Subcellular Biochemistry

Volume **14**
Artificial and Reconstituted
Membrane Systems

SUBCELLULAR BIOCHEMISTRY

SERIES EDITOR

J. R. HARRIS, North East Thames Regional Transfusion Centre, Brentwood, Essex, England

ASSISTANT EDITORS

H. J. HILDERSON, University of Antwerp, Antwerp, Belgium
J. J. M. BERGERON, McGill University, Montreal, Canada

Recent Volumes in this Series:

Volumes 5–11	Edited by Donald B. Roodyn
Volume 12	**Immunological Aspects** Edited by J. R. Harris
Volume 13	**Fluorescence Studies on Biological Membranes** Edited by H. J. Hilderson
Volume 14	**Artificial and Reconstituted Membrane Systems** Edited by J. R. Harris and A.-H. Etémadi
Volume 15	**Virally Infected Cells** Edited by J. R. Harris

A Continuation Order Plan is available for this series. A continuation order will bring delivery of each new volume immediately upon publication. Volumes are billed only upon actual shipment. For further information please contact the publisher.

Subcellular Biochemistry

Volume 14
Artificial and Reconstituted Membrane Systems

Edited by
J. R. Harris
North East Thames Regional Transfusion Centre
Brentwood, Essex, England

and

A.-H. Etémadi
University of Paris VI
Paris, France

PLENUM PRESS • NEW YORK AND LONDON

The Library of Congress cataloged the first volume of this title as follows:

Sub-cellular biochemistry.

   London, New York, Plenum Press.
      v. illus. 23 cm. quarterly.
      Began with Sept. 1971 issue. Cf. New serial titles.
   1. Cytochemistry – Periodicals. 2. Cell organelles – Periodicals.
QH611.S84   574.8'76   73-643479

ISBN-13:978-1-4613-9364-1   e-ISBN-13:978-1-4613-9362-7
DOI: 10.1007/978-1-4613-9362-7

This series is a continuation of the journal *Sub-Cellular Biochemistry*, Volumes 1 to 4 of which were published quarterly from 1972 to 1975

© 1989 Plenum Press, New York
Softcover reprint of the hardcover 1st edition 1989

A Division of Plenum Publishing Corporation
233 Spring Street, New York, N.Y. 10013

All rights reserved

No part of this book may be reproduced, stored in a retrieval system, or transmitted in any form or by any means, electronic, mechanical, photocopying, microfilming, recording, or otherwise, without written permission from the Publisher

INTERNATIONAL ADVISORY EDITORIAL BOARD

J. L. AVILA, Institutode Biomedicina, Caracas, Venezuela
B. B. BISWAS, Bose Institute, Calcutta, India
N. BORGESE, CNR Center for Pharmacological Study, Milan, Italy
M. J. COSTELLO, Duke University Medical Center, Durham, North Carolina, USA
N. CRAWFORD, Royal College of Surgeons, London, England
C. de DUVE, International Institute of Cellular and Molecular Pathology, Brussels, Belgium
A.-H. ETÉMADI, University of Paris VI, Paris, France
W. H. EVANS, National Institute for Medical Research, London, England
H. GLAUMANN, Karolinska Institute, Huddinge, Sweden
D. R. HEADON, University College Galway, Galway, Ireland
P. L. JØRGENSEN, University of Aarhus, Aarhus, Denmark
J. KIM, Osaka University, Osaka, Japan
J. B. LLOYD, University of Keele, Keele, England
J. A. LUCY, Royal Free Hospital School of Medicine, London, England
A. H. MADDY, University of Edinburgh, Edinburgh, Scotland
A. MONNERON, Institut Pasteur, Paris, France
D. J. MORRÉ, Purdue University, West Lafayette, Indiana, USA
M. OSBORNE, Max Planck Institute for Biophysical Chemistry, Göttingen, FRG
P. QUINN, King's College London, London, England
G. RALSTON, The University of Sydney, Sydney, Australia
S. ROTTEM, The Hebrew University, Jerusalem, Israel
M. R. J. SALTON, New York University Medical Center, New York, New York, USA
G. SCHATTEN, University of Wisconsin–Madison, Madison, Wisconsin, USA
F. S. SJOSTRAND, University of California–Los Angeles, Los Angeles, California, USA
T. TAKAHASHI, Aichi Cancer Center, Nagoya, Japan
G. B. WARREN, Imperial Cancer Research Fund, London, England
F. WUNDERLICH, University of Düsseldorf, Düsseldorf, FRG
G. ZAMPIGHI, University of California–Los Angeles, Los Angeles, California, USA
I. B. ZBARSKY, Academy of Sciences of the USSR, Moscow, USSR

Contributors

Q. F. Ahkong Department of Biochemistry and Chemistry, Royal Free Hospital School of Medicine, University of London, London NW3 2PF, United Kingdom

Ludwig Brand Biology Department and McCollum–Pratt Institute, Johns Hopkins University, Baltimore, Maryland 21218

Heinz Breer Department of Biology and Biophysics, Osnabrück University, Osnabrück D-4500, Federal Republic of Germany

Lesley Davenport Department of Chemistry, Brooklyn College of the City University of New York, Brooklyn, New York 11210

Nejat Düzgüneş Cancer Research Institute and Department of Pharmaceutical Chemistry, University of California, San Francisco, California 94143-0128

Abol-Hassan Etémadi Laboratory of Biochemistry and Molecular Biology of Bacterial Lipids, Faculty of Sciences of the University of Paris VI, Paris, France

Bruce Grasberger Section on Chemical Immunology, Arthritis and Rheumatism Branch, National Institute of Arthritis and Musculoskeletal and Skin Diseases, National Institutes of Health, Bethesda, Maryland 20892

Gregory Gregoriadis Medical Research Council Group, Academic Department of Medicine, Royal Free Hospital School of Medicine, London NW3 2QG, United Kingdom

Wolfgang Hanke Department of Biology and Biophysics, Osnabrück University, Osnabrück D-4500, Federal Republic of Germany

Dick Hoekstra   Laboratory of Physiological Chemistry, University of Groningen, 9712 KZ Groningen, The Netherlands

Jean-Pierre Kinet   Section on Chemical Immunology, Arthritis and Rheumatism Branch, National Institute of Arthritis and Musculoskeletal and Skin Diseases, National Institutes of Health, Bethesda, Maryland 20892

Jay R. Knutson   Laboratory of Technical Development, National Institutes of Health, Bethesda, Maryland 20892

J. A. Lucy   Department of Biochemistry and Chemistry, Royal Free Hospital School of Medicine, University of London, London NW3 2PF, United Kingdom

Darrell R. McCaslin   Department of Chemistry, Rutgers University, Newark, New Jersey 07102

P. J. Quinn   Department of Biochemistry, King's College London, Kensington Campus, London W8 7AH, United Kingdom

Jacqueline A. Reynolds   Department of Physiology, Duke University Medical Center, Durham, North Carolina 27701

Nobuhito Sone   Department of Biochemistry, Jichi Medical School, Minamikawachi-machi, Tochigi-ken 329-04, Japan

H. Ti Tien   Membrane Biophysics Laboratory, Department of Physiology, Michigan State University, East Lansing, Michigan 48824-1101

Peter M. Vassilev   Membrane Biophysics Laboratory, Department of Physiology, Michigan State University, East Lansing, Michigan 48824-1101

Preface

This collection of 11 chapters is devoted to a survey of artificial and reconstituted membrane systems. These are fundamental themes and areas of great current importance in membrane biochemistry. They also relate well to the founding concept of this series, namely, to present studies that progressively work toward and provide us with an "integrated view of the cell." In this volume, it is the application of a wide range of physiochemical and biochemical techniques to the study of membrane lipids and proteins which serves to demonstrate the significant progress that has been made in this field over the past 25 years. From the understanding of *simplified* artificial systems, it is hoped that it will ultimately be possible to gain a more accurate understanding of natural biological membranes, in all their diversity.

This book is an appropriate successor to Volume 13 of the series, which deals with fluorescence studies on biological membranes. Indeed, the present chapter by Lesley Davenport and colleagues was originally due for inclusion in Volume 13, but has been held over for inclusion in this volume, where it integrates remarkably well with the other topics.

The extremely varied and interesting contents of this volume are now briefly outlined. In Chapter 1, Jacqueline A. Reynolds and Darrell R. McCaslin present a pertinent survey of the interaction of detergents with membrane lipids and proteins, together with an assessment of the reconstitution process. This chapter sets the technical scene for much of what is to follow, since the use of detergents occupies a principal role in many of the studies to be described. P. J. Quinn then contributes a very thorough and extensively illustrated account of membrane lipid phase behavior and lipid–protein interactions, and in Chapter 3 a consideration of the reconstitution of physiological membrane molecular mechanisms (such as ion transport and receptors) in bilayer lipid membranes and patch-clamp bilayers is presented by Peter M. Vassilev and H. Ti Tien. A detailed biophysical assessment of the application of fluorescence studies to membrane dynamics and heterogeneity is then offered by Lesley Davenport, Jay R. Knutson, and Ludwig Brand. This is followed by an excellent account of membrane fusion, fusogenic agents, and osmotic forces from J. A. Lucy and Q. F. Ahkong. In Chapter 6, Dick Hoekstra and Nejat Düzgüneş present a survey of lectin–carbohydrate interactions in model and biological membrane systems. This is an impressively thorough account of what is currently an extremely important topic in membrane biochemistry.

Nobuhito Sone then gives us an interesting review of the energy-transducing complexes in bacterial respiratory chains, and this is followed by a somewhat personal description of the reconstitution of the high-affinity receptor for immunoglobulin E by Jean-Pierre Kinet and Bruce Grasberger. Wolfgang Hanke and Heinz Breer then discuss the reconstitution of acetylcholine receptors in planar lipid bilayers and a brief, yet pertinent, survey of the use of liposomes as carriers of drugs by Gregory Gregoriadis follows. Finally, an extremely thorough description of the impact of reconstitution experiments on the understanding of physiological protein translocation processes is contributed by Abol-Hassan Etémadi.

It is hoped that the diversity and scientific depth of the material included in this volume of *Subcellular Biochemistry* will enable it to be of widespread interest and value to membrane biochemists and others drawn to this fascinating field of study.

J. R. Harris
Brentwood, Essex, United Kingdom
Abol-Hassan Etémadi
Paris, France

Contents

Chapter 1
The Role of Detergents in Membrane Reconstruction
Jacqueline A. Reynolds and Darrell R. McCaslin

1.	Introduction	1
2.	The Membrane Environment	2
	2.1. Lipid Components	2
	2.2. Protein Components	3
3.	Detergents	5
	3.1. Types and Properties of Detergents	5
	3.2. Modes of Interaction with Proteins	7
	3.3. Interaction with Diacyl Lipid Bilayers	8
4.	Reconstitution	9
	4.1. Detergent–Lipid–Protein Interactions: A Reversible Process	10
	4.2. Kinetics	11
	4.3. Experimental Requirements for Lipid Vesicles	11
	4.4. Experimental Procedures for Detergent-Induced Reconstitution	14
	4.5. Kinetic Control of Vesicle Size	19
	4.6. Incorporation of Intrinsic Membrane Proteins	22
5.	Some Concluding Comments	23
6.	References	23

Chapter 2
Membrane Lipid Phase Behavior and Lipid–Protein Interactions
P. J. Quinn

1.	Introduction	25
2.	Polymorphism of Membrane Polar Lipids	26
	2.1. Biophysical Characterization of Membrane Lipid Polymorphism	27

2.2.	Factors Responsible for Determining Lipid Phase	38
2.3.	Phases of Pure Polar Lipids	42
2.4.	Hexagonal Phases	55
2.5.	Cubic and Other Nonlamellar Phases	56
3.	Phase Behavior of Mixed Lipid Systems	57
3.1.	Phase Behavior of Lipid Enantomers	58
3.2.	Lateral Miscibility of Bilayer-Forming Lipids	61
3.3.	Bilayer–Nonbilayer Phase Separations	64
4.	Protein–Lipid Interactions	66
4.1.	Exclusion of Proteins from Bilayer Gel Phases	69
4.2.	Phase Separation of Nonbilayer Lipid Phases in Membranes	73
5.	Conclusions	82
6.	References	83

Chapter 3
Reconstitution of Membrane Molecular Mechanisms in Bilayer Lipid Membranes and Patch-Clamp Bilayers
Peter M. Vassilev and H. Ti Tien

1.	Introduction	97
2.	Techniques for Formation and Characterization of Reconstituted Bilayers	98
2.1.	Formation of Planar and Patch-Clamp Bilayers	98
2.2.	Electrophysiological Methods for Analyzing the Properties of Reconstituted Membranes	101
3.	Reconstitution of Membrane Molecular Mechanisms Related to Ion Transport and Excitability	107
3.1.	Reconstitution of Sodium Ion Transport Mechanisms	108
3.2.	Calcium Channels Functioning in Bilayers	112
3.3.	Mechanisms of Potassium Ion Transport	119
3.4.	Chloride Channels	126
3.5.	Reconstitution of Active Transport Mechanisms in Planar Bilayers	128
4.	Reconstitution of Synaptic Events	129
4.1.	Acetylcholine-Receptor Mechanisms	129
4.2.	Dopamine Receptors	132
5.	Reconstitution of Sensory Mechanisms	132
5.1.	Visual Receptor Membranes	132
5.2.	Olfactory Receptors	132
6.	Physiological Processes in Nonexcitable Membranes	133
6.1.	Erythrocyte Plasma Membranes	133

6.2.	Epithelial Membranes	133
6.3.	Kidney Plasma Membranes	134
6.4.	Ciliary Membranes	135
6.5.	Outer Mitochondrial Membrane	136
7.	Concluding Remarks and Perspective	136
8.	References	137

Chapter 4
Fluorescence Studies of Membrane Dynamics and Heterogeneity
Lesley Davenport, Jay R. Knutson, and Ludwig Brand

1.	Introduction	145
2.	Do Domains Exist?	147
	2.1. Thermal Effects	149
	2.2. Chemically Induced Domain Formation	157
3.	What Are the Relaxation Times for Lipid Domains?	167
4.	Future Trends	174
5.	References	177

Chapter 5
Membrane Fusion: Fusogenic Agents and Osmotic Forces
J. A. Lucy and Q. F. Ahkong

1.	Introduction	189
2.	Fusogenic Lipids	190
	2.1. Osmotic Effects	190
	2.2. Cellular Proteases	192
	2.3. Fusion Driven by Cell Swelling	195
3.	Fusion Induced by Polyethylene Glycol	197
	3.1. Purity of the Polymer	197
	3.2. Dehydration	199
	3.3. Rehydration	201
4.	Electrically Induced Cell Fusion	204
5.	Molecular Models for Membrane Fusion	208
6.	Virally Induced Membrane Fusion	210
7.	Exocytosis	212
	7.1. Model Systems	212
	7.2. Biological Systems	214
8.	References	218

Chapter 6
Lectin–Carbohydrate Interactions in Model and Biological Membrane Systems
Dick Hoekstra and Nejat Düzgüneş

1.	Introduction	229
2.	Lectins: Abundance and Properties	230
	2.1. Structural Features of Lectins	231
	2.2. Carbohydrate Binding Properties	234
3.	Glycolipids and Glycoproteins: Biological Receptors for Lectins	237
4.	Lectin–Carbohydrate Interactions in Model Systems	240
	4.1. Properties of Glycolipid-Containing Membranes	240
	4.2. Lectin–Glycolipid Interactions	242
	4.3. Interactions of Lectins with Glycoprotein-Containing Membranes	256
5.	Lectin–Carbohydrate Interactions in Biological Systems	259
	5.1. Exogenous Lectins: Applications	259
	5.2. Exogenous Lectins: Biological Effects	261
	5.3. Endogenous Lectins	262
6.	References	267

Chapter 7
Energy-Transducing Complexes in Bacterial Respiratory Chains
Nobuhito Sone

1.	Introduction	279
2.	Electron Transfer Chain and Energy Coupling	280
	2.1. Bacterial Respiratory Chains	280
	2.2. Two Approaches Using Whole Cells or Membrane Vesicles	282
3.	Reconstitution into Liposomes	287
	3.1. A Tool to Study Energy-Transducing Enzymes	287
	3.2. Purification of Respiratory Complexes	289
	3.3. Reconstitution Methods	289
4.	Cytochrome Oxidase	291
	4.1. Bacterial Cytochrome Oxidase	291
	4.2. Reconstitution of Proton Pump Activity	293
	4.3. Properties of Cytochrome aa_3-Type Oxidases	296
	4.4. Cytochrome o-Type and d-Type Oxidases	297
5.	Cytochrome bc_1 Complexes	299
	5.1. Properties	299
	5.2. Reconstitution	303
6.	NADH Dehydrogenases and Complex I	305
	6.1. Membrane-Bound NADH Dehydrogenases	305

	6.2. Evidence for Proton Pumping NADH : Quinone Oxidoreductase	306
	6.3. Na$^+$ Pumping NADH : Quinone Oxidoreductase	307
7.	Energy-Transducing Components Other Than Complexes I–IV	308
	7.1. Anaerobic Electron Transfer Systems	308
	7.2. Transhydrogenase	310
8.	Concluding Remarks and Outstanding Problems	311
9.	References	313

Chapter 8
Reconstitution of the High-Affinity Receptor for Immunoglobulin E
Jean-Pierre Kinet and Bruce Grasberger

1.	Introduction	321
2.	Purification of the Receptor	322
	2.1. Sensitivity to Detergent	322
	2.2. Protective Effect of Lipids	324
	2.3. Protocol for Purification	325
3.	Mechanisms of Action of the Receptor: Possible Functions to Reconstitute	327
	3.1. Role of Aggregation	327
	3.2. Functions Associated with the Receptor-Mediated Triggering of Mast Cells	327
4.	Reconstitution Studies	328
	4.1. Planar Bilayers	328
	4.2. Vesicles	330
5.	Conclusions	334
6.	References	335

Chapter 9
Reconstitution of Acetylcholine Receptors into Planar Lipid Bilayers
Wolfgang Hanke and Heinz Breer

1.	Introduction	339
2.	Technical Aspects of Acetylcholine-Receptor Reconstitution	342
	2.1. Biochemical Procedures	342
	2.2. Formation of Planar Lipid Bilayers	344
	2.3. Incorporation of Proteins into Planar Lipid Bilayers	346
	2.4. How to Collect and Process Data	347
3.	Properties of Reconstituted Acetylcholine-Receptor Channels	350
4.	Gating Models and Theoretical Aspects	356
5.	Data from Other Experiments	358

6. Principal Requirements for a Functional Acetylcholine-Receptor Channel 358
7. Summary and Conclusion.................................. 359
8. References... 359

Chapter 10
Liposomes as Carriers of Drugs: Observations on Vesicle Fate after Injection and Its Control
Gregory Gregoriadis

1. Introduction... 363
2. Retention of Drugs by Liposomes in Contact with Blood 364
3. Circulation of Liposomes in the Blood 368
4. Distribution of Liposomes in Tissues 373
5. Conclusions... 375
6. References.. 375

Chapter 11
Reconstitution and Physiological Protein Translocation Processes
Abol-Hassan Etémadi

1. Introduction... 379
2. Membrane Proteins and Protein Translocation Processes......... 386
   2.1. Microsomal Membrane Proteins 386
   2.2. Bacterial Membrane Proteins.......................... 391
   2.3. Organellar Membrane Proteins 393
3. Energy Dependence of Protein Translocation Processes 398
   3.1. Secretory and Transmembrane Integral Proteins 398
   3.2. Migratory Proteins 400
4. Translocation and Membrane-Anchoring Signals 403
   4.1. Signal and Membrane-Anchoring Sequences for Proteins Translocated through the Endoplasmic Reticulum and the Bacterial Cytoplasmic Membrane 403
   4.2. Signal Sequences for Migratory Proteins 412
5. Post-Translational Protein Translocation 427
   5.1. Post-Translational Translocation of Secretory and Transmembrane Integral Proteins 427
   5.2. Post-Translational Translocation of Migratory Proteins..... 429
   5.3. Bound Ribosomes and Post-Translational Translocation of Migratory Proteins 430
   5.4. The Case of Endoproteins 433
6. Protein Translocation through or Insertion into Lipid Aggregates.. 436

Contents

    6.1. Experiments Related to Translocation of Secretory and
           Integral Membrane Proteins........................... 436
    6.2. Experiments on Translocation of Migratory Proteins....... 440
7. General and Concluding Remarks 445
8. References.. 450

Index 487

Chapter 1

The Role of Detergents in Membrane Reconstitution

Jacqueline A. Reynolds and Darrell R. McCaslin

1. INTRODUCTION

The study of amphiphilic molecules has a long and distinguished history, including Traube's initial investigation of surface activity (1891), the introduction of the concept and term "micelle" by McBain (1913), and the first rigorous description of the solution properties of these important structures by Hartley (1936). It is unlikely that early workers in this field could have foreseen the impact their research would eventually have in the study of living systems. Consider, for example, a number of currently widespread experimental procedures.

1. The use of polyacrylamide gel electrophoresis in the presence of sodium dodecylsulfate for the separation and categorization of proteins (Weber and Osborne, 1969) is based on a well-characterized interaction between two amphiphilic molecules, an ionic detergent, and a polypeptide chain (Fish *et al.*, 1970; Reynolds and Tanford, 1970a,b).
2. The ability to solubilize proteins associated with biological membranes rests entirely on decades of investigation of amphiphilic molecules in general and

---

Jacqueline A. Reynolds Department of Physiology, Duke University Medical Center, Durham, North Carolina 27701. **Darrell R. McCaslin** Department of Chemistry, Rutgers University, Newark, New Jersey 07102.

of protein–amphiphile interactions in particular (reviewed by Helenius and Simons, 1975; Reynolds, 1982; Steinhardt and Reynolds, 1969; Tanford, 1980; Tanford and Reynolds, 1976).
3. The *in vitro* investigation of enzymes that act on hydrophobic substrates has been facilitated by the recognition that relatively water-insoluble substances can be incorporated into mixed micelles, thus increasing their "concentration" in an aqueous environment—a process so familiar to colloid chemists that it was already part of textbook material in 1950 (McBain, 1950).

> The colloidal micelles of soaps and detergents have the remarkable property of incorporating in or upon themselves large quantities of molecules which are otherwise insoluble in water. This process leads to a reversible equilibrium.

This chapter addresses yet another application of detergent chemistry to the field of biological science—formation of lipid bilayers and reconstitution of membrane-bound proteins from solutions containing detergent–lipid–protein mixed micelles. Membrane proteins that have been solubilized in detergent solutions are amenable to study by the vast repertoire of tools of the protein chemist. However, the properties of such a protein, which are dependent on the vectorial nature of their normal membrane environment (e.g., transport), can only be examined by incorporation into a lipid vesicle. The goals of this chapter are to provide an overview of our current state of knowledge in the context of the biological sciences and to discuss complexities inherent (and often unrecognized) in these multicomponent systems. The emphasis is on concepts, methodology, and new developments rather than on a comprehensive review of the current literature.

2. THE MEMBRANE ENVIRONMENT

2.1. Lipid Components

The fundamental structural unit of the biological membrane is the lipid bilayer that provides the permeability barrier isolating the cytoplasm from the external fluid or the lumen of an organelle from the cytoplasm. The lipid components are heterogeneous in composition with respect to both polar head groups and hydrophobic regions, and the biological functions served by this heterogeneity are still not completely understood. The bilayer structure itself is not symmetrical with respect to distribution of head group types between the inner and outer leaflet—the minor components such as phosphatidylinositol and phosphatidylserine being located predominantly, if not entirely, in the inner leaflet of plasma membranes. Whether a similar distributional asymmetry exists for the hydrocarbon moieties is not established, although net compositional differences in the fatty acid moieties among different types of membrane is well known.

Bilayers composed of naturally occurring lipids with approximately equal amounts of saturated and unsaturated fatty acyl chains are "fluid" at physiological temperatures; that is, they do not form two-dimensional crystals at normally encountered temperatures. However, the frequently cited analogy between the interior of the bilayer and an organic solvent such as hexane is misleading. Organic solvents are usually isotropic fluids; lipid bilayers are intrinsically anisotropic perpendicular to the plane of the membrane because one end of each hydrocarbon tail must remain at the bilayer–solution interface due to its covalent attachment to the polar head group. Also, bilayers are not necessarily isotropic within the plane of the membrane since lateral segregation of different types of lipid can occur.

It is also intellectually misleading to envision the boundary between hydrophobic and hydrophilic regions of a lipid bilayer as equivalent to the well-defined Gibbsian surface of discontinuity. In fact, the phase boundary between these two regions of lipid molecules is complex and not strictly comparable to the interface between two immiscible isotropic fluids.

These descriptions of the physical properties of a biological membrane may seem self-evident, but, nonetheless, misleading models continue to abound in the literature—not only in the form of spatially incorrect pictorial cartoons but also in discussions of the functional and structural interactions between proteins and lipids.

2.2. Protein Components

The description of a biological membrane as a *mosaic* of lipophilic and hydrophilic regions was postulated by Collander and Bärlund (1933), and the concept that the hydrophilic regions were generated by protein components had received a wide degree of acceptance by the 1940s. Höber (1945), for example, in discussing an incorrect postulate of the structure of plasma membranes, points out: "The picture seems to ignore the *rather well established concept of a mosaic* of lipoid and protein areas, as a background for the solvent-sieve theory of permeability."

Today, the proteins associated with biological membranes are usually classified according to their physical location (Green, 1972; Singer, 1971). Extrinsic proteins are those found associated with the membrane surface, while intrinsic proteins interact directly with the hydrocarbon portion of the lipid bilayer. The latter interaction may well involve a small protein domain that does not span the bilayer (e.g., cytochrome b_5; Spatz and Strittmatter, 1971) or, alternatively, the polypeptide chain may repeatedly cross the entire bilayer structure (e.g., bacteriorhodopsin; Henderson and Unwin, 1975). Intrinsic membrane

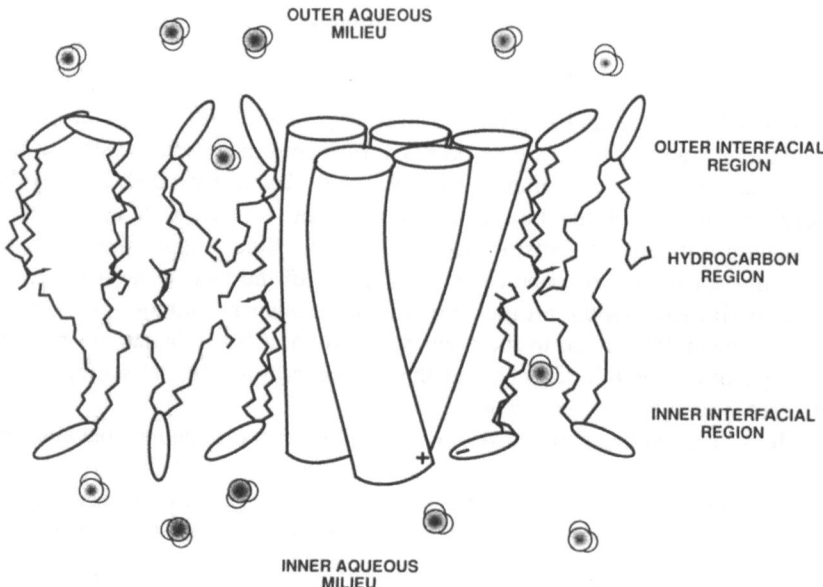

FIGURE 1. Schematic diagram drawn to scale of a phosphatidylcholine bilayer and a hypothetical intrinsic membrane protein. The bilayer is shown as a two-dimensional cross section with the head group and hydrophobic region scaled correctly for naturally occurring phosphatidylcholines. The transmembrane part of the protein molecule is shown as a bundle of helices, all of which should be imagined as part of the same molecule, joined together outside the membrane by other segments of the protein.

proteins then are multidomain in nature, consisting of portions interacting with two potentially different aqueous milieus and a region interacting with a hydrocarbon-like medium. An additional domain is clearly present as well, although it is frequently ignored in discussion of protein structure and function. This is the domain that penetrates the lipid head-group region in which the water activity is significantly lower than that of bulk water and in which is found a diversity of densely packed, polar structures.

Figure 1 shows a correctly scaled diagram of a hypothetical structure of an intrinsic membrane protein embedded in a bilayer of naturally occurring lipids. Unfortunately, one can only portray pictorially an approximation to the variety of diverse motions and configurations accessible to the lipid molecules, but the geometrical relationship between the different solvent environments to which the protein is exposed is apparent.

If one removes an intrinsic membrane protein from its native environment by solubilizing it in detergent–protein mixed micelles, the different domains must be kept in solvent milieus similar to those of the bilayer system or structural alterations ranging from minor to catastrophic can occur. There is little problem in matching either the hydrophobic interior of the bilayer or the bulk aqueous solvent surrounding the hydrophilic domains of the protein. This merely requires the proper choice of detergent with respect to the length of its hydrocarbon tail and an appropriate buffer solution. (Obviously, since there is only one bulk solvent in this system, one can match only one of the two aqueous milieus.) However, little thought is given by most investigators to the interfacial domain that is normally approximately 8–10 Å in thickness and that contains tightly packed, hydrated lipid head groups of varying chemical composition.

An additional complication in the system shown in Figure 1 arises from the fact that many biological membranes are under the influence of an electrical potential gradient. Since proteins are dipolar structures, their overall conformation is dependent on the potential to which they are exposed—a potential that cannot exist when cells are ruptured to obtain microsomal fractions or when membrane proteins are solubilized in detergent systems. The importance of this parameter is not known in general, but we are all familiar with the voltage-gated ion channels in which a change in membrane potential results directly in a conformational change leading to a flow of ions through the protein structure. Less dramatic, but equally important structural dependencies on membrane potential will undoubtedly be found as our technical abilities to observe them improve.

3. DETERGENTS

Detergents are amphiphilic compounds generally containing a single hydrophobic region attached to a polar head group. Most of the familiar detergents will spontaneously self-associate to form micelles which are small, water-soluble particles containing a hydrophobic core and a polar surface. The condition required for this self-association is that the negative free energy due to hydrophobic association and the positive free energy of repulsion between polar head groups be balanced in the aggregated structure, and the concentration at which the self-association occurs is referred to as the *critical micelle concentration*, CMC.

In this section we discuss the properties of the types of detergent most frequently used in biological applications, and we recommend to the interested reader more detailed and general treatments of this subject (Israelachvilli, 1985; Tanford, 1980).

3.1. Types and Properties of Detergents

In Table I is presented a brief list of common commercial detergents together with their relevant physical properties. In general, these compounds can

Table I
Properties of Some Common Commercial Detergents

Detergent[a]	CMC (M × 10³)	Aggregation number	$\bar{v}$ (cm³/g)	Ionic strength
$C_{12}OSO_3^-\,Na^+$	8.13	62	0.864	0
	2.30	84		0.05
	0.92	103		0.20
$C_{12}NMe_3^+\,Br^-$	14.6	61	0.97	0
	5.71	72		0.05
	2.54	83		0.25
$C_8\phi E_{9.6}$ (Triton X-100)	0.3	140	0.908	
$C_9\phi E_{10}$ (Triton N-101)	0.097	100	0.922	
$C_9\phi E_{15}$	0.087	52	0.965	
$C_{12}E_8$	0.087	120	0.973	
$C_{16}E_{20}$ (Brij 58)	0.0039	70	0.919	
C_{12}Sorbitan E_{20} (Tween 20)	0.059		0.869	
C_{18}Sorbitan E_{20} (Tween 80)	0.012	60	0.896	
C_8 Glucoside	25.0			
Deoxycholate	2.0	1.7	0.778	0 (pH 9)
	1.6	2.2		0.01
	0.91	22.0		0.15
Cholate	10.0	2	0.771	0 (pH 8–9)
	5.0	2.8		0.01
	3.0	4.8		0.15

[a]Data from Reynolds (1982). C_n indicates the number of carbon atoms in the hydrophobic region; E_m indicates the number of (O—CH₂—CH₂) groups in the hydrophilic region.

be divided into two classes—those that possess ionic or strongly dipolar head groups and those that have nonionic head groups. The most commonly used members of the latter class included polyoxyethylene or carbohydrate derivatives. The bile salts represent a unique class of detergents in that the hydrophobic portion consists of a fused ring system rather than an alkyl chain, and furthermore, they contain an extremely small polar region. Note that these latter compounds form extremely small self-aggregated structures quite unlike the ellipsoidal or nearly spherical forms assumed by the other amphiphiles listed in the table.

The CMC and average aggregation number of ionic detergents are dependent on the ionic strength, while these properties of nonionic amphiphiles are relatively insensitive to this parameter in the range of ionic concentrations normally used by biological scientists. However, it should be kept in mind that any large reduction in the activity of water such as occurs at high concentrations of solutes (e.g., 1 M NaCl, 30% sucrose) will affect the physical properties of all amphiphilic compounds as a consequence of changes in the degree of preferential hydration.

The aggregation number given in Table I is a weight average association number, the distribution in micellar sizes about the mean being approximately Gaussian and a function of the free energy of head-group interactions. Both the mean micellar size and the distribution can be altered by incorporation of an amphiphilic molecule of different chemical composition (e.g., a protein, a different detergent, a lipid, a diacylglycerol), and the perturbations induced by such incorporation will obviously increase as the number of "foreign" molecules relative to the principal detergent increases.

It is apparent from Table I that the CMC is a function of both the size of the hydrophobic domain and the chemical nature of the hydrophilic head group. Thus, the long alkyl chain detergents that most closely match naturally occurring phospholipids in their hydrophobic moieties have very low CMCs, a property that as we shall see in the next section is inconvenient for reconstitution purposes. On the other hand, short alkyl chain detergents with much higher CMCs have a disadvantage when used for solubilization of membrane-bound proteins in that their hydrophobic domain is significantly smaller than that found in a normal lipid bilayer.

Additional cautionary words regarding these amphiphilic compounds are in order at this point. (1) Many commercial detergents contain impurities remaining from the synthetic process—impurities such as transition metal ions and preservatives against oxidation. This is a particularly severe problem with polyoxyethylene derivatives. (2) These compounds are also subject to oxidation reactions (e.g., Donbrow, 1987), leading to the formation of reactive peroxides, and special precautions are needed to protect solutions containing these amphiphiles from oxygen and ultraviolet radiation. (3) A frequently unappreciated problem is the heterogeneity in chemical structure of many commercial products. As examples, sodium dodecylsulfate may well contain significant amounts of octyl and decyl derivatives; polyoxyethylene head-group lengths are usually polydisperse because they are synthesized by a polymerization process that generates an approximately Poisson distribution of head-group sizes; Triton X-100 contains both para and ortho derivatives of the phenyl ring as well as heterogeneity in the length of the polyoxyethylene chain. (4) Temperature effects on the physical properties of detergents are usually not well characterized in physiological media, the majority of the reported data having been obtained in pure water or at very low ionic strength.

3.2. Modes of Interaction with Proteins

Most water-soluble proteins in their native state do not have specific binding sites for amphiphilic compounds—serum albumin and beta lactoglobulin being notable exceptions. However, all water-soluble proteins investigated to date interact with ionic detergents containing n-alkyl chains of 12 carbons or more in a highly cooperative process which involves a concomitant conformational change. This latter phenomenon has been known for many years and has been exploited in polypeptide separations using sodium dodecyl sulfate as a "denaturant."

Intrinsic membrane proteins, on the other hand, interact with detergents by forming a mixed micellar system in which the amphiphiles shield the hydrophobic domain of the protein. There is no reason to believe that the size and organization of the detergent aggregate in this case is necessarily identical to that of a pure detergent micelle. If the polypeptide in question is small in size as in the case of cytochrome b_5 or the fd coat protein, the perturbation of the normal micelle may be minimal; larger proteins can induce significant changes, however, in the number of detergent monomers incorporated into the mixed micellar system (Reynolds, 1982).

3.3. Interaction with Diacyl Lipid Bilayers

The following mechanistic sequence has been suggested as a description of the reversible interaction of detergents and bilayer-forming lipids (Reynolds, 1982) and has recently received experimental support in a detailed study of the interactions between egg yolk lecithin and octyl glucoside (Ollivon et al., 1988).

At low ratios of detergent to lipid, single tail amphiphiles partition into lipid aggregates without disruption of the normal bilayer structure.

$$PL_{(\text{bilayer})} + iD \leftrightarrow PL - D_{j(\text{bilayer})} + (i-j)D \tag{1}$$

This partitioning may or may not obey the law of ideal mixing (Reynolds et al., 1985) and is frequently kinetically limited by the barrier to flip-flop presented by the hydrophobic region of lipid bilayers. As the ratio of detergent to lipid is increased, the curvature of the bilayer increases due to the larger number of amphiphilic head groups present relative to the volume of hydrocarbon domain, and, at some point, fragmentation occurs.

$$PL - D_{j(\text{bilayer})} + kD \leftrightarrow PL - D_{(j+1)(\text{fragment})} + (k-1)D \tag{2}$$

A further increase in partitioned detergent then leads to the formation of mixed, soluble micelles.

$$PL - D_{(j+1)(\text{fragment})} + mD \leftrightarrow PL - D_{n(\text{mixed micelle})} + (m-n)D_{(\text{pure micelles})} \tag{3}$$

At this point, some further discussion of the partitioning of amphiphiles into lipid bilayers at low detergent to lipid ratios is warranted. The equilibration of a detergent molecule between the outer and inner leaflet of a lipid bilayer is affected by a number of factors. From the standpoint of thermodynamic considerations alone, the standard chemical potential of a detergent molecule dissolved in a lipid bilayer is a function of the interaction energies with adjacent molecules. These interaction energies are in turn a function of the chemical compositions of the neighboring molecules and of the radii of curvature at the inner and outer surfaces of the vesicle, the latter parameter affecting the molecular packing at both the surface and the interior of the bilayer (e.g., see Huang and Mason, 1978). Thus, the equilibrium constant for detergent association with the inner leaflet may not necessarily be equal to that for association with the outer leaflet since both the chemical composition and the radii of curvature may differ between the two halves. As a corollary, it is apparent that the rate constants for "flip-flop" will be unequal if the two equilibrium constants are unequal (i.e., the detergent has a thermodynamic preference for one or the other of the bilayer leaflets).

The final state in the solubilization scheme consists of pure detergent micelles and mixed micelles containing both lipid and detergent. In the absence of specific interactions, the lipid molecules will be distributed randomly in the micellar particles according to normal statistical laws. Thus, an *average* of i lipids per micelle *does not* mean that each micelle contains exactly i lipid molecules, but rather that there is a distribution around the mean that can be described by a standard statistical distribution function. For example, if an average of two diacylglycerol molecules per micelle are added to a detergent solution, the Poisson distribution predicts that the fraction of micelles containing i copies of diacylglycerol is as follows: $i=0$, 0.135; $i=1$, 0.271; $i=2$, 0.271; $i=3$, 0.181; $i=4$, 0.092; $i=5$, 0.036; $i=6$, 0.012; $i=7$, 0.002. There are a number of examples in the literature in which mixed micellar systems such as this are used in an attempt to demonstrate the number of lipid molecules required for "activation" of a particular protein. In these reported experiments lipid is incorporated in a detergent system at a sufficient concentration to provide the desired *average* number of lipids per micelle. The protein under investigation is added to this system at a molar concentration usually one to three orders of magnitude less than that of the mixed micelles, and inferences are drawn regarding the number of lipid molecules necessary for "activation" based on changes in "activity" as a function of the *average* number of lipids per micelle. Under these conditions, no valid conclusion can be reached.

4. RECONSTITUTION

All lipids do not form bilayer type structures when suspended in an aqueous solution (e.g., see Israelachvilli, 1985, for a recent review). Those that do are characterized by relatively large polar head groups such as phosphatidylcholine, and, hence, many reconstitution studies have been carried out with diacyl derivatives of this species as the *principal* lipid. Unfortunately for the biological scientist, simple suspension of diacyl phosphatidylcholines in an aqueous solvent leads to multilamellar bilayer structures—not single-wall vesicles. This phenomenon has motivated scientists to devise experimental procedures that will favor formation of the latter structures. Among the methods that produce at least a partial population of single-wall vesicles are sonication, freeze–thaw cycles, dilution from an organic solvent such as diethyl ether, and transient exposure of lipids containing negatively charged head groups to pH 11.

The experienced biochemist, however, is less than delighted to subject a functioning protein to these manipulations, all of which can lead to protein *denaturation* (e.g., see Tanford, 1968, 1970; Lapanje, 1978). Consequently, methods have been sought for the formation of single-wall lipid vesicles and concomitant incorporation of intrinsic membrane proteins that minimize the possibility of alterations in the native structure of proteins. It was recognized by the protein chemist that substitution of detergent molecules for the normal lipid milieu was a rational approach to solubilization and purification of membrane proteins (Helenius and Simons, 1975; Tanford and Reynolds, 1976) and furthermore, that the reversal of this procedure could be used to reconstitute the *in vivo* asymmetric bilayer system. It is in this context that a knowledge of colloid chemistry in general and the physicochemical properties of detergents in particular has become essential to the biological scientist.

4.1. Detergent–Lipid–Protein Interactions: A Reversible Process

The partitioning of detergents into lipid bilayers and the formation of mixed, soluble micelles were discussed in Section 3.3. Here, we are concerned with the thermodynamic reversibility of this process such that removal of detergent from the mixed micellar system leads to the formation of the preferred physical state of the pure lipid. The first detailed characterization of vesicles formed by this process was carried out by Brunner *et al.* (1976), in which sodium cholate was used to solubilize egg yolk lecithin and the detergent was subsequently removed by gel filtration chromatography. Since that time, other studies have appeared using different detergent systems (Mimms *et al.*, 1981; Ueno *et al.*, 1984), and in all cases single-wall phospholipid vesicles are formed as long as the mixed micellar system contains sufficient detergent to totally disperse the lipid and disrupt all direct lipid–lipid interactions. (It has been demonstrated experimentally that incomplete solubilization of lipids will lead to mixed populations of multi- and single-wall vesicles on detergent removal (Mimms *et al.*, 1981).)

Investigation of more complex systems consisting of protein in addition to lipid and detergent has demonstrated that here too detergent solubilization is a reversible process. Thus, intrinsic membrane proteins are reincorporated into lipid bilayers when the solubilizing amphiphile is sufficiently depleted. [Helenius *et al.* (1977) published what is probably the first study of true reconstitution of detergent solubilized, purified, membrane-associated proteins. Previous claims of reconstitution from detergent-containing systems are unsubstantiated experimentally; that is, complete solubilization of lipid and protein components in detergent micelles such that no "memory" of the original bilayer structure was retained was not demonstrated and, in fact, was highly unlikely given the low concentrations of detergent used.]

The interaction of lipids with aposerum lipoproteins has also been demonstrated to be under thermodynamic control in that removal of detergent from mixed micellar systems containing these proteins and naturally occurring lipids leads to the formation of water-soluble, lipid–protein particles similar to those found *in vivo* (Dhawan and Reynolds, 1983; Reynolds, 1982, 1984; Watt and Reynolds, 1981)—not to protein-containing lipid bilayers.

4.2. Kinetics

While it is reassuring to demonstrate experimentally that intrinsic membrane proteins and bilayer-forming lipids reassemble into the same physical structure found *in vivo*, complacency on the subject of thermodynamic control during reconstitution studies is not warranted. As is well known, the *rate* at which an equilibrium state is approached is not readily predictable, nor are the number of quasiequilibrium states that may exist along a particular reaction pathway. The kinetics of each step in the various molecular rearrangements that occur when detergent is removed from a mixed micellar system containing dispersed lipid (and protein) can have a profound effect on the "temporally" stable structures that are obtained—structures that are not necessarily at true equilibrium.

Let us consider as an example a system we have already discussed—diacyl phosphatidylcholine in an aqueous solvent. If we suspend the crystalline lipid in an aqueous buffer solution, multilamellar vesicles are formed. If we totally solubilize it in detergent mixed micelles and then remove the detergent, we obtain a population of only single-wall vesicles. Which of the two lipid structures is the true equilibrium state? The answer to this question is not known, but it is obvious that a kinetic barrier for interconversion between multiwall and single-wall vesicles must exist since both structures, once formed, are stable for weeks. Furthermore, the size of the single-wall vesicles formed by detergent depletion of a micellar system depends critically on the kinetics of detergent removal (Ueno *et al.*, 1984; see also Section 4.5).

4.3. Experimental Requirements for Lipid Vesicles

Reconstitution is of little interest for its own sake. We know that intrinsic membrane proteins are intercalated into lipid bilayers *in vivo*, so we have learned very little by simply being able to reinsert them under artificial conditions. The scientist is interested in reconstitution of *purified* membrane proteins so that he or she can study the mechanism of individual functions such as transport and transmembrane signaling. In order to do this, the reconstituted system must meet specific, minimal requirements.

1. The vesicles must be surrounded by a *single* bilayer.
2. The vesicles must be impermeable to ions and water-soluble macromolecules.
3. The internal volume must be sufficiently large to allow the experimental determination of transport rates.
4. The bilayer must be free of impurities that will affect the function and/or structure of the incorporated protein.
5. The average number of protein molecules per vesicle must be controlled and known.
6. The vectorial orientation of incorporated protein molecules must be known.

These requirements are qualitatively apparent to the experimentalist investigating the functional characteristics of purified, intrinsic membrane proteins that have been incorporated into a lipid bilayer. However, the quantitative implications as they relate to experimental design are often appreciated only after the fact, if at all. For example, in Table II are presented passive permeability and internal volume characteristics of pure egg yolk lecithin vesicles as a function of the external diameter. The half-life, $t_{1/2}$, for passive equilibration of Cl^- between the internal vesicle volume and the bulk solution is significantly faster than that for Na^+, and in small vesicles $t_{1/2}(Cl^-)$ is sufficiently low that this may become a limiting factor in designing experiments in which asymmetry in the anion concentration must be maintained. It will be obvious to those who follow the literature on reconstitution that few published studies have been re-

Table II
Size-Dependent Characteristics of Lipid Vesicles

Vesicle diameter (Å)	Internal volume (ml × 10^{15})	$t_{1/2}(Cl^-)$ hr[a]	$t_{1/2}(Na^+)$ hr[b]	Number of molecules at 10^{-3} M
600	0.065	3.2	160	39
1000	0.408	5.8	292	245
2000	3.71	12.0	602	2226
2600	8.38	16.0	802	5028

[a] Assuming a passive permeability coefficient of 5×10^{-11} cm/sec.
[b] Assuming a passive permeability coefficient of 10^{-12} cm/sec.

Table III
Membrane Potentials in Reconstituted Systems

Vesicle diameter (Å)	$d\psi/dt$ (mV/sec)[a]
600	92.3
1000	31.1
2000	7.5
2400	5.2

[a] Initial rate of formation of membrane potential assuming one copy of an electrogenic pump protein per vesicle and a molecular turnover number of 58 ions/sec per pump.

ported in which the "leak rates" are as slow as those shown in the table. Indeed, one commonly finds reported half-lives that are two to three orders of magnitude greater.

The last column in Table II shows the number of molecules present in the internal vesicle volume at a concentration of 10^{-3} M. Note that one copy of an active ion pump incorporated into a 600 Å vesicle with a molecular turnover number of 100 sec^{-1} will increase the intravesicular concentration of the transported ion to 0.0026 M in 1 sec—a concentration that is inhibitory for some ion transporting proteins (e.g., the CA^{2+} pump from sarcoplasmic reticulum).

A particularly troublesome problem arises when electrogenic transport proteins are incorporated into lipid bilayers since each kinetic cycle of such a protein will produce a membrane potential. Table III indicates the *initial* rate of formation of a membrane potential as a function of vesicle size when one copy of an electrogenic pump transporting one charge per cycle (molecular turnover number = 58 sec^{-1}) is incorporated. It is instructive to consider that many of the studies of the Na^+,K^+ protein pump have been carried out in 600 Å vesicles under conditions where approximately 200–300 copies of the protein have been incorporated per vesicle. Since the kinetic parameters of any pump protein carrying a net charge across the membrane are a function of the membrane potential, interpretation of data obtained in systems such as this is clearly impossible.

These cautionary words should not lead the reader to conclude that mechanistic studies are impossible in small reconstituted vesicular systems, but rather should emphasize the need for a few simple calculations *before* the experiment is carried out. Induced membrane potentials and concentration gradients, for example, can be collapsed by the use of appropriate, lipid-soluble ionophores; the time course of an experiment can often be adjusted to meet limitations inherent in the system being studied; the size of reconstituted vesicles can be altered by a variety of procedures and the effect of internal volume on a particular process can then be tested directly.

4.4. Experimental Procedures for Detergent-Induced Reconstitution

Solubilization. Solubilization of lipids in detergent mixed micelles is carried out by depositing the appropriate amount of lipid (normally obtained as a solution in organic solvent such as chloroform–methanol) on the surface of a glass tube. The organic solvent is evaporated by purging with an inert gas such as argon, and a suitable volume of aqueous buffer containing the detergent of choice is added to the tube. The final molar concentration of detergent must be above the CMC of the pure compound and, furthermore, must exceed that of the lipid by a factor of 8–10 if total dispersion of the lipid is to be assured. The solution is blanketed with argon and protected from exposure to ultraviolet light. Detergents with relatively high CMCs solubilize rapidly, but those with low CMCs are slow to equilibrate. For example, 24 hr are required to solubilize a film of egg yolk lecithin (1.2×10^{-5} mol) with 1 ml of 0.15 M $C_{16}E_{20}$. If, however, this detergent is codeposited with the lipid on the surface of a glass tube, addition of an aqueous solution will result in total solubilization of both components within 1 hr.

Detergent Removal. If the CMC of the detergent is sufficiently high that a reasonable monomer concentration gradient can be established across a dialysis membrane (i.e., $> 10^{-4}$ M), the detergent can be depleted by dialysis against large volumes of an appropriate buffer solution. A more generally applicable procedure, however, is the use of hydrophobic beads that irreversibly adsorb amphiphilic monomers. Since the available monomer concentration of detergents is significantly higher than that of diacyl lipids, the detergent is adsorbed more rapidly and lipid loss can be kept to a minimum by appropriate manipulative procedures. Furthermore, the *rate* at which detergent is removed from the mixed micellar system is readily controlled when hydrophobic adsorbents are used for detergent depletion.

A complete characterization with respect to both the number of adsorption sites and the rate constants applicable to the process must be obtained for each detergent–(lipid)–hydrophobic adsorbent system, if experiments are to be designed with rationality rather than guesswork. Hydrophobic beads are macroreticular structures with varying nominal pore size and discrete adsorption sites for amphiphilic molecules, the absolute number of such sites being a function of both the surface area of the bead and the size of the amphiphilic molecule interacting with this surface. The rate of adsorption is determined by the product of the monomer concentration of the amphiphile (adsorbate) and the unimolecular rate constant specific for this irreversible process. This latter parameter reflects the rate of diffusion to the sites, orientation factors, and the ease of access of the adsorbate through the macroreticular pores.

For an irreversible adsorption process with the detergent monomer concentration (i.e., CMC) constant, the rate of interaction of a homogeneous amphiphile with a hydrophobic surface containing i sets of sites is described generally by the following equation:

$$\text{unoccupied sites/total sites} = \Sigma_i (S_i/S_T) \exp[-k_i(\text{CMC})t] \tag{4}$$

where S_i = total number of sites of type i, S_T = total number of all sites, and k_i = forward rate constant for association with sites of type i.

Table IV contains data obtained in this laboratory for the interaction of three amphiphiles with SM-2 and SM-4 hydrophobic beads (Bio-Rad Laboratories). As might be intuitively expected, the total number of sites/gram of moist beads decreases as the hydrocarbon chain length increases, and the rates of adsorption are related to the relative sizes of the adsorbate and the average macroreticular pore size. (Note that variation in these parameters occurs with different lots of beads, temperature, and solvent composition. Hence, data must be obtained by the individual investigator under his or her own experimental conditions.)

The association of octyl glucoside with the hydrophobic beads is too rapid to be conveniently measured, and that of $C_{12}E_8$ with SM-2 is difficult to deter-

Table IV
Adsorption Properties of Hydrophobic Beads[a]

	SM-2	SM-4
Pore size (Å)	90	40
C_8 Glucoside		
Extended length = 15 Å		
CMC = 0.025 M		
Moles sites/g moist beads	3×10^{-4}	
$C_{12}E_8$		
Extended length = 27 Å		
CMC = 9×10^{-5} M		
Moles sites/g moist beads	1.7×10^{-4}	
$k_1 \times$ CMC (M/sec)	0.0405	
$C_{16}E_{20}$[b]		
Extended length = 40 Å		
CMC = 4×10^{-6} M		
Moles sites/g moist beads	1.5×10^{-4}	3.1×10^{-4}
$k_1 \times$ CMC (M/sec)	3.45×10^{-4}	2.92×10^{-4}
$k_2 \times$ CMC (M/sec)	9.25×10^{-5}	3.5×10^{-5}
S_1/S_T	0.33	0.14
S_2/S_T	0.67	0.86

[a] $T = 20°C$, 0.05 M TES, 0.236 M Na$^+$, 0.2 M Cl$^-$.
[b] $C_{16}E_{20}$ is a heterogeneous detergent with respect to head group, and the reported CMC is an average value.

mine with high accuracy. This latter, homogeneous detergent, however, appears within the limits of experimental error to interact with a single set of sites with a forward rate constant of approximately 450 sec^{-1}. On the other hand, the commercial detergent, Brij 58 (nominally $C_{16}E_{20}$) contains head groups of varying lengths, the distribution of which can be approximately described by a Poisson probability function. The degree of homogeneity of the hydrocarbon tail is not reported and will depend on the purity of the long-chain alcohol used in the manufacturing process. This detergent interacts with both SM-2 and SM-4 in a multiexponential fashion. Based on Equation (4), the simplest explanation for the multiexponential nature of the kinetics of adsorption of $C_{16}E_{20}$ will require two sets of sites with differing unimolecular rate constants (and therefore differing affinities), the values of which are given in Table IV. It should be appreciated, however, that this is not the only mechanism that will produce the experimental result. A decrease in the effective monomer concentration of this heterogeneous amphiphile as the result of selective depletion of specific chemical species during the adsorption process would also generate multiphasic kinetics. This latter explanation is considered unlikely since the data shown in Table IV were obtained under conditions where the *total* concentration of detergent was such that 60% of the original amphiphile remained in the solution at the point at which significant deviation from monoexponential kinetics was noted. Such a high proportion of species with a significantly lower free monomer concentration cannot be accounted for by a normal Poisson distribution of head groups nor by expected contamination levels of hydrocarbon chains greater than 16 carbons in length. For example, the species $C_{16}E_9$ has a CMC only a factor of 2 lower than that of $C_{16}E_{20}$ and should be present at approximately 0.29% of the total detergent; but the two pseudo-first-order rate constants in Table IV differ by a factor of 3.7 (SM-2) and 8.3 (SM-4) with a total detergent concentration that is still 60% of the original adsorbate.

For practical purposes it is not necessary for the biological scientist engaged in the *use* of hydrophobic beads to ascertain the molecular level explanation for a multiexponential adsorption process. It is only essential that he or she have available the pseudounimolecular rate constants that describe the process and the values of the preexponential terms that provide a measure of the apparent distribution of sites. This information is sufficient to allow a rational design of an experimental protocol to form lipid vesicles from detergent mixed micellar solutions.

Since the presence of foreign molecules in a micellar system (e.g., lipid dissolved in detergent micelles) leads to an alteration in the detergent monomer concentration, it is also necessary to determine the kinetics of adsorption to hydrophobic beads under these conditions. Figure 2 provides examples of such measurements using mixed micelles of $C_{16}E_{20}$ and egg yolk lecithin. The open circles represent the number of moles of detergent bound per gram of hydro-

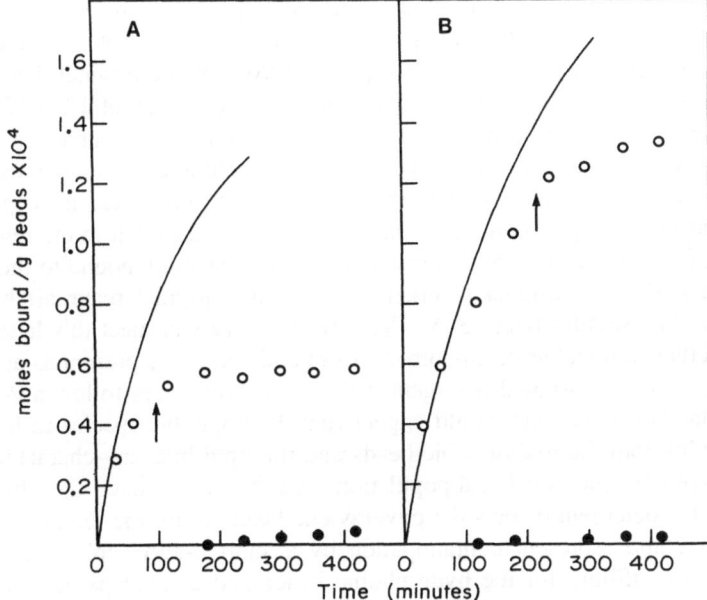

FIGURE 2. Kinetics of adsorption of $C_{16}E_{20}$ from egg yolk lecithin mixed micelles: (A) 1.5 g SM-2 hydrophobic beads, 4 ml 0.0026 M egg yolk lecithin, 0.023 $MC_{16}E_{20}$, 0.05 M TES, pH 7.5, 0.236 M Na^+, 0.2 M Cl^-; (B) 1 g SM-4 hydrophobic beads, 0.0364 M $C_{16}E_{20}$, 0.0035 M egg yolk lecithin, buffer as in (A). The arrows designate a detergent to lipid ratio in the solution of 1. ○, $C_{16}E_{20}$; ●, egg yolk lecithin.

phobic beads and are obtained by subtracting the concentration of nonadsorbed detergent (determined by chemical analysis) from the starting concentration. The filled circles represent the number of moles of bound lipid per gram of hydrophobic beads. The solid lines in the figure are the calculated rates of removal of this detergent in the absence of lipid (see Table IV). In these experiments the total number of sites available is 2.5 times greater than the total number of moles of detergent—a condition that one would normally use for rapid and complete removal of the detergent.

The rate of adsorption becomes extremely slow when the detergent has been depleted by 90% (marked by arrows in Figure 2—detergent to lipid ratio $\cong$ 1) despite the fact that fewer than 40% of the sites are occupied and the total detergent concentration is two to three orders of magnitude higher than the measured average CMC. Also, apparent lipid adsorption to the beads (filled circles) occurs only after this level of detergent depletion has been achieved.

It is of some interest to determine whether the detergent remaining in the solution at this point is actually associated with the lipid or occupies a separate phase and, furthermore, whether this residual detergent is representative of the original material or is a subpopulation. To this end, the solution was removed from the hydrophobic beads after adsorption of 96% of the original detergent. Chemical analysis revealed an egg yolk lecithin concentration of 1.8×10^{-3} M and a $C_{16}E_{20}$ concentration of 6×10^{-4} M. This solution was subjected to ultrafiltration through a Centricon 30 containing a membrane that allows passage of species with a radius less than 35 Å. The filtrate contained no egg yolk lecithin but the detergent concentration was nearly equal to that in the retentate indicating (1) little or none of this residual $C_{16}E_{20}$ is actually bound to the lipid vesicles and (2) the residual detergent (4% of the original material) has an effective radius smaller than 35 Å. $C_{16}E_x$ with $x \leq 12$ can meet this latter criterion whether in micellar or monomeric form and should be present as approximately 3.8% of the original detergent if the head group sizes follow a Poisson distribution. However, one would expect such hydrophobic species to interact strongly with both the hydrophobic beads and the lipid bilayer—characteristics not displayed by this small subpopulation. Alternatively, since the chemical assay for this detergent detects the oxyethylene head group, the remaining subpopulation could represent contamination by a short n-alkyl polyoxyethylene with very low affinity for the hydrophobic beads and a small partition coefficient for egg yolk lecithin.

These experiments with $C_{16}E_{20}$ have been presented in some detail because they are indicative of the experimental problems that must be addressed when heterogeneous detergents are used, and they emphasize the need for detailed studies of effects of differing hydrophobic and hydrophilic domains on each of the processes involved in solubilization and reconstitution. In light of the scarcity of homogeneous, nonionic detergents that are commercially available, most scientists will inevitably find themselves in the position of coping with the problems arising from polydispersity—problems that include, as we have shown here, differences in the properties and behavior of subpopulations and/or contaminants of the purchased material.

Experimental Protocols for Use of Hydrophobic Beads. For systems in which the pseudounimolecular rate constant, k_i(CMC), is high (e.g., octyl glucoside or $C_{12}E_8$), it is convenient to prepare small columns, 0.7×10 cm, filled with an appropriate weight of the hydrophobic beads. The beads are washed exhaustively with buffer solution, and excess liquid is removed by applying gentle pressure with a syringe attached to a Luer fitting on the column reservoir cap. The sample containing mixed micelles of detergent and lipid is then applied to the top of the column and rapidly forced through the moist beads by using the attached syringe. Removal of most of the aqueous solution from the beads prior to addition of the sample prevents undesirable dilution, and rapid passage of the sample minimizes lipid losses. Multiple passes are needed to reduce the detergent to the desired level since equilibrium with all sites is not accomplished by this method. Using this procedure and a two- to threefold excess of adsorption sites, intact vesicles can be formed in less than 10 min from detergent systems with CMCs greater than 5×10^{-5} M (e.g., octyl glucoside or $C_{12}E_8$).

The slow rates of adsorption exhibited by detergents with low CMCs (and the attendant long hydrophobic tails) preclude the column procedure we have just described. An alternative method involves sealing the mixed micellar solution and an appropriate weight of hydrophobic beads in a tube and rotating gently until the detergent is depleted. Customarily, we have used a 2- to 2.5-fold excess of adsorption sites and a volume of mixed micellar solution at least two to three times that of the moist beads. Following detergent depletion and formation of lipid vesicles, the contents of the tube are poured into a Bio-Rad disposable column containing a polyethylene support and gentle pressure is applied from a syringe until no further liquid emerges from the column exit line. (All manipulations, of course, are carried out under an argon blanket and in the absence of ultraviolet light.)

4.5. Kinetic Control of Vesicle Size

Early studies on vesicle formation from detergent mixed micelles were carried out using dialysis to deplete the single tail amphiphile—a procedure that is not readily adaptable to the study of kinetic factors in the determination of vesicle size. Thus, dialysis from octyl glucoside mixed micelles produced vesicles with an average external diameter of 2400 Å while the same procedure led to vesicles of 600 Å when deoxycholate was used as the solubilizing detergent. In neither case were the rates of detergent depletion followed, and it was assumed that these differences in size were in some way related to the type of detergent used.

The data presented in Table V show that, in fact, small vesicles are formed from three chemically distinct detergents when the rate of detergent depletion is rapid compared to that achieved in the customary dialysis procedures. Furthermore, if detergent depletion is halted at specific ratios of detergent to lipid and the solution allowed to stand for variable periods of time, the final population of vesicles is increased in average size. An example of this is shown in Table V in which $C_{12}E_8$ was removed from a mixed micellar solution containing egg yolk lecithin to a level of 1.4 mol of detergent/mol of lipid. The solution was incubated for 24 hr and the remainder of the detergent removed rapidly using hydrophobic beads. The average vesicle size was increased by this procedure from 560 to 880 Å (Ueno et al., 1984).

These observations have been investigated in more detail by preparing bath sonicated single-wall vesicles of pure egg yolk lecithin and studying the effects of incubation in the presence of added detergent on vesicle diameter.

Table V
Size of Vesicles Prepared from Mixed Micellar Solutions

Detergent	Method of preparation	Diameter (Å)	Residual detergent (m/m)
C_8 Glucoside	30 sec on hydrophobic beads	520	<0.01
	24 hr dialysis	2400	<0.005
	Dialysis to 4 m detergent/m lipid followed by rapid removal on hydrophobic beads	2400 and 850 (broad bimodal distribution)	<0.01
$C_{12}E_8$	5 min on hydrophobic beads	560	0.3
	3 min on hydrophobic beads followed by incubation for 24 hr at 1.4 m detergent/m lipid	880	0.3
$C_{16}E_{20}$	3 hr on hydrophobic beads	630	<0.01

The results are presented in Table VI. All the sonicated vesicle preparations were characterized with respect to permeability to Cl^- and/or Na^+, internal volume, and hydrodynamic diameter as determined by gel filtration chromatography on an S-1000 column (Reynolds et al., 1983). In all cases, the diameters determined from internal, trapped volume and from gel chromatography agreed within 10% indicating no significant population of multilamellar structures. (Permeability to Cl^- ranged from 2 to 7×10^{-11} cm/sec and to Na^+ was 2×10^{-12} cm/sec.)

Detergent was added and incubation of the solution carried out for the times indicated, following which the detergent was removed by adsorption on hydrophobic beads. The effect of incubation with the various single-tail amphiphiles on the permeability characteristics of the vesicles was tested, and in all but two of the instances cited in Table VI the added detergent resulted in a total loss of trapped volume suggesting that the original structure was fragmented. Incubation of these detergent-containing fragments led to a significant increase in the diameter of the vesicles formed following detergent removal.

In two instances, fragmentation did not occur. At a bound octyl glucoside to lipid ratio of 0.5 mol/mol (total octyl glucoside concentration below the CMC of the pure detergent), the vesicles remained intact by the criterion of permeability to ions, and incubation did not lead to an increase in vesicle size. The addition of 0.1 mol $C_{16}E_{20}$/mol egg yolk lecithin, on the other hand, also left the vesicles intact, but did lead to a significant increase in the average diameter. It therefore appears that more than one mechanism is operative in the alterations in the vesicle populations induced by added detergent.

Let us consider the possibilities. The mechanistic sequence of solubilization and reconstitution shown in Section 3.3 indicates that rupture of the intact lipid vesicle leads to the formation of fragments of detergent-containing bilayers. All but one of the experiments shown in Table VI as well as previous work by Ueno *et al.* (1984) are consistent with a mechanism in which the length of the detergent-containing lipid fragment is the determinant of the ultimate diameter of the vesicle. Thus, incubation of these fragments leads to progressive growth in fragment length, and subsequent detergent depletion leads to fusion to form closed vesicles, the diameter of which is a direct function of the fragment length.

However, it is clear that this mechanism cannot be operative in the case of $C_{16}E_{20}$, which caused an increase in average vesicle diameter at a mole ratio of 0.1 *without* rupturing the original vesicle. In this case, one must invoke some inherent property of the detergent itself that, even at low levels, can induce *vesicle* fusion, and the most likely property is the unique nature of the extremely large polyoxyethylene head group. It is not conceivable that this head group of 20 oxyethylenes sufficiently reduces the water activity at the

Table VI
Detergent-Induced Size Alterations in Preformed Phospholipid Vesicles[a]

Detergent	Detergent/Lipid (mol/mol)	Average diameter (Å)[b]		
		No detergent	$t = 3$ hr	$t = 22$ hr
$C_{16}E_{20}$	0.1	640		810
Deoxycholate	3.0	580	710	740
$C_{12}E_8$	1.0	600	900	840
	2.5	500	770	810
Octyl glucoside	0.5	600	600	600
	1.4	500	1800	2420

[a] Egg yolk lecithin at a concentration of 0.01 M was suspended in 0.05 M TES, 0.236 M Na$^+$, 0.2 M Cl$^-$, pH 7.5 and bath sonicated for 30 min at 4°C. The ratio of detergent to lipid is calculated as (total detergent − CMC)/total lipid except in the case of octyl glucoside at a ratio of 0.5. In this latter instance the total octyl glucoside concentration is below the CMC and the moles of associated detergent were calculated from the previously determined partition coefficient (Reynolds *et al.*, 1985). Addition of detergent at the tabulated ratio led to total loss of internal volume as measured by trapped Na$^+$ or Cl$^-$ in all cases except $C_{16}E_{20}$(0.1 mol/mol) and octyl glucoside (0.5 mol/mol).
[b] Average diameter refers to the diameter measured after incubation and subsequent detergent removal.

local surface of the bilayer to allow close approach of a similar region of another vesicle and subsequent fusion. [See Israelachvilli (1985) for a discussion of fusion mechanisms as they are currently envisioned.]

4.6. Incorporation of Intrinsic Membrane Proteins

Equilibration of the components of a mixed micellar system occurs by dissolution of single molecules from the aggregated state followed by diffusion through the aqueous solvent. This is a rapid process as long as the effective thermodynamic monomer concentration that can be sustained is equal to or greater than 10^{-6} M. Most detergents equilibrate rapidly, but phospholipids with attainable monomer concentrations less than 10^{-12} M do not. Intrinsic membrane proteins, for the most part, are totally insoluble in native form in aqueous media and are effectively trapped in a specific mixed micelle after solubilization in detergent. Attempts at reconstitution from micellar systems are therefore fraught with experimental difficulties, since depletion of detergent must be accompanied by local rearrangements of lipid and protein molecules at a sufficient rate to prevent exposure of the hydrophobic domain of the membrane protein to water. If lipid is present in the same micelle as the protein, there is less danger of inadvertent "denaturation" by this mechanism. However, if the lipid and the protein are present in separate micelles, collision and subsequent molecular rearrangement may be significantly slower than the loss of "protective" detergent.

One procedure that can be used to avoid these difficulties is to assure the incorporation of several moles of phospholipid into the protein–detergent mixed micelle by designing the experiment in such a manner that *no* pure micelles are present during dissolution of a phospholipid film (containing 5 or 6 mol lipid/mol micelle in solution) from the surface of a glass tube. Sufficient additional lipid is then prepared in a mixed micellar system as described in Section 4.4 to bring the total lipid concentration to that required for the desired final mole ratio of protein to lipid. The final solution will then contain two types of mixed micelle—one containing lipid alone and the other containing lipid and protein.

The use of hydrophobic beads in membrane-protein-containing systems has not been extensively investigated. However, since the rate of adsorption is rapid for detergents and extremely slow for either lipid of intrinsic-membrane proteins due to the low solubility of the latter two components in water (see Equation (4)), appropriate experimental design should prevent significant protein loss on the beads. Preliminary studies in this laboratory using the Ca^{2+} ATPase from sarcoplasmic reticulum substantiate this view.

5. SOME CONCLUDING COMMENTS

Our understanding of the mechanistic processes involved in lipid–protein–detergent interactions is far from complete, and there are, unfortunately, no "cookbook recipes" for reconstitution of lipid vesicles and intrinsic membrane proteins. However, years of research in fields such as colloid and surface chemistry, protein structure and function, chemical equilibria, and kinetics provide a fundamental body of knowledge on which one can build. This fundamental framework was provided by individuals indulging their intellectual curiosity about the world around them, and it behooves us to remember, particularly in this period of overwhelming emphasis on "applied science," that technological advances *follow* basic discoveries. Rarely can the applicability of truly intellectual endeavors be predicted.

6. REFERENCES

Brunner, J., Skrabal, P., and Hauser, H., 1976, Single bilayer vesicles prepared without sonication. *Biochim. Biophys. Acta* **455**:322–331.

Collander, R., and Bärlund, P., 1933, Permeabilitatstudien an chara ceratophylla. *Acta Bot. Fenn.* **11**:1–114.

Dhawan, S., and Reynolds, J. A., 1983, Interaction of apolipoprotein B from human serum low density lipoprotein with egg yolk phosphatidylcholine and cholesterol. *Biochemistry* **22**:3660–3664.

Donbrow, M., 1987, Stability of the polyoxyethylene chain. In *Nonionic Surfactants, Physical Chemistry* (M. J. Schick, ed.), pp. 1011–1072, Marcel Dekker, New York.

Fish, W. W., Reynolds, J. A., and Tanford, C., 1970, Gel chromatography of proteins in denaturing solvents. *J. Biol. Chem.* **245**:5166–5168.

Green, D. E., 1972, Membrane proteins: A perspective. *Ann. N.Y. Acad. Sci.* **195**:150–172.

Hartley, G. S., 1936, *Aqueous Solutions of Paraffin Chain Salts*, Hermann et Cie, Paris.

Helenius, A., and Simons, K., 1975, Solubilization of membranes by detergents. *Biochim. Biophys. Acta* **415**:29–79.

Helenius, A., Fries, E., and Kartenbeck, J., 1977, Reconstitution of Semliki Forest virus membrane. *J. Cell Biol.* **75**:866–880.

Henderson, R., and Unwin, P., 1975, Three dimensional model of purple membrane obtained by electron microscopy. *Nature* **257**:28–32.

Höber, R., 1945, *Physical Chemistry of Cells and Tissues*, Blakiston, Philadelphia.

Huang, C., and Mason, J. T., 1978, Geometric packing constraints in egg phosphatidyl choline vesicles. *Proc. Natl. Acad. Sci. USA* **75**:308–310.

Israelachvilli, J. N., 1985, *Intermolecular and Surface Forces*, Academic Press, Orlando.

Lapanje, S., 1978, Physicochemical Aspects of Protein Denaturation, Wiley, New York.

McBain, J. W., 1913, Mobility of highly charged micelles. *Trans. Faraday Soc.* **9**:99–101. (Translated from Beweglichkeit von Hochbeladenen Mizellen, *Kolloid Z.* **12**:256–257.)

McBain, J. W., 1950, *Colloid Science*, Heath, Boston.

Mimms, L. T., Zampighi, G., Nozaki, T., Tanford, C., and Reynolds, J. A., 1981, Phospholipid vesicle formation and transmembrane protein incorporation using octyl glucoside. *Biochemistry* **20**:833–840.

Ollivon, M., Eidelman, O., Blumenthal, R., and Walther, A., 1988, Micelle-vesicle transition of egg phosphatidyl choline and octylglucoside. *Biochemistry* **27**:1695–1703.

Reynolds, J. A., 1982, Interactions between proteins and amphiphiles. In *Lipid–Protein Interactions* (P. Jost and O. H. Griffith, eds.), Vol. 2, pp. 193–224, Wiley, New York.

Reynolds, J. A., 1984, Interaction of apolipoprotein A1 from human serum high density lipoprotein with egg yolk phosphatidylcholine. *Biochemistry* **23**:1124–1129.

Reynolds, J. A., and Tanford, C., 1970a, Binding of dodecylsulfate to proteins at high binding ratios. *Proc. Natl. Acad. Sci. USA* **66**:1002–1007.

Reynolds, J. A., and Tanford, C., 1970b, The gross conformation of protein–sodium dodecylsulfate complexes. *J. Biol. Chem.* **245**:5161–5165.

Reynolds, J. A., Nozaki, Y., and Tanford, C., 1983, Gel exclusion chromatography on S-1000 Sephacryl: Application to phospholipid vesicles. *Anal. Biochem.* **130**:471–474.

Reynolds, J. A., Tanford, C., and McCaslin, D. R., 1985, Detergents and the Na,K pump protein. In *The Sodium Pump* (I. M. Glynn and C. Ellory, eds.), pp. 1–9, The Company of Biologists, Ltd., Cambridge.

Singer, S. J., 1971, The molecular organization of biological membranes. In *Structure and Function of Biological Membranes* (L. I. Rothfield, ed.), pp. 145–222, Academic Press, Orlando.

Spatz, L., and Strittmatter, P., 1971, A form of cytochrome b_5 that contains an additional hydrophobic sequence of 40 amino acid residues. *Proc. Natl. Acad. Sci USA* **68**:1042–1046.

Steinhardt, J., and Reynolds, J. A., 1969, *Multiple Equilibria in Proteins*, Academic Press, Orlando.

Tanford, C., 1968, Protein denaturation. *Adv. Protein Chem.* **23**:121–282.

Tanford, C., 1970, Protein denaturation. *Adv. Protein Chem.* **24**:2–93.

Tanford, C., 1980, *The Hydrophobic Effect*, Wiley, New York.

Tanford, C., and Reynolds, J. A., 1976, Characterization of membrane proteins in detergent solutions. *Biochim. Biophys. Acta* **457**:133–170.

Traube, J., 1891, Ueber die Capillaritatsconstanten organischer Stoffe in wassenigen Losungen. *Liebig's Ann. Chem.* **265**:27–55.

Ueno, M., Tanford, C., and Reynolds, J. A., 1984, Phospholipid vesicle formation using nonionic detergents with low monomer solubility. Kinetic factors determine vesicle size and permeability. *Biochemistry* **23**:3070–3076.

Watt, R. M., and Reynolds, J. A., 1981, Interaction of apolipoprotein B from human serum low density lipoprotein with egg yolk phosphatidylcholine. *Biochemistry* **20**:3897–3901.

Weber, K., and Osborn, M., 1969, The reliability of molecular weight determinations by dodecyl sulfate–polyacrylamide gel electrophoresis. *J. Biol. Chem.* **244**:4406–4412.

Chapter 2

Membrane Lipid Phase Behavior and Lipid–Protein Interactions

P. J. Quinn

1. INTRODUCTION

One of the most distinguishing features of nearly all biological membranes is that they contain a highly complex assortment of polar lipids. Only a few major lipid classes can be recognized but there is a whole spectrum of molecular species within each of these major lipid classes which differ in the type, length, and number of unsaturated residues of their hydrophobic component. There is a general consensus, based on observations that have been obtained by a variety of biophysical methods, that the lipid constituents of all biological membranes are arranged in a bilayer configuration in which the polar groups are located on the outside, in contact with the aqueous medium, and the hydrocarbon substituents are oriented toward the interior to form a hydrophobic domain that excludes water. It has been argued on the basis of the hydrophilic to hydrophobic balance within the molecules that membrane lipids have a relatively low critical micellar concentration and a discrete distribution of domains within the molecule and this is responsible for creating a stable bilayer structure.

Although the bilayer arrangement appears to be the dominant phase of lipids in biological membranes (Blaurock, 1983), this is not necessarily the phase preferred by individual molecular species of lipid isolated from particular

---

P. J. Quinn Department of Biochemistry, King's College London, Kensington Campus, London W8 7AH, United Kingdom.

biological membranes. If polar lipids, for example, extracted from biological membranes are separated into molecular species and then dispersed in water or dilute solutions of inorganic salts, they are found to form one of a number of well-characterized phases. The structures include bilayer phases in which the hydrocarbon components are arranged in either a crystal lattice, a gel phase, or a liquid-crystal configuration. In addition, a number of nonbilayer phases, such as hexagonal II or cubic phases, are also found at temperatures approximating to the growth temperature of the organisms from which the lipid was extracted. The composition of the molecular species, even of membranes that perform very simple functions, is highly complex, and individual molecular species of lipid may number more than 100 distinct chemical entities. In general, individual molecular species of lipids that will form any one of these particular phases are likely to be constituents of most biological membranes. Of particular interest, therefore, is the phase behavior of complex lipid mixtures of the type encountered in biological membranes since this will provide information that will be helpful in understanding the factors that govern the structure and stability of the lipid matrix of membranes. It is largely these factors that determine the limits within which organisms will grow and ultimately survive.

The importance of the phase behavior of mixed lipid systems is that it is now widely believed that the characteristics of membrane lipid phase behavior underlie many physiological processes. Included among these are the fusion of membranes, the transduction of signals across membranes, and the stability of membranes when organisms are subjected to physiological stresses. Furthermore, the processes of phase transition and phase separation are central to problems associated with the successful low-temperature storage of living cells and tissues.

This review will provide a summary of the polymorphic behavior of membrane polar lipids. An understanding of the phases adopted by biological membrane lipids in aqueous systems will form the basis for a discussion of the phase behavior of mixed lipid systems and the factors that underlie mixing and phase separation of lipids in binary mixtures. Finally, this information will be used to describe evidence for phase mixing and phase separations of lipids and lipid–protein complexes within biological membranes. From this discussion some indications of the way membrane proteins interact with the lipid constituents may be attempted.

2. POLYMORPHISM OF MEMBRANE POLAR LIPIDS

The macroscopic structures formed by membrane polar lipids when dispersed in aqueous systems depend on the precise chemistry of the lipid molecules and vary according to the temperature, salt concentration, pH, hydrostatic pressure, and amount of water relative to lipid in the system. In this section, methods used to determine lipid phase characteristics and the types of macroscopic structures observed in these lipids will be examined.

2.1. Biophysical Characterization of Membrane Lipid Polymorphism

The polymorphic behavior of membrane polar lipids in aqueous systems has been extensively investigated using a variety of biophysical techniques. Each technique has its own particular advantages and drawbacks and, in general, the combination of several approaches to lipid phase determination is required to provide unambiguous phase assignments. In this section some applications of specific methods will be described and their usefulness in characterizing phase behavior of lipid–water systems will be discussed in the context of relevant examples.

2.1.1. Static X-Ray Diffraction Methods

The classical technique for structural assignments of the macroscopic organization of lipid aggregates is small-angle X-ray diffraction (for reviews see Luzzati *et al.*, 1968; Shipley, 1973). Where these lipid systems possess some form of long-range order, an analysis of the relationship between the diffraction orders can be used to provide a reliable method for lipid phase determination. Thus, in multibilayer systems the spacings of the first- and higher-order Bragg reflections correspond as $1 : \frac{1}{2} : \frac{1}{3} : \cdots$ which is in good agreement with theoretical predictions (Harlos and Eibl, 1980; Janiak *et al.*, 1976; Luzzati *et al.*, 1968; Rand *et al.*, 1971; Reiss-Husson, 1967; Shipley, 1973). In practice, the relative intensity of the various Bragg reflections is highly variable and depends to a large extent on the nature of the sample. In some multilamellar systems, for example, only the first-order reflection can be observed (Harlos and Eibl, 1980; Marsh and Seddon, 1982), in which case the unambiguous assignment of the phase cannot be made. The first-order low-angle reflection represents the lamellar repeat distance or d spacing, which is the dimension of the bilayer thickness plus the thickness of the intervening water layer between successive lipid bilayer sheets.

Other types of macroscopic structure can be recognized from the ratio of the higher-order Bragg reflections. Thus, sharp reflections in ratio $1 : 1/\sqrt{3} : 1/2 : 1/\sqrt{7}$ is another commonly observed combination in some lipid dispersions (Luzzati *et al.*, 1968; Marsh and Seddon, 1982; Seddon *et al.*, 1984). This order is consistent for a two-dimensional periodicity of hexagonally packed cylinders and the orientation of the molecules within these cylinders for naturally occurring polar lipids is the hexagonal-II arrangement. This is illustrated in Figure 1, which shows the polar head groups of the molecules oriented

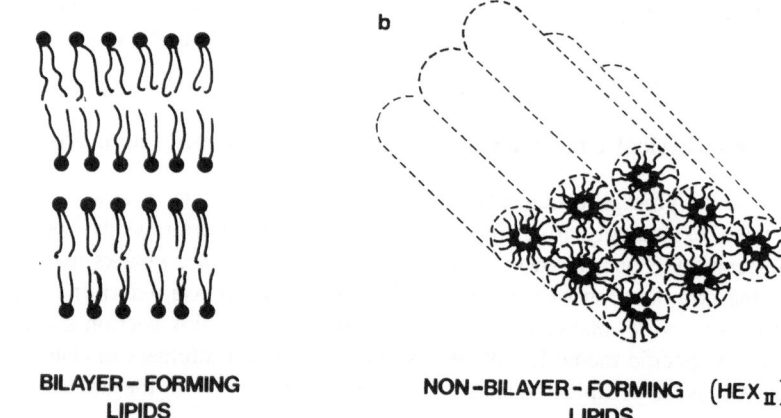

FIGURE 1. Molecular arrangements of membrane polar lipids in (a) bilayer and (b) hexagonal-II structures. The circles represent the polar group and the lines are the hydrocarbon chains.

toward the aqueous core of the cylinder with the hydrocarbon chains radiating out into the interior of the structure. An arrangement of the hexagonal-I type is not typical of membrane lipids but is formed by certain detergents and membrane destabilizing agents such as the lysophosphatides. In hexagonal-I structure, the polar head groups reside at the periphery of the cylinders and the hydrocarbon chains project into the core. The highest-order reflection of the hexagonal phase, d, is related to the diameter of the cylinders, a, in the form $d = (\sqrt{3}/2)a$.

Other three-dimensional phases are also recognized and are of the types indexing rectangular, rhombohedric, and cubic structures (Luzzati and Tardieu, 1974). In such arrangements there are often many reflections and, provided enough of these are observed, it is possible to characterize the space group and lattice type to which the structure can be assigned. Cubic phases are common among membrane lipids (Larsson *et al.*, 1980; Lindblom *et al.*, 1979), one of which consists of a structure based on deformed bilayer units that are attached to create two different continuous aqueous networks.

More recently, direct methods have been adapted to obtain information on the precise structural rearrangements that are associated with transitions between phases. Conventional X-ray diffraction methods using highly focused beams require sample exposure times of at least minutes to record useful diffraction patterns (Akiyama *et al.*, 1982) and are suitable only for relatively slow transitions (Ruocco and Shipley, 1982a). X-rays from synchrotron sources, however, are several orders of magnitude more intense than can be achieved by these machines (Kirkland *et al.*, 1983; Nave *et al.*, 1985), hence diffraction measurements can be made on time scales of 10–50 msec without serious radiation damage to the specimen (Caffrey, 1984). Since the time for transition between phases of membrane lipid–water systems is generally on the order of seconds, the method can be used to determine the kinetics as well as the mechanism associated with transition from one phase to another (Lis and Quinn, 1987).

Time-resolved X-ray diffraction methods have been used to study conformational transitions in membranes and lipid extracts. Conformational transitions in membranes and lipid extracts from an unsaturated fatty acid auxotroph of *Escherichia coli* and in lipids containing a preponderance of branched chain substituents extracted from *Bacillus subtilis* have been reported using synchrotron radiation X-ray diffraction methods (Ranck *et al.*, 1984). Other studies have concentrated on characterizing lamellar to hexagonal-II transition kinetics in pure phosphatidylethanolamine–water systems (Caffrey, 1985). The mechanism of the transition is said to be a nucleation and growth process initiated by bilayer fusion. An example of the use of time-resolved X-ray diffraction methods to investigate the kinetics of phase transitions is shown in Figure 2. This shows an experiment in which a saturated monogalactosyldiacylglycerol–water system is subjected to a temperature jump from 25°C at the rate indicated in

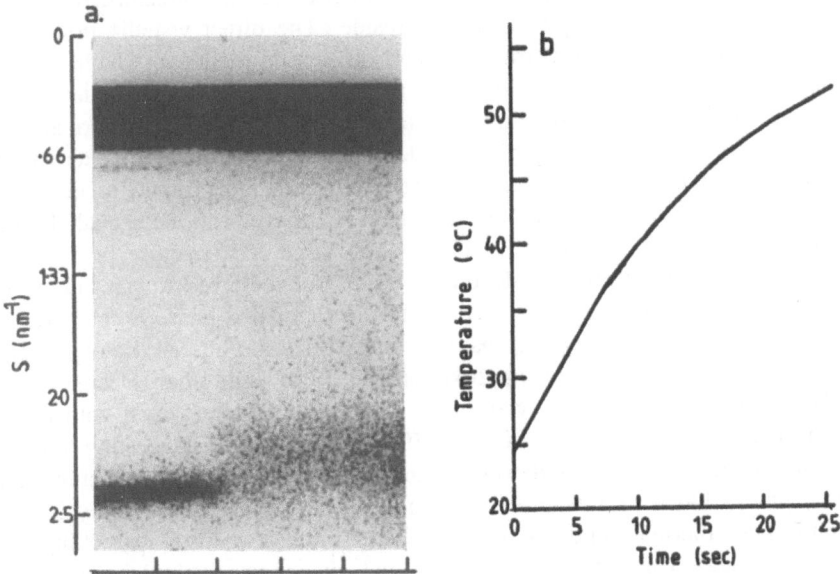

FIGURE 2. Kinetics of the gel to liquid-crystalline phase transition in a saturated monogalactosyldiacylglycerol–water system recorded by time-resolved X-ray diffraction methods. (a) X-ray scattering intensity as a function of diffraction spacing and (b) corresponding sample temperature recorded as a function of time.

Figure 2b. The intense diffraction maxima at wide angle corresponding to the gel phase packing of the acyl chains gives way to a diffuse scattering consistent with disordered hydrocarbon chains as the phase undergoes a transition from the gel to liquid-crystalline arrangement. At the same time the higher-order lamellar repeat observed in the gel phase shows that the lamellar phase is disordered at the transition temperature.

2.1.2. Electron Microscopic Methods

A direct method of examination of lipid polymorphism is the use of electron microscopic techniques. Possibly the most useful of these techniques has been freeze–fracture electron microscopy. The reasons for this are that reliable thermal quenching procedures have been developed to overcome relaxation phenomena associated with thermotrophic events. Replicas prepared from deposition of platinum on the fractured surface are relatively durable in the electron beam and the phase can be examined directly from the surface features at a resolution of about 3 nm (Verkleij and Ververgaert, 1978; Zingsheim, 1972). The planes of weakness through which the fracture passes are generally believed to be the hydrocarbon interior of structures where the terminal ends of the hydrocarbon chains reside. The direct visualization of the structure contrasts with spectroscopic and diffraction methods, which generate only averaged information of structure throughout the specimen under examination. Thus, the freeze–fracture technique has enabled the characterization of phase separations such as lipidic particles interpreted as inverted spherical micelles sandwiched within a bilayer structure which are difficult to determine unambiguously by other methods. Examples of bilayer and hexagonal-II structure are illustrated in Figure 3. These show a smooth fracture plane in the multibilayer structure that is devoid of water and believed to pass along the central domain of the bilayer structure. The hexagonal-II phase is characterized by fracture faces exposing long parallel lines which are observed along at least two fracture planes at angles approximately 120° to each other (Deamer et al., 1970). The periodicity observed within the structure agrees well with the repeat distances observed by X-ray diffraction methods, suggesting that the phases are preserved during thermal quenching. Relaxation phenomena in mixed lipid phases have been observed in large lipid aggregates presumably because of the poor thermal conductivity and slower rates of thermal quenching (Sen et al., 1982a). Current methods of thermal quenching, including jet-freezing (Moor et al., 1976), allow sample preparations in the absence of cryopreservatives that are known to alter lipid polymorphic behavior (Sen et al., 1982b).

Lipid Phase Behavior and Lipid–Protein Interactions

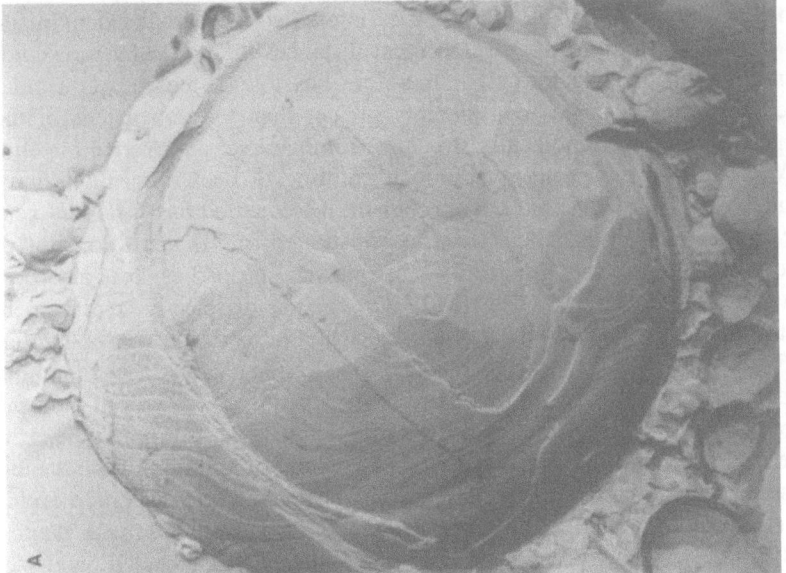

FIGURE 3. Freeze–fracture electron micrographs of replicas prepared from aqueous lipid dispersions forming (A) multibilayer liposomes, (B) hexagonal-II structure.

2.1.3. Nuclear Magnetic Resonance Methods

While X-ray and freeze–fracture techniques yield mainly static information, NMR methods can provide both static and dynamic information. Thus, apart from characteristic average structural properties of lipid–water phases, it is possible to derive rotational and translational diffusion coefficients and correlation times for molecular motions. In respect of the structural properties of these systems, information on the average conformation of the lipid acyl chains and polar head groups can be obtained as well as an indication of specific and nonspecific interactions between molecules in mixed lipid systems. An important parameter that can also be obtained is the order parameter that is a measure of the angular distribution of molecules about a preferred molecular orientation. In bilayer arrays, for example, there is effective axial symmetry about the normal to the bilayer surface referred to as the director (McLaughlin *et al.*, 1975; Seelig and Niederberger, 1974; Stockton *et al.*, 1974).

Translational diffusion of lipids in lipid–water systems can be measured by pulsed field gradient NMR techniques (Bull and Lindman, 1974; Charvolin and Rigny, 1971; Kuo and Wade, 1979; Lindblom and Wennerstrom, 1977; Roeder *et al.*, 1976; Stilbs, 1987; Stilbs *et al.*, 1984). The major advantages of the NMR methods over other methods are, first, a direct measurement of diffusion coefficients can be obtained and no probes or model-dependent assumptions are required. Second, the time during which measurements are made can be varied according to the displacement distances the particular molecules experience within the structure. It is therefore possible to measure molecular displacements over distances considerably greater than the dimensions of the components of the macromolecular structure, for example, an inverted micelle.

NMR methods exploiting quadrupolar splitting of deuterium nuclei have been widely used to study lipid polymorphism. The method has wide and very specific applications in that the ^2H can be substituted for ^1H in all domains of the lipid without significantly perturbing the physical properties. It is also possible to substitute deuterated water for ^1H_2O. The use of selective replacement techniques in such studies has been the subject of a number of reviews (Seelig, 1977; Seelig and Seelig, 1980). In anisotropic arrangements the deuterium quadrupolar resonance is split to an extent that depends on the size of the lipid structures and decreases as motional averaging tends to an isotropic type. It is possible to derive an order parameter from the extent of splitting of the deuterium resonances for particular residues of lipids in the different polymorphic phases (Gally *et al.*, 1980). The spectral features of phospholipid with deuterium substituted in the hydrocarbon chains in bilayer, hexagonal-II, and isotropic phases are presented in Figure 4.

Distinct types of cubic phases can be characterized from a comparison of the lipid translational diffusion coefficient in cubic and lamellar liquid crystal

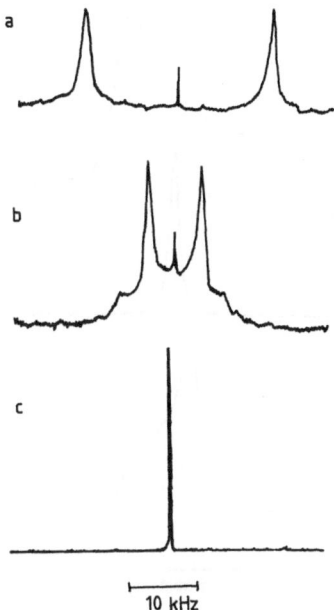

FIGURE 4. Typical ^{2}H-NMR spectra of aqueous dispersions of phospholipid-containing deuterium substitution at the [9,10] position of the acyl chains in (a) bilayer, (b) hexagonal-II, and (c) micellar configurations.

phases and from the diffusion coefficient of water in cubic phases and in pure water (Lindblom and Wennerstrom, 1977). Structures with continuous regions of both hydrocarbon chains and water (Lindblom et al., 1979; Scriven, 1976) and either discontinuous hydrocarbon regions but continuous water domains (Bull and Lindman, 1974; Eriksson et al., 1982) or with discontinuous water domains but with continuous hydrocarbon regions (water in oil emulsions) can all be distinguished. In the case of bicontinuous cubic systems, if diffusion of lipid molecules can take place over macroscopic distances without polar groups entering the hydrocarbon phase or exposure of hydrocarbons to water, the diffusion coefficient is found to be in the same order as in liquid-crystal bilayers. Similarly, if water diffusion is in the same order as in bulk water, then the cubic phase is of an oil in water emulsion type of phase (Rilfors et al., 1986).

^{31}P-NMR has been extensively used to characterize phospholipid polymorphism (Cullis and de Kruijff, 1978a,b; Cullis et al., 1985; de Kruijff et al., 1985; Seelig, 1978; Tilcock et al., 1986). The method provides a convenient and quantitative estimate of macromolecular structures found in most of the major lipid–water phases. This is derived from line shapes of proton decoupled spectra, which indicate chemical shielding anisotropy of the lipid phosphate moiety. Thus, different orientations of the phosphate segment yield characteristic resonances at different frequencies. In the case of bilayers (Figure 5a) the ^{31}P-NMR spectrum is typical of a shielding tensor that is axially symmetric

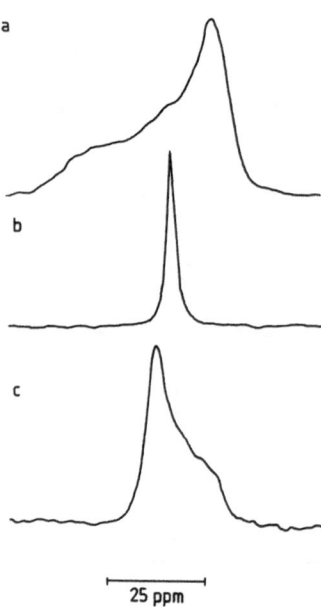

FIGURE 5. Typical ^{31}P-NMR spectra of aqueous dispersions of phospholipids in (a) bilayer, (b) micellar, and (c) hexagonal-II configurations.

25 ppm

around a director axis consisting of a low-field shoulder and high-field peak. Oriented bilayers give rise to individual narrow resonances with an angular dependent chemical shift (Hemminga and Cullis, 1982; McLaughlin et al., 1975, 1981; Seelig and Gally, 1976). Such studies indicate that molecules with their long axis perpendicular to the magnetic field contribute to the high-field peak and molecules with a parallel orientation correspond to resonances at a position in the shoulder region, and that the director axis coincides with the bilayer normal about which the phosphate segment rotates.

The rate of reorientation resulting from Brownian motion of lipid molecules in bilayers can also be determined from the line shape (Burnell et al., 1980). When the rate of reorientation of the phosphate segment is fast the line shape narrows until eventually, when motion is effectively isotropic on the time scale of the measurement (<10 μsec), motional averaging leads to a narrow and symmetric resonance line (Figure 5b). Motion of this type is experienced by phospholipid molecules in small, rapidly tumbling bilayer vesicles, micelles, inverted micelles, and a variety of cubic phases. Another common phase structure that can readily be identified is hexagonal-II structure. Although there is virtually no difference in the motion experienced by the phosphate segment of phospholipid in bilayer and hexagonal II, the macroscopic orientation of the lipid cylinders within the spectrometer magnetic field gives rise to additional averaging of the chemical shift anisotropy. The resulting spectrum of molecules in hexagonal-II arrangement is illustrated in Figure 5c.

2.1.4. Absorption Spectroscopy

Various forms of vibrational spectroscopy have been used to characterize the polymorphic behavior of lipids. Among the earliest reports of hydrated lipids was that of Chapman *et al.* (1967), who examined lipids at low water contents by conventional infrared spectroscopy. One of the major problems with the method, however, is that the absorption of water tends to dominate the spectrum. This can be overcome by using Raman spectroscopy, and many studies have been undertaken using this technique (Cary, 1982; Lord and Mendelsohn, 1981; Wong, 1984). The adaptation of conventional infrared methods, however, has helped to overcome the problem of water absorption, including the use of attenuated total reflection methods (Fringeli and Gunthard, 1981). Furthermore, the introduction of interoferometric Fourier transform infrared spectroscopy together with computer data analysis has proved very useful for characterizing lipid polymorphism (Casal and Mantsch, 1984). Absorption bands corresponding to CH_2 antisymmetric and symmetric stretching modes can be used to determined *trans/gauche* isomerizations of the acyl chains. Acyl chain packing and conformation also affects absorption bands associated with CH_2 bending or scissoring modes. C—C stretching and low-frequency vibrational modes can be detected only by Raman spectroscopy and have also been used to assign phases to acyl chain domains. As in ^{2}H-NMR, deuteration at specific sites in the molecule causes modification of the infrared spectrum and provides useful information on motion and structure of lipids (Mendelsohn and Koch, 1980; Smith and Mantsch, 1979) and biomembranes (Cameron *et al.*, 1981; Casal *et al.*, 1979, 1980, 1982, 1983; Sunder *et al.*, 1978). Absorption bands corresponding to residues found in the polar head group of lipid molecules have also been identified (Arrondo *et al.*, 1984).

2.1.5. Thermal Analysis

Transitions between phases are usually characterized on the basis of enthalpy changes measured by some form of calorimetry. Although differential thermal analysis has been used for this purpose, the most useful system is differential scanning calorimetry (McElhaney, 1982; Silvius, 1982; Small, 1986). This method relies on the accurate measurement of heat required to sustain a constant change in temperature within the specimen compared with a reference of similar heat capacity. While the method does not provide any structural information, it gives an accurate temperature at which conversion from one

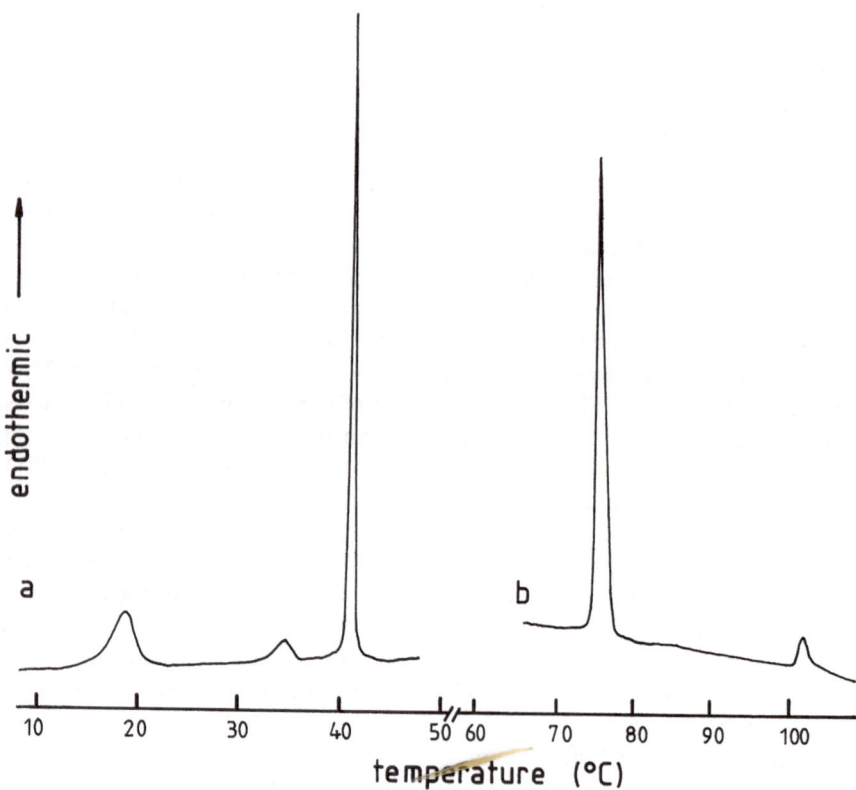

FIGURE 6. Differential scanning calorimetric heating scans of (a) 1,2-dipalmitoylphosphatidylcholine dispersed in excess water and equilibrated for 3 days at 0°C and (b) 1,2-distearoylphosphatidylethanolamine in excess water after several cycles of heating and cooling.

phase to another takes place as well as the enthalpy of the particular transition. Useful information on phase mixing and compound formation can also be derived. Thus, the relative cooperativity of a phase transition can be evaluated directly from the thermogram since it is related to the sharpness of the excess specific heat absorption curve of an endothermic transition. This is generally expressed as the temperature width at half height of the heat absorption peak or as the temperature difference between the onset or low-temperature boundary of the phase transition and the completion or upper limit of the transition temperature. Very pure, single-component lipids exhibit highly cooperative chain melting endotherms with temperature widths at half height of less than 0.1°C, but this may extend to 10–15° for complex mixtures and biological membranes. Differential scanning calorimetry can be used on the basis of the perturbing effects of membrane proteins on lipid phase transitions and phase separation processes to investigate the interaction between protein and lipid components (McElhaney, 1986).

The transition enthalpy of mesomorphic phase changes depends on the type of transition involved and, with the exception of transitions between certain bilayer phases, heating transitions are almost invariably endothermic processes. The thermogram shown in Figure 6a is a differential scanning calorimetric heating scan of dipalmitoylphosphatidylcholine equilibrated for 3 days at 0°C, in which the relative enthalpies (peak areas) of the subtransition (18°C), the pretransition (34°C), and the main transition (41°C) can be clearly seen. The relative enthalpies of bilayer to nonbilayer transitions are illustrated by the thermogram of distearoylphosphatidylethanolamine (Figure 6b), in which the main gel to liquid-crystalline phase transition at 74°C precedes the lamellar to hexagonal-II transition at 101°C. It should be emphasized that the structural transitions in each of these thermal events must be verified by independent methods such as X-ray diffraction and freeze–fracture electron microscopy.

2.1.6. Dilatometry

The specific volume of each particular mesophase is characteristic of that phase, and transition between phases results in a change in specific volume. This change as a function of temperature can be monitored accurately by direct forms of dilatometry (Wilkinson and Nagle, 1978). The method can be adapted to measure changes in specific volume during temperature scans of lipid dis-

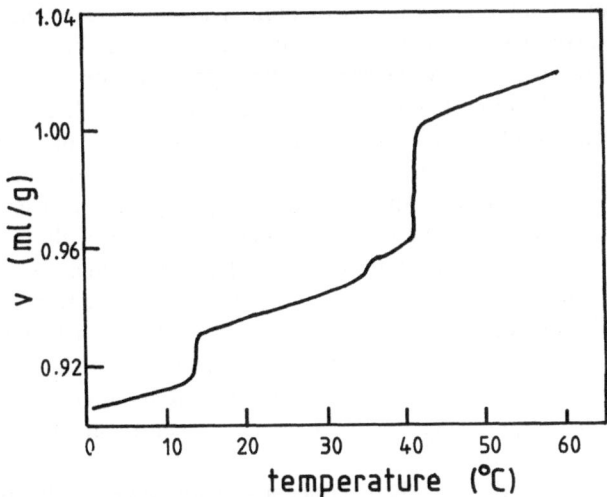

FIGURE 7. Relationship between absolute specific volume, v, of aqueous dispersions of 1,2-dipalmitoylphosphatidylcholine and temperature.

persions or in a thermal quench mode in which the temperature is lowered as rapidly as possible to a designated quenching temperature where it is subsequently held while changes in volume are recorded as a function of time. The relationship between absolute specific volume of 1,2-dipalmitoylphosphatidylcholine dispersed in excess water as a function of temperature is shown in Figure 7. It can be seen that there are changes in specific volume associated with all the transitions observed in the bilayer phase of this lipid by calorimetry and the temperatures at which the transitions occur are coincident. Again, however, the structural transitions cannot be determined from the measurements and it is necessary to assign these using direct structural methods.

2.2. Factors Responsible for Determining Lipid Phase

It is a general observation that some species of membrane lipid form one particular phase under a given set of conditions while another molecular species will form a different phase under the same conditions. A large amount of data have accumulated over the years on both lipids of natural origin as well as on pure synthetic materials which have helped to establish the principles on which the phase behavior of these membrane components depend.

One way that the phase characteristics have been described is in terms of the so-called molecular shape concept. According to this idea, those lipids with a relatively large polar head group relative to the cross-sectional area of the hydrocarbon chains are defined as inverted cone-shaped molecules. Lipids that would fall into this category are primarily lysophospholipids and detergents, which are not common constituents of biological membranes and, in general, will have a disruptive effect on membrane stability. The form in which these lipids are found in excess water are micellar aggregates or a hexagonal-I type of structure. On the other hand, lipids with a smaller polar head group relative to the hydrocarbon chains are referred to as cone-shaped lipids and in excess water will tend to form inverted structures such as the hexagonal-II phase or inverted micellar structures. An intermediate lipid type, which is said to have a cylindrical structure, is where the cross-sectional area of the acyl chains is comparable to that of the polar head group of the molecules. Such lipids will tend to form smectic mesophases or, on vigorous dispersion, bilayer vesicles. In some of the cubic phases that have been described and that appear to be intermediate between bilayer and hexagonal-II phases, the shape of the molecules is also said to be intermediate between that required for formation of bilayers and that required for formation of hexagonal-II phase (Larsson et al., 1980).

Despite the fact that the shape concept appears to be overly simplistic, the stability of different phases and transition between phases, such as between bilayer and nonbilayer phases, can be reasonably well understood in these terms.

The temperature dependence, for example, of the transition from a bilayer to hexagonal-II phase can be described in terms of an increase in hydrocarbon area with increasing temperatures as the chains become more disordered. The effect of this is an increased cone shape of the molecule eventually achieving a state that is able to transform from a lamellar to a hexagonal-II phase. Likewise, the introduction of *cis* unsaturated double bonds in the hydrocarbon chains tends to increase the area swept out by the chains by imposing a kink. The kink resulting from a restricted rotation about the bond prevents the close-packing alignment of the chains and strong cohesion by van der Waals interactions, thus reducing the temperature at which the cone shape is adopted. Similar arguments can be advanced to explain why monohexosyldiacylglycerols prefer a hexagonal-II phase when the hydrocarbon components are unsaturated compared to similar molecules with two hexose residues which have a clear preference for a lamellar structure. In this case, however, there is evidence that the hexose residues are linear in orientation and hence the basic cylindrical structure is unchanged by addition of further hexoses.

To overcome such problems, the tendency to form particular phases can also be expressed in terms of domain size as well as spatial arrangement. A schematic illustrating the concept of amphipathic balance is shown in Figure 8. This model relies more on the affinity of the molecule for the respective polar and nonpolar environments and has the implicit assumption that the molecules are to a certain extent flexible and able to adopt a configuration that represents the lowest free energy in the system. In this way the dynamic properties of the molecules such as their rotation about the long axis resulting in a time-averaged configuration, interactions both within the molecule and between neighboring molecules involving van der Waals interactions, dipolar interactions associated with the polar head group, and finally the hydration properties of the lipid head groups can all be taken into account. In effect these ideas are related to the critical micellar concentration of the polar lipids. This in turn is related to the solubility of the molecules in an aqueous medium and, in the case of bilayer-forming lipids, the critical micellar concentration is generally in the range of nanomolar concentrations. According to Figure 8 the hydrophilic to hydrophobic balance referred to as a normal membrane lipid polarity is that in which a bilayer phase would tend to form in excess water. Where the hydrophilic to hydrophobic balance is relatively high, micelles or hexagonal-I structure would be the preferred phase. On the other hand, where the hydrophilic–hydrophobic balance is relatively low, inverted micelles in the form of hexagonal-II or some form of cubic structure would form in excess water.

It should be emphasized that the system as a whole should be considered rather than the individual molecules. Thus, in certain mixed lipid dispersions consisting of components with a tendency to form hexagonal II and hexagonal I when dispersed alone may form a stable bilayer phase under certain condi-

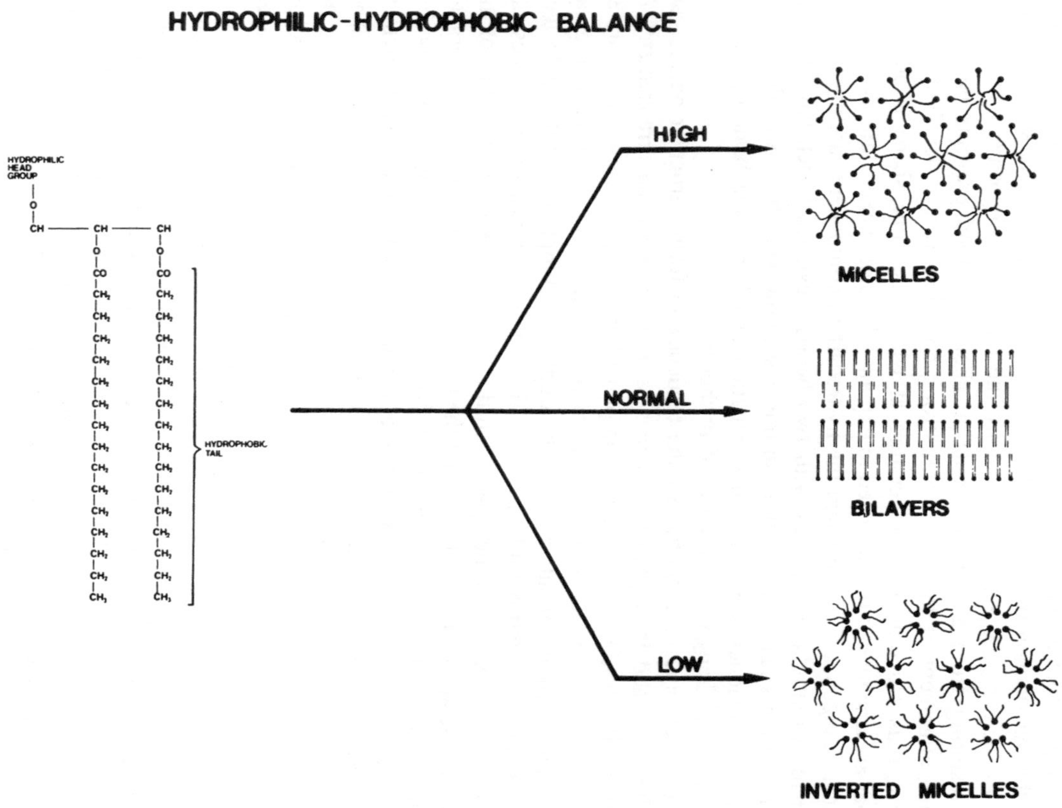

FIGURE 8. The amphipathic balance model to explain the phase behavior of membrane polar lipids.

tions (e.g., see Madden and Cullis, 1982). Another example is the case where two bilayer-forming lipids such as phosphatidylcholines and fatty acids, which, when mixed in certain molar proportions, will form a hexagonal-II phase at high temperatures (Koynova et al., 1987; Marsh and Seddon, 1982).

It has been pointed out by Gruner (1985) that the literal acceptance of the molecular volume distribution is determined by the phase, on the other. To overcome this problem, certain refinements to the shape concept are required. molecular volume distribution is determined by the phase on the other. To overcome this problem, certain refinements to the shape concept are required. The ideas introduced by Kirk et al. (1984), for example, are based on an evaluation of the free energy per molecule when it occupies a certain molecular volume of a given shape. To achieve this, it is assumed that the shape-dependent free energy per molecule may be divided into those components which are concerned with the necessary bending of lipid monolayers into configurations dictated by nonlamellar arrangements (elastic bending factor) and those concerned with the packing energies involved in occupying the hydrocarbon domain within the structure. In addition, it was found necessary to include an allowance for the lattice-specific factors such as electrostatic and hydration potentials.

The elastic bending factor accounts for the fact that in a bilayer configuration the individual monolayers are flat but are required to roll up into cylinders of radii about 3 nm in the hexagonal-II phase. This suggests that there is a minimum free energy of curvature of monolayers associated with an intrinsic curvature of each structure that prevails at particular temperatures. Balanced against this is the fact that the packing energies associated with the constraint of the polar head groups into geometric space lead to variation in accessibility of the nonaqueous domains to the hydrocarbon chains. Thus, the density of hydrocarbon in hexagonal-II structure is greatest at the aqueous interaction and decreases with distance from the aqueous cylinders, reaching a minimum at a point equidistant from the adjacent cylinders of the hexagonal-II matrix. This requires that some hydrocarbon chains be extended into energetically unfavorable configurations in order to occupy those hydrophobic domains within the structure which are relatively remote from the polar domain of the system. In practice, this means a combination of *trans/gauche* chain isomerizations that differ depending on the precise location of the molecule in the hexagonal-II phase; no such constraints operate in the bilayer phase where there are equal probabilities of occupancy of the hydrocarbon domain. Finally, it is well known (see Parsegian et al., 1979) that there are strong repulsive forces operating between opposed lipid polar groups across thin water layers. This is one factor that determines the thickness of the water layers betwen multilamellar structures (Rand, 1981). The repulsive force decreases exponentially with increasing separation of the bilayer surfaces with a decay length of approximately 0.2–

0.3 nm. This repulsive force cannot be explained in terms of standard electrostatic theory (Derjaguin and Landau, 1941; Verwey and Overbeek, 1948) and has been accounted for according to a surface-induced polarization density in the water (Gruen and Marcelja, 1983; Israelachvili, 1982; Israelachvili *et al.*, 1980).

The theory has certain predictive qualities, some of which have been tested. It has been shown (Kirk and Gruner, 1985), for example, that the radius of curvature of the lipid tubes of the hexagonal-II phase can be varied according to the proportion of high-curvature (dioylylphosphatidylethanolamine) and low-curvature (dioylylphosphatidylcholine) lipids present in binary mixtures of the two components. Moreover, the limit to curvature was found to be eventually constrained by the packing of the hydrocarbon into the structure. This could be relieved by the addition of an alkane such as dodecane to the polar lipid mixture, which was manifested as a considerable reduction in temperature at which the phase was converted from a lamellar to a hexagonal-II structure and an expansion of the hexagonal-II lattice due to an increase in the diameter of the water channels.

2.3. Phases of Pure Polar Lipids

Our understanding of the phase behavior of lipids found in biological membranes has been greatly assisted by studies of pure synthetic lipids. Methods are available for the synthesis of a wide variety of membrane polar lipids containing well-defined hydrocarbon substituents. Many of these are also available commercially in the relatively abundant quantities required for many biophysical studies. In this section we examine the structures and behaviors of the predominant lipid phases observed in molecular species of lipid isolated from biological membranes.

2.3.1. Bilayer Phases

When water or dilute electrolyte solutions are added to dry preparations of certain membrane lipid classes, the polar head group becomes hydrated and forms a phase with two-dimensional order. This order consists of stacked bilayers of lipid with intervening layers of water. The thickness of the respective lipid and water layers depends on the area occupied by the lipid at the lipid–water interface and, in excess water, the extent of hydration of the polar groups. The dimensions and arrangements of the molecules in smectic mesophases of this type can be determined from stained preparations under the electron microscope or from X-ray diffraction measurements. The latter method is considered more reliable because measurements can be formed with preliminary treatments that might alter the dimensions of the structure. In simple terms, X-rays are scribed reasonably accurately in terms of an equilibrium between the repulsive forces of the double-layer potential and the attractive van der Waals forces between opposing bilayers. It has been observed that the addition of sufficient cations causes shrinkage of the swollen phase of acidic phospholipids, presumably because they are able to screen the electrostatic charges partially on the lipid. Indeed, the packing of phosphatidylserine in excess NaCl, or in an excess of a salt in general, is similar to that observed with phosphatidylcholines in that the lipid undergoes only limited swelling, and the two-phase system consists of multilamellar aggregates dispersed in excess water (Hauser, 1984). It was also reported that the addition of NaCl (0.5 mol/dm^3) to phosphatidylserine vesicles in excess water induced aggregation and fusion of the vesicles, leading to the formation of a multilamellar dispersion. Reversible phase changes that occur on adding or removing Ca^{2+} from dispersions of phosphatidylserine have also been described (Papahadjopoulos et al., 1975).

The swollen phase consisting of single bilayer vesicles has a characteristic X-ray scattering profile (Wilkins et al., 1971). The relationship between X-ray scattering diffraction intensities from single-layered vesicles in water is illustrated in Figure 11. This shows broad continuous diffraction bands centered about the original diffraction maxima. The most intense scattering band is centered at a spacing corresponding to a dimension of 3.6 nm and weaker bands corresponding to one-half and one-third are also observed. The first order can therefore be interpreted as the thickness of a single lipid layer which represents the repeating structure within the phase.

Apart from low-angle X-ray reflections arising from long-range order within membrane lipid–water mesophases, scattering at wide angles (approximately 10°) is common to all lipids and is due to scattering from the hydrocarbon

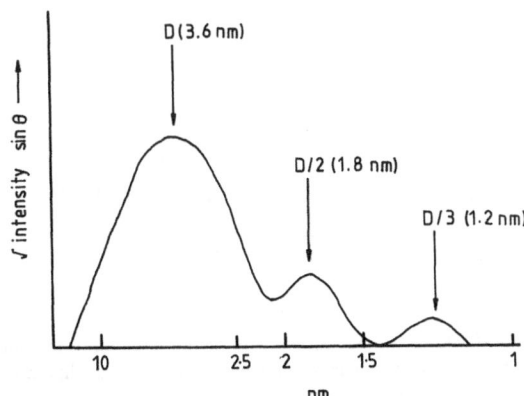

FIGURE 11. X-ray scattering intensity as a function of lattice spacing recorded from an aqueous dispersion of single bilayer vesicles of phosphatidylcholine.

component. Thus, information from the scattering characteristics of the acyl chains can provide information on the state of order and packing of the hydrocarbon chains into the mesophase and different packing arrangements characterize different bilayer phases. The characteristic features of these bilayer phases are examined in turn.

2.3.2. Lamellar Crystalline Phases

Lamellar crystalline phases, designated L_c phases, in membrane lipid–water systems were first reported by Chen et al. (1980) from calorimetric studies of phosphatidylcholines equilibrated for several days at about 0°C. Subsequently, the structural features of this phase and its transition to high-temperature phases have been characterized (Cameron and Mantsch, 1982; Finegold and Singer, 1984, 1986; Kodama et al., 1985; Nagle and Wilkinson, 1982; Serrallach et al., 1984; Silvius et al., 1985; Singer and Finegold, 1985; Ter-Minassian-Saraga and Madelmont, 1984; Wilkinson and McIntosh, 1986). The structural features have been examined by X-ray (Fuldner, 1981; Ruocco and Shipley, 1982a, b) in which characteristic sharp reflections in the wide-angle region index a unit cell of hydrocarbon chain packing of the dimensions illustrated in Figure 12a. Other x-ray and calorimetric studies have established the presence of L_c phases in isobranched lipids (Church et al., 1986), phosphatidylethanolamines (Sakurai et al., 1983; Seddon et al., 1983; Tenchov et al., 1984), phosphatidylglycerol (Blaurock and McIntosh, 1985), and galactolipids (Lis and Quinn, 1986; Sen et al., 1983). The long spacing of the L_c phase is relatively short in dipalmitoylphosphatidylcholine (5.9 nm), leading to the conclusion that the water layer is reduced (Cameron and Mantsch, 1982) compared to other multibilayer phases (see Figure 12b). Other characteristics of the L_c phase of this lipid are that the hydrocarbon chains are fully extended in the all-*trans* configuration

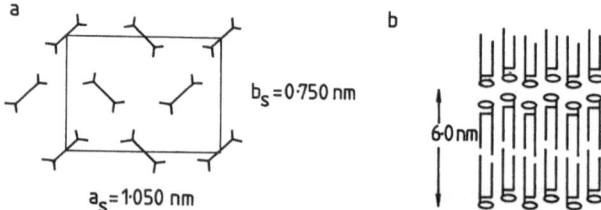

FIGURE 12. (a) Proposed cross section parallel to the bilayer plane of the acyl chain packing and (b) lamellar arrangement of the crystal phase of dipalmitoylphosphatidylcholine. Data from Tenchov et al. (1986).

with their long axes tilted to the bilayer normal. The phosphate head groups are relatively immobile (Lewis et al., 1984) and are thought to lie parallel to the bilayer plane (Griffin et al., 1978; Pearson and Pascher, 1979).

The formation of the L_c phase from the high-temperature lamellar gel phase during equilibration at around 0°C has been studied by conventional x-ray and differential scanning calorimetry (Ruocco and Shipley, 1982a). Two successive stages in the transition were observed over a period of about 15 hr. The first involved a rapid shift over the first hour of storage of the two characteristic $L_{\beta'}$ chain spacings and a slower change in the spacings to their position in the L_c phase accompanied by a progressive decrease in the lamellar repeat spacing. The latter change was found to be the result of an expulsion of water from the interlamellar space. The characteristic transition in diacylphosphatidylcholines from the crystal phase to higher-temperature phases has been examined in an homologous series of fully saturated molecular species differing in length of the acyl chains (Lewis et al., 1987). All the homologues studied ranging in chain length from 10 to 22 carbon atoms were found to form L_c phases on equilibration at appropriate low temperatures. The formation of the stable crystal phase appears to proceed by a complex metastable intermediate phase(s) and is subject to considerable hysteresis. The appearance of metastable L_c phase intermediates in the formation of the apparently stable L_c phases is associated particularly with odd-numbered phosphatidylcholines (carbon atoms 13, 15) while even-numbered homologues (carbon atoms 12, 14) exhibit only one metastable L_c phase intermediate. It was suggested by Lewis et al. (1987) that the formation of stable L_c phases of odd- and even-numbered acyl chain phosphatidylcholines may proceed via different mechanistic pathways. Behavior of this type is thought to arise from differences in the end-group interactions of crystalline or quasicrystalline odd- and even-numbered homologues of long-chain paraffinic compounds in which the hydrocarbon chains are tilted to the end-group planes (Broadhurst, 1962).

The transition from the L_c phase of phosphatidylcholine to higher temperature phases also proceeds by pathways that are chain-length dependent. Data summarized from the differential scanning calorimetric study of Lewis et al. (1987) are shown in Table I. The mechanism and kinetics of the transition of dipalmitoylphosphatidylcholine from the L_c phase to higher temperature phases have been examined by time-resolved x-ray diffraction methods (Tenchov et al., 1987). During the subtransition, two prominent wide-angle reflections, characteristic of the low-temperature L_c phase, gradually change such that a sharp peak at a spacing of 0.430 nm decreases in intensity and ultimately disappears while a broader peak, initially located at 0.375 nm, progressively shifts to an eventual spacing of 0.410 nm. These changes are shown in Figure 13a. This behavior is interpreted as a lateral deformation of the acyl chain packing

Table I
Thermal Transition of L_c Phases of Saturated Phosphatidylcholine Homologues Differing in Acyl Chain Length to High-Temperature Bilayer Phases

Chain length (carbon atoms)	Transition pathway
10–13	$L_c \longrightarrow L_\alpha$
14–15	$L_c \longrightarrow P_{\beta'} \longrightarrow L_\alpha$
16–22	$L_c \longrightarrow L_{\beta'} \longrightarrow P_{\beta'} \longrightarrow L_\alpha$
22	$L_c \longrightarrow L_{\beta'} \longrightarrow L_\alpha$

Data from Lewis *et al.* (1987).

subcell as the chains begin to rotate until a state is reached where the chains pack on a regular hexagonal array characteristic of the lamellar gel phase. There is also an increase in the lamellar repeat spacings from 6.0 to 6.4 nm which takes place simultaneously with the acyl chain rearrangement (Figure 13b). The relationship between the structural changes and thermal changes represented by the excess specific heat variation recorded by differential scanning calorimetry

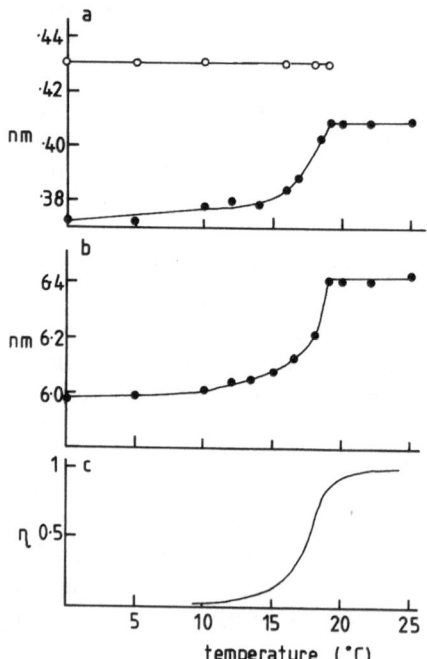

FIGURE 13. Structural transitions associated with the thermal decomposition of the L_c phase of dipalmitoylphosphatidylcholine showing the relationship between temperature and (a) wide-angle spacing, (b) low-angle spacing, and (c) integral of excess specific heat. Data from Tenchov *et al.* (1987).

is shown in Figure 13c. The transition mechanism is consistent with a continuous (second-order) phase transition in which a presumably orthorhombic subcell is transformed into a hexagonal subcell in a gradual process. The lamellar gel phase into which the L_c phase is transformed is characterized by an acyl chain orientation that is vertical to the bilayer plane rather than tilted at an angle. Temperature jump experiments through the subtransition indicated a relaxation time of about 5 sec.

2.3.3. Lamellar Gel Phases

These phases are designated with a subscript β and are characterized by a hexagonal packing of the hydrocarbon chains. The chains may be oriented normal with respect to the bilayer plane, L_β, inclined at an angle to the bilayer plane, $L_{\beta'}$, and a third phase in which the chains are oriented perpendicular to the bilayer plane but the bilayer is subject to a rippled conformation. The ripple phase, designated $P_{\beta'}$, is a phase characteristic of the phosphatidylcholines. The molecular packing of the hydrocarbon chain subcells and the character of the lamellar arrangement in the $L_{\beta'}$ and $P_{\beta'}$ phases in dipalmitoylphosphatidylcholine are illustrated in Figure 14.

The formation of the L_β phase from the L_c phase is metastable and is

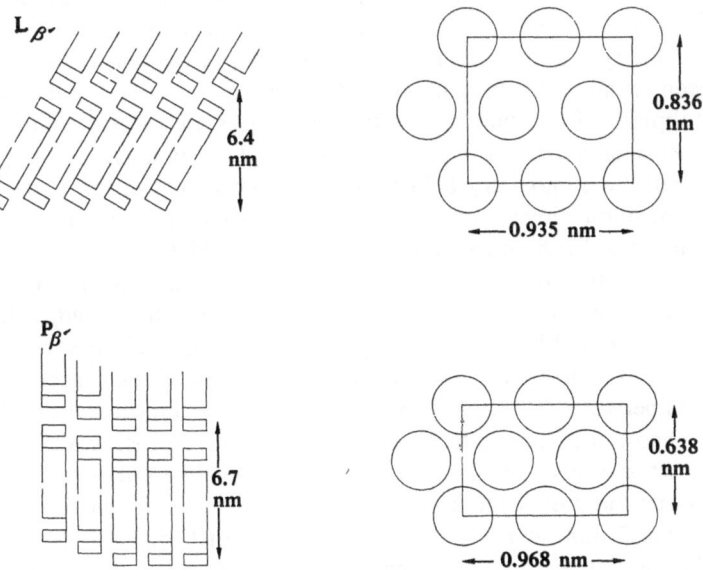

FIGURE 14. Molecular packing arrangements of multibilayers of dipalmitoylphosphatidylcholine in excess water in the $P_{B'}$ and $L_{B'}$ phases.

transformed to the $L_{\beta'}$ phase with a relaxation time of the order of 12 min (Tenchov et al., 1987). This is associated with a change in the wide angle reflection observed at a spacing of 0.418 nm consistent with a hexagonal arrangement of the hydrocarbon chains into an asymmetric peak which has a pronounced shoulder at 0.408 nm. It is assumed that these two reflections represent diffraction from (2,0) and (1,2) lattice planes of a two-dimensional rectangular lattice of the hydrocarbon chains. In this arrangement the hydrocarbon chains are tilted at an angle to the normal to the plane of the bilayer and it is unlikely that the lateral packing arrangements correspond to any of the precise simple and hybrid subcells (Abrahamsson et al., 1978; Hitchcock et al., 1974). The angle of tilt with respect to the bilayer normal has been reported as 23° (Pope et al., 1981).

Hawton and Doane (1987) measured the water order using ^{2}H-NMR methods in saturated phosphatidylcholine and phosphatidylethanolamine in the region of the chain order–disorder transition and found, like earlier studies, a strong motional narrowing in the deuterium quadrupolar splitting (Salisbury et al., 1971; Strenk et al., 1985). They concluded that the marked reduction of water order is the result of pretransitional fluctuations in the phospholipid molecules.

The polymorphic behavior of 1,2-dilauroyl-sn-glycero-3-phosphoethanolamine (Chang and Epand, 1983; Seddon et al., 1983; Wilkinson and Nagle, 1981) and 1,2-dimeristoyl-sn-glycero-3-phosphoethanolamine (Mantsch et al., 1986; Wilkinson and Nagle, 1984) in water has been studied, using calorimetry, dilatometry, infrared spectroscopy, and x-ray diffraction methods, and both have been found to exhibit a number of distinct gel phases. The data from both phospholipids indicate that at least two gel phases are formed, similar to those of the phosphatidylcholines, such that an L_β phase is a metastable phase and relaxes with time into a crystalline phase L_c. When heated, the L_c phase undergoes a transition directly to the L_β phase, at a temperature that is somewhat higher than that of the transition between the L_β and L_α phases.

The transition between the $L_{\beta'}$ and $P_{\beta'}$ phases is generally referred to as the pretransition. The kinetics of the pretransition has been studied in multilamellar phospholipid bilayers using differential scanning calorimetry (Cho et al., 1981), and the half-time of the pretransition becomes longer near the pretransition temperature. The reverse process, that is, a transition from the $P_{\beta'}$ to the $L_{\beta'}$ phase, does not conform to a single relaxation process. Conventional x-ray diffraction studies of dimyristoylphosphatidylcholine (Akiyama et al., 1982) showed that the pretransition from the $P_{\beta'}$ phase to the $L_{\beta'}$ phase is 90% complete within 7 min after a temperature shift and that imperfections in the multilamellar structure cause an extension of the relaxation time. Defects in the ripple structure have been studied using freeze–fracture methods by Ruppel and Sackmann (1983), who classified defects in the $P_{\beta'}$ phase topologically and analyzed them in terms of a homotopy theory. Freeze-fracture studies of the $P_{\beta'}$ phase transformation to the $L_{\beta'}$ phase in dipalmitoylphosphatidylcholine bilayers have shown that when the temperature is reduced from 38 to 30°C, the ripple structure disappears in a manner in which the intervals between ripples begin to expand with the decrease in ripple density following the temperature shift in a process that continues for several tens of minutes. Complete disappearance of the ripples requires a duration of about 3 hr following the temperature shift (Tsuchida *et al.*, 1987). Relaxation times measured by electron spin resonance probe methods (Tsuchida *et al.*, 1985) correlate a fast relaxation process with an expansion of the intervals between ripples. X-ray diffraction studies of the $P_{\beta'}$ phase in dimyristoyl- and dipalmitoylphosphatidylcholine have been reported by Alecio *et al.* (1985). In oriented multibilayers, two-dimensionally resolved patterns were observed near the meridian, which were analyzed to reveal a ripple wavelength of 165 nm and with widths in excess of 100 nm in a direction perpendicular to this.

2.3.4. Lamellar Liquid-Crystalline Phases

The high-temperature stable phase of phosphatidylcholines, dihexosyldiacylglycerols, and many of the acidic phospholipids is a lamellar phase in which the hydrocarbon chains are disordered and characterized by a diffuse x-ray scattering band centered at 0.46–0.47 nm. The hydration of bilayers on undergoing a transition from L_β phase to L_α phase changes very significantly as does the bilayer thickness. This is illustrated in Figure 15, which shows the lamellar repeat spacings for dimyristoylphosphatidylcholine and dipalmitoylphosphatidylcholine as a function of the concentration of water in the dispersions in the respective phases. In the gel phase (L_β, at less than 10°C), the addition of water to both phospholipids results in an increase in lamellar repeat distance, reaching a limiting value of 6.16 and 5.93 nm for the dimyristoyl and dipalmitoyl derivatives, respectively. This limiting value is slightly greater when the lipid is maintained at temperatures greater than the phase transition temperature in which they exist in the L_α phase. The amount of water in the system that is required to hydrate the lipid fully is increased proportionately for each of the lipids. This is reflected in a marked decrease in thickness of the bilayer on transition from a gel to a liquid-crystalline phase.

Electron density profiles of the type shown in Figure 9 are also able to provide information on the thickness of bilayers and the method has led to the identification of interdigitated phases that distinguish them from purely bilayer phases. Furthermore, the electron density profiles can indicate the extent of chain tilting. A number of phospholipids are known to form chain interdigitated phases either spontaneously or by altering the level of hydration and so on. Interdigitated phases have been described in dipalmitoylphosphatidylglycerol

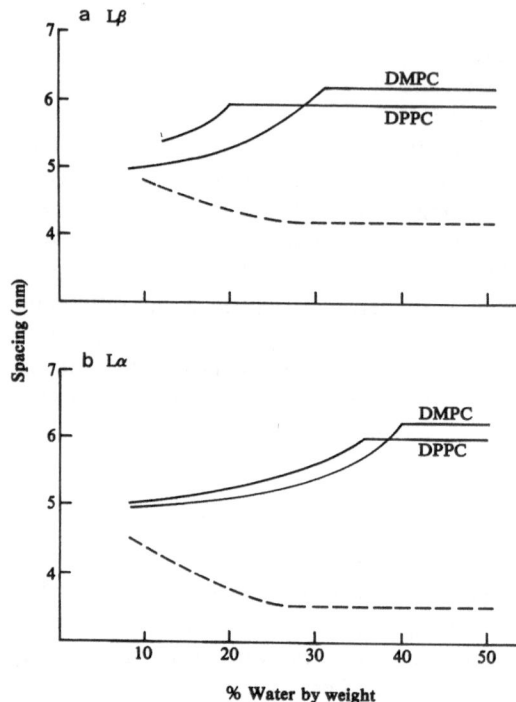

FIGURE 15. The dependence of lamellar repeat spacing of dimyristoyl (DMPC) and dipalmitoyl-(DPPC) phosphatidylcholines in the (a) L_β and (b) L_α phases on the content of water in the dispersion. Dashed line represents the derived value of bilayer thickness for DMPC. Data from Janiak et al. (1979) and Ruocco and Shipley (1982a).

(Ranck et al., 1977), 2-deoxylysophosphatidylcholine (Hauser et al., 1980), 1,3-dipalmitoylglycerol-2-phosphocholine (Serralach et al., 1983), dipalmitoylphosphatidylcholine in nonaqueous solvents (McDaniel et al., 1983), and in the presence of anesthetics (McIntosh et al., 1983) and high hydrostatic pressures (Braganza and Worcester, 1986). Electron density profiles have been used to determine bilayer thickness in the ether analogue of dipalmitoylphosphatidylcholine and to distinguish bilayer phases in the liquid-crystal phase and two gel phases, one of which is formed at relatively high levels of hydration and is characterized by tilted hydrocarbon chains and the other at low hydration at which the hydrocarbon chains are interdigitated (Kim et al., 1987). The three structures are shown schematically in Figure 16.

One of the features of liquid-crystalline bilayers is the relatively rapid rate of lateral diffusion of molecules in the plane of the bilayer. Measurements of the rate of lateral diffusion have been made using fluorescence quenching and excimeration of pyrene analogues of phospholipids by Galla et al. (1979). They reported that the rate of lateral mobility in bilayers of saturated phospholipids in the liquid-crystal phase was independent of the chain length or presence of cis-unsaturated bonds; however, jump frequencies within the lattice were two-

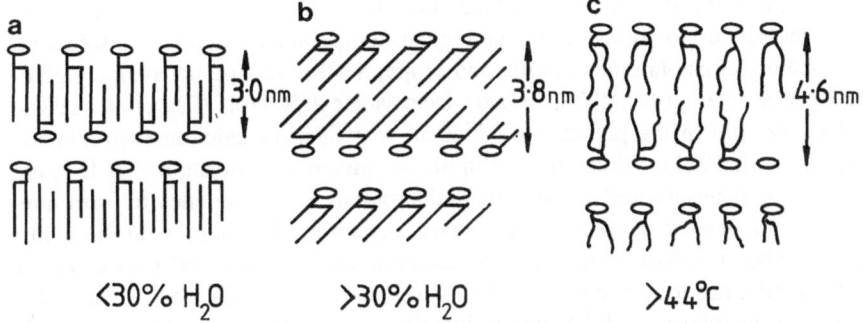

FIGURE 16. Structure and transitions of dihexadecylphosphatidylcholine from (a) an interdigitated bilayer phase to (b) a gel phase with tilted chains on hydration and to (c) a liquid-crystalline phase on heating above 44°C.

fold greater for the unsaturated phospholipids. Essentially similar rates of diffusion were also observed in pure synthetic bilayers and in biological membranes. Detailed studies of the temperature dependence of lateral diffusion by pyrene excimer formation and a fluorescence photobleaching recovery method (Vaz et al., 1984, 1985) have shown that the rate does not conform to a simple exponent of an Arrhenius form but is related more precisely to a free volume model (King and Marsh, 1986). According to this model the creation of a critically sized free volume adjacent to a molecule and of sufficient size to accommodate it must be formed in order for lateral diffusion to occur (Cohen and Turnbull, 1959). While lateral diffusion of lipid molecules can take place with lateral diffusion coefficients in the region of 10–11 m^2/sec, the diffusion of molecules from one side of the bilayer to the other is a relatively slow process (Kornberg and McConnell, 1971; Renooij et al., 1976; Rousselet et al., 1976; Seigneuret and Devaux, 1984; Verkleij et al., 1973). The rate of transmembrane movement is related to the acyl chain length of the phospholipid and decreases with increasing chain length (Fujii et al., 1985; Tamura et al., 1986).

2.4. Hexagonal Phases

As indicated in Figure 1 one of the nonbilayer phases adopted by membrane lipids and observed also in particular synthetic lipid analogues is hexagonal-II phase. Hexagonal-I phases are not observed with membrane lipid analogues and are associated with lysophospholipids and surfactants that, in general, destabilize bilayer membranes. The hydrocarbon chains of lipids in bilayer arrangement are invariably in the disordered configuration.

Lipids from biological membranes that form hexagonal-II structures in dilute electrolyte solutions at physiological temperatures include unsaturated phosphatidylethanolamines (Cullis and Hope, 1978), monoglucosyldiacylglycerols (Wieslander et al., 1978), monogalactosyldiacylglycerols (Sen et al., 1981), and cardiolipin in the presence of divalent ions such as calcium (van Venetie and Verkleij, 1981). When charged lipids are mixed with hexagonal-II-forming lipids such as unsaturated phosphatidylethanolamine, the tendency to form nonbilayer structures is suppressed. However, screening the charges on the acidic lipid in phosphatidylserine/phosphatidylethanolamine (Tilcock and Cullis, 1980), phosphatidylinositol/phosphatidylethanolamine (Nayer et al., 1982), phosphatidylglycerols/phosphatidylethanolamine (Farren and Cullis, 1980), and cardiolipin/phosphatidylethanolamine (de Kruijff and Cullis, 1980a) mixtures by bivalent ions like calcium or pH cause transition to the hexagonal-II phase. Similar inductions of hexagonal-II phases in cardiolipin-containing bilayers can be induced by polypeptides such as poly(lysine) (de Kruijff and Cullis, 1980b) and proteins like cytochrome c (de Kruijff and Cullis, 1980a).

Hydration factors appear to be in part responsible for inducing hexagonal-II phases. The lipids that tend to form such structures, in general, have a low hydration. Chaotropic agents, for example, tend to convert unsaturated phosphatidylethanolamines from hexagonal-II to lamellar phases (Sen and Yeagle, 1986). It was argued that, in the hexagonal-II phase, exposure of the polar groups to water is reduced because of the close packing of these groups in the structure.

2.5. Cubic and Other Nonlamellar Phases

The phases described above are all two-dimensional but other phases with three-dimensional symmetry have been described in lipid–water systems (Luzzati and Reiss-Husson, 1966). These, most generally, are characteristic of lipid mixtures rather than pure membrane lipids; however, many polar lipids such as monoglycerides (Larsson et al., 1980; Lindblom et al., 1979) and glucolipids (Wieslander et al., 1981) are believed to form these so-called cubic phases. Attempts have been made to characterize these phases by freeze–fracture electron microscopy (Hui and Boni, 1981) but the precise structure requires X-ray and other evidence for definition (Williams et al., 1981). One feature that can be regularly distinguished by freeze–fracture electron microscopy, however, is lipidic particles sandwiched within a bilayer of lipid (Sen et al., 1982a). The structure of these particles is thought to be an inverted micelle of lipid. In some cases these lipidic particles often appear to form an interwoven network with particles present predominantly at the intersections of bilayers to form honeycomb (Cullis et al., 1980) or sponge (Noordam et al., 1981). As well as coexisting with bilayer phases of lipids, inverted micelle structures have also been shown to be present in mixtures with hexagonal-II structure and the relative amounts of lipid in the various phases can be quantitated by ^{31}P-NMR (De Grip et al., 1979; Vasilenko et al., 1982). Most notable of the lipids which form three-dimensional phases are the membrane lipids of the extreme halophiles such as *Halobacterium halobium* and the thermoacidophiles like *Sulpholobus sulphataricus*. These membrane lipids are characterized by the presence of phytanyl linked hydrocarbon residues in place of the fatty acyl chains (Kates, 1978), and in the case of the thermoacidophile, the phytanyl chains are long and connect polar groups located at each end (De Rosa et al., 1980a,b, 1983). Freeze–fracture electron microscopic studies of the diphytanyl derivatives of phosphatidylglycerol phosphate from *Halobacterium cutirubrum* (Quinn et al., 1986) showed that the tendency of the lipid to adopt nonbilayer structures increased as the charges on the two phosphate groups were screened by inorganic ions, especially in concentrations found in the habitat of the living organism. The bipolar isoprenyl ether lipids from the archaebacteria also displayed polymorphic behavior under physiological conditions as judged by X-ray diffraction methods (Gulik et al., 1985). In one phase the unsubstituted glycerol ends of the lipids were found to segregate in the hydrocarbon region away from the other polar groups. Two cubic phases were identified each consisting of two intertwined and unconnected three-dimensional networks of rods.

3. PHASE BEHAVIOR OF MIXED LIPID SYSTEMS

Unlike pure molecular species of lipid, combinations of different lipids in dispersions do not usually form ideal mixtures and there is a tendency for the components to separate into distinct phases. Phase separations in multicomponent mixtures of lipids can be driven by a variety of factors, including differences in melting point (gel phase to liquid-crystalline phase transition temperature), binding of multivalent ions, amphipathic balance within the molecules, and steric configuration. It has been pointed out that great care is required to determine the precise boundary conditions about which phase separations occur. The reason for this is that diffusion of molecules within the system is required in order to effect the phase separation as this may require very extensive periods of equilibration under defined conditions. Furthermore, phase separation may result in the formation of other phases, which in turn may have hysteritic properties (Mansourian and Quinn, 1986). For these reasons it is essential to determine the composition of lipid mixtures that are as close as possible to their equilibrium conditions, which means, for the most part, that prolonged incubation or storage at designated temperatures is required before an equilibrium state is achieved.

These considerations place rather stringent conditions on the methods of preparation of samples and of analysis of data required for accurate thermodynamic analysis for two-component or, more generally, multicomponent lipid–water systems. At present, the thermodynamic parameters by which the mixing of lipids in liquid-crystalline lamellar phases may be described are usually evaluated from data that describe equilibria between liquid and solid phases of each lipid. For example, the liquidus and solidus curves in an experimental phase diagram may be fitted simultaneously by using an appropriate thermodynamic model (Lee, 1977; 1978; Van Dijck et al., 1977). This approach requires careful attention to details of the solid phase mixing of lipids, including the possibility of multiple solid phases, even in cases where the mixing of the lipids that occurs in the liquid phase may be of great interest from a biological viewpoint. The literature, unfortunately, is replete with examples of experimental conditions which have failed to account for the complicating factors of multiple solid phases in the construction of phase diagrams, and caution must be exercised in deriving thermodynamic values from such data.

Mathematical modeling of two-component lipid bilayer systems has been attempted, and the value of such models in predicting thermodynamic parameters is likewise subject to achieving an equilibrium condition (Jan et al., 1984). Some models that purport to simulate the behavior of two-component lipid bilayer systems (Snyder and Freire, 1980) have been found to be deficient in this respect.

3.1. Phase Behavior of Lipid Enantomers

Molecular species of lipids of biological origin are generally of the L-enantomeric form. Comparison of the L-enantomer with the DL racemic modification have shown that phases formed in the two cases are often different. An example is the formation of the lamellar crystalline phase in 1,2-dipalmitoylglycero-3-phosphocholine (Boyanov et al., 1986). It has been shown that only the pure enantomers will form the crystal phase upon equilibration of the fully hydrated lipid at around 0°C. Figure 17 shows a study of the effect of D-1,2-dipalmitoylglycero-3-phosphocholine on the subtransition of L-1,2-dipalmitoylglycero-3-phosphocholine. This shows that adding increasing amounts of the D-enantomer prevents the formation of the crystal phase and that the mixing of 1 mol of either enantomer with 15 mol of the other is sufficient to prevent formation of the crystalline lamellar phase. It may be noted that the pre-endothermic and main endothermic transitions of the phospholipid are unaffected by the enantomeric composition. It can be seen from Figure 17 that as the mole fraction of the D-enantomer increases the subtransition of the L-enantomer of 1,2-dipalmitoylglycero-3-phosphocholine there is no change in the temperature at which the subtransition is observed but the enthalpy of the transition decreases as the proportion of D-1,2-dipalmitoylglycero-2-phosphocholine in-

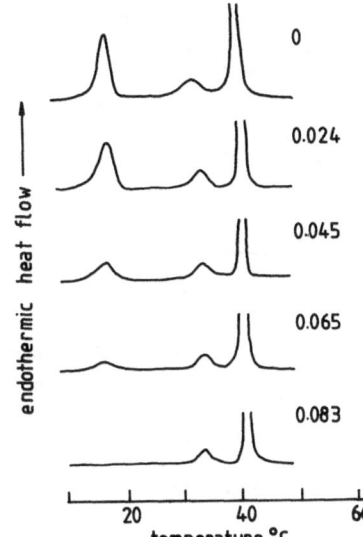

FIGURE 17. Differential scanning calorimetric heating scans of L-1,2-dipalmitoylglycero-3-phosphocholine codispersed in the mole fractions of D-enantomer indicated and equilibrated for 4 days at 4°C. Data from Boyanov et al. (1986).

creases; the transition ultimately disappears when the mole fraction of the D-enantomer reaches about 0.08. This effect was found to be less dramatic in the case of mixtures of D-enantomer with the L-enantomer of 1,2-dipalmitoylglycero-3-phosphoethanolamine; if the mole fraction of the D-enantomer is 0.15, the transition does not occur. A major difference, however, can be seen in the thermograms of mixtures that contain either of the 1,2-dipalmitoylglycero-3-phosphoethanolamines; the pre-endothermic transition is considerably perturbed and merges with the main endothermic transition. The effect of the L-2-dipalmitoylglycero-3-phosphoethanolamine on the subtransition of the L-1,2-dipalmitoylglycero-3-phosphocholine is shown in Figure 18. In this mixture there is an obvious reduction in the enthalpy of the pretransition; it may be concluded from this observation that the tilting of the acyl chains in the gel phase is perturbed.

It is known that racemic substances can exist in three principally different states depending on the relative strength of the interaction between antipode and similar enantomers (Eliel, 1962). Racemic mixtures and solution on the one hand and racemic compounds on the other are characterized by different, easily discernible phase diagrams. It has been suggested that the greater thermal stability of the L_c phase of D-dipalmitoylphosphatidylethanolamine compared with the L_c phase of the pure L-enantomer is that a racemic compound is formed at low hydration in which there is a stronger attraction in the DL pairs than in the LL and DD pairs (Tenchov et al., 1984). These intermolecular complexes

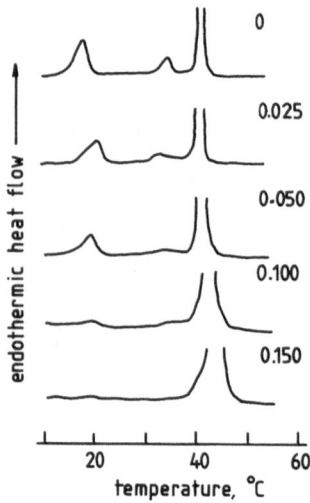

FIGURE 18. Differential scanning calorimetric heating scans of L-1,2-dipalmitoylglycero-3-phosphocholine showing the effects of L-1,2-dipalmitoylglycero-3-phosphoethanolamine on the subtransition and pretransition endotherms. Values represent mole fractions of ethanolamine phosphatide in the choline phosphatide. Data from Boyanov et al. (1986).

appear to dissociate after equilibration in the L_α phase and, in the hydrated state, the racemic dipalmitoylphosphatidylethanolamine forms a nearly ideal solution that is characterized by gel to liquid crystal phase transitions similar in character and temperature to that of the L-dipalmitoylphosphatidylethanolamine. The origins of the effects described above on the subtransition

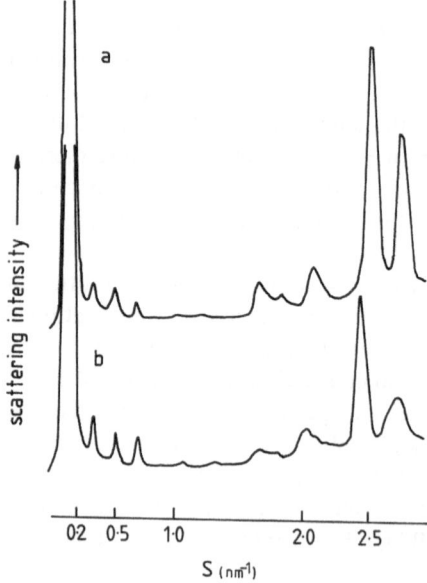

FIGURE 19. X-ray diffraction intensity versus reciprocal spacing of (a) L-dipalmitoylphosphatidylethanolamine and (b) DL-dipalmitoylphosphatidylethanolamine.

appear to reside in the particularly close and cooperative packing arrangement of the lipid molecules in the lamellar phase; this is perturbed by racemic mixtures (or other phospholipids, cholesterol, fatty acids, etc.) that prevent the close-packing arrangement within the phase. Direct evidence for this has been obtained in the case of racemic dipalmitoylphosphatidylethanolamines as shown in Figure 19. It can be seen that the x-ray diffraction patterns of the L_c phase recorded at 20°C show slightly different arrangements of lipid molecules in the unit cells of L- and DL-dipalmitoylphosphatidylethanolamine. The Bragg reflections at low angle show a lamellar repeat distance of 5.7 nm that is identical in the two lipids. The wide-angle scattering patterns are typical powder patterns of the lipid crystal phases with two intense scattering peaks and at least two weaker intensity bands at longer spacings. This suggests that although the lamellar repeat is not dependent on the particular enantiomer, the lateral interactions within the bilayer are different. As inferred previously from differential scanning calorimetric studies (Tenchov et al., 1984), it may be expected that the observed differences are due to formation of DL pairs in a lateral direction and reflect different interactions in these pairs compared to LL and DD pairs.

3.2 Lateral Miscibility of Bilayer-Forming Lipids

One of the earliest demonstrations of the phase behavior of mixed lipid systems was the interaction between binary mixtures of saturated and unsaturated bilayer-forming lipids (Phillips et al., 1972). Thus, calorimetric studies, shown in Figure 20, illustrate the effect of cooling a binary mixture of 1,2-distearoylphosphatidylcholine and 1,2-dioleoylphosphatidylcholine from a temperature above the gel to liquid-crystalline phase transition of the fully saturated lipid. The heating thermogram (curve c) obtained from such a mixture shows two endothermic transitions arising from a transition from the gel to liquid-crystalline phase of the dioleoyl constituent at about −23°C and the saturated component at about 45°C. The pure components are seen to undergo transitions at −23 and 56°C, respectively. Curve c is simply the result of phase separations that are created by cooling the sample, first below the phase transition temperature of the saturated component, which causes it to separate into a distinct gel phase domain and separate from the remaining liquid-crystalline unsaturated component. The slight perturbation of this endotherm indicates that complete phase separation is not achieved. Cooling below the liquid-crystalline to gel phase transition of the unsaturated component occurs in a separate domain. Subsequent heating of the mixture therefore induces two separate phase transitions corresponding to the cooperative melting of the two domains that were created during the cooling process. Curve d of Figure 20 is a heating scan of a single species of phospholipid which contains an oleoyl hydrocarbon chain in the 2-position of the phosphatidylcholine and a steroyl chain in the carbon atom 1-position. Since there is no phase separation in this single molecular

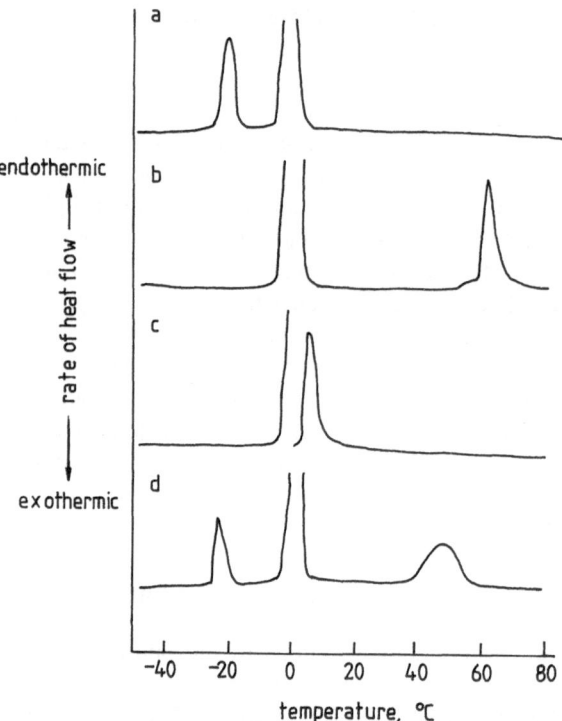

FIGURE 20. Thermograms obtained from differential scanning calorimetry of (a) 1,2-dioleoylphosphatidylcholine, (b) 1,2-distearoylphosphatidylcholine, (c) equimolar mixture of (a) and (b), and (d) 1-stearoyl,2-oleoylphosphatidylcholine. All lipids were dispersed in an equal weight of water and heat exchange recorded during heating scans. Data from Phillips et al. (1972).

species of phospholipid there is a single melting endotherm intermediate between those of the two components of the binary mixture. Similar data have been obtained using freeze–fracture electron microscopic techniques (Verkleij et al., 1972; Ververgaert et al., 1973) and fluorescence probe methods (Sklar et al., 1979). In general, lamellar-forming membrane polar lipids exhibit complete miscibility in the fluid phase, and their ability to phase separate upon cooling depends on the difference in phase-transition temperatures of the individual components, the rate of cooling, and the interaction between the polar groups. With more sophisticated methods of measuring phase separations, systems that were hitherto regarded as miscible are now known to form phase-separated systems, at least over some of the regions of the phase diagram (Tenchov, 1985).

Studies of mixtures of phosphatidylcholines with differing hydrocarbon chains by differential scanning calorimetry (Lentz *et al.*, 1976; Mabrey and Sturtevant, 1976; Van Dijck *et al.*, 1977), electron paramagnetic resonance spectroscopy (Luna and McConnell, 1978; Wu and McConnell, 1975), and fluorescent probe methods (Shimshick and McConnell, 1973) have shown that when the chain length differs by four or more carbon atoms there is solid-state immiscibility. Continuous mixtures are observed if the components differ by no more than two carbon atoms. Other studies of mixtures of phospholipids with the same acyl chains but differing polar head groups (Arnold *et al.*, 1981; Blume and Ackerman, 1974; Chapman *et al.*, 1973; Mendelsohn and Koch, 1980; Silvius and Gagne, 1984a,b) have led to the conclusion that phosphatidylserines and phosphatidylethanolamines are completely miscible while mixtures of phosphatidylcholines with phosphatidylserines and phosphatidylethanolamines, in general, display solid-state immiscibility. Furthermore, any mixture of lipids differing in both chain length or saturation and polar group structure exhibit solid–solid phase separation and in some cases fluid–fluid phase separation can also be observed. In general, the deviations from ideal mixing are greater in the solid state of the mixture and at low concentrations of the higher melting component.

Many studies have been reported of phase separations induced in mixtures of zwitterionic and acidic phospholipids mediated by charged screening by counterions in the aqueous phase. Ca^{2+}-induced phase separations between dieladoyl- and dimyristoyl-containing phosphatidylserines and similar derivatives of phosphatidylcholine in binary mixtures have been reported (Silvius and Gagne, 1984b). It is known that the addition of phosphatidylcholine to bilayer vesicles of phosphatidylserine strongly antagonizes the tendency of phosphatidylserine to mediate the fusion of vesicles in the presence of calcium ions. This is related to the ability of calcium to induce lateral phase separations in the mixture of the two lipids. It was found that dimyristoyl- or dieladoyl-containing phosphatidylcholines were completely miscible with similar derivatives of phosphatidylserine in the fluid phase, in the absence of calcium. Limited lateral separation into phases, however, could be detected if calcium were added. At higher temperatures, mixtures that contained less than 40% or more than 70% phosphatidylserine in phosphatidylcholine did not show evidence of phase separation when they were at equilibrium in the presence of calcium, while two domains could be distinguished at intermediate concentrations. If the mixture were cooled, observations indicated that extensive immiscibility of solid phases exists in this system. Other studies of mixtures of phosphatidylcholines with phosphatidic acid, which differs from phosphatidylserine in that there is a net potential negative charge of 2 on the lipid, have been reported by Graham *et al.* (1985) using differential scanning calorimetry and by Kouaouci *et al.* (1985) using Raman spectroscopy. The studies have shown that the acyl chains of dimyristoylglycerophosphoric acid in the ionized state possess more *gauche* conformers than occur in bilayers of phosphatidylcholine in which the chains are of equivalent length. The phospholipids display complete miscibility in the liquid-crystalline lamellar phase, but if calcium ions are added they undergo a rapid and extensive phase separation. Domains of phase-separated dimyristoylglycerophosphocholine adopt packing characteristics that are typical of the pure lipid, but a small proportion of the phosphatidylcholine that is trapped within the phosphatidic-acid-rich domain (about 20 mol%) displays quite different features. This component is packed more tightly since, depending on the temperature, it is incorporated into either a solid solution in the dimyristoylglycerophosphoric acid or a complex of the phospholipid with calcium, in a so-called cochleate phase. The latter phase is highly ordered and does not appear to undergo a cooperative thermotropic phase transition between 5 and 100°C. The clusters of trapped dimyristoylphosphatidylcholine do not melt cooperatively if the phase-separated mixture is heated but become fluid as the dimyristoylglycerophosphoric acid that is not complexed with calcium melts, at about 50°C.

Lateral phase separations between phospholipids and sulfatides (Viani *et al.*, 1986) and gangliosides (Masserini and Freire, 1986) have been reported. Likewise, transbilayer phase separations in highly sonicated vesicle systems consisting of phospholipids and sphingomyelin have been observed and ascribed to the precise chemical configuration of the head group of the respective molecules (Kumar and Gupta, 1985).

3.3. Bilayer–Nonbilayer Phase Separations

When lipids that form bilayer phases are mixed with lipids that form hexagonal-II phases under the same conditions in aqueous systems, there is a tendency for the components to phase separate. One of the best characterized of such systems are glycolipid mixtures consisting of mixtures of monoglycolipid and diglycolipid. The galactolipids representing major lipid components of photosynthetic membranes of higher plant chloroplasts have been well characterized in this respect. Phase equilibria of each of the major galactolipids of this membrane, when dispersed in pure form, have been examined by x-ray diffraction, freeze–fracture electron microscopy, and differential scanning calorimetry (Sen *et al.*, 1983). The native lipids from chloroplasts, in general, contain polyunsaturated fatty acyl chains. The monogalactolipid forms a reverse hexagonal phase and the digalactosyl-containing lipid forms a lamellar phase in excess water at temperatures between -10 and 80°C. The phase behavior of binary mixtures of monogalactosyl- and digalactosyl-containing lipids has recently been studied, using ^{2}H-NMR and low-angle X-ray diffraction methods (Brentel *et al.*, 1985). The mixtures that were examined consisted of monogalactosyl- and digalactosyldiacylglycerols, codispersed in water, in mole ratios of 1:2, 1.2:1, and 2:1. The phase properties over the temperature range 10–

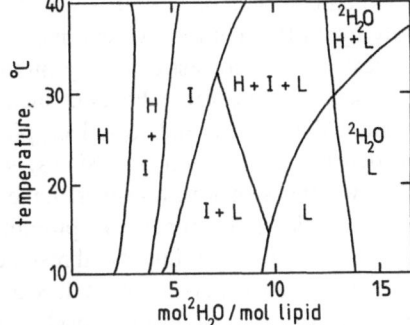

FIGURE 21. A composition–temperature phase diagram of a 2 : 1 mole ratio mixture of mono- and digalactosyl-diacylglycerols from chloroplasts dispersed in ^2H_2O. H = hexagonal-II; I = isotropic; L = lamellar. Data redrawn from Brentel et al. (1985).

40°C and for water contents up to 14 mol of water/mol of lipid in the dispersion in which the mole ratio of lipids was 2:1 are shown in Figure 21. This mole ratio is approximately equivalent to that in the photosynthetic membrane. At low water content, the hexagonal-II phase dominates the system. As the concentration of water is increased, a reverse cubic–liquid-crystalline phase gradually replaces the hexagonal-II phase and, in turn, is replaced by a liquid-crystalline lamellar phase in excess water. With increasing proportions of monogalactolipid in the mixture, the nonlamellar regions of the phase diagram increase. The reverse cubic–liquid-crystalline phase was characterized by x-ray diffraction methods and its space group was found to be Ia3d; some deuterium NMR measurements of translational diffusion of molecules in this phase indicated that the structure of the cubic phase was bicontinuous. As shown in the deuterium NMR study, the native lipid of chloroplasts, which contains almost exclusively linolenoyl fatty acid substituents, forms hexagonal-II structures in excess water but on saturation of the hydrocarbon chains, as in distearoylmonogalactosyldiacylglycerol, a lamellar gel phase forms at temperatures less than about 50°C. When the saturated and the unsaturated lipids are mixed in excess water, extensive phase separations of the gel phase lamellar and hexagonal-II phase can be observed by freeze–fracture and other methods (Mansourian and Quinn, 1986). In experiments in which the extent of saturation of the monogalactolipid was varied in mixtures with bilayer-forming lipids (Gounaris et al., 1983b), it was found that the tendency to phase separate was determined by the level of unsaturation of the hydrocarbon chains and mimicked the effect on the tendency of the lipid to form hexagonal-II phase.

Electrostatic Effects

The effects of charge shielding on the phase behavior of mixtures of nonbilayer and acidic lipids have been studied extensively as a model of the fusion between cell membranes. It is well known that polycations such as polylysine, polymyxin B, mellitin, cytochrome c, and polyamines are able to induce fusion of vesicles that consist of binary mixtures of a zwitterionic and an acidic lipid. The same is true for polyvalent metal ions, but the potency (in terms of the amount that is required to induce phase separation of the type that would be needed to cause fusion of vesicles) is much less than that of the polycations. The two types of ion, however, have been shown to act in concert, such that, for example, polyamines can promote the calcium-induced fusion of bilayer membranes by lowering the threshold concentration of metal ion that is required to induce fusion. In studies of bilayer vesicles consisting of equimolar mixtures of dioleoylglycerophorylcholine and bovine heart cardiolipin, polylysine has a synergistic effect on calcium-induced fusion; its effect is believed to be mediated by inducing aggregation between vesicles (Gad et al., 1985). In systems that contain subthreshold concentrations of calcium, the limiting factor for fusion of vesicles was found to be the capacity to induce aggregation of vesicles, which can be largely overcome by interaction of polylysine at the vesicle surface (Wilschut et al., 1985).

Other systems that have been examined are binary mixtures of phosphatidylethanolamines with dioleoyl-, dieladoyl-, and dimyristoylglycerophorylserines (Silvius and Gagne, 1984a). Phase diagrams of these systems indicate that mixtures that contain less than 20 mol% or more than 80 mol% phosphatidylserine in phosphatidylethanolamine do not separate into phases in the presence of calcium, but the formation of domains of immiscible phases can be detected in all mixtures in which the mole ratio is between these values. The particular composition of the fatty acyl groups does not appear to affect the kinetics of separation of lipid phases significantly, as judged by calcium-ion-induced membrane fusion, as long as the mixed lipid system was in the liquid-crystalline state. The rate of fusion, however, was found to be significantly affected by temperature; the rate markedly increased with increasing temperature. Fusion was also observed to occur between vesicles in which the proportion of phosphatidylserine was so low that calcium did not induce a phase separation of the acidic lipid, indicating that it is the regions of phosphatidylethanolamine that are involved in the fusion of bilayers and that the role of phosphatidylserine may be to stabilize the phosphatidylethanolamine in a bilayer, rather than allowing the lipid to assume some nonbilayer configuration within the system.

4. PROTEIN–LIPID INTERACTIONS

The presence of protein in biological membranes has a significant influence on the phase behavior of the constituent membrane lipids. A number of examples are known in which the structures formed by total polar lipid extracts of biological membranes differ greatly from the bilayer arrangement that is believed to represent the membrane lipid matrix. These membrane systems include the retinol rod outer-segment disc membranes (De Grip et al., 1979) and the photosynthetic membrane of higher plant chloroplasts (Quinn et al., 1982). In other membranes, such as the erythrocyte, the lipids have been shown to exist in bilayer configuration in both the intact membrane (Cullis and Grathwohl, 1977) and in dispersions of total polar lipids (Cullis and de Kruijff, 1979). Liver microsomal membranes isolated from rat (de Kruijff and Cullis, 1980b), rabbit (Stier et al., 1978), or cattle (de Kruijff et al., 1978) at physiological temperatures show lipids in an isotropic arrangement but the structure is transformed to a bilayer if the temperature is reduced. The isolated lipids by contrast, when dispersed in buffer, formed bilayer structures. Stier et al. (1978) have suggested that the interaction of proteins like cytochrome P450 with lipid may be responsible for the isotropic structure in the membrane. Rapid transbilayer motion of the phospholipids could also give rise to the observed isotropic structure (Van den Bresselaar et al., 1978), but this would have to be mediated by the membrane proteins as no structures are observed in the lipid dispersions.

Other indications that the proteins affect the phase behavior of lipids have been derived from reconstituted systems in which it has been shown that intrinsic membrane proteins cause a disordering of the lipids with which they are in contact compared with bulk lipids in the fluid phase (Rice and Oldfield, 1979). This is explained by a need for the hydrocarbon chains of the lipids to fill the voids created between and around the amino acid side chains of the intrinsic protein in contact with the hydrophobic domain of the lipid bilayer. It has been suggested that non-bilayer-forming lipids may best be suited to occupy these protein–lipid boundary domains because the relative bulk of the hydrocarbon residues of these lipids is greater than those of the bilayer-forming lipids (Quinn and Williams, 1983). It has also been reported that intrinsic membrane proteins intercalated into bilayer-forming lipids cause an increased leakage of ions across the membrane (Van der Steen et al., 1982) which can be prevented by including the non-bilayer-forming lipid, phosphatidylethanolamine, in the membrane lipid mixture (Van der Steen et al., 1983). This effect of phosphatidylethanolamine could be due to its preferred interaction with the hydrophobic domain of the intrinsic membrane protein, thus effectively sealing the protein into the lipid bilayer matrix.

Proteins have also been shown to induce or suppress phase separations and lamellar to hexagonal-II phase transitions in phospholipid dispersions. The addition of cytochrome c to cardiolipin model membranes, for example, specifically induces a hexagonal-II phase and possibly an inverted micellar structure of the phase-separated lipids in these systems (de Kruijff and Cullis, 1980a). It has also been reported that calcium is able to induce nonlamellar phases, but the effect of calcium is prevented by polypeptides such as poly(L-lysine) which

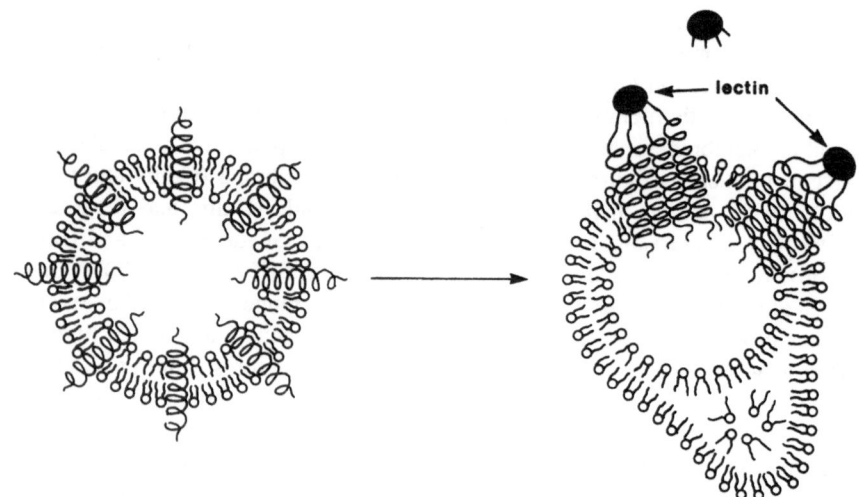

FIGURE 22. The suppression of nonbilayer lipid phase separation by intrinsic membrane proteins and the creation of phase-separated nonbilayer lipid structures induced by lectin-mediated lateral aggregation of the proteins.

can be regarded as models of extrinsic membrane proteins (de Kruijff and Cullis, 1980b). The suppression of nonbilayer lipid phase separations by the incorporation of intrinsic membrane proteins into hexagonal-II-forming lipids such as dioleoylphosphatidylethanolamine has also been noted (Taraschi et al., 1982). This suppression occurs at molar ratios of polypeptide to lipid on the order of 1 : 200, indicating that the influence of the protein extends well beyond the lipids in direct contact with the hydrophobic domain of the polypeptide. Figure 22 shows what happens when the proteins are aggregated within the plane of such membranes; for example, by the addition of multivalent lectins, domains rich in non-bilayer-forming lipids are created and the membranes are rendered leaky to ions presumably by the formation of inverted micellar structures within the bilayer (Quinn et al., 1982).

We may assume from these model studies that phase separations of membrane lipids may have profound consequences on the phase behavior of the lipids in biological membranes. There are a number of ways in which both intrinsic and extrinsic proteins may be phase separated and we shall examine how exposure of biological membranes to low temperatures, high salt concentrations, and so on may induce phase separations or lateral aggregation of membrane proteins.

4.1. Exclusion of Proteins from Bilayer Gel Phases

The processes associated with the cooling of biological membranes can be simulated by random two-dimensional arrays of discs approximating the relative cross-sectional areas of lipid hydrocarbon chains and helical polypeptide sequences (Chapman *et al.*, 1977; Quinn and Chapman, 1980). The process likely to accompany cooling of biological membranes using this model approach is illustrated in Figure 23. The events that take place depend to some extent on the proportions of lipid to polypeptide components but at relatively high lipid contents cooling the system to temperatures below the phase-transition temperature of the lipid results in a disordering of the gel structure in the region of the polypeptide chains. Packing faults are seen to radiate out along the hexagonal axis of the closely packed hydrocarbon chains. If the lipid is cooled throughout this transition so as to avoid thermal gradients in the sample, the hydrocarbon chains will crystallize and the protein will become trapped within the dislocations that form spontaneously due to thermal fluctuations that develop during crystallization. The so-called zone refining of the protein into these lateral aggregates is likely to be prevented by entrapment in packing faults until sufficient protein is present to produce a dislocation density in excess of that which occurs spontaneously in the pure lipid. This is clearly seen in Figure 23 where regions of membrane most affected by crystallizing lipid are those that spontaneously possess very few protein–protein or protein–lipid–protein contacts. The crystallizing lipid excludes the protein and at least part of the noncrystalline lipid surrounding the protein into regions in which the protein

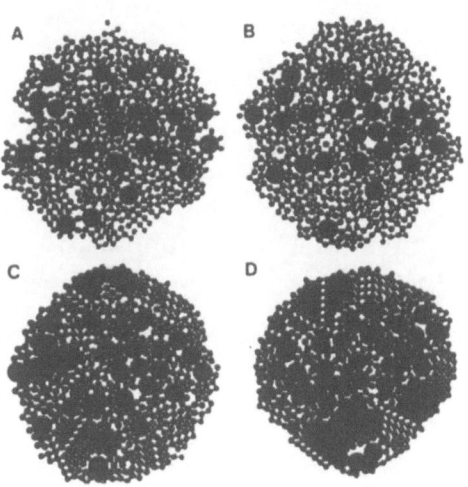

FIGURE 23. Simulation of lipid crystallization from (A) a random array of lipid (represented by the small circles) and polypeptide components (the large circles are of relative dimensions of a helical coil cross section) leading to (B) and (C) nucleation, packing faults, and finally in (D) to aggregation of the polypeptides. From Chapman *et al.* (1977).

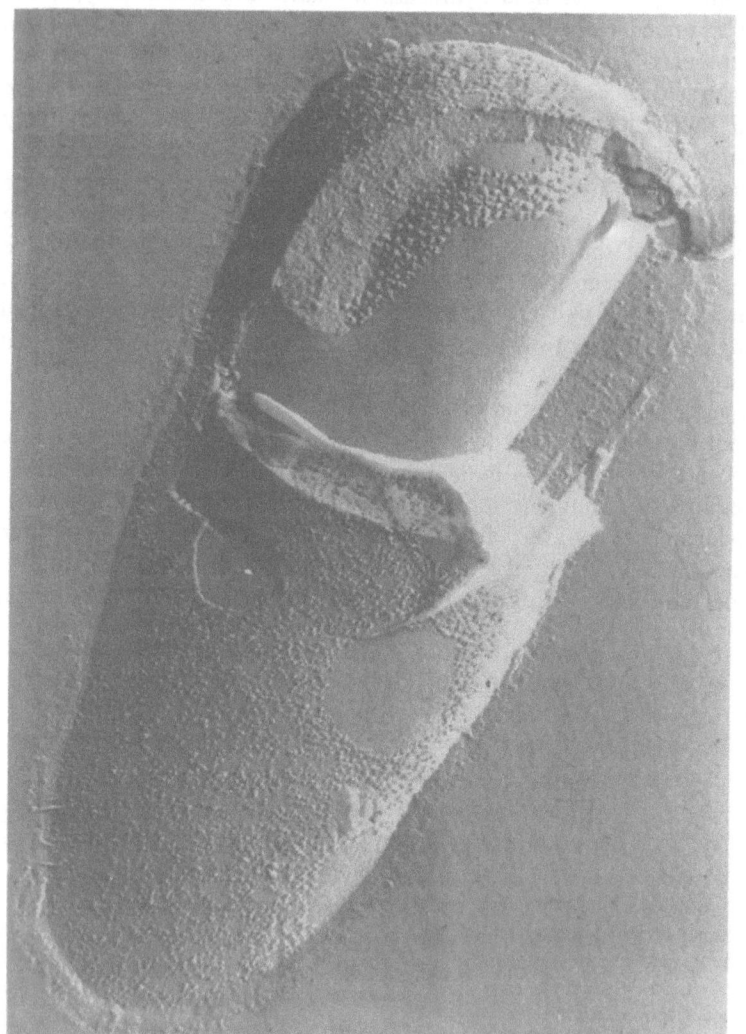

FIGURE 24. Electron micrograph of a freeze–fracture replica prepared from *Anacystis* cells cultured at 38°C and thermally quenched from 15°C. The particle-free domains of the plasma membrane and photosynthetic membrane are believed to represent domains of the gel phase lipid from which the remaining fluid phase lipid and protein have been excluded.

concentration is relatively high. The degree to which these protein-rich areas coalesce depends to some extent on the rate at which the system is cooled. Cooling the fluid phase in which the distribution of the components is random produces areas of lipid, representing nucleation sites, which ultimately form domains of crystalline lipid intersected by packing faults. As crystallization of the lipid proceeds, protein molecules are forced to aggregate, eventually achieving the phase separation depicted in the final segment of the illustration shown in Figure 23.

It is obvious that the situation in biological membranes is considerably more complex than the simple process illustrated by the model depicted in Figure 23. Nevertheless, the similarity in behavior of many biological membranes is striking. An experiment using the blue-green alga *Anacystis nidulans* is relevant to this point (Furtado *et al.*, 1979). Thus freeze–fracture electron microscope investigations of *Anacystis* cells cultured at a temperature of 38°C and thermally quenched for freeze–fracture at 35°C compared with those equilibrated at 15°C before thermal quenching show that at the growth temperature membrane-associated particles are randomly distributed in the plane of the membrane. Cooling the cells to 15°C causes a lateral phase separation of the membrane-associated particles and the creation of large domains, both in the plasma membrane and in the photosynthetic membranes of these organisms, to smooth fracture face. This phase separation is illustrated in the electron micrograph shown in Figure 24. The smooth regions are interpreted to be gel phase lipid domains from which the intrinsic membrane proteins have been excluded. In control experiments, where cells are grown at low temperatures and thermally quenched from 15°C, a random distribution of membrane-associated particles is observed and this correlates with changes in the lipid composition of the membrane. Thus, the membrane lipids possess fatty acyl residues that are considerably less saturated than when the organism is grown at high temperature.

The question as to which lipids are located in the phase-separated domains and which lipids phase separate together with the proteins has been investigated by differential scanning calorimetry, low-angle x-ray diffraction, and freeze–fracture electron microscopy (Mannock *et al.*, 1985). The results of the calorimetric study are illustrated in Figure 25. This shows that organisms cultured at 28°C exhibit a broad endotherm in their heating scans and a midpoint temperature of 14°C which presumably corresponds with the transition of gel phase lipid domains into a liquid-crystalline configuration. A similar endotherm is observed in cells cultured at 38°C, although judging from the enthalpy of these transitions the amount of gel-phase-separated lipid in cells cultured at 38°C is greater than in the corresponding membranes of cells cultured at 28°C. There is, however, a dramatic change in the midpoint temperatures of the endotherms observed in heating thermograms of total polar lipid extracts of organisms cul-

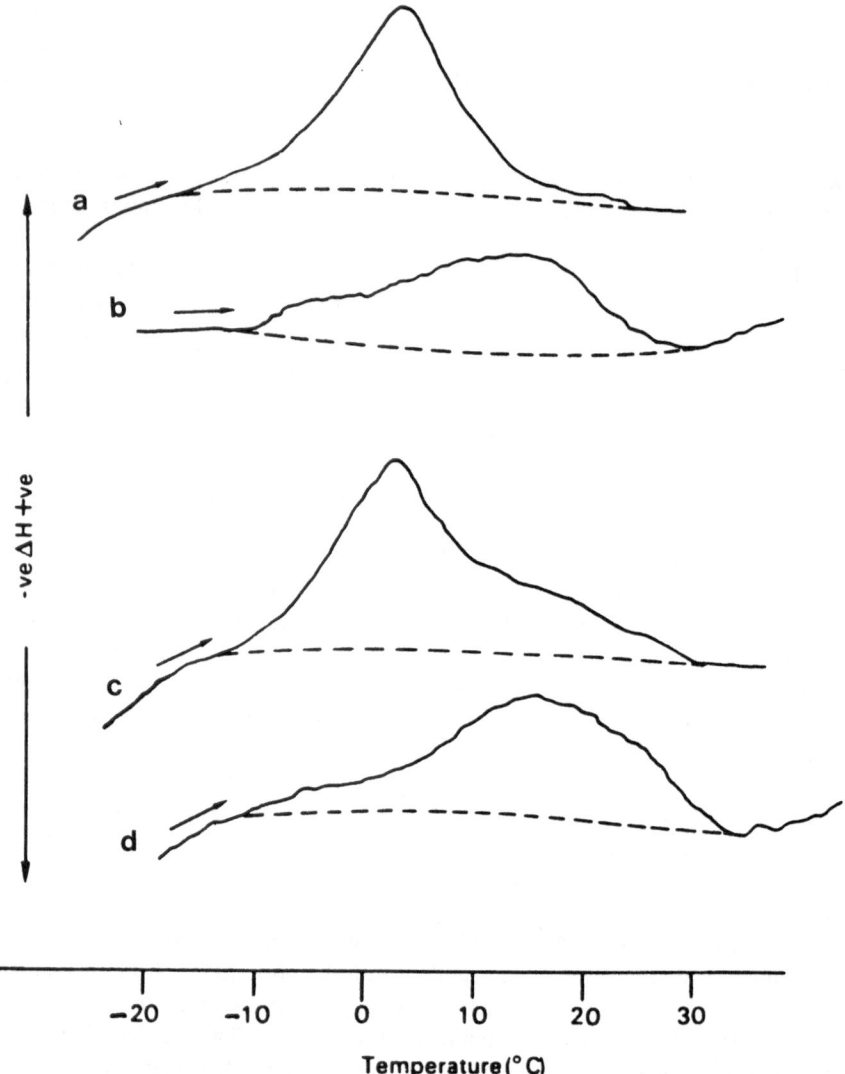

FIGURE 25. Differential scanning calorimetric heating curves of total polar lipid extracts of *Anacystis nidulans* cultured at (a) 28°C and (c) 38°C. Thermograms of whole cells grown at (b) 28°C and (d) 38°C are also shown. Data from Mannock *et al.* (1985).

tured at 20 and 38 °C which occur at 2 and 4°C, respectively. This observation clearly shows that it is the high-melting-point lipids that phase separate from the membrane proteins in the intact biological membrane. The question remains as to which lipids have higher melting points and which lipids have low melting points. This has important implications with regard to the ability of the membrane to restore a random distribution of components after thermally induced phase separations. It is well known, for example, that the phase separations of the type observed in *Anacystis nidulans* result in irreversible changes and loss in viability of the cells which are unable to repair their leaky membranes (Brand *et al.*, 1979). It is clear that the lipid composition of *Anacystis* is relatively simple, and appreciable proportions of the membrane lipids will undergo phase transitions over a relatively narrow range of temperatures. With the more complex mixtures, typical of many biological membranes, the transition endotherms observed on heating membranes previously cooled to low temperatures exhibit transitions that are invariably broad and extend over tens of degrees. Attempts have been made to resolve these broad transitions in erythrocyte membranes and lipid mixtures simulating the membrane lipid composition into components contributed to by each of the major lipid classes present in the membrane (Van Dijck *et al.*, 1976). The results of these experiments using human erythrocyte phospholipids have shown that the phase separation of the sphingolipids dominated the higher-temperature phase behavior of the outer leaflet of the membrane, and the phosphatidylethanolamines appear to phase separate initially in the inner leaflet upon cooling the membrane.

4.2. Phase Separation of Nonbilayer Lipid Phases in Membranes

The phase separation of nonbilayer lipid structures has been observed in biological membranes that contain a high proportion of non-bilayer-forming lipid under certain conditions. The data in Table II show that some subcellular

Table II
Nonbilayer Lipids of Rat Liver Subcellular Membranes and Photosynthetic Membranes and the Respective Ratios of Protein to Lipid

Membrane	Nonbilayer lipid[a] (%)	Protein/lipid (w/w)
Rat liver		
Plasma membrane	18.5	1.4
Nuclear membrane	22.7	1.6
Endoplasmic reticulum	18.6	0.9
Inner mitochondrial membrane	42.7	3.2
Outer mitochondrial membrane	26.6	1.1
Plants and algae		
Spinach chloroplasts	38	1.5
Anacystis nidulans	54	1.9

[a] Phospholipids for rat liver; total polar lipid of plants and algae.

membranes of mammalian tissues and those of photosynthetic membranes have relatively high proportions of lipids that form nonbilayer structures under physiological conditions. The proportion of such lipids does not necessarily correlate with the amount of membrane protein, but it may be expected that the membrane proteins differ considerably in the way they interact with the lipid component and the effects they exert on membrane lipid phase behavior.

Under normal conditions, nonbilayer structures are not found in chloroplasts or other biological membranes. They are, however, observed when such membranes are subjected to stress. Nonbilayer structures are formed in the membranes of heat-stressed chloroplasts (Gounaris et al., 1983a) and chloroplast thylakoids treated with 6 M guanidinethiocyanate (Machold et al., 1977). Similar structures have also been observed in mitochondria following the addition of high concentrations of manganese (Van Venetie and Verkleij, 1982) and dehydrated retinal rod outer-segment disc membranes (Corless and Costello, 1982; Gruner et al., 1982) and sarcoplasmic reticulum (Crowe and Crowe, 1982). The factors responsible for these phase separations and their implications with regard to protein–lipid interactions will now be examined.

4.2.1. High-Temperature Stress in Plant Cell Membranes

The biophysical and functional properties of plant cell organelles exposed to high temperatures reveal characteristic damage to membranes. It appears that the upper limit of temperatures at which particular plant species survive is a function of the stability of the chloroplast membrane. Leaves and isolated chloroplasts, for example, show marked reductions in the photosynthetic activity following exposures to temperatures above 40–50°C (Quinn and Williams, 1983). Measurements of changes in chlorophyll a fluorescence emission under such conditions (Krause and Santarius, 1975; Schreiber and Berry, 1977; Schreiber et al., 1976) indicate that the light-harvesting apparatus of photosystem-II is particularly susceptible to thermal damage. Thus, heating briefly to relatively high temperatures results in irreversible functional reorganization of the photosynthetic apparatus, particularly photosystem-II (Schreiber and Armond, 1978). Freeze–fracture studies (Armond et al., 1980; Gounaris et al., 1984) have shown that incubation of isolated chloroplast suspensions at elevated temperatures leads to a progressive dissociation of the supramolecular complex corresponding to the photosynthetic light-harvesting unit of photosystem-II and the consequent destacking of the thylakoid membrane. Differential scanning calorimetric studies (Cramer et al., 1981) have indicated a series of endothermic transitions corresponding to order–disorder transitions of different structural domains within the chloroplast membrane. The lowest temperature transition, with a transition maximum of 42–44°C, correlates with the release of manganese from the thylakoid membrane, loss of oxygen evolution ability by the chloroplasts, and a decrease in the redox potential of high-potential cytochrome b-559. It appears to correspond to the thermal disruption of a protein component on the donor side of photosystem-II and, as such, probably reflects part of the dissociation process observed in freeze–fracture studies.

The structural changes in thylakoid membranes of higher plant chloroplasts subjected to thermal stress have been examined in detail by freeze–fracture electron microscopy (Gounaris et al., 1983). It was reported that a normal morphology was preserved during brief exposure (5 min) of the chloroplast suspension to temperatures up to 35°C. Incubation between 35 and 45°C caused complete destacking of the grana and higher temperatures caused a phase separation of nonbilayer lipids into stable aggregates of cylindical inverted micelles. Interpretation of the effects of temperatures greater than 45°C is based on phase conditions that result in a release of the constraints imposed by interaction of the non-bilayer-forming lipid, monogalactosyldiacylglycerol, with other membrane components and its segregation into domains of nonbilayer lipid structure. Gross phase separations of this type require that the shift in thermal stability of the stacked membrane is relatively large because three-dimensional aggregates of lipid are not observed if the chloroplast membrane is destacked by manipulation of the ionic environment before heat treatment.

The explanation of the effect of heat treatment on destacking the native membrane is not so clear-cut. The factors responsible for membrane stacking have been postulated in a model proposed by Barber (1980). According to this model thylakoid stacking and related phenomena are explained in terms of the effect of cations on electrostatic charges on the membrane surface. A difference in the surface charges of light-harvesting chlorophyll a/b–protein complexes associated with photosystem-II (believed to carry little or no charge) and P700-chlorophyll a/b–protein complexes associated with photosystem-I (possessing a relative abundance of negative charge) is said to result in a randomization of the complexes laterally in the membrane in conditions of low salt concentration. It is argued that addition of cations screens the charges and reduces electrostatic repulsion between photosystem-I complexes, allowing a reorganization of the membrane system in which the two photosystems become spatially segregated. The formation of the grana stacks is explained by a reduction in Coulombic repulsion forces between opposing membrane surfaces (Mullet and Arntzen, 1980; Ryrie et al., 1980).

A careful analysis of the ultrastructural changes associated with temperature-induced destacking of thylakoid membranes has been carried out using electron microscopy (Gounaris et al., 1984), where it was shown that normal granal stacks are progressively replaced by modified thylakoid attachment sites in which contact between opposing membranes becomes restricted to regions of focal contact. Changes in the distribution of particles observed in the exoplasmic phase of stacked and unstacked regions and corresponding regions of

Table III
Size and Density of Membrane-Associated Particles Observed on the Exoplasmic (EF) and Protoplasmic (PF) Fracture Faces of Heat-Treated Chloroplast Membranes

Fracture face[a]	Particle diameter (nm)	Particle density (μm^{-2})
Unheated control (25°C)		
EF_s	11.2 ± 2.4	1238 ± 163
EF_u	9.3 ± 2.1	348 ± 59
PF_s	7.8 ± 2.0	4583 ± 322
PF_u	9.2 ± 2.1	3580 ± 263
Heated (45°C, 5 min)		
EF_s	—	—
EF_u	8.9 ± 3.0	818 ± 38
PF_s	6.6 ± 1.7	7290 ± 258
PF_u	9.3 ± 2.4	4346 ± 478

[a] s and u represent stacked and unstacked regions, respectively.

the protoplasmic phase due to heat temperature are presented in Table III. The most notable difference is a disappearance of the intramembranous particles in the exoplasmic profiles of the attachment sites formed during exposure to high temperature. The reverse occurs outside the stacked regions of the exoplasmic face, where the density of the particles increases significantly. Changes in particle size and regions of contact between membranes in profiles of exposed exoplasmic and protoplasmic fracture faces are shown in the form of difference histograms in Figure 26. These data are interpreted as a dissociation of light-harvesting units of photosystem-II in which the antenna complexes cluster together, maintaining regions of membrane adhesion, while excluding the core complexes of photosystem-II and light-harvesting units of photosystem-I. This process is illustrated schematically in Figure 27. In this model the electrostatic interactions maintaining contact between the membranes are not perturbed by the brief exposure to high temperature, but a shift in the phase behavior of the membrane lipids leads to dissociation of oligomeric complexes. This is consistent with functional changes observed by Schreiber and Armond (1978) in photosystem-II of chloroplasts subjected to heat stress. It has been suggested that a shift in the phase of the nonbilayer lipid, monogalactosyldiacylglycerol, underlies this structural change and that the functional role of this nonbilayer lipid is to package the light-harvesting chlorophyll a/b–protein complexes together with the photosystem-II core protein complex into an efficient functional unit localized within the grana stack (Quinn and Williams, 1983). A similar functional role of nonbilayer lipids, including phosphatidylethanolamines and monogalactosyldiacylglycerols, has been proposed on the basis of reconstitution studies of Ca^{2+}-ATPase in which the efficiency of pumping is markedly improved by the presence of non-bilayer-forming lipids (Navarro et al., 1984).

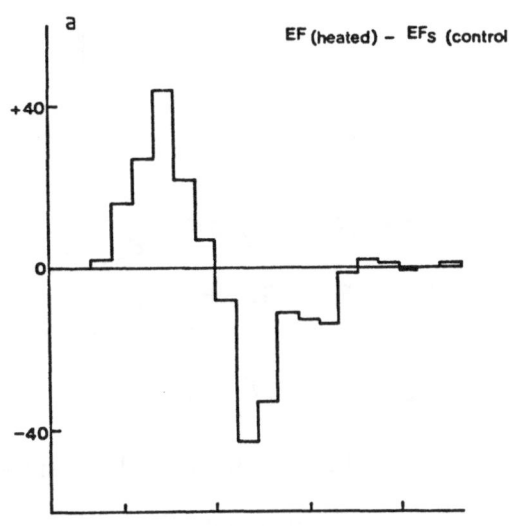

FIGURE 26. Particle size difference histograms (heated minus control) of the intramembranous particles found in the (a) exoplasmic and (b) protoplasmic fracture faces of chloroplast thylakoid membrane. EF_s and PF_s refer to the exoplasmic and protoplasmic fracture faces, respectively, of the stacked region of the membrane.

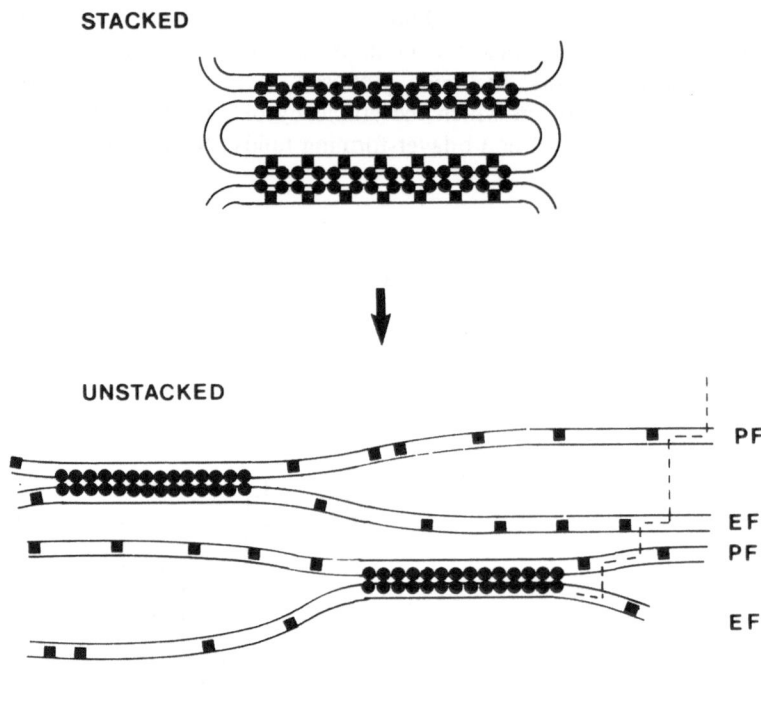

FIGURE 27. Schematic model illustrating the effect of the dissociation of the light-harvesting apparatus on the organization of the photosynthetic membrane. ●, chlorophyll a/b light-harvesting proteins; ■, photosystem-II core proteins only are shown.

4.2.2. Subzero Temperatures

Biological membranes are subject to damage at subzero temperatures resulting generally in a destruction of the permeability barrier properties. A lipid phase separation model has been proposed to account for the irreversible damage to membranes (Quinn, 1985). The model takes into account the phase behavior of membrane lipids and the likely events that occur during cooling and subsequent reheating of the membrane. Figure 28 is a schematic illustration of the molecular redistributions that probably take place during these processes. Thus, when membranes are cooled from the growth temperature, the first transition that the membrane lipids will be subjected to will be a hexagonal-II to liquid-crystalline lamellar phase transition of those lipids that tend to form nonbilayer phases at the growth temperature. It has been shown that intrinsic membrane proteins impose a lamellar configuration on these lipids at the growth temperature; the role of these lipids in packaging of proteins has already been discussed. When this constraint is no longer imposed and in the absence of any specific interactions between the proteins and the nonlamellar lipids, it may be predicted that the lipids will diffuse to form a homogeneous fluid bilayer of lipids in the absence of any small component of lipid which is below the gel to liquid-crystalline phase transition temperature. Such lipids may form small domains of gel phase lipid in the fluid membrane matrix. There is, as yet, no direct information on the extent to which bilayer and non-bilayer-forming lipids may be heterogeneously distributed within membranes but where this is due to a tendency of the particular lipids to form nonbilayer structures, this constraint would be expected to be removed on cooling below the hexagonal-II to lamellar phase transition temperature. With the possible exception of membranes subjected to chilling damage, such as those described for *Anacystis nidulans*, this type of phase separation is not likely to cause irreversible damage

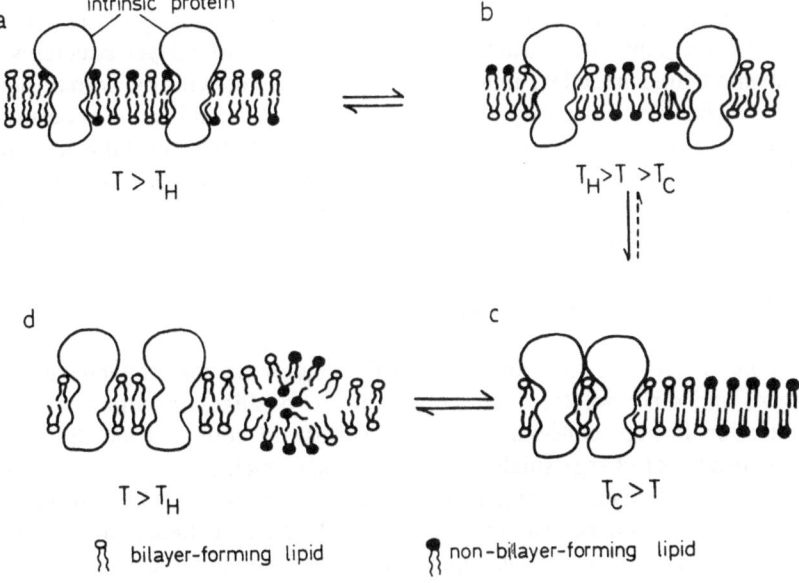

FIGURE 28. Schematic representation of the phase separations in biological membranes associated with cooling from the growth temperature (a) to a temperature below the hexagonal-II to liquid-crystalline phase transition temperature of the non-bilayer-forming lipids (b) and subsequently below the gel to liquid-crystalline phase transition temperature of all the membrane lipids (c). The effect of reheating to the growth temperature is shown in (d).

to membranes and any heterogeneity may be restored on reheating to the growth temperature. Further cooling results in transitions from liquid-crystalline to gel phase beginning first with the higher-melting-point lipids, which as we have seen will segregate into domains of pure gel phase lipid from which the intrinsic membrane proteins and low-melting-point lipids are excluded.

The central argument of the lipid phase separation model of irreversible damage to biological membranes is that the higher-melting-point lipids present in gel phase domains will be rich in the hexagonal-II-forming lipids. This assumption is based on the fact that, with equivalent hydrocarbon substituents, lipids that tend to form hexagonal-II structures have liquid-crystalline to gel phase transitions at temperatures that are considerably higher than those corresponding to bilayer-forming lipids (see Quinn, 1985). Where phase separations between lipid classes of this type are created, the changes are not likely to be reversed on reheating to temperatures above the hexagonal-II to lamellar phase transition temperature. In practice, it would be improbable for there to be a single temperature where all the nonlamellar lipids would coexist together with the bilayer-forming lipids in the fluid phase, allowing the original distribution of the lipids and the proteins to be restored. Damage to the membrane in these circumstances would be expected to result when the membrane is reheated to temperatures where the domains of the phase-separated hexagonal-II-rich lipids tend to form nonbilayer structures. The creation of nonbilayer structures such as inverted lipid micelles, unless suppressed or dealt with by normal homeostatic mechanisms operating within the membrane, would serve to destroy the permeability barrier properties of the membrane. If this breakdown of membrane barrier was of sufficient duration to permit loss of essential components or irreversible alterations in the intracellular compartmentation, loss of cell viability would result.

4.2.3. The Role of Acidic Lipids

One of the complicating effects of freezing is the zone refinement of electrolytes and other solutes from ice as water crystallizes, resulting in their accumulation in high concentration at the surface of cells and membranes. The consequences of charge shielding of the usually small proportion of membrane lipids is well known in mixed lipid systems (see above). Other studies have also shown that charge shielding or removal of acidic lipids can also induce phase separation of nonbilayer lipids from other membrane components.

As already mentioned, temperature of mitochondria with high concentrations of manganese (Van Venetie and Verkleij, 1982) and chloroplasts with high concentrations of magnesium (Carter and Staehelin, 1980) causes phase separation of hexagonal-II lipid structures. Similar types of phase separation

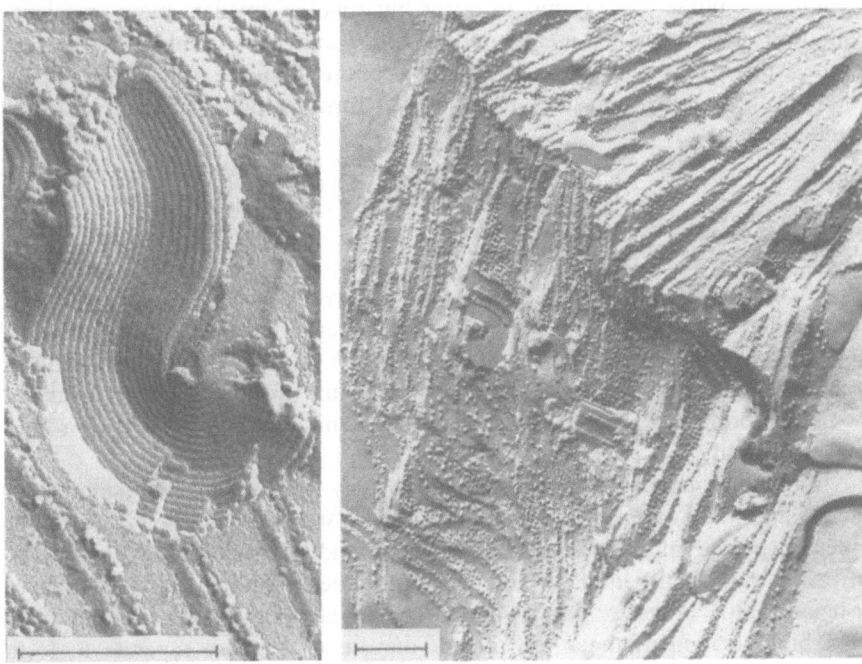

FIGURE 29. Electron micrographs of a freeze–fracture replica prepared from pea chloroplasts exposed to a medium of pH 4.5 for 5 min before being restored to pH 7.6 prior to thermal quenching. Data from Thomas *et al.* (1985). Bars = 200 nm.

have been observed when chloroplasts are exposed briefly to low pH (4.5, 5 min) or subjected to digestion with phospholipase A_2 (Thomas *et al.*, 1985). Figure 29 shows an electron micrograph of a freeze–fracture replica prepared from a suspension of pea chloroplasts exposed to pH 4.5 where a large domain of hexagonal phase lipid from which the protein appears to be excluded can be seen embedded in the stacked thylakoid region. It should be emphasized that phase separations such as those observed in this figure only appear in stacked chloroplasts. No nonbilayer lipid aggregates are found in chloroplasts treated with phospholipase A_2, for example, under conditions in which stacking is prevented. This suggests that the formation of such structures is dependent on the existence of two or more bilayers containing "destabilized" nonbilayer lipids in close proximity to each other. A similar conclusion has been drawn from studies of freeze-stressed protoplasts (Gordon–Kam and Steponkus, 1984). Failure to observe such aggregates does not necessarily imply the absence of any phase separation of nonbilayer lipids in unstacked membranes. It is quite possible that phase separations do occur but involve smaller, more transient, nonbilayer structures such as inverted micelles sandwiched within the bilayer. All these effects are interpreted as a removal of the influence of the acidic lipids on imposing a bilayer phase on non-bilayer-forming lipids of the biological membrane.

5. CONCLUSIONS

Membrane lipids comprise an extremely diverse group of molecules and most biological membranes contain a large number of individual molecular species of lipid. The purpose of this diversity is not yet known and is puzzling in view of the general observation that apparently full functional activity of many membrane proteins can be reconstituted with a minimal requirement of a fluid bilayer of lipids.

Individual molecular species of lipid display rich structural polymorphism depending on the temperature, water and electrolyte concentrations, and so on. When dispersed in excess aqueous medium under physiological conditions, however, most of the lipid species will form either a fluid or gel phase bilayer or hexagonal-II structure. When bilayer-forming and non-bilayer-forming lipids from biological membranes are mixed, they tend to phase separate to form intermediate structures such as micelles sandwiched between leaflets of bilayers. In some membranes such as the erythrocytes the proportion of non-bilayer-forming lipid is low and only bilayers are observed when dispersed in dilute salt solutions. In other membranes, such as photosynthetic membranes and membranes of halobacteria, large-scale phase separations are observed in membrane lipids dispersed under physiological conditions. Phase separations within the biological membrane, however, are not observed, leading to the conclusion that interaction of the non-bilayer-forming lipids with other membrane components prevents phase separation.

Subjecting biological membranes to extreme environmental conditions or unphysiological stresses such as freeze–thawing, dehydration, and so on often provides evidence of lipid–lipid and lipid–protein phase separations. Where phase separations cannot be reversed, it is likely that irreversible damage to the membrane will result. This can be caused by leakage of solutes either at the interface between intrinsic membrane protein and lipid interface or mediated by the transient formation of nonbilayer lipid structures. These conclusions imply that there may be some preference to interaction of intrinsic membrane proteins with non-bilayer-forming lipids which helps to seal them into a bilayer matrix and also to package particular oligomeric complexes into efficient functional units.

6. REFERENCES

Abrahamsson, S., Dahlen, B., Lofgren, H., and Pascher, I., 1978, Lateral packing of hydrocarbon chains. *Prog. Chem. Fats Other Lipids* **16**:125-143.
Akiyama, M., Terayama, Y., and Matsushima, N., 1982, Kinetics of pretransition in multilamellar dimyristoylphosphatidylcholine vesicle. X-ray diffraction study. *Biochim. Biophys. Acta* **687**:337-339.
Alecio, M. R., Miller, A., and Watts, A., 1985, Diffraction of X-rays by rippled phosphatidylcholine bilayers. *Biochim. Biophys. Act* **815**:139-142.
Armond, P. A., Bjorkman, O., and Staehelin, L. A., 1980, Dissociation of supramolecular complexes in chloroplast membranes. A manifestation of heat damage to the photosynthetic apparatus. *Biochim. Biophys. Acta* **601**:433-441.
Arnold, K., Losche, A., and Gawisch, K., 1981, ^{31}P-NMR investigations of phase separation in phosphatidylcholine/phosphatidylethanolamine mixtures. *Biochim. Biophys. Acta* **645**:143-148.
Arrondo, J. L. R., Goni, F. M., and Macarulla, J. M., 1984, Infrared spectroscopy of phosphatidylcholines in aqueous suspension. A study of the phosphate group vibrations. *Biochim. Biophys. Acta* **794**:165-168.
Barber, J., 1980, An explanation for the relationship between salt-induced thylakoid stacking and the chlorophyll fluorescence changes associated with changes in spillover of energy from photosystem II to photosystem I. *FEBS Lett.* **117**:1-10.
Blaurock, A. E., 1983, Evidence of bilayer structure and of membrane interactions from X-ray diffraction analysis. *Biochim. Biophys. Acta* **650**:167-207.
Blaurock, A. E., and McIntosh, T. J., 1985, Structure of the crystalline bilayer in the subgel phase of dipalmitoylphosphatidylglycerol. *Biochemistry* **25**:299-305.
Blume, A., and Ackerman, Th., 1974, A calorimetric study of the lipid phase transitions in aqueous dispersions of phosphorylcholine-phosphonylethanolamine mixtures. *FEBS Lett.* **43**:71-74.
Boyanov, A. I., Koynova, R. D., and Tenchov, B. G., 1986, Effect of lipid admixtures on the dipalmitoylphosphatidylcholine subtransition. *Chem. Phys. Lipids* **39**:155-163.
Braganza, L. F. and Worcester, D. L., 1986, Hydrostatic pressure induces hydrocarbon chain interdigitation in single-component phospholipid bilayers. *Biochemistry* **25**:2591-2596.
Brand, J. J., Kirchanski, S. J., and Ramirez-Mitchell, R., 1979, Chill-induced morphological alterations in *Anacystis nidulans* as a function of growth temperature. *Planta* **145**:63-68.
Brentel, I., Selstam, E., and Lindblom, G., 1985, Phase equilibria of mixtures of plant galactolipids. The formation of a bicontinuous cubic phase. *Biochim. Biophys. Acta* **812**:816-826.
Broadhurst, M. G., 1962, Analysis of the solid-phase behaviour of the normal paraffins. *J. Res. Natl. Bur. Stand. Sect A* **66A**:241-249.
Browning, J. L., 1981, Motions and interactions of phospholipid headgroups at the membrane surface. 3. Dynamic properties of amine-containing head groups. *Biochemistry* **20**:7144-7151.
Bull, T., and Lindman, B., 1974, Amphiphile diffusion in cubic lyotropic mesophases. *Mol. Cryst. Liq. Cryst.* **28**:155-160.
Burnell, E. E., Cullis, P. R., and de Kruijff, B., 1980, Effects of tumbling and lateral diffusion on phosphatidylcholine model membrane ^{31}P-NMR lineshapes. *Biochim. Biophys. Acta* **603**:63-69.
Caffrey, M., 1984, X-radiation damage of hydrated lecithin membranes detected by real-time X-ray diffraction using wiggler-enhanced synchrotron radiation as the ionizing radiation source. *Nucl. Instrum. Methods Phys. Res.* **222**:329-338.
Caffrey, M., 1985, Kinetics and mechanism of the lamellar gel/lamellar liquid-crystal and lamellar/inverted hexagonal phase transition in phosphatidylethanolamine: A real-time X-ray diffraction study using synchrotron radiation. *Biochemistry* **24**:4826-4844.

Cameron, D. G., and Mantsch, H. H., 1982, Metastability and polymorphism in the gel phase of 1,2-dipalmitoyl-3-*sn*-phosphatidylcholine. A Fourier transform infrared study of the subtransition. *Biophys. J.* **38**:175–184.

Cameron, D. G., Casal, H. L., Mantsch, H. H., Boulanger, Y., and Smith, I. C. P., 1981, The thermotropic behaviour of dipalmitoylphosphatidylcholine bilayers. *Biophys. J.* **35**:1–16.

Cary, P. R., 1982, *Biochemical Applications of Raman and Resonance Raman Spectroscopy*, pp. 208–233, Academic Press, Orlando.

Carter, D. P., and Staehelin, L. A., 1980, Proteolysis of chloroplast thylakoid membranes. II. Evidence for the involvement of the light-harvesting chlorophyll a/b–protein complex in thylakoid stacking and for the effects of magnesium ions on photosystem II–light-harvesting complex aggregates in the absence of membrane stacking. *Arch. Biochim. Biophys* **200**:374–386.

Casal, H. L., and Mantsch, H. H., 1984, Polymorphic phase behaviour of phospholipid membranes studied by infrared spectroscopy. *Biochim. Biophys. Acta* **779**:381–401.

Casal, H. L., Smith, I. C. P., Cameron, D. G., and Mantsch, H. H., 1979, Lipid reorganization in biological membranes. A study by Fourier transform infrared difference spectroscopy. *Biochim. Biophys. Acta* **550**:145–149.

Casal, H. L., Cameron, D. G., Smith, I. C. P., and Mantsch, H. H., 1980, *Acholeplasma laidlawii* membranes: A Fourier transform infrared study of the influence of protein on lipid organisation and dynamics. *Biochemistry* **19**:444–451.

Casal, H. L., Cameron, D. G., Jarrell, H. C., Smith, I. C. P., and Mantsch, H. H., 1982, Lipid phase transitions in fatty acid-homogeneous membranes of *Acholeplasma laidlawii* B. *Chem. Phys. Lipids* **30**:17–26.

Casal, H. L., Mantsch, H. H., Cameron, D. G., and Gaber, B. P., 1983, On the subtransitions of deuterated derivatives of 1,2-dipalmitoyl-*sn*-glycero-3-phosphocholine. *Chem. Phys. Lipids* **33**:109–112.

Chang, H. and Epand, R. M., 1983, The existence of a highly ordered phase in fully hydrated dilauroylphosphatidylethanolamine, *Biochem. Biophys. Acta,* **728**:319–324.

Chapman, D., Williams, R. M., and Ladbrooke, B. D., 1967, Physical studies of phospholipids. VI. Thermotropic and lyotropic mesomorphism of some 1,2-diacylphosphatidylcholines (lecithins). *Chem. Phys. Lipids* **1**:445–475.

Chapman, D., Urbina, J., and Keough, K. M., 1973, Biomembrane phase transitions. Studies of lipid–water systems using differential scanning calorimetry. *J. Biol. Chem* **249**:2512–2518.

Chapman, O., Cornell, B. A., Eliasz, A. W., and Perry, A., 1977, Interactions of helical polypeptide segments which span the hydrocarbon region of lipid bilayers. Studies of the gramicidin A lipid–water system. *J. Mol. Biol.* **113**:517–538.

Chapman, D., Cornell, B. A., and Quinn, P. J., 1977, Phase transitions, protein aggregation and a new method for modulating membrane fluidity. In *Biochemistry of Membrane Transport* (G. Semenza and E. Carafoli, eds.), pp. 72–85, Springer-Verlag, Berlin.

Charvolin, J., and Rigny, P., 1971, NMR study of molecular motions in the mesophases of potassium laurate-water-d2 system. *J. Magn. Reson.* **4**:40–46.

Chen, S. C., Sturtevant, J. M., and Gaffney, B. J., 1980, Scanning calorimetric evidence for a third phase transition in phosphatidylcholine bilayers. *Proc. Natl. Acad. Sci. U.S.A.* **77**:5060–5063.

Cho, K. C., Choy, C. L., and Young, K., 1981, Kinetics of the pretransition of synthetic phospholipids. A calorimetric study. *Biochim. Biophys. Acta* **662**:14–21.

Church, S. E., Griffiths, D. J., Lewis, R. N. A. H., McElhaney, R. N., and Wickman, H. H., 1986, X-ray structure study of thermotropic phases in isoacylphosphatidylcholine multibilayers. *Biophys. J.* **49**:597–605.

Cohen, M. H., and Turnbull, D., 1959, Molecular transport in liquids and gases. *J. Chem. Phys.* **31:**1164–1169.
Corless, J. M., and Costello, M. J., 1982, Isolation, rapid freezing and freeze–fracture methods for frog retinal photoreceptors. *Exp. Eye Res.* **32:**217–228.
Cornell, B. A., and Separovic, F., 1983, Membrane thickness and acyl chain length. *Biochim. Biophys. Acta* **733:**189–193.
Cramer, W. A., Whitmarsh, J., and Low, P. S., 1981, Differential scanning calorimetry of chloroplast membranes: Identification of an endothermic transition associated with the water-splitting complex of photosystem II, *Biochemistry* **20:**157–162.
Crowe, L. M., and Crowe, J. H., 1982, Hydration-dependent hexagonal phase lipid in a biological membrane *Arch. Biochem. Biophys.* **217:**582–587.
Cullis, P. R., and de Kruijff, B., 1978a, Polymorphic phase behaviour of lipid mixtures as detected by ^{31}P-NMR. *Biochim. Biophys. Acta* **507:**207–218.
Cullis, P. R., and de Kruijff, B., 1978b, Lipid polymorphism and the functional role of lipids in biological membranes. *Biochim. Biophys. Acta* **559:**399–420.
Cullis, P. R. and de Kruijff, B., 1979, Lipid polymorphism and the functional role of lipids in biological membrane. *Biochim. Biophys. Acta* **559:**399–420.
Cullis, P. R., and Grathwohl, Ch., 1977, Hydrocarbon phase transitions and lipid–protein interactions in the erythrocyte membrane. *Biochim. Biophys. Acta* **473:**213–226.
Cullis, P. R., and Hope, M. J., 1978, Effects of fusogenic agent on membrane structure of erythrocyte ghosts and the mechanism of membrane fusion. *Nature* **271:**672–674.
Cullis, P. R., de Kruijff, B., Hope, M. J., Nayar, R., and Schmid, S. L., 1980, Phospholipids and membrane transport. *Can. J. Biochem.* **58:**1091–1100.
Cullis, P. R., Hope, M. J., de Kruijff, B., Verkeij, A. J., and Tilcock, C. P. S., 1985, Lipid polymorphism and function of biological membranes, In *Phospholipids and Cellular Regulation* (J. F. Kuo, ed.), pp. 1–60, CRC Press, Boca Raton, FL.
Deamer, D. W., Leonard, R., Tardieu, A., and Branton, D., 1970, Lamellar and hexagonal lipid phases visualized by freeze-etching. *Biochim. Biophys. Acta* **219:**47–60.
De Grip, W. J., Drenthe, E. H. S., Van Echfeld, C. J. A., de Kruijff, B., and Verkleij, A. J., 1979, A possible role of rhodopsin in maintaining bilayer structure in the photoreceptor membrane. *Biochim. Biophys. Acta* **558:**330–337.
de Kruijff, B., and Cullis, P. R., 1980a, Cytochrome c specifically induces non-bilayer structures in cardiolipin-containing model membranes. *Biochim. Biophys. Acta* **602:**477–490.
de Kruijff, B., and Cullis, P. R., 1980b, The influence of poly(L-lysine) on phospholipid polymorphism. Evidence that electrostatic polypeptide–phospholipid interactions can modulate bilayer–non-bilayer transitions. *Biochim. Biophys. Acta* **601:**235–240.
de Kruijff, B., Van den Besselaar, A. M. H. P., Cullis, P. R., Van den Bosch, H., and Van Deenen, L. L. M., 1978, Evidence for isotropic motion of phospholipids in liver microsomal membranes. A ^{31}P-NMR study. *Biochim. Biophys. Acta* **514:**1–8.
de Kruijff, B., Cullis, P. R., Verkleij, A. J., Hope, M. J., Van Echfeld, C. J. A., and Tarashi, T. F., 1985, Lipid polymorphism and membrane function. In *The Enzymes of Biological Membranes* (A. N. Martonosi, ed.), Vol. 1, pp. 131–204, Plenum Press, New York.
Derjaguin, B. V., and Landau, L., 1941, Theory of the stability of strongly charged lyophobic sols and of the adhesion of strongly charged particles in solutions of electrolytes. *Acta Physicochem. USSR* **14:**633–692.
De Rosa, M., Gambarcorta, A., Nicolaus, B. and Bu'lock, J. D., 1980a, Complex lipids of *Caldaviella acidophila*, a thermoacidophile archaebacterium. *Phytochemistry* **19:**821–825.
De Rosa, M., Esposito, E., Gambarcorta, A., Nicolaus, B., and Bu'lock, J. D., 1980b, Effects of temperature on ether lipid composition of *Caldaviella acidophila*. *Phytochemistry* **19:**827–831.

De Rosa, M., Gambacorta, A., Nicolaus, B., Chappe, B., and Albrecht, P., 1983, Isoprenoid ethers; backbone of complex lipids of the archaebacterium *Sulfolobus solfataricus*. *Biochim. Biophys. Acta* **753**:249–256.

Eliel, E. G., 1962, *Stereochemistry of Carbon Compounds*, Chap. IV, McGraw-Hill, New York.

Eriksson, P.-O:, Khan, A., and Lindblom, G., 1982, Nuclear magnetic resonance studies of molecular motion and structure of cubic liquid crystalline phases. *J. Phys. Chem* **86**:387–393.

Farren, S. B., and Cullis, P. R., 1980, Polymorphism of phosphatidylglycerol–phosphatidylethanolamine model membrane systems. A ^{31}P-NMR study. *Biochem. Biophys. Res. Commun.* **97**:182–191.

Finegold, L., and Singer, M. A., 1984, Phosphatidylcholine bilayers: Subtransitions in pure and mixed lipids. *Chem. Phys. Lipids* **35**:291–297.

Finegold, L., and Singer, M. A., 1986, The metastability of saturated phosphatidylcholines depends on the acyl chain length. *Biochim. Biophys. Acta* **855**:417–420.

Fringeli, U. P., and Gunthard, Hs. H., 181, Infrared membrane spectroscopy. In *Membrane Biology, Biochemistry and Biophysics* (E. Grell, ed.), Vol. 31, pp. 270–332, Springer-Verlag, Berlin.

Fujii, T., Tamura, A., and Yamane, T., 1985, Trans-bilayer movement of added phosphatidylcholine and lysophosphatidylcholine species with various acyl chain lengths in plasma membrane of intact human erythrocytes. *J. Biochem* **98**:1221–1227.

Fuldner, H. H., 1981, Characterization of a third phase transition in multilamellar dipalmitoyllecithin liposomes. *Biochemistry* **20**:5707–5710.

Furtado, D., Williams, W. P., Brain, A. P. R., and Quinn, P. J., 1979, Phase separations in membranes of *Anacystis nidulans* grown at different temperatures. *Biochim. Biophys. Acta* **555**:352–357.

Gad, A. E., Bental, M., Elyashiv, G., Weinberg, H., and Nir, S., 1985, Promotion and inhibition of vesicle fusion by polylysine. *Biochemistry* **24**:6277–6282.

Galla, H.-J., Hartmann, W., Theilen, U., and Sackmann, E., 1979, On two-dimensional passive random walk in bilayers and fluid pathways in biomembranes. *J. Membr. Biol* **48**:215–236.

Gally, H. U., Pluschke, G., Overath, P., and Seelig, J., 1980, Structure of *Escherichia coli* membranes. Fatty acyl chain order parameters of inner and outer membranes and derived liposomes. *Biochemistry* **19**:1638–1643.

Gordon-Kam, W. J., and Steponkus, P. L., 1984, Lamellar-to-hexagonal-II phase transitions in the plasma membrane of isolated protoplasts after freeze-induced dehydration. *Proc. Natl. Acad. Sci. U.S.A.* **81**:6373–6377.

Gounaris, K., Brain, A. P. R., Quinn, P. J., and Williams, W. P., 1983a, Structural and functional changes associated with heat-induced phase-separations of non-bilayer lipids in chloroplast thylakoid membranes. *FEBS Lett.* **153**:47–52.

Gounaris, K., Mannock, D. A., Sen, A., Brain, A. P. R., Williams, W. P., and Quinn, P. J., 1983b, Polyunsaturated fatty acyl residues of galactolipids are involved in the control of bilayer/non-bilayer lipid transitions in higher plant chloroplasts. *Biochim. Biophys. Acta* **732**:229–242.

Gounaris, K., Brain, A. P. R., Quinn, P. J., and Williams, W. P., 1984, Structural reorganization of chloroplast thylakoid membranes in response to heat stress. *Biochim. Biophys. Acta* **766**:198–208.

Graham, I., Gagne, J., and Silvius, J. R., 1985, Kinetics and thermodynamics of calcium-induced lateral phase separations in phosphatidic acid containing bilayers. *Biochemistry* **24**:7123–7131.

Griffin, R. G., Powers, L., and Persham, R. S., 1978, Head-group conformation phospholipids: A phosphorus-31 nuclear magnetic resonance study or oriented monodomain dipalmitoylphosphatidylcholine bilayers. *Biochemistry* **17**:2718–2722.

Gruen, D. W. R., and Marcelja, S., 1983, Spatially varying polarization in water. A model for the electric double layer and the hydration force. *J. Chem. Soc. Faraday Trans. II* **79**:225–242.

Gruner, S. M., 1985, Intrinsic curvature hypothesis for biomembrane lipid composition: a role for nonbilayer lipids, *Proc. Natl. Acad. Sci., U.S.A.*, **82**:3665–3669.

Gruner, S. M., Rothschild, K. J., and Clark, N. A., 1982, X-ray diffraction and electron microscope study of phase separation in rod outer segment photoreceptor membrane multilayers. *Biophys. J.* **39**:241–245.

Gulik, A., Luzzati, V., De Rosa, M., and Gambacorta, A., 1985, Structure and polymorphism of bipolar isopranyl ether lipids from archaebacteria. *J. Mol. Biol.* **182**:131–149.

Gulik-Krzywicki, T., Rivas, E., and Luzzati, V., 1967, Structure et polymorphisme des lipides: Etude par diffraction des rayons X du systeme forme de lipides de mitochondries de coeur de boeuf et d'eau. *J. Mol. Biol.* **27**:303–322.

Gulik-Krzywicki, T., Tardieu, A., and Luzzati, V., 1969, Smectic phase of lipid–water systems: Properties related to the nature of the lipid and to the presence of net electrical charges. *Mol. Cryst. Liq. Cryst.* **8**:285–291.

Harlos, K., and Eibl, H., 1980, Influence of calcium on phosphatidylglycerol. Two separate lamellar structures. *Biochemistry* **19**:896–899.

Hauser, H., 1984, Some aspects of the phase behaviour of charged lipids. *Biochim. Biophys. Acta* **772**:37–50.

Hauser, H., Pascher, I., and Sundell, S., 1980, Conformation of phospholipids. Crystal structure of a lysophosphatidylcholine analogue. *J. Mol. Biol.* **137**:249–264.

Hawton, M. H., and Doane, J. W., 1987, Pretransitional phenomena in phospholipid/water multilayers. *Biophys. J.* **52**:401–404.

Hemminga, M. A., and Cullis, P. R., 1982, Phosphorus-31-NMR studies of oriented phospholipid multilayers. *J. Magn. Reson.* **47**:307–323.

Hitchcock, P. B., Mason, R., Thomas, K. M., and Shipley, G. G., 1974, Structural chemistry of 1,2-dilauroyl-DL-phosphatidylethanolamine: Molecular conformation and intermolecular packing of phospholipids. *Proc. Natl. Acad. Sci. U.S.A.* **71**:3036–3040.

Hui, S. W., and Boni, L. T., 1981, Lipidic particles and cubic phase lipids. *Nature* **296**:175–176.

Israelachvili, J. N., 1982, Forces between surfaces in liquids. *Adv. Colloid Interface Sci.* **16**:31–47.

Israelachivili, J. N., Marcelja, S., and Horn, R. G., 1980, Physical principles of membrane organization, *Q. Rev. Biophys* **13**:121–200.

Jan, N., Lookman, T., and Pink, D. A., 1984, On computer simulation methods used to study models of two-component lipid bilayers. *Biochemistry* **23**:3227–3231.

Janiak, M. S., Small, D. M., and Shipley, G. G., 1976, Nature of the thermal pretransition of synthetic phospholipids, dimyristryl- and dipalmitoyllecithin. *Biochemistry* **25**:4575–4580.

Janiak, M. J., Small, D. M., and Shipley, G. G., 1979. Temperature and compositional dependence of the structure of hydrated dimyristoyl lecithin. *J. Biol. Chem.* **254**:6068–6078.

Kates, M., 1978, Phytanylether-linked polar lipids and neutral lipids of extremely halophilic bacteria. *Prog. Chem. Fats Other Lipids* **15**:301–342.

Kim, J. T., Mattai, J., and Shipley, G. G., 1987, Gel phase polymorphism in ether-linked dihexadecylphosphatidylcholine bilayers. *Biochemistry* **26**:6592–6598.

King, M. D., and Marsh, D., 1986, Free volume model for lateral diffusion coefficients. Assessment of the temperature dependence in phosphatidylcholine and phosphatidylethanolamine bilayers. *Biochim. Biophys. Acta* **862**:231–234.

Kirk, G. L., and Gruner, S. M., 1985, Lyotropic effects of alkanes and headgroup composition on the L_α-H-II lipid liquid crystal phase transition: Hydrocarbon packing versus intrinsic curvature. *J. Phys.* **46**:761–769.

Kirk, G. L., Gruner, S. M., and Stein, D. L., 1984, A thermodynamic model of the lamellar to inverse hexagonal phase transition of lipid membrane–water systems. *Biochemistry* **23**:1093–1102.

Kirkland, J. P., Nagel, D. J., and Cowan, P. L., 1983, The Naval Research Laboratory materials analysis beam line at the National Synchrotron Light Source. *Nucl. Instrum. Methods* **208**:49–54.

Kodama, M., Hashigami, H., and Seki, S., 1985, Static and dynamic calorimetric studies on the three kinds of phase transition in the systems of L- and DL-dipalmitoylphosphatidylcholine/water. *Biochim. Biophys. Acta* **814**:300–306.

Kornberg, R. D., and McConnell, H. M., 1971, Inside–outside transitions of phospholipids in vesicle membranes. *Biochemistry* **10**:1111–1120.

Kouaouci, R., Silvius, J. R., Graham, I., and Pezolet, M., 1985, Calcium-induced lateral phase separations in phosphatidylcholine–phosphatidic acid mixtures. A Raman spectroscopic study. *Biochemistry* **24**:7132–7140.

Koynova, R. D., Boyanov, A. I., and Tenchov, B. G., 1987, Gel-state metastability and nature of the azeotropic points in mixtures of saturated phosphatidylcholines and fatty acids. *Biochim. Biophys. Acta* **903**:186–196.

Krause, G. H., and Santarius, K. A., 1975, Relative thermostability of the chloroplast envelope. *Planta* **127**:285–299.

Kumar, A., and Gupta, C. M., 1985, Transbilayer phosphatidylcholine distributions in small unilameller sphingomyelin-phosphatidylcholine vesicles: effect of altered polar head group, *Biochemistry* **24**:5157–5163.

Kuo, A.-L., and Wade, C. G., 1979, Lipid lateral diffusion by pulsed nuclear magnetic resonance. *Biochemistry* **18**:2300–2308.

Larsson, K., Fontell, K., and Krog, N., 1980, Structural relationships between lamellar, cubic and hexagonal phases in monoglyceride–water systems. Possibility of cubic structures in biological systems. *Chem. Phys. Lipids* **27**:321–328.

Lee, A. G., 1977, Lipid phase transitions and phase diagrams. II. Mixtures involving lipids. *Biochim. Biophys. Acta* **472**:285–344.

Lee, A. G., 1978, Calculation of phase diagrams for non-ideal mixtures of lipids, and a possible non-random distribution of lipids in lipid mixtures in the liquid crystalline phase, *Biochim. Biophys. Acta* **507**:433–444.

Lentz, B. R., Barenholz, Y., and Thompson, T. E., 1976, Fluorescence depolarization studies of phase transitions and fluidity in phospholipid bilayers. 2. Two-component phosphatidylcholine liposomes. *Biochemistry* **15**:4529–4537.

Levine, Y. K. and Wilkins, M. H. F., 1971, Structure of oriented lipid bilayers, *Nature New Biol.* **230**:69–72.

Lewis, B. A., DasGupta, S. K., and Griffin, R. G., 1984, Solid-state NMR studies of the molecular dynamics and phase behaviour of mixed-chain phosphatidylcholines. *Biochemistry* **23**:1988–1993.

Lewis, R. N. A. H., Mak, N., and McElhaney, R. N., 1987, A differential scanning calorimetric study of the thermotropic phase behaviour of model membranes composed of phosphatidylcholines containing linear saturated fatty acyl chains. *Biochemistry* **26**:6118–6126.

Lindblom, G., and Wennerstrom, H., 1977, Amphile diffusion in model membrane systems studied by pulsed NMR. *Biophys. Chem.* **6**:167–171.

Lindblom, G., Larsson, K., Johansson, L., Fontell, K., and Frosen, S., 1979, The cubic phase of monoglyceride–water systems. Arguments for a structure based upon lamellar bilayer units. *J. Am. Chem. Soc.* **101**:5465–5470.

Lis, L. J., and Quinn, P. J., 1986, A time-resolved synchrotron X-ray study of a crystalline phase bilayer transition and packing in a saturated monogalactosylglycerol–water system. *Biochim. Biophys. Acta* **862**:81–86.

Lis, L. J., and Quinn, P. J., 1987, Kinetics of lyotropic liquid crystal phase transitions: A time resolved X-ray diffraction study. *Mol. Cryst. Liq. Cryst* **146**:35–39.

Lord, R. C., and Mendelsohn, R., 1981, Application of spectroscopic methods to study membrane structure. In *Membrane Spectroscopy* (E. Grell, ed.), pp. 377–436, Springer-Verlag, Berlin.

Luna, E. J., and McConnell, H. M., 1978, Multiple phase equilibria in binary mixtures of phospholipids. *Biochim. Biophys. Acta* **509**:462–473.

Luzzati, V., 1986, X-ray diffraction studies of lipid–water systems. In *Biological Membranes, Physical Facts and Functions* (D. Chapman, ed.), pp. 71–123, Academic Press, London.

Luzzati, V., and Reiss-Husson, F., 1966, Structure of the cubic phase of lipid–water systems. *Nature* **20**:1351–1352.

Luzzati, V., and Tardieu, A., 1974, Lipid phases; structure and structural transitions. *Annu. Rev. Phys. Chem.* **25**:79–94.

Luzzati, V., Gulik-Kryzwicki, T., and Tardieu, A., 1968, Polymorphism of lecithine. *Nature (London)* **218**:1031–1034.

Mabrey, S., and Sturtevant, J., 1976, Investigation of phase transitions of lipids and lipid mixtures by high sensitivity differential scanning calorimetry. *Proc. Natl. Acad. Sci. U.S.A.* **73**:3862–3866.

Machold, O., Simpson, D. J., and Hayes-Hanson, G., 1977, Correlation between the freeze-fracture appearance and polypeptide composition of thylakoid membranes in barley. *Carlsberg Res. Commun.* **42**:499–516.

Madden, T. D., and Cullis, P. R., 1982, Stabilization of bilayer structure for unsaturated phosphatidylethanolamines by detergents. *Biochim. Biophys. Acta* **684**:149–153.

Mannock, D. A., Brain, A. P. A., and Williams, W. P., 1985, Phase behaviour of the membrane lipids of the thermophilic blue-green alga *Anacystis nidulans*. *Biochim. Biophys. Acta* **821**:153–164.

Mansourian, A. R., and Quinn, P. J., 1986, Phase properties of binary mixtures of monogalactosyldiacylglycerols differing in hydrocarbon chain substituents dispersed in aqueous systems. *Biochim. Biophys. Acta* **855**:169–178.

Mantsch, H. H., Casal, H. L., and Jones, R. N., 1986 Infrared spectroscopy of membrane lipid systems. In *Spectroscopy of Biological Systems* (R. J. H. Clark and R. E. Hester, eds.), pp. 1–46, Wiley, Chichester.

Marsh, D., and Seddon, J. M., 1982, Gel-to-inverted hexagonal (L-H11) phase transitions in phosphatidylethanolamine and fatty-acid phosphatidylcholine mixtures, demonstrated by ^{31}P-NMR spectroscopy and X-ray diffraction. *Biochim. Biophys. Acta* **690**:117–123.

Masserini, M., and Freire, E., 1986, Thermotropic characterization of phosphatidylcholine vesicles containing ganglioside G_{M1} with homogenous ceramide chain length, *Biochemistry* **25**:1043–1049.

McDaniel, R. V., McIntosh, T. J., and Simon, S. A., 1983, Nonelectrolyte substitution for water in phosphatidylcholine bilayers. *Biochim. Biophys. Acta* **731**:97–108.

McElhaney, R. N., 1982, The use of differential scanning calorimetry and differential thermal analysis in studies of model and biological membranes. *Chem. Phys. Lipids* **30**:229–259.

McElhaney, R. N., 1986, Differential scanning calorimetric studies of lipid–protein interactions in model membrane systems. *Biochim. Biophys. Acta* **864**:361–421.

McIntosh, T. J., McDaniel, R. V., and Simon, S. A., 1983, Induction of an interdigitated gel phase in fully hydrated phosphatidylcholine bilayers. *Biochim. Biophys. Acta* **731**:109–114.

McLaughlin, A. C., Cullis, P. R., Hemminga, M. A., Hoult, D. I., Radda, G. K., Ritchei, G. A., Seeley, P. J., and Richards, R. E., 1975, Application of ^{31}P-NMR to model and biological membrane systems. *FEBS Lett.* **57**:213–218.

McLaughlin, A. C., Herbette, L., Blasie, J. K., Wang, C. T., Hymel, L., and Fleischer, S., 1981, ^{31}P-NMR studies of oriented multilayers formed from isolated sarcoplasmic reticulum and reconstituted sarcoplasmic reticulum. Evidence that "boundary-layer" phospholipid is not immobilized. *Biochim. Biophys. Acta* **643:**1-16.

Mendelsohn, R., and Koch, C. C., 1980, Deuterated phospholipids as Raman spectroscopic probes of membrane structure. Phase diagrams for the dipalmitoylphosphatidylcholine (and its d62 derivative)—dipalmitoylphosphatidylethanolamine system. *Biochim. Biophys. Acta* **598:**260-271.

Moor, H., Kistler, J., and Muller, M., 1976, Freezing in a propane jet. *Experientia* **32:**805-815.

Mullet, J. E., and Arntzen, C. J., 1980, Simulation of grana stacking in a model membrane system, *Biochim. Biophys. Acta* **589:**100-117.

Nagle, J. F., and Wilkinson, D. A., 1982, Dilametric studies of the sub-transition in dipalmitoylphosphatidylcholine. *Biochemistry* **21:**3817-3821.

Navarro, J., Tovio-Kunnuncan, M., and Racker, E., 1984, Effect of lipid composition on the calcium/adenosine 5′-triphosphate coupling ratio of the Ca^{2+}-ATPase of sarcoplasmic reticulum. *Biochemistry* **23:**130-135.

Nave, C., Helliwell, J. R., Moore, P. R., Tompson, A. W., Worgan, J. S., Greenall, R. J., Miller, A., Burley, S. K., Bradshaw, J., Pigram, W. J., Fuller, W., Siddons, D. P., Deutsch, M., and Tregear, R. T., 1985, Facilities for solution scattering and fibre diffraction at the Daresbury SRS. *J. Appl. Crystallogr.* **18:**396-403.

Nayer, R., Schmid, S. L., Hope, M. J., and Cullis, P. R., 1982, Structured preferences of phosphatidylinositol and phosphatidylinositol–phosphatidylethanolamine model membranes. Influence of Ca^{2+} and Mg^{2+}. *Biochim. Biophys. Acta* **688:**169-176.

Noordam, P. C., van Echteld, C. J. A., DeKruijff, B., and DeGier, J., 1981, Rapid transbilayer movement of phosphatidylcholine in unsaturated phosphatidylethanolamine-containing model systems. *Biochim. Biophys. Acta* **646:**483-487.

Papahadjopoulos, D., Vail, W. J., Jacobson, K., and Poste, G., 1975, Cochleate cylinders: Formation by fusion of unilamellar vesicles. *Biochim. Biophys. Acta* **394:**483-491.

Parsegian, V. A., Fuller, N., and Rand, R. P., 1979, Measured work of deformation and repulsion of lecithin bilayers. *Proc. Natl. Acad. Sci. U.S.A.* **76:**2750-2754.

Pearson, R. H., and Pascher, I., 1979, Molecular structure of lecithin dihydrate. *Nature* **281:**499-451.

Phillips, M. C., Hauser, H., and Paltauf, F., 1972, Inter and intra-molecular mixing of hydrocarbon chains in lecithin/water systems. *Chem. Phys. Lipids* **8:**127-133.

Pope, J. M., Walker, L., Cornell, B. A., and Francis, G. W., 1981, NMR study of synthetic lecithin bilayers in the vicinity of the gel-liquid-crystalline transition. *Biophys. J.* **35:**509-520.

Quinn, P. J., 1985, A lipid phase separation model of low-temperature damage to biological membranes. *Cryobiology* **22:**28-46.

Quinn, P. J., and Chapman, D., 1980, The dynamics of membrane structure. *Crit Rev. Biochem.* **8:**1-117.

Quinn, P. J., Gounaris, K., Sen, A., and Williams, W. P., 1982, Structural configuration of plant membrane lipids and their role in the organisation of chloroplast thylakoid constituents. In *Biochemistry and Metabolism of Plant Lipids* (J. F. G. M. Wintermans and P. J. C. Kuiper, eds.), pp. 327-330, Elsevier Biomedical Press, Amsterdam.

Quinn, P. J., and Williams W. P., 1983, The structural role of lipids in photosynthetic membranes. *Biochim. Biophys. Acta* **737:**223-266.

Quinn, P. J., Brain, A. P. R., Stewart, L. C., and Kates, M., 1986, The structure of membrane lipids of the extreme halophile, *Halobacterium cutirubrum*, in aqueous systems studied by freeze-fracture. *Biochim. Biophys. Acta* **863:**213-223.

Ranck, J. L., Keira, T., and Luzzati, V., 1977, A novel packing of the hydrocarbon chains in lipids. The low temperature phases of dipalmitoylphosphatidyl-glycerol. *Biochim. Biophys. Acta* **488:**432-441.

Ranck, J. L., Letellier, L., Shechter, E., Krop, B., Pernot, P., and Tardieu, A., 1984, X-ray analysis of the kinetics of *Escherichia coli* lipid and membrane structural transitions. *Biochemistry* **23**:4955–4961.

Rand, R. P., and Luzzàti, V., 1968, X-ray diffraction study in water of lipids extracted from human erythrocytes. The position of cholesterol in the lipid lamellae. *Biophys. J.* **8**:125–137.

Rand, R. P., Tinker, D. A., and Fast, P. G., 1971, Polymorphism of phosphatidylethanolamines from two natural sources. *Chem. Phys. Lipids* **6**:333–342.

Reiss-Husson, F., 1967, Structure des phases liquides-cristallines de differentes phospholipides, monoglycerides, sphingolipides, anhydres ou en presence d'eau. *J. Mol. Biol.* **25**:363–382.

Renooij, W., Van Golde, M. G., Zwall, R. F. A., and van Deenen, L. L. M., 1976, Topological asymmetry of phospholipid metabolism in rat erythrocyte membranes. Evidence for flip-flop of lecithin. *Eur. J. Biochem.* **61**:53–58.

Rice, D., and Oldfield, E., 1979, Deuterium nuclear magnetic resonance studies of the interaction between dimyristoylphosphatidylcholine and gramicidin A. *Biochemistry* **18**:3272–3279.

Rilfors, L., Eriksson, P.-O., Arvidson, G., and Lindblom, G., 1986, Relationship between three-dimensional arrays of "lipidic particles" and bicontinuous cubic lipid phases. *Biochemistry* **25**:7702–7710.

Roeder, S. B. W., Burnell, E. E., Kuo, A.-L., and Wade, C. G., 1976, Determination of the lateral diffusion coefficient of potassium oleate in the lamellar phase. *J. Chem. Phys.* **64**:1848–1849.

Rousselet, A., Guthmann, C., Matricon, J., Bienvenue, A., and Devaux, P. F., 1976, Study of the transverse diffusion of spin-labelled phospholipids in biological membranes. I. Human red blood cells. *Biochim. Biophys. Acta* **426**:357–371.

Ruocco, M. J., and Shipley, G. G., 1982a, Characterisation of the sub-transition of hydrated dipalmitoylphosphatidylcholine bilayers. Kinetic, hydration and structural study. *Biochim. Biophys. Acta* **691**:309–320.

Ruocco, M. J., and Shipley, G. G., 1982b, Characterisation of the subtransition of hydrated dipalmitoylphosphatidylcholine bilayers. *Biochim. Biophys. Acta* **684**:59–66.

Ruppel, D., and Sackmann, E., 1983, On defects in different phases of two-dimensional lipid bilayers, *J. Phys.* **44**:1025–1034.

Ryrie, I. J., Anderson, J. M., and Goodchild, D. J., 1980, The role of the light-harvesting chlorophyll a/b-protein complex in chloroplast membrane stacking, *Eur. J. Biochem.* **107**:345–354.

Sakurai, I., Sakurai, T., Seto, T., and Iwayanagi, S., 1983, Lyotropic phase transitions in single crystals of L- and DL-dipalmitoylglycerophosphocholines. *Chem. Phys. Lipids* **32**:1–11.

Salisbury, N. J., Dark, A., and Chapman, D., 1971, Deuteron magnetic resonance studies of water associated with phospholipids. *Chem. Phys. Lipids* **8**:142–151.

Schreiber, U., and Armond, P. A., 1978, Heat-induced changes in chlorophyll fluorescence in isolated chloroplasts and related heat damage at the pigment level. *Biochim. Biophys. Acta* **502**:138–151.

Schreiber, U., and Berry, J., 1977, Heat-induced changes in chlorophyll fluorescence in intact leaves correlated with damage to the photosynthetic apparatus. *Planta* **136**:233–238.

Schreiber, U., Colbow, K., and Vidaver, W., 1976, Analysis of temperature jump chlorophyll fluorescence induction in plants. *Biochim. Biophys. Acta* **423**:249–263.

Scriven, L. E., 1976, Equilibrium bicontinuous structure. *Nature* **263**:123–125.

Seddon, J. M., Harlos, K., and Marsh, D., 1983, Metastability and polymorphism in the gel and fluid bilayer phases of dilauroylphosphatidylethanolamine. Two crystalline forms in excess water. *J. Biol. Chem* **258**:3850–3854.

Seddon, J. M., Cevc, G., Kaye, R. D., and Marsch D., 1984, X-ray diffraction study of the polymorphism of hydrated diacyl- and dialkylphosphatidylethanolamines. *Biochemistry* **23**:2634–2644.

Seelig, J., 1977, Deuterium magnetic resonance. Theory and application to lipid membranes. *Q. Rev. Biophys.* **10**:353–418.

Seelig, J., 1978, ^{31}P nuclear magnetic resonance and the head group; structure of phospholipids in membranes. *Biochim. Biophys. Acta* **515**:105–140.

Seelig, J., and Gally, H. U., 1976, Investigation of phosphatidylethanolamine bilayers by deuterium and phosphorus-31 nuclear magnetic resonance. *Biochemistry* **15**:5199–5204.

Seelig, J., and Niederberger, W., 1974, Deuterium-labelled lipids as structural probes in liquid-crystalline bilayers. A deuterium magnetic resonance study. *J. Am. Chem. Soc.* **96**:2069–2072.

Seelig, J., and Seelig, A., 1980, Lipid conformation in model membranes and biological membranes. *Q. Rev. Biophys.* **13**:19–61.

Seigneuret, M., and Devaux, P. F., 1984, ATP-dependent asymmetric distribution of spin-labelled phospholipids in the erythrocyte membrane: Relation to shape changes. *Proc. Natl. Acad. Sci. U.S.A.* **81**:3751–3755.

Sen, A., and Yeagle, P. L., 1986, Hydration and the lamellar to hexagonal II phase transition of phosphatidylethanolamine. *Biochemistry* **25**:7518–7522.

Sen, A., Williams, W. P., and Quinn, P. J., 1981, The structure and thermotropic properties of pure 1,2-diacylgalactosylglycerols in aqueous systems. *Biochim. Biophys. Acta* **663**:380–389.

Sen, A., Williams, W. P., Brain, A. P. R., and Quinn, P. J., 1982a, Bilayer and non-bilayer transformation in aqueous dispersions of mixed sn-3-galactosyldiacylglycerols isolated from chloroplasts. A freeze–fracture study. *Biochim. Biophys. Acta* **685**:297–306.

Sen, A., Brain, A. P. R., Quinn, P. J., and Williams, W. P., 1982b, Formation of inverted lipid micelles in aqueous dispersions of mixed sn-3-galactosyldiacylglycerol, induced by heat and ethylene glycol. *Biochim. Biophys. Acta* **686**:215–224.

Sen, A., Mannock, D. A., Collins, D. J., Quinn, P. J., and Williams, W. P., 1983, Thermotropic phase properties and structure of 1,2-distearoylgalactosylglycerols in aqueous systems. *Proc. R. Soc. London B* **218**:349–364.

Serrallach, E. N., Dijkman, R., de Haas, G. H., and Shipley, G. G., 1983, Structure and thermotropic properties of 1,3-dipalmitoyl-glycero-2-phosphocholine. *J. Mol. Biol.* **170**:155–174.

Serrallach, E. N., de Haas, G. H., and Shipley, G. G., 1984, Structure and thermotropic properties of mixed chain phosphatidylcholine bilayer membranes. *Biochemistry* **23**:713–720.

Shimshick, E. J., and McConnell, H. M., 1973, Lateral phase separation in phospholipid membranes, *Biochemistry* **12**:2351–2360.

Shipley, G. G. 1973, Recent X-ray diffraction studies of biological membranes and membrane components. In *Biological Membranes* (D. Chapman and D. F. H. Wallach, eds.), Vol.2, pp. 1–89, Academic Press, London.

Silvius, J. R., 1982, Lipid–protein interactions in biological membranes. In *Lipid–Protein Interactions* (P. C. Jost and O. H. Griffith, eds.), Vol. 2, pp. 239–281, Wiley, New York.

Silvius, J. R., and Gange, J., 1984a, Lipid phase behaviour and calcium-induced fusion of phosphatidylethanolamine–phosphatidylserine vesicles. Calorimetric and fusion studies. *Biochemistry* **23**:3232–3240.

Silvius, J. R., and Gagne, J., 1984b, Calcium-induced fusion and lateral phase separations in phosphatidylcholine–phosphatidylserine vesicles: Correlation by calorimetric and fusion measurements. *Biochemistry* **23**:3241–3247.

Silvius, J. R., Lyons, M., Yeagle, P. L., and O'Leary, T. J., 1985, Thermotropic properties of bilayers containing branched-chain phospholipids. Calorimetric, Raman, and ^{31}P-NMR studies. *Biochemistry* **24**:5388–5395.

Singer, M. A., and Finegold, L., 1985, Permeability and morphology of low temperature phases in bilayers of single and of mixtures of phosphatidylcholines. *Biochim. Biophys. Acta* **816**:303–312.

Sklar, L. A., Milijanich, G. P., and Dratz, E. A., 1979, Phospholipid lateral phase separation and the partition of *cis*-parinaric acid and *trans*-parinaric acid among aqueous, solid lipid and fluid lipid phases. *Biochemistry* **18:**1707–1716.
Small, D. M., 1986, Phase behaviour of lipid–water systems. In *Handbook of Lipid Research* (D. J. Hanahan, ed.), Vol. 4, pp. 1–672, Plenum Press, New York.
Smith, I. C. P., and Mantsch, H. H., 1979, A look at membranes by Fourier transform NMR and IR of deuterated lipids. *Trends Biochem. Sci.* **4:**152–154.
Snyder, B., and Freire, E., 1980, Compositional domain structure in phosphatidylcholine–cholesterol and sphingomyelin–cholesterol bilayers. *Proc. Natl. Acad. Sci. U.S.A.* **77:**4055–4059.
Stier, A., Finch, S. A. E., and Bosterling, B., 1978, Non-lamellar structure in rabbit liver microsomal membranes. *FEBS Lett.* **91:**109–122.
Stilbs, P., 1987, Fourier transform pulsed-gradient spin-echo studies of molecular diffusion. *Prog. Nucl. Magn. Reson. Spectrosc.* **19:**1–45.
Stilbs, P., Arvidson, G., and Lindblom, G., 1984, Vesicle membrane–water partition coefficients determined from Fourier transform pulsed-gradient spin-echo NMR based self-diffusion data. Application to anesthetic binding in tetracaine–phosphatidylcholine–water systems. *Chem. Phys. Lipids* **35:**309–314.
Stockton, G. W., Polnaszek, C. F., Leitch, L. C., Tulloch, A. P., and Smith, I. C. P., 1974, A study of mobility and order in model membranes using ^{2}H-NMR relaxation rates and quadrupole splittings of specifically deuterated lipids. *Biochem. Biophys. Res. Commun.* **60:**844–850.
Strenk, L. M., Westerman, P. W., Vaz, N. P. A., and Doane, J. W., 1985, Spatial modulation in lecithin bilayers. *Biophys. J.* **48:**355–359.
Sunder, S., Cameron, D. G., Mantsch, H. H., and Bernstein, H. J., 1978, Infrared and laser Raman studies of deuterated model membranes: Phase transition in 1,2-perdeuterodipalmitoyl-*sn*-glycero-3-phosphocholine. *Can. J. Chem.* **56:**2121–2126.
Tamura, A., Yoshikawa, K., Fujii, T., Ohki, K., Nozawa, Y., and Sumida, Y., 1986, Effect of fatty acyl chain length of phosphatidylcholine on their transfer from liposomes to erythrocytes and transverse diffusion in the membranes inferred by TEMPO-phosphatidylcholine spin probes. *Biochim. Biophys. Acta* **855:**250–256.
Tanford, C., 1973, *The Hydrophobic Effect: Formation of Micelles and Biological Membranes*, Wiley-Interscience, New York.
Taraschi, T. F., de Kruijff, B., Verleij, A. J., and van Echfeld, C. J. A., 1982, Effect of glycophorin on lipid polymorphism. A ^{31}P-NMR study. *Biochim. Biophys. Acta* **685:**153–161.
Tenchov, B. G., 1985, Nonuniform lipid distribution in membranes. *Prog. Sur. Sci.* **20:**273–340.
Tenchov, B. G., Lis, L. J., and Quinn, P. J., 1987, Mechanism and kinetics of the subtransition in hydrated L-dipalmitoylphosphatidylcholine, *Biochim. Biophys. Acta* **897:**143–151.
Tenchov, B. G., Boyanov, A. J., and Koynova, R., 1984, Lyotropic polymorphism of racemic dipalmitoylphosphatidylethanolamine. A differential scanning calorimetric study. *Biochemistry* **23:**3553–3558.
Tenchov, B. G., Lis, L. J., and Quinn, P. J., 1987, Mechanism and kinetics of the subtransition in hydrated L-dipalmtoylphosphatidylcholine. *Biochim. Biophys. Acta* **897:**143–151.
Ter-Minassian-Saraga, L., and Madelmont, G., 1984, Subtransition and hydration studies of fully hydrated DPPC gel-phase. *J. Colloid Interface Sci.* **99:**420–426.
Thomas, P. G., Brain, A. P. R., Quinn, P. J., and Williams, W. P., 1985, Low pH and phospholipase A$_2$ treatment induce the phase separation of non-bilayer lipids within pea chloroplast membranes. *FEBS Lett.* **183:**161–166.
Tilcock, C. P. S., and Cullis, P. R., 1980, The polymorphic phase behaviour of mixed phosphatidylserine–phosphatidylethanolamine model systems as detected by ^{31}P-NMR. Effects of divalent cations and pH. *Biochim. Biophys. Acta* **641:**189–201.

Tilcock, C. P. S., Cullis, P. R., and Gruner, S. M., 1986, On the validity of ^{31}P-NMR determinations of phospholipid polymorphic phase behaviour. *Chem. Phys. Lipids* **40**:47–56.
Tsuchida, K., Hatta, I., Imaizumi, S., Ohki, K., and Nozawa, Y., 1985, Kinetics near the pretransition of a multilamellar phospholipid studied by ESR. *Biochim. Biophys. Acta* **812**:249–254.
Tsuchida, K., Ohki, K., Sekiya, T., Nozawa, Y., and Hatta, I., 1987, Dynamics of appearance and disappearance of the ripple structure in multilamellar liposomes of dipalmitoylphosphatidylcholine. *Biochim. Biophys. Acta* **898**:53–58.
Van den Bresselaar, A. M. H. P., de Kruijff, B., Van den Bosch, H., and van Deenen, L. L. M., 1978, Phosphatidylcholine mobility in liver microsomal membranes. *Biochim. Biophys. Acta* **510**:242–255.
Van der Steen, A. T. M., De Kruijff, B., and De Gier, J., 1982, Glycophorin incorporation increases the bilayer permeability of large unilamellar vesicles in a lipid-dependent manner. *Biochim. Biophys. Acta* **691**:13–23.
Van der Steen, A. T. M., Taraschi, T. F., Voorhout, W. F., and de Kruijff, B., 1983, Barrier properties of glycophorin–phospholipid systems prepared by different methods. *Biochim. Biophys. Acta* **733**:51–64.
Van Dijck, P. W. M., Van Zoelen, E. J. J., Seldenrijk, R., van Deenen, L. L. M., and de Gier, J., 1976, Calorimetric behaviour of individual phospholipid classes from human and bovine erythrocyte membranes. *Chem. Phys. Lipids* **17**:336–343.
Van Dijck, P. W. M., Kaper, A. J., Oonk, H. A. J., and de Gier, J., 1977, Miscibility properties of binary phosphatidylcholine mixtures. A calorimetric study. *Biochim. Biophys. Acta* **470**:58–69.
Van Venetie, R., and Verkleij, A. J., 1981, Analysis of the hexagonal-II phase and its relations to lipidic particles and the lamellar phase. A freeze–fracture study. *Biochim. Biophys. Acta* **645**:262–269.
Van Venetie, R., and Verkleij, A. J., 1982, Possible role of non-bilayer lipids in the structure of mitochondria. A freeze–fracture electron microscopy study. *Biochim. Biophys. Acta* **692**:397–405.
Vasilenko, I., de Kruijff, B., and Verkleij, A. J., 1982, The synthesis and use of thiophospholipids in ^{31}P-NMR studies of lipid polymorphism. *Biochim. Biophys. Acta* **685**:144–152.
Vaz, W. L. C., Goodsaid-Zalduondo, F., and Jacobson, K., 1984, Lateral diffusion of lipids and proteins in bilayer membranes. *FEBS Lett.* **174**:199–207.
Vaz, W. L. C., Clegg, R. M., and Hallmann, D., 1985, Translational diffusion of lipids in liquid crystalline phase phosphatidylcholine multibilayers. A comparison of experiment with theory. *Biochemistry* **24**:781–786.
Verkleij, A. J., and Ververgaert, P. H. J. Th., 1978, Freeze–fracture morphology of biological membranes. *Biochim. Biophys. Acta* **515**:303–327.
Verkleij, A. J. Ververgaert, P. H. J. Th., van Deenen, L. L. M., and Elbers, P. F., 1972, Phase transitions of phospholipid bilayers and membranes of *Acholeplasma laidlawii* B visualized by freeze fracturing electron microscopy. *Biochim. Biophys. Acta* **288**:326–332.
Verkleij, A. J., Zwaal, R. F. A., Roelofsen, B., Comfurius, P., Kastelijn, D., and van Deenen, L. L. M., 1973, The asymmetric distribution of phospholipids in the human red cell membrane. A combined study using phospholipases and freeze-etch electron microscopy. *Biochim. Biophys. Acta* **323**:178–193.
Ververgaert, P. H. J. Th., Verkleij, A. J., Elbers, P. F., and van Deenen, L. L. M., 1973, Analysis of the crystalline process in lecithin liposomes: A freeze-etch study. *Biochim. Biophys. Acta* **311**:320–329.
Verwey, E. J. W., and Overbeek, J. Th. G., 1984, *Theory of the Stability of Lyophobic Colloids*, Elsevier, Amsterdam.

Viani, P., Cervato, G., Marchesini, S., and Cestaro, B., 1986, Fluorospectroscopic studies of mixtures of distearoylphasphatidylcholine and sulphatides with defined fatty acid compositions, *Chem. Phys. Lipids* **39**:41–51.

Wieslander, A., Rilfors, L., Johansson, B. A., and Lindblom, G., 1981, Reversed cubic phase with membrane glucolipids from *Acholeplasma laidlawii*. ^{1}H, ^{2}H and diffusion nuclear magnetic resonance measurements. *Biochemistry* **20**:730–735.

Wieslander, A., Ulmius, J., Lindblom, G., and Fontell, K., 1978, Water binding and phase structures for different *Acholeplasma laidlawii* membrane lipids studied by deuteron magnetic resonance and X-ray diffraction. *Biochim. Biophys. Acta* **512**:241–253.

Wieslander, A., Rilfors, L., Johansson, B. A., and Lindblom, G., 1981, Reversed cubic phase with membrane glucolipids from *Acholeplasma laidlawii*. ^{1}H, ^{2}H and diffusion nuclear magnetic resonance measurements. *Biochemistry* **20**:730–735.

Wilkins, W. H. F., Blaurock, A. E., and Engelman, D. M., 1971, Bilayer structure in membranes. *Nature New Biol.* **230**:72–76.

Wilkinson, D. A., and McIntosh, T. J., 1986, A subtransition in a phospholipid with a net charge, dipalmitoylphosphatidylglycerol. *Biochemistry* **25**:295–298.

Wilkinson, D. A., and Nagle, J. F., 1978, A differential dilatometer. *Anal. Biochem.* **84**:263–271.

Wilkinson, D. A., and Nagle, J. F., 1981, Dilatometry and calorimetry of saturated phosphatidylethanolamine dispersions, *Biochemistry* **20**:187–192.

Wilkinson, D. A., and Nagle, J. F., 1984, Metastability in the phase behavior of dimyristoylphosphatidylethanolamine bilayers, *Biochemistry* **23**:538–541.

Williams, W. P., Sen, A., Brain, A. P. R., Quinn, P. J., and Dickens, M. J., 1981, Lipidic particles and cubic phases. *Nature* **296**:175–176.

Wilschut, J., Nir., Scholma, J., and Hoekstra, D., 1985, Kinetics of Ca^{2+}-induced fusion of cardiolipin–phosphatidylcholine vesicles: Correlation between vesicle aggregation, bilayer destabilization and fusion. *Biochemistry* **24**:4630–4636.

Wong, P. T. T., 1984, Raman spectroscopy of thermotropic and high-pressure phases of aqueous phospholipid dispersions. *Annu. Rev. Biophys. Bioeng.* **13**:1–24.

Wu, S. H. and McConnell, H. M., 1975, Phase separations in phospholipid membranes, *Biochemistry* **14**:847–854.

Zingsheim, H. P., 1972, Membrane structure and electron microscopy. The significance of physical problems and technics (freeze etching). *Biochim. Biophys. Acta* **265**:339–366.

Chapter 3

Reconstitution of Membrane Molecular Mechanisms in Bilayer Lipid Membranes and Patch-Clamp Bilayers

Peter M. Vassilev and H. Ti Tien

1. INTRODUCTION

A variety of important cellular processes such as excitability, ion transport, and neurohormonal regulation depend on the operation of macromolecular complexes in biomembranes. In view of these diverse and intricate processes, it is not surprising that biomembranes are very complex functionally. The molecular organization of biomembranes, as revealed by physical chemical studies, includes a lipid bilayer core, associated proteins, glycolipids, and other nonlipid materials. The bilayer is about 50 Å thick in a relatively liquid state with proteins and nonlipid materials arranged on each side as well as transversely across the lipid bilayer. Because of this delicate structure and its dimensions, direct experimental investigation of the biomembrane is exceedingly difficult at the membrane level. Thus, to overcome constraints imposed by the complexity of biomembrane, a number of artificial membrane systems have been developed. In this survey we are concerned with the experimental bilayer lipid membrane (BLM) of planar configuration as well as with patch-clamp bilayers. The review on BLM is in no sense complete; it is not possible to cite a great number

---

Peter M. Vassilev and H. Ti Tien Membrane Biophysics Laboratory, Department of Physiology, Michigan State University, East Lansing, Michigan 48824-1101.

of excellent papers in this expanding field of BLM research owing to space limitation. References may be consulted on the topics not covered here (Blumenthal and Klausner, 1982; Etemadi, 1983; Miller, 1986; Shamoo and Tivol, 1980; Tien, 1974).

2. TECHNIQUES FOR FORMATION AND CHARACTERIZATION OF RECONSTITUTED BILAYERS

2.1. Formation of Planar and Patch-Clamp Bilayers

Achievements in the field of electrophysiology were significantly promoted by the development of voltage and current-clamp techniques. Although the higher resistance and the size of BLMs imposed some specific requirements concerning the measurement of membrane electrical parameters, a tendency always existed to apply the most recently developed electrophysiological methods to BLM experiments. On the other hand, the achievements in BLM research also influenced, to some extent, electrophysiological developments, so that presently it is possible to state that the electrophysiological and the BLM experimental approaches are complementary.

BLM can be formed on a hole in a Teflon partition separating two compartments filled with aqueous solution by using a brush technique or other modified techniques (Mueller *et al.*, 1962; Tien, 1974). Ionic currents through single ion channels in BLM were observed in the late 1960s (Bean *et al.*, 1969). Discrete conductance fluctuations mediated by polypeptide antibiotics were also characterized in BLM systems (Gordon and Haydon, 1975). The use of improved voltage-clamp techniques and of folded bilayers (Montal and Mueller, 1972; Tancrede *et al.*, 1983) later permitted single-channel recordings to be obtained with a better resolution. The small membrane surface area of the BLMs formed from monolayers by the folding method is one of the main prerequisites for improvement of the signal-to-noise ratio.

A BLM with increased mechanical stability was obtained by using a Langmuir–Blodgett technique for passing a hydrostatically closed Teflon chamber with an aperture of 2 mm through a monolayer (Tancrede *et al.*, 1983; Tien, 1974; Vodyanoy *et al.*, 1982). Records with good signal-to-noise ratio were obtained with this method by using a voltage-clamp circuit (Vodyanoy and Murphy, 1983). Techniques for studying single-channel activities in native membranes were developed in the late 1970s. The voltage-clamp of whole cells could not give information about single-channel events due to the large surface of the whole cells. Efforts were made to develop a patch-clamp method in which only a small region of the membrane is clamped (Neher and Sakmann, 1976; Neher *et al.*, 1978). Initially, the resolution of the single-channel events under this method was not quite satisfactory due to the low resistance of the seals between the tip of the glass micropipette and the membrane patch. Later, the conditions for obtaining seals with high resistance (gigaseals) were determined and the patch-clamp electrical circuits were improved (Hamill *et al.*, 1981).

The patch-clamp technique became one of the most important and promising electrophysiological methods and permits quick progress toward an understanding of the molecular mechanisms of excitability and other membrane phenomena. However, the complex composition and structure of the natural membrane in some cases makes interpretation of the experimental results in this field difficult. Therefore, a quite useful approach is related to the perfor-

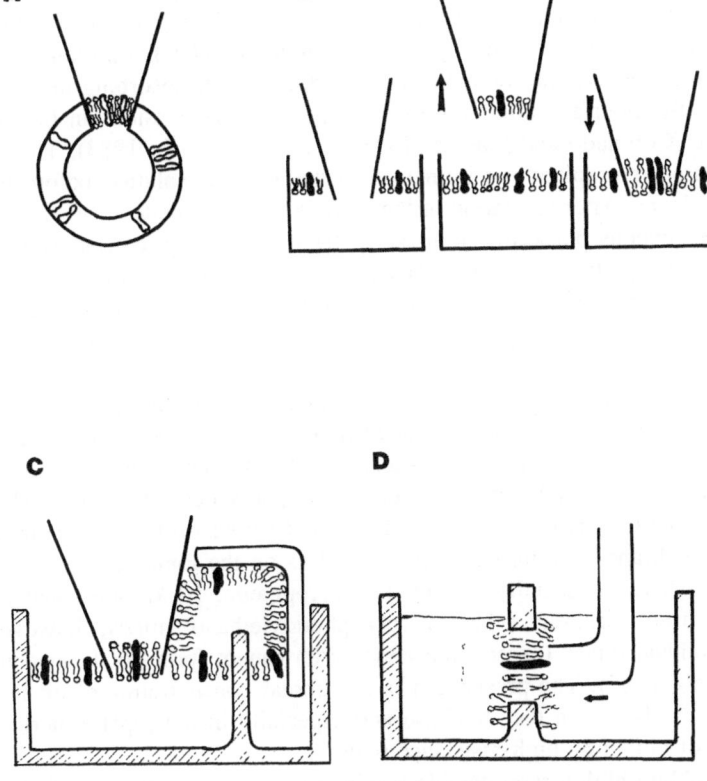

FIGURE 1. Techniques for formation of bilayers at the tips of patch-clamp pipettes. (A) Formation of bilayers from liposomal membranes. (B) Formation of bilayers from monolayers using the dip-tip technique. (C) Bilayers formed by micropipette guided contact of two monolayers. (D) Formation of bilayers by introducing micropipette tips into preformed planar BLMs.

mance of patch-clamp studies on the channel molecular mechanisms by using a model system with well-defined components and characteristics such as BLM. Soon after the elaboration of the improved patch-clamp technique (Hamill et al., 1981), this method was successfully applied in BLM research (Tank et al., 1982). Subsequently, several different ways for forming BLMs on the tips of patch-clamp micropipettes have been proposed (Figure 1). In one of them (Tank et al., 1982) (Figure 1A), macromolecules were inserted in small liposomes (Higashi et al., 1987), followed by enlargement of the lipid vesicles to about 10 μm by using the freeze–thaw method. The tip of a macropipette is brought near one of the blebs formed by the liposomal membrane and, by applying negative pressure, an excised membrane patch is formed on the tip of the pipette. The difficulties of this method are related to preparation of liposomes (Higashi et al., 1987) with defined size and form as well as the delicate manipulations of the liposomal membranes required of the micropipette. Another method is the so-called dipping technique as demonstrated in Figure 1B. In this case, a monolayer is formed either by spreading lipid and channel protein molecules dissolved in pentane on the air–water interface, or by absorbing these components on the interface after introducing liposomal suspensions in the aqueous subphase (Coronado and Latorre, 1983; Suarez-Isla et al., 1983). The tip of a patch-clamp micropipette is immersed in the aqueous solution before forming the monolayer. After formation of the monolayer, the pipette is slowly removed from the aqueous phase to the air and later is cautiously reintroduced into the solution through the monolayer. Depending on different conditions, a BLM is frequently formed on the tip of the pipette by using this simple procedure. A monolayer is formed on the tip when the pipette is lifted to the air after passing it through the interface. When reimmersing the pipette into the solution, the phospholipid hydrocarbon chains of the monolayers on the tip and on the air–water interface interact when apposed against each other and a BLM is formed on the pipette tip. If the first trial is unsuccessful, the procedure can be repeated several times (Coronado and Latorre, 1983). The success rate of BLM formation is usually high, but it depends to a great extent on the lipid composition of the membrane-forming solution, as well as on the ionic composition of the aqueous phases, especially that in the micropipette. BLMs are usually easily formed from different types of phosphotidylethanolamines; however, the phyosphotidylcholines (except diphytanoyl-PC) cannot form good gigaohm seals (Coronado, 1985). The presence of a millimolar concentration of divalent cations (Ca^{2+}, Ba^{2+}) in the subphase and especially in the pipette increases the chances of obtaining high-resistance seals.

To obtain highly reproducible results with a success rate of 80% for BLM formation, the surface pressure of the monolayers must be controlled. The optimal surface pressure during the removal and reintroduction of the pipette should be about 30 dyn/cm (Suarez-Isla et al., 1983). A special automatic control system is needed to maintain the surface pressure at such a constant value. Both fire polished (Suarez-Isla *et al.*, 1983) or nonfire polished (Coronado and Latorre, 1983) pipettes can be used. A success rate of 90% for forming gigaseals in the range of 1–10 GΩ can be achieved when controlling the surface pressure of the monomolecular film (Suarez-Isla *et al.*, 1983).

A micropipette guided method which differs from the dipping technique was recently described (Schuerholz and Schindler, 1983). In this case, the top of the micropipette remains on the interface and the BLM forms by micropipette guided contact of the two monolayers (Figure 1C). One of the advantages of this method is the facility for investigating channel activities as a function of the surface pressure. The dependence of transport processes on changes in surface pressure in the range of 28–48 mN/m has been studied. This important advantage, however, has not been largely exploited, probably due to the rather difficult experimental procedure.

A method that differs from the previously described techniques includes the preformation of a standard planar BLM into which the tip of a patch-clamp micropipette is introduced forming a BLM on the pipette tip (Andersen, 1978, 1984) (Figure 1D). In this case, the BLM may contain some solvent, while in the previously described methods the BLMs are solvent-free or virtually solvent-free.

Presently, the dipping technique is one of the most convenient and promising methods that have recently been introduced in connection with the modern electrophysiological developments. It is particularly useful when only small amounts of purified proteins and membrane-forming materials are available. All the bilayer techniques are valuable especially for investigating channel activities in reconstituted intracellular membranes, which so far cannot be studied directly in isolated single cells by the standard patch-clamp technique.

2.2. Electrophysiological Methods for Analyzing the Properties of Reconstituted Membranes

The patch clamp technique may be used for whole-cell current as well as for single-channel current measurements. In the former case, the micropipette interior has free access to the intracellular space, while in the latter case a hemispherically or cylindrically shaped membrane patch with a small diameter is sealed into the micropipette tip by applying negative hydrostatic pressure. Inside-out or outside-out patches can be formed depending on the sealed membrane configuration. One, two, or more channels may be functioning in the patch as a result of varying density of channel molecules in the membranes. The permeation of ions through the open channels can be influenced by electrical potentials applied across the membrane as well as by concentration gradients. The ion transport process can be examined by measuring the single-

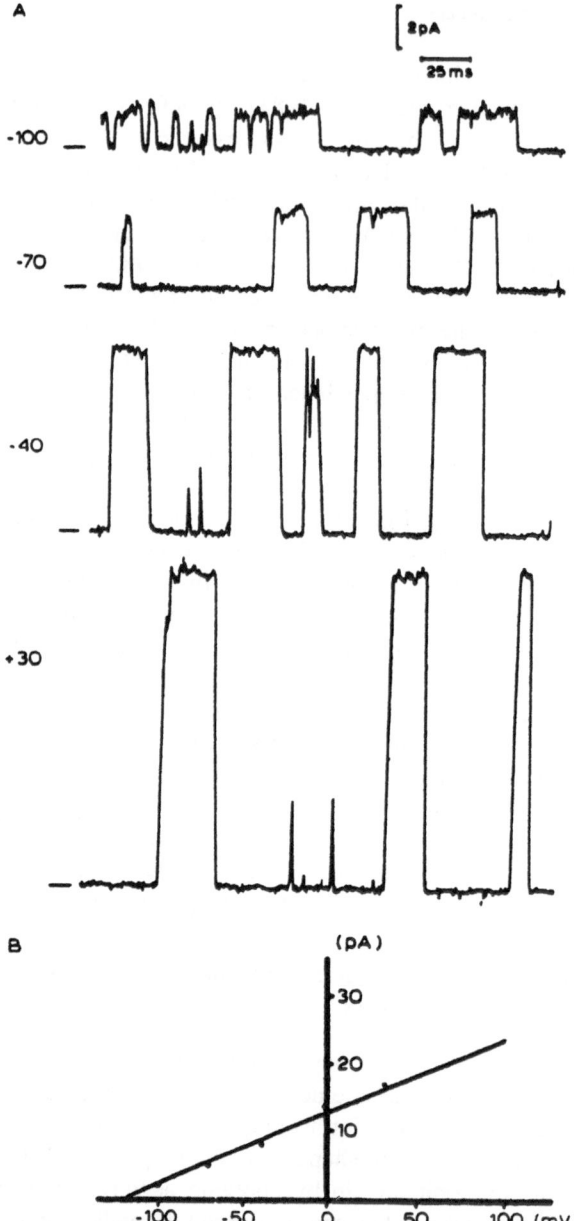

FIGURE 2. Single Ca^{2+} channels from brain microsomal membranes reconstituted in bilayers that were made at the tips of patch-clamp micropipettes in 50 mM Ca^{2+} as a current carrier. (A) Representative single-channel current traces at four different applied voltages, their magnitude being indicated on the left side of each trace. The horizontal bars on the left side indicate the channel closed state for each record. The vertical and the horizontal calibration bars on the right side above the top current trace are valid for all the records shown. (B) Single-channel current–voltage relations. The equilibrium potential was -122 mV and the conductance determined from the slope of the regression line was 107 pS. From Vassilev *et al.* (1987).

channel current events, which appear as square-shaped current fluctuations during the transitions of the channels between the different conducting states (Figure 2A). For example, a single-channel current displacement with an amplitude of 1 pA and 1 ms duration may be caused by the flow of 10,000 ions through the open channel across the membrane. So the single-channel current fluctuation may be characterized by determining its amplitude and its duration. Channels with varying amplitudes and durations may be present in the patch. The amplitude distributions are analyzed by plotting current amplitude histograms, while the open and closed time duration distributions may be estimated after obtaining frequency histograms. An amplitude histogram for brain microsomal Ca^{2+} channels is shown in Figure 3 in which the number of channel-opening events as a function of single-channel current amplitudes is plotted. A peak amplitude of 2.2 pA fitted by a Gaussian distribution curve is defined from the diagram. After determining the peak amplitudes at more than two applied potentials, the single-channel current–voltage relationship may be defined as shown in Figure 2B. If this relationship is linear, as is the case for this type of channel for this range of applied potentials, the single-channel conductance (i.e., slope conductance) of the channels may be determined by taking into account the slope of the current–voltage (I–V) curve. In this case, the slope conductance was defined to be 107 pS in 50 mM Ca^{2+} as a current carrier.

The single-channel (unit) conductance varies as a function of the concentration of the current carrier. A linear dependence may be observed at a defined ion concentration range, but it usually becomes nonlinear and saturates at high concentrations. For some channels, like the plasma membrane Ca^{2+} channels, the saturating ion concentration can be relatively high (~150 mM), while for other types of channels it is much lower, for example, 18 mM Cs^+ for the K^+ channels of the skeletal muscle sarcoplasmic reticulum membranes (Cukierman et al., 1985). The single-channel conductance may also vary depending on the charge carrier. For example, the unit conductance of the plasma membrane Ca^{2+} channels is 23 pS in 100 mM Ba^{2+} and 7 pS in 100 mM Ca^{2+} (Rosenberg et al., 1985), which means that these channels are significantly more permeable for Ba^2 than for Ca^{2+}. The permeability ratio for any type of channel may be determined after defining the current–voltage relationship and the equilibrium (Nernst) potential under biionic conditions. For example, in the presence of 50 mM Ba^{2+} on one side of the membrane and 80 mM Cs^+ on the other side, the permeability ratio $P(Ba)/P(Cs)$ of the muscle sarcoplasmic reticulum Ca^{2+} channels was defined to be 11.4 (Smith et al., 1985).

By determining the reversal potential under biionic conditions in the presence of defined concentrations of monovalent and divalent ions, the ionic selectivity and specificity may be characterized.

The kinetic characteristics of the channels may be analyzed by taking into account the time-dependent transitions of the channels between their closed and

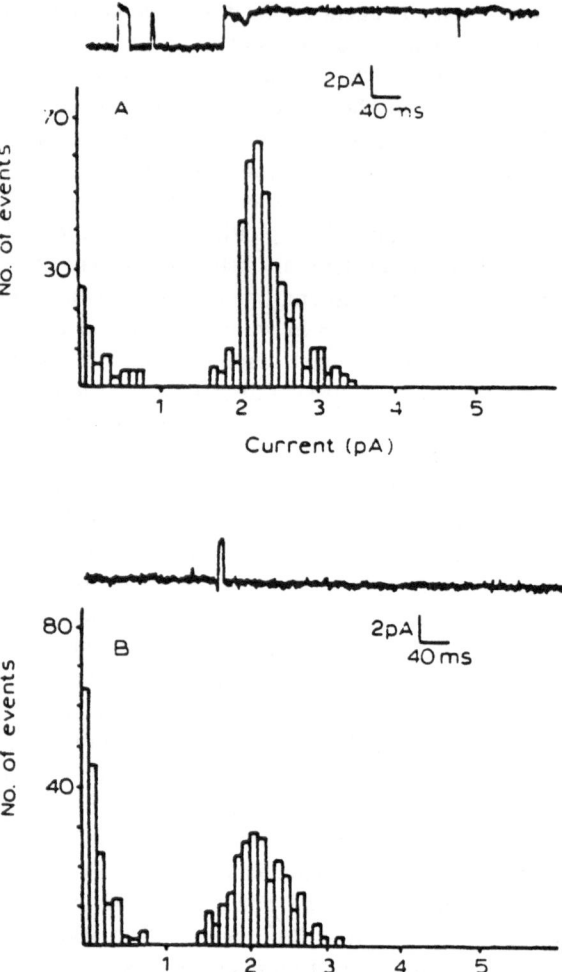

FIGURE 3. Ruthenium red-induced inhibition of the brain microsomal Ca^{2+} channels. The patch contains one functioning channel. (A) Current amplitude histogram at -100 mV in the presence of 1 mM ATP before the addition of ruthenium red, and a representative single-channel current trace showing that most of the time the channel resides in the open state. (B) Current amplitude histogram and a representative current trace at -100 mV in the presence of 1 μM ruthenium red. The data were taken 8–12 sec after addition of the agent by analyzing a 4 sec segment of the single-channel record. From Vassilev et al. (1987).

Reconstituted Bilayer Membrane

open states. The transitions of the channel between only two conductance states, one closed and one open state, can be described by the reaction

$$\text{closed} \underset{k_{-1}}{\overset{k_1}{\rightleftharpoons}} \text{open} \tag{1}$$

where k_1 and k_{-1} are the rate constants for opening and closing of the channel.

Some channels, like acetylcholine and other agonist-activated channels, can be opened only after binding of the activating ligands to the receptor region of the channel. In this case, two closed and one open channel states are available and the transitions between them are described as follows:

$$\underset{\text{(unbound ligand)}}{\text{closed}} \underset{k_{-i_1}}{\overset{k_1}{\rightleftharpoons}} \underset{\text{(bound ligand)}}{\text{closed}} \underset{k_{-1}'}{\overset{k_1'}{\rightleftharpoons}} \text{open} \tag{2}$$

The sodium channels are also fluctuating between two closed states and one open state. However, in this case the second closed state is related to the inactivation mechanism of the Na channels. Transitions may occur between all three different channel states:

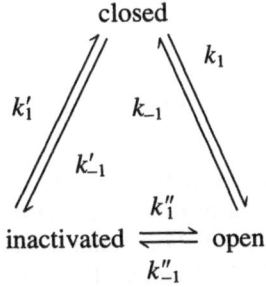

The characteristics of the simple channel mechanism with transitions between only one closed and one open state will be briefly described. The relation between the rate constants for channel opening (K_1) and closing (K_{-1}) and the difference in energy between the closed and open channel states (ΔE) are given by

$$\frac{k_1}{k_{-1}} = \exp\left(\frac{-\Delta E}{RT}\right) \tag{4}$$

where R is the gas constant and T is the absolute temperature.

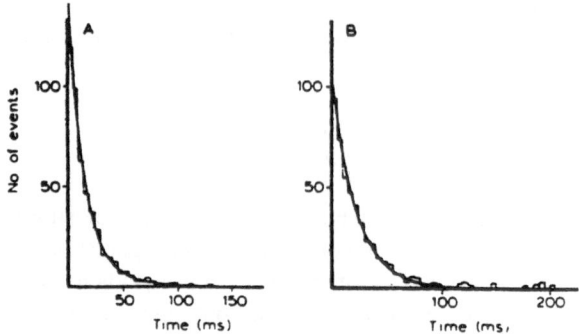

FIGURE 4. Stimulation of the brain microsomal Ca^{2+} channel activity by addition of 1 mM ATP to the cis side of the bilayer. (A) Open time histogram of data recorded at 100 mV in the absence of ATP. A sum of two exponentials with time constants $\tau_1 = 14$ msec and $\tau_2 = 39$ msec fitted the obtained data. (B) Open time histogram of data recorded at -100 mV in the presence of 1 mM ATP. The two time constants in this case were $\tau = 21$ msec and $\tau_2 = 108$ msec. From Vassilev et al. (1987).

Information about k_1 and k_{-1} can be obtained by taking into consideration the expressions for the distributions of open times, $P_o(t)$, and of closed times, $P_c(t)$, of the channels:

$$P_a(t) = k_{-1}\exp(-k_{-1}t) \qquad (5)$$
$$P_a(t) = k_1\exp(k_1 t) \qquad (6)$$

For determining the time distributions and k_1 and k_{-1}, the number of opening or closing events as a function of their time duration is estimated. As a result, frequency histograms are plotted. For example, an open state frequency histogram for brain microsomal Ca^{2+} channels is shown in Figure 4. It is clear that the number of opening events decreases exponentially as a function of the open time duration. In this particular case, the data are fitted by a sum of two exponentials with time constants τ_1 and τ_2, which may be due to the presence of channels with two different mean open times. Channels with one or more than two time constants are also described in the literature (Vassilev et al., 1987).

Closed time histograms may be obtained in a similar way by plotting the number of closing events as a function of the closed time durations. The data in this case may be fitted by exponentials, which may differ substantially from those of the open time histograms and may be characteristic of defined types of channel as well as defined blocking effects. When the channel resides in more than one closed state, as shown for some agonist-activated channels, a sum of several exponentials may fit the data from closed time histograms.

In addition to the mean open times, which are determined by analyzing the frequency histograms, the channel kinetics are also characterized by their defined open state probabilities. If among a total number of N channels in the membrane patch only n channels are open, then the probability of finding simultaneously open channels $P_{(n)}$ is given by

$$P_{(n)} = \frac{N!}{n!(n-n)!} P_o^n [1 - P_o^n]^{N-n} \qquad (7)$$

The probability of the open state (P_o) may be determined by using the equation

$$P_o = \frac{I}{N_i} \qquad (8)$$

where I is the time-averaged current flowing through the open channels for a defined period of time, N is the number of channels functioning independently in the membrane patch, and i is the single-channel current.

The relation between P_o and the mean open time, t_o, that is, the average time the channel resides in the open state, is given by

$$t_o = \frac{P}{n/T} \qquad (9)$$

where n is the number of the channel openings and T is the time period during which the parameters were analyzed.

It should be noted that one of the prerequisites of an adequate analysis of the single-channel parameters is to work with patches containing only one functioning channel. This is a considerable limitation, and recently efforts to develop more versatile methods were reported by Liebovitch and Fishbarg (1985). It should be pointed out that many of the theoretical considerations of the membrane transport mechanisms are based on BLM experimental background (Eisenman and Horn, 1983).

3. RECONSTITUTION OF MEMBRANE MOLECULAR MECHANISMS RELATED TO ION TRANSPORT AND EXCITABILITY

Considerable progress toward elucidating the molecular aspects of transport mechanisms in excitable membranes has been achieved by incorporating some polypeptides with defined molecular structures, such as gramicidin, alamethicin, monazomycin, and suzukacillin, into BLMs (Andersen, 1984; Mountz and Tien, 1976; Mueller and Rudin, 1968). These agents were found to induce voltage-dependent changes in the membrane conductance as well as channel transitions between the closed and open states with kinetics and multiple conductance states typical of the open channels. These effects and their relation to the excitable processes have been considered in detail in several reviews (Blumenthal and Klausner, 1982; Laeuger, 1982; Vassilev and Tien, 1985) and they will not be discussed here.

Although the experiments with alamethicin and other excitability-inducing compounds provide valuable information about the excitable processes, the relationships between the structure as well as the mechanism of functioning of these agents and those of the native channel proteins are not well known.

Another important approach for studying the membrane physiological processes relates to the incorporation of functionally active channel macromolecules isolated from native membranes into BLMs. Due to the difficulties concerning the purification of channel proteins, so far mostly crude extracts or membrane vesicles containing enriched fractions of channel macromolecules have been used in the reconstitution experiments. After incorporation of these channel molecules into the BLM, the channel activities have been studied under conditions similar to those used in studies on transport mechanisms in native membranes.

Owing to space limitations a certain number of excellent papers in this field will not be discussed here. Previous review articles may be consulted for more information in this field (Blumenthal and Klausner, 1982; Laeuger, 1982; Miller, 1986; Tien, 1974; Vassilev and Tien, 1985). BLM studies with substantial correlations with mechanisms in native membranes are considered in more detail.

3.1. Reconstitution of Sodium Ion Transport Mechanisms

3.1.1. Nerve Membranes

The investigation of Na channel activity is one of the most challenging tasks in electrophysiology, owing to the relatively fast kinetics of activation of Na^+ channels (Hille, 1984; Hodgkin and Huxley, 1952; Oosting, 1979; Sigworth and Neher, 1980). They enter an open state during the first several milliseconds after applying depolarizing pulses. Their inactivation also occurs in the range of milliseconds. A special electrical circuit must be used to compensate for the large capacitance transient appearing in the beginning of the pulse. When using the patch-clamp method, the capacitance transient can also be decreased by coating the glass micropipette with a Sylgard layer about 10–40 μm from the tip.

Reconstitution of Na^+-channel activity in a BLM is a difficult task mainly due to the inactivation mechanism. Krueger et al. (1983) succeeded in incorporating brain membrane components containing Na^+-channel molecules into BLMs and recorded a voltage-dependent Na^+-channel activity by using an inhibitor of the Na$^+$-channel inactivation, batrachotoxin. The use of batrachotoxin was essential for obtaining Na$^+$-channel activity. The rat brain membrane vesicles containing the channels were incorporated into the BLM by using the osmotic gradient fusion procedure (Cohen *et al.*, 1980, 1982; Miller and Racker, 1976) which so far has been the most successful method for reconstitution of channel activities in BLM. The authors observed several similarities between the characteristics of the reconstituted and native channel activities. Nanomolar concentrations of saxitoxin (STX) blocked the channels in the presence of 0.5 M NaCl on both sides of the BLM. The same concentrations of STX inhibit the functioning of the Na channels in native membranes. When the concentration of STX was increased, the mean duration of the Na$^+$-channel openings was shortened. Another interesting observation was the voltage dependence of the STX-blocking effect, which was more pronounced at hyperpolarizing potentials. This effect was confirmed and analyzed in detail later (French *et al.*, 1984). The findings concerning the voltage dependence of the neurotoxin effects in native membranes are rather controversial (Baer *et al.*, 1976; Cohen and Strichartz, 1977; Heggeness and Starkus, 1986; Vassilev *et al.*, 1986a). A voltage-dependent block of Na channels by TTX was found in the heart (Baer *et al.*, 1976) which has not been confirmed in later experiments in multicellular preparations (Cohen and Strichartz, 1977; Cohen *et al.*, 1981). Recently, a TTX-induced voltage shift of the inactivation curve of Na channels was obtained by using the patch-clamp single-channel recording technique (Vassilev *et al.*, 1986a), indicating that under some conditions the effect of TTX may be voltage dependent. All these experiments were performed in different types of preparation by using different techniques and therefore it is possible to suppose that the neurotoxins are acting by various mechanisms in defined types of cell. Further experiments in this direction by using other types of reconstituted BLM system should be particularly helpful in solving this problem, which might facilitate the elucidation of the molecular mechanisms of excitability.

The Na$^+$ channels reconstituted in BLM (French *et al.*, 1984; Krueger *et al.*, 1983) shared other properties typical of the sodium channels in native membranes such as similar unit conductance and selectivity for sodium over potassium. The unit conductance was 30 pS, which is actually about two times higher than that of the Na$^+$ channels in native membranes (15 pS), but in the model system a much higher concentration of NaCl was used (0.5 M) to obtain a good current resolution. The permeability ratio (P_K/P_{Na}) of 0.06 and the lack of Cs permeation are characteristics that also demonstrate the similarities between the properties of the model and native systems.

It should be pointed out, however, that these comparisons are valid for the case when the sodium inactivation is removed by batrachotoxin which modifies the channel kinetics. Further efforts should be made to reconstitute and characterize Na$^+$-channel activity in BLMs without influencing the inactivation mechanisms. Well-defined purified preparations should be used for the incorporation procedures. Pure synthetic lipids must be employed for preparing the BLM-forming solution. We found channel activities in BLMs prepared from most of the commercially available lipids isolated from brain or other tissues. Although their origin is not known it is possible to suppose that they are due to some proteolipid molecules remaining after the isolation procedures. They may, however, cause misleading interpretations of the results from the reconstitution experiments.

Substantial progress has recently been achieved toward an understanding of the molecular structure of the sodium channels (Hartshorne and Caterall, 1981, 1984). Radiolabeled neurotoxins, TTX and especially STX, have been used successfully as molecular probes for purification and reconstitution of brain sodium channel activity in phospholipid vesicles. Homogeneous fractions of sodium channels were purified from rat brains and their molecular composition was defined. Three polypeptide subunits were separated; α, with a molecular weight of 260,000 Da; β_1, 39,000; and β_2, 37,000. The β_1 subunit covalently binds scorpion toxin. β_2 and α subunits are covalently linked by disulfide bonds. This trimeric complex has been incorporated into planar BLMs (Hartshorne et al., 1985). The voltage-dependent Na^+-channel activity was reversibly blocked by tetrodotoxin ($K_i = 8.3$ nM at -50mV). The unit conductance was 25 pS in 0.5 M NCl, 3.5 pS in 0.5 M KCl, and 1.5 pS in 0.5 M RbCl.

More information about the structure and the properties of the Na^+ channels, and especially about the region of the selectivity filter, was obtained in a recent study on Na^+ channels incorporated into BLM from rat forebrain membranes (Worley et al., 1986).

It has been shown that an agent binding to the selectivity filter, trimethyloxonium (TMO), significantly modified the sensitivity of the channel to the blocking effects of Ca^{2+} and of STX. TMO also reduced the single Na^+ channel conductance from 25 to 15.8 pS. Ca^{2+} used at 10 mM "extracellular" concentration reduced the single-channel current by 60% and induced a decrease in the blocking effect of STX. STX and Ca^{2+} were able to prevent the modification of the Na^+ channels by TMO, which indicates that all these agents may compete for binding to a common site in the channel molecule. It was suggested that this blocking site may include a specific carboxyl group that is methylated by TMO. The carboxyl group may be important for the toxin binding but may not substantially affect the access of Na^+ and Ca^{2+} to the pore.

3.1.2. Muscle Membranes

Many similarities exist between the physiological processes in muscle and in nerve membranes. The maintenance of the resting potentials and the ion channel activation and inactivation mechanisms underlying the action potentials share similar characteristics. Substantial differences, however, also exist: for example, much longer action potentials and large Ca^+ currents are found in muscle. In most types of muscle cell the membrane processes are particularly complicated in relation to the excitation–contraction coupling process. This key physiological mechanism is dependent on the coupling between the ionic fluxes across the plasma membrane, which has a complex morphology involving the transverse tubular system, and ion transport across the sarcoplasmic reticulum membrane.

Progress in the field of muscle electrophysiology was previously slow due to the lack of single-cell preparations. Interpretation of the results from experiments on multicellular muscle fiber preparations was more difficult than that of the data on squid giant axon used in neurophysiology.

Later, single muscle cell preparations were obtained. The patch-clamp method was improved and combined with the whole-cell voltage clamp. The BLM method has also been quite helpful for characterization of the muscle membrane ion channels under well-defined conditions. BLM remains the best technique for characterizing the single-channel activities of intracellular SR membranes and of plasma membrane formations which protrude deeply inside the cell, such as the transverse tubules. The channel mechanisms in these membranes so far cannot be analyzed by direct application of the conventional patch-clamp method.

Sodium channels from skeletal muscle membranes were incorporated into BLM by Moczydlowski et al. (1984a,b) soon after the experiments on reconstituted nerve membrane sodium channels (French et al., 1984; Krueger et al., 1983). The gating, selectivity neurotoxin modulation, and other properties of the channels were then analyzed and an elaborate model for the voltage-dependent neurotoxin effect was proposed. This model may serve as a clue for understanding the molecular structure of the skeletal muscle sodium channel.

The reconstituted Na^+ channels showed many of the characteristics typical of the batrachotoxin-treated native channels.

1. A steep voltage-dependent activation assessed by measuring the changes in the opening probability as a function of the applied voltage and a negative shift of about 50 mV of the activation curve were observed similar to the batrachotoxin-modified channels in native membranes. It was clearly demonstrated that at potentials more positive than -60 mV, the channel is almost continuously open, while at -120 mV the openings are very infrequent.
2. The skeletal muscle Na^+ channels showed relatively low unit conductance (20 pS at 0.2 M NaCl), coinciding with that of the Na^+ channels in rat myotubes (Patlak and Horn, 1982) after taking into account the difference in the temperatures (Horn et al., 1984; Yamamoto et al., 1984).
3. The sequence of ionic selectivity was as that of the batrachotoxin-treated and of the normal Na^+ channels. $Li^+ \simeq Na^+ > K^+ > Rb^+ > Cs^+ >> Cl^-$. The permeability ratio, P_{Na^+}/P_{K^+}, was close to normal; 14.3 : 1.

4. The channels show specific blocking effects induced by TTX, STX, and other neurotoxin analogues.

A detailed analysis of the voltage dependence of the block was performed. It was clearly demonstrated that the inhibitory reactions were voltage dependent. The changes of the neurotoxins and their analogues, which varied from 0 to +2, did not influence the block. It seems evident also that the charges of the surrounding lipids do not exert any effect. The applied potential affects both the association and the dissociation rates of TTX. Na^+ competes with TTX, but it competitively inhibits only the association rate. This effect is observed only when Na^+ and TTX are added to the same side of the membrane.

The aforementioned authors reconsidered the concept of the selectivity filter of the Na^+ channel on the basis of the above data. Krueger and his colleagues (1983) suggested that the TTX-receptor site and a Na^+ site are located at the outer mouth of the pore. The Na^+ site operates as a prefilter, accelerating Na^+ entry into deeper pore regions.

The data on the voltage dependence of the toxic effects are better explained by a mechanism in which the neurotoxin molecule does not interact directly with the deeper regions of the selectivity filter; rather, it binds to the receptor site at the outer pore mouth and induces a voltage-dependent conformational change of a polypeptide segment of the protein molecule controlling Na^+ conduction.

Thus, it must be pointed out that most of the data concerning neurotoxin action on Na^+ channels incorporated into BLMs support the supposition that neurotoxin block of Na^+ channel activity is voltage dependent (French *et al.*, 1984; Krueger *et al.*, 1983; Moczydlowski *et al.*, 1984a,b). This property may be characteristic of different kinds of Na^+ channels, because similar results were obtained from experiments on BLMs reconstituted with Na^+ channels isolated from muscles (Moczydlowski *et al.*, 1984a,b), as well as from nerve membranes (French *et al.*, 1984; Krueger *et al.*, 1983). On the other side, the findings in the native membranes are contradictory, especially those in cardiac muscle (Baer *et al.*, 1976; Cohen and Strichartz, 1977; Cohen *et al.*, 1981; Heggeness and Starkus, 1986; Vassilev *et al.*, 1986a).

Further detailed studies examining, in parallel, native and reconstituted model membranes are needed to clarify this question.

3.2. Calcium Channels Functioning in Bilayers

3.2.1. Nerve Membranes

Calcium ions entering the cells mainly through voltage-dependent channels and Na^+–Ca^{2+} exchange are involved in the regulation and triggering of many physiological functions. Calcium ions play an important role in nerve cells, triggering the release of neurotransmitters into the synaptic gaps, as well as in the secretion of neurohypophysins. Many pharmacological agents were found to exert their effects through modulating the Ca^{2+} currents, which show specific characteristics for different types of cells. Resolution of the single-channel events related to Ca^{2+} transport meets some difficulties owing to the very low unit conductance of the Ca^{2+} channels. By using high calcium or barium concentrations (110 or 96 mM in the micropipette solution), however, high-quality single-Ca^{2+}-channel recordings can be obtained in different types of cell (Hagiwara and Byerly, 1981; Nilius et al., 1985; Nowycky et al., 1985; Reuter, 1983; Tsien, 1983). Nelson et al. (1984) recently incorporated brain membrane vesicles containing Ca^{2+} channels into BLMs formed from phosphatidylethanolamine (PE) (3.3%) and phosphatidylserine (PS) (2.6%) and studied their characteristics at high concentrations of divalent ions (250 mM). The reconstituted channels shared several common properties with Ca^{2+} channels in native membranes:

1. The channel activity was voltage dependent. It increased at more positive potentials, but showed saturation at very high voltages.
2. A selectivity for divalent over monovalent cations was demonstrated.
3. A dose-dependent blocking effect by the following ions was observed: $La^{3+} > Cd^{2+} >> Mn^{2+}$. However, even concentrations higher than 0.5 mM did not completely inhibit the channel activity.
4. There was a selectivity among several divalent ions: Ca^{2+}, Ba^{2+}, and Sr^{2+}.
5. The unit single-channel conductances were very small, 5–8.5 pS. Clearly distinguished differences between the unit conductances in the presence of different divalent cations as current carriers were also found. The single-channel conductance carried by Ba^{2+} was much higher, 8.5 pS, than that carried by Ca^{2+} or Sr^{2+}, 5 pS. This means that these two ions are moving through the channel 1.7 times slower than Ba^{2+}. Another peculiar effect induced by Ba^{2+} is the shortening of the mean channel open time, 127 msec, in comparison to those of Ca^{2+} and Sr^{2+}, 454 and 385 msec, respectively, at +100 mV.

Thus, the order of the mean open times is $Ca^{2+} > Sr^{2+} > Ba^{2+}$. It is noteworthy that it coincides with the order of affinities of the same ions for calcium channels in brain synaptosomes. The data can be interpreted in favor of the supposition that when the cation permeates the channel, it prevents its closure. A similar behavior is exhibited by some K^+ channels in squid axons.

It should be pointed out that the reconstituted Ca^{2+} channels were also functioning in the absence of complete inactivation, similar to the Na^+ channels reconstituted in BLM as previously described.

The characteristics of Ca^{2+} channels cannot properly be analyzed in excised patches of cellular membranes due to the rapid "rundown" of the Ca^{2+}-channel activity. The loss of functionally important phosphorylating enzymes is probably responsible for this phenomenon.

So far the BLM technique has proved adequate for measuring single-channel activity in synaptosomes, which, due to their small size (1 μm in diameter), cannot be studied directly by using the patch-clamp technique. Synaptosomal fractions from rat brain tissue have been used as a source for reconstitution of voltage-dependent Ca^{2+} channels (Nelson, 1986). Several correlations were found between the properties of channels characterized by radioisotopic methods and those of the reconstituted fractions:

1. Same order of maximum flux rate of permeating divalents ($Ba^{2+} > Ca^2 = Sr^{2+} > Mn^{2+}$).
2. Permeation and blocking effect of Mn^{2+}.
3. Blockade by organic ions with an order of potency: $La^{3+} >^2 Cd^{2+} \gg Mg^{2+} > Na^+ > K^+$.
4. The permeant and the blocking ions compete for a single site, the order of their apparent affinities for the site being the same as the orders of maximum flux rate and of blocking potency, respectively.
5. Same inhibitory effects by micromolar concentrations of D-600 and verapamil.
6. Lack of complete inactivation.

One of the typical features of these channels was the order of the unit conductances for the permeant ions, which was the inverse of the order of their mean open times: $Mn > Ca = Sr > Ba$. The blockers reduced the single-channel conductance and affected the mean open times. The blocking effect of Cd^{2+}, which is not a permeable ion, may be explained by its binding to a single site. The mechanism of the Mn^{2+}-induced block, however, is more complicated. In the absence of permeant ions Mn^{2+} can permeate through the channels, the single-channel conductance (4 pS) being lower than that for other divalents (Ba^{2+}, 7.5 pS; Sr^{2+}, 6 pS). When added in the presence of permeant ions, Mn^{2+} can reduce the amplitude of the single-channel fluctuations in a voltage-dependent manner. The existence of more than one binding site is compatible with the data on the Mn^{2+} effect.

The reconstituted Ca^{2+} channels show some deviations from the normal behavior in native membranes. For instance, a shift of their activation curve toward more positive is observed. It has been suggested that this phenomenon is due to the lack of phosphorylating enzymes in the BLM system. These enzymes may be involved in the control of the Ca^{2+}-channel activity *in vivo*.

We have studied the properties of single Ca^{2+} channels from brain endoplasmic reticulum (ER) membranes reconstituted in BLMs made on the tips of patch-clamp micropipettes (Vassilev et al., 1987). They were similar to those of sarcoplasmic reticulum (SR) membranes from skeletal muscle, as recently described (Smith et al., 1985). We observed large current displacements accompanying the Ca^{2+} channel openings at 50 mM Ca^{2+} (Figure 2A) and even as low as $[Ca^{2+}] = 3.6$ mM. Two exponentials described the open channel kinetics (Figure 4), the first and the second time constants being 14 and 39 ms, respectively. The Ca^{2+} channel activity was inhibited by $1 \mu M$ ruthenium red (Figure 3) and was stimulated by nucleotides, 1 mM ATP (Figure 4). These Ca^{2+} channels are probably responsible for Ca^{2+} transport across the ER membranes and for the triggering of different processes related to excitability in nerve cells and the release of neuromediators in the synaptic terminals. The similarities between the unit conductances as well as the kinetics of the Ca^{2+} channels in SR and in ER membranes are interpreted in terms of a common Ca^{2+}-channel mechanism controlling the excitation–contraction coupling in muscle cells, as well as the excitation–secretion coupling in nerve and other cells.

In a separate series of experiments, we found that in the presence of brain tubulin (spread on the surface as a component of the monolayer or added to the aqueous subphase underlying the surface film), the brain ER Ca^{2+} channel activity reconstituted in patch-clamped bilayers was substantially increased. The opening probability (p_o) was almost two times higher when tubulin was added. We suggested that tubulin, which is available as peripheral or even as an integral protein in the brain microsomal membranes (Soifer and Czosnek, 1980), may be related in some way to the regulation of their Ca^{2+}-channel activity.

We found in previous studies that microtubules may mediate interactions between two apposed BLMs (Vassilev et al., 1985, 1986c). Colchicine was shown to inhibit this effect (Vassilev et al., 1985). Dielectric breakdown measurements were carried out which confirmed the evidence that the microtubules can physically interconnect two membranes (Vassilev et al., 1986c). Furthermore, brain tubulin may influence the current–voltage relationships, the dielectric breakdown voltage, and other characteristics of brain lipid BLMs.

In further experiments we tried to verify that channel activity is available in BLMs containing only microtubule proteins and defined lipid molecules. Some lipid molecular species remain associated to microtubule preparations even after a careful isolation of the proteins. The presence of these lipids may be one of the factors determining the assembly of the microtubule protein fractions into membranelike sheets and vesicular formations under defined conditions (Feit and Shay, 1980). This finding allowed us to suggest that a hydrophobic fraction containing membrane tubulin and associated lipids may be prepared from isolated microtubule materials. We succeeded in isolating such a hydrophobic fraction from a pellet obtained after two cycles of tubulin poly-

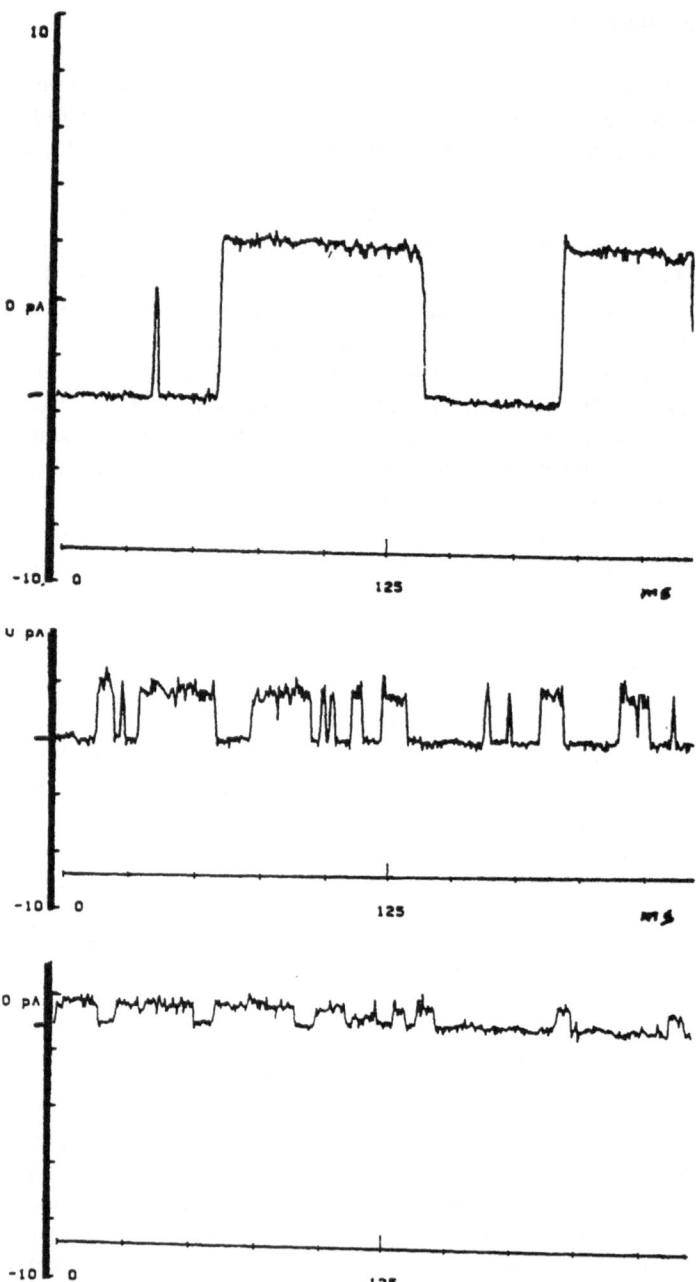

FIGURE 5. Single-channel current displacements in patch-clamp bilayers containing microtubule proteins. The magnitudes of the applied potentials are indicated on the left side of each current trace. The horizontal bars on the left side show the channel closed state for each record. Pipette solution: 47 mM $CaCl_2$, 3 mM $Ca(OH)_2$, 10 mM Hepes (pH 7.4). Bath solution: 10 mM Tris, 0.1 mM EGTA, 0.1 mM $Ca(OH)_2$, 10 mM Hepes (pH 7.4).

merization and depolymerization. The BLMs prepared from these materials on the tips of patch-clamp micropipettes were stable and gigaohm seals formed readily. We then studied single-channel activity in these membranes (Figure 5) under the same conditions used in previous investigations on reconstituted Ca^{2+} channels from brain microsomal membranes. Striking similarities were found between the characteristics of the Ca^{2+} channels in BLMs made from materials containing microtubule proteins and those in BLMs from brain microsomal components. The current–voltage relationships, single-channel conductances, and the time constants describing the channel kinetics in the two types of membrane were similar. One of the differences was related to the stimulating effects of nucleotides. Guanosine 5′-triphosphate (GTP) exerted a more prominent stimulating effect on the Ca^{2+} channel activity in the BLM containing microtubule components. This may be attributed to the specific role of GTP in tubulin structure and functioning. From these findings we suggest that some microtubule protein fractions may be involved in the regulation of the Ca^{2+}-channel mechanism in intracellular membranes. Further detailed studies shall be needed to define the structural relationships between channel molecules and tubulin subunits as well as those of microtubule-associated proteins (MAPs). It is also noteworthy that low concentrations of Ca^{2+} exert a dramatic influence on microtubule structure (Margolis, 1983). Micromolar $[Ca^{2+}]$ can induce disassembly of the microtubules. Thus, low- as well as high-affinity Ca^{2+}-binding sites are available in the tubulin molecules.

The Ca^{2+}-modulated microtubule depolymerization, integration of tubulin molecules into membranes, and defined Ca^{2+}-transport mechanisms may be interrelated events. One of the physiologically important Ca^{2+}-binding proteins, calmodulin, was found tightly associated with microtubule proteins (Margolis, 1983). Calmodulin-dependent protein kinases can interact with the MAPs (Schulman, 1984), and calmodulin antagonists, such as trifluoperazine, were found to influence Ca^{2+}-transport events. Recently, we observed a trifluoperazine-induced inhibition of the activity of brain microsomal Ca^{2+} channels which confirms the importance of the cytoskeletal proteins for the Ca^{2+} channel mechanisms.

3.2.2. Muscle Membranes

Highly functional Ca^{2+} channels from cardiac sarcolemma were incorporated into BLM in a recent study by Rosenberg et al. (1986). The reconstituted channels shared most of the typical Ca^{2+} channel properties of the native cardiac cellular membranes:

1. Slope conductances of 23 pS in 100 mM Ba^{2+} and 7 pS in 100 mM Ca^{2+} were typical for L-type Ca^{2+} channels in cardiac sarcolemmal membranes (Nowycky et al., 1985).

2. Characteristic kinetics of Ca^{2+} channel gating in the presence of Ba^{2+} and the dihydropyridine Ca^{2+} agonist Bay K8644 were found. The mean open time was 20 msec, with long-lasting opening events being promoted by the agonist.
3. A much faster inactivation in the presence of Ca^{2+} was found.
4. Steepness and voltage dependence of activation was typical for the L-type Ca^{2+} channels.
5. A full block was produced by 6 μM nimodipine, a dihydropyridine Ca^{2+} antagonist added to both sides of the bilayer in the presence of Bay K8644.
6. Block produced by Cd^{2+} induced a noisy open state current. Cd^{2+} (30 μM) exhibited a 50% inhibition of the single-channel current in the presence of 100 mM Ba^{2+}, which coincides with the findings in cell-attached patches.
7. The channels exhibited perfect divalent cation selectivity, the permeability ratio $P_{Ba^{2+}}/P_{K^+}$ being 2800 : 1.

The channel was found to be functionally symmetric under symmetric ionic conditions. Several substantial differences also exist between the characteristics of Ca^{2+} channels derived from cardiac and skeletal muscle sarcolemma (Rosenberg et al., 1986):

1. The unit conductance of the skeletal muscle Ca^{2+} channel (10.6 pS) is less than half that of the cardiac membrane channel (22.7 pS).
2. The cardiac and skeletal muscle membrane Ca^{2+} channels display different gating behavior.
3. The skeletal muscle Ca^{2+} channel shows slower activation and longer mean open times.

On the basis of these findings a structural diversity among Ca^{2+} channels in different membranes may be suggested. Sarcolemmal Ca^{2+} channels from porcine left ventricle were incorporated into patch-clamped bilayers in a study by Ehrlich et al. (1986) and it was found that, similar to the channels in the native membrane, the reconstituted Ca^{2+} channels are: (1) regulated by voltage, (2) stimulated by Bay K8644, (3) reversibly blocked by nitrendipine (5 μM), (4) selective for divalents with a permeability ratio of $Ba^{2+}/Ca^{2+}/Mg^{2+}$ = 1:0.45:0.08, and (5) with a slope conductance of 6–9 pS. An interesting effect concerning the blocking action of two stereoisomers of the verapamil analogue, D-600, was also found. One of the stereoisomers, (−) D-600, induced a more prominent block of the Ca^{2+}-channel activity than the other, (+) D-600, which correlates with the differences in the binding abilities of the two agents to the channel receptor sites as well as with their physiological actions in heart muscle. Cardiac action potentials are inhibited to a much greater extent by (−) D-600 than by (+) D-600. The former agent also induces a 100-fold greater negative inotropic effect in heart muscle.

The effect of D-600 on Ca^{2+} channels reconstituted from skeletal muscle transverse tubules was studied by Affolter and Coronado (1986). This agent as well as its charged membrane-impermeable derivative, D-890, substantially reduced the Ca^{2+}-channel activity. Their effects on the reconstituted channels were similar to those found in experiments on cellular preparations:

1. D-600 blocked the reconstituted channels with a half maximal blocking dose of 5 μM. In whole skeletal frog muscles this dose was 13.4 μM.
2. D-600 exerted similar blocking effects when added on either side of the bilayer.
3. D-890 effectively blocked the Ca^{2+} current only when added to the "cytoplasmic" side, which correlates with findings in isolated cardiac cells (Hescheler et al., 1982). Its half maximal blocking dose was 3 μM.

The mechanism of action of D-890 remains little understood, but obviously the positive charge of its molecule plays some important role.

A different type of Ca^{2+} channel was also shown to be present in skeletal muscle SR membranes by Smith et al. (1985). Similar to their native membrane correlates the SR Ca^{2+} channels incorporated into BLM were activated by 1mM ATP and blocked by 1 μM ruthenium red and Mg^{2+}. Their unit conductance, 170 pS in 50 mM Ba^{2+} and 125 pS in 50 mM Ca^{2+}, is much larger than that of the sarcolemmal Ca^{2+} channels. Ca^{2+} channels with similar properties were also characterized in cardiac SR membranes reconstituted in BLM (Vassilev and Hume, 1986; Vassilev et al., 1986b). The channels were activated by micromolar $[Ca^{2+}]$ on the same "cytoplasmic" side of the membrane and inhibited by ruthenium red and ryanodine (Vassilev and Hume, 1986).

3.3. Mechanisms of Potassium Ion Transport

3.3.1. Cytoplasmic Membranes

K^+ channels show considerably more variety than other channels depending on the types of cell, specific regulatory mechanisms, and so on. In addition to several main types, such as inward rectifier, delayed rectifier, and Ca^{2+}-dependent K^+ channels, other channels with differing unit conductance in various cells were described.

K^+ channels, which are probably responsible for the resting K^+ conductance of brain synaptosomes, were reconstituted in BLMs formed from PE and DPG or PE and PS by Nelson and Reinhardt (1984). A rectification of the current–voltage relationship was observed, which can be described by the Goldman–Hodgkin–Katz equation (Tien, 1988a). A selectivity of K^+ over Cl^- and of K^+ over Na^+ with a permeability ratio, K^+/Na^+, of 3:1–5:1 was demonstrated. K^+-channel blockers were tested and it was found that Ba^{2+} (20 mM) provoked a voltage-dependent block, but tetraethylammonium (TEA) did not influence the channel activity. Despite the differences in the actions of inhibitors, the other properties of these reconstituted channels may be interpreted in favor of their specificity for K^+.

An important type of K^+ channel which differs from those involved in resting conductance is *the delayed rectifier channel*. The single-channel analysis of its activity in native membranes using the patch-clamp technique is relatively difficult, but BLM was successfully used for the characterization of this channel. Coronado *et al.* (1984) incorporated lobster axon membrane vesicles into BLMs by using an osmotic gradient (Cohen *et al.*, 1980; Miller, 1978) and observed the appearance of single-K^+-channel events which share some of the characteristics typical for the delayed rectifier channels in a natural environment:

1. Bursting behavior with quiescent long duration interbursting intervals.
2. Voltage inactivation.
3. Lack of Cs^+ permeability and block by Cs^+, TEA, and nonyltrimethylammonium.
4. The effects of these blocking agents are exerted only when added to a defined side of the BLM, which is coincidental with their side-dependent inhibitory effects in nerve membranes.

Some differences, however, were also observed.

1. The single-channel conductance, 28 pS, was almost three times larger than that of the squid axon membrane K^+ channel. It was suggested that in lobster axon the delayed-rectifier conductance may be lower if a different lipid environment were available. On the basis of data about the dependence of conductance on the lipid composition of BLMs, it was concluded that decreased amounts of PS or increased amounts of cholesterol in the membranes may cause a lowering of the unit channel conductance.
2. There are similarities, but some differences as well, in the bursting kinetics of the reconstituted channels in relation to those of the native channels in squid axons. Temperature was suggested as one of the possible factors influencing the bursting channel lifetimes.
3. At high depolarizing voltages a current decrease is observed which remains unexplained.

In addition to the single channels with 6–7 pA current amplitude, which are qualified as delayed rectifiers, another type of channel with larger current amplitude, 12 pA, is observed in the recordings, but its behavior is clearly distinguished from that of the delayed rectifier opening events.

Data about the ionic selectivity of these channels were obtained on the basis of the reversal potentials and relative permeabilities calculated by using the Goldman–Hodgkin–Katz equation. The permeability ratio P_{Na}/P_K was defined to be 1 : 30. Only four ions are highly permeant: $Tl^+>K^+>Rb^+>NH_4$.

The kinetics of these channels may be described by a scheme that includes one open *(O)* and two closed states $(C_1$ and $C_2)$:

$$C_2 \underset{k_{-2}}{\overset{k_2(V)}{\rightleftarrows}} C_1 \underset{k_{-1}}{\overset{k_1}{\rightleftarrows}} O \tag{10}$$

The voltage-dependent step is supposed to be between the two closed states designated by $K_2(V)$. This scheme may be typical for the channels showing bursting activity.

Channels with similar characteristics have also been reconstituted from lobster axonal membranes into BLMs made on the tips of patch-clamp micropipettes by Coronado and Latorre (1983).

A variety of K^+ channels have been characterized in muscle cells. By using the patch-clamp and the BLM technique, K^+ channels have been found not only in the muscle plasma membranes but also in the intracellular membranes, such as those of the sarcoplasmic reticulum (Shamoo and Tivol, 1980).

Coronado and Latorre (1982) incorporated K^+ channels from calf cardiac sarcolemma into BLMs from PE/PS and characterized their activity in detail. They found three different types of K^+ channels with various unit conductances, 15, 28, and 95 pS. The 15 pS channel was shown to be independent of Ca^{2+} concentration. Its selectivity was not very high, the permeability ratio being $P_{K^+}/P_{Na^+} = 3$. The 95 pS channel, on the contrary, was influenced by changes in Ca^{2+} concentration. At Ca^{2+} concentrations lower than 1 mM, the long closing events disappeared. The 28 pS channel activity was found to be voltage dependent. At -80 mV, the mean open time was 6 msec, but at more depolarizing voltage, -40 mV, it was increased 10 times. The voltage dependence of this channel is quite similar to that of the repolarizing current, i_x, in cardiac muscle. The probability histogram of this channel reveals one open state and two closed states. When a fivefold K^+ concentration gradient is created across the BLM, the reversal potential was -34 mV which coincides with the predicted potential for a K^+-selective channel according to the Nernst equation. The permeability ratio, P_{K^+}/P_{Na^+}, of this channel was defined to be relatively low, 5. This low selectivity is also similar to that of the cardiac repolarizing current, i_x.

The Ca^{2+}-activated K^+ channels, which were found in a variety of cells (Adams *et al.*, 1982; Barret *et al.*, 1982; Marty, 1981; Methfessel and Boheim, 1982; Pallota *et al.*, 1981), differ substantially from other types of K^+ channels such as the delayed rectifier and the inward rectifier channels. Their opening and closing are regulated mainly by [Ca^{2+}], but the voltage across the membrane also plays a controlling role. Another typical feature of these channels is their very high unit conductance, 200–230 pS, which is substantially higher than that of the other types of channel (Latorre et al., 1984). The Ca^{2+}-activated K^+ channels incorporated into BLMs (Latorre et al., 1982; Moczydlowski and Latorre, 1983; Vergara and Latorre, 1983; Vergara et al., 1984) shared most of the characteristics of the native channels (Adams et al., 1982; Barret et al., 1982; Marty, 1981; Methfessel and Boheim, 1982; Pallota et al., 1981) of this type. A summary of these studies is as follows:

1. A high unit conductance.
2. Ca^{2+} dependence. Calcium, however, does not influence the conductance, but it does induce substantial changes in the mean open time. The latter increases as a linear function of [Ca^{2+}], whereas the mean closed time is linearly proportional to the reciprocal [Ca^{2+}]. The Ca^{2+} effect depends on the side of addition of the cation.
3. Voltage dependence. The voltage also influences the mean open time. It is interesting that in the experiments on the reconstituted transverse tubule membrane channels (Latorre et al., 1982), where one open and two closed (fast and slow) conductance states were described, the dwelling time of the channel in the slow state increased when applying more positive potentials. An opposite voltage dependence for the fast state was observed.
4. Typical ionic selectivity.
5. Block by TEA, depending on the side of its addition. This effect is interestingly voltage dependent. The apparent dissociation constant, K_d, decreases as a function of the applied potential.
6. Block by Ba^{2+}. The effect of Ba^{2+} shows some similarities, but also some significant differences in relation to that of the inward- and delayed-rectifier K^+ channels. Analysis of the data concerning Ba^{2+} block (Vergara and Lattorre, 1983) led to a valuable conclusion about the mechanisms of the ion conduction through the different types of K^+ channel. The results indicate that the Ca^{2+}-activated K^+ channel behaves as a single-ion pore, while the delayed-rectifier channel behaves as a multi-ion pore.

After analyzing the interrelations between the Ca^{2+} and the voltage dependence of the channels, Vergara and Latorre (1983) proposed a model explaining the activation kinetics by two voltage-dependent Ca^{2+}-binding reactions. The channel transitions between the open and the closed states are due to a conformational change of the channel protein molecule which is not controlled by the voltage.

3.3.2. Intracellular Membranes

The reconstitution and characterization of SR membranes in BLM are so far the most adequate approaches for obtaining information about their single-channel properties. This opportunity was exploited during the first steps of the BLM-reconstitution experiments (Shamoo and Tivol, 1980). The fusion method, which was later used as a reproducible method for reconstitution of various channel activities, has been elaborated to some extent on the basis of studies on sarcoplasmic reticulum components inserted into BLM (Miller and Rosenberg, 1979; Miller and Racker, 1976). Discrete fluctuations of BLM conductance were observed only when three main conditions for fusion between the sarcoplasmic reticulum vesicles and BLM were available: high $[Ca^{2+}]$, not less than 0.5 mM; osmotic gradient; and the presence of negatively charged phospholipids in the BLM-forming solution (Miller and Racker, 1976). It has been found that the cation channel conductance induced by sarcoplasmic reticulum vesicles in BLM is gated by voltage (Miller and Rosenberg, 1979). This could be related to the physiological mechanism of the excitation–contraction coupling, bearing in mind that the permeability of the sarcoplasmic reticulum membrane is controlled in some way by voltage changes during the coupling of the events occurring in the transverse tubule and the sarcoplasmic reticulum membranes. The voltage-gated K^+ conductance was found to be modulated by some transition metal ions, for example, Ag^+, Hg^+, Cu^{2+} (Miller and Rosenberg, 1979). They affect the opening probability rather than the channel conductance. The inhibition effect of one of the agents, mersalyl, is voltage dependent. The results indicated that a sulfhydryl group may be involved in the functioning of the channel. The unit conductance of the channel was 140 pS in 0.1 M K^+ (Labarca et al., 1980; Miller, 1979; Miller and Rosenberg, 1979).

Later, many other properties of the channel were analyzed in detail (Coronada and Miller, 1979, 1982; Garcia and Miller, 1984; Labarca and Miller, 1981; Miller, 1982a,b; Miller et al., 1984). It has been found that the channel to be perfectly selective for K^+ over Cl^- and for K^+ over $Ca^{2+} - Cs^+$ asymmetrically blocked the channel. Increasing $[Cs^+]$ induced several effects, such as reduced channel conductance increased mean open time and voltage-dependent changes of the open state conductance (Coronado and Miller, 1979; Coronado et al., 1980). Two cholinergic drug agents, decamethonium and hexamethonium, were found to induce blocking effects similar to Cs^+ (Coronado and Miller, 1980; Miller, 1982a,b). However, there was some difference between the influences of the two compounds. Decamethonium provoked flickerings from the open state, their rate increasing at higher drug concentrations. The flickering was voltage dependent, which is found only at negative potentials when the drug molecules are driven inside the channels. Hexamethonium, on the contrary, did not induce flickering, but it provoked a decrease in the unitary conductance. The different actions of the two blocking agents may be explained by differences in the lifetime of the open state. These cholinergic blockers were found to induce similar effects on channels isolated from another source, cardiac SR (Gray *et al.*, 1985; Tomlins *et al.*, 1984). Some differences, however, were also observed. In the cardiac SR, decamethonium caused flickering from either side of the BLM, while in BLM reconstituted with channels from skeletal muscle sarcoplasmic reticulum, it was effective only when added to the trans side (Coronado and Miller, 1980; Miller, 1982a,b). Nevertheless, in the former type of membrane the voltage dependence of the block was less pronounced when the agent was present on the same side. In both cases, decamethonium probably binds in a bent configuration to a site about 55% of the voltage drop across the channel from the opposite side of the membrane.

In the cardiac membranes there were also channels with differing conductance. Two open states were described, one of them (β) being a fully open state with a unit conductance of 157.2 pS, and the other (α) a substate with a conductance of 100.7 pS. Both states can be blocked by the cholinergic drugs and both of them have the same ionic selectivity sequence that is identical with that of the channels from skeletal muscle sarcoplasmic reticulum. The positive potentials favor the channel opening in both types of membrane.

Bell and Miller (9184) found the activity of the K^+ channel of sarcoplasmic reticulum to be dependent on the charges of the surrounding membrane phospholipids. The channel conductance is lower in BLMs made from neutral phospholipids than BLMs from negatively charged phospholipids at a given $[K^+]$, the effect being more pronounced at lower ionic strength. The maximum unit conductance at a saturating K^+ concentration is 220 pS. The charge of the phospholipids does not influence the ionic selectivity, but it affects the apparent rate constant of the block induced by the divalent channel blocker bis Q11. When a positively charged PC analogue was present in the BLM, the K^+ conductance was lower than that of neutral BLMs. Based on these results, it was suggested that the entry way of the channel is located 1–2 nm from the lipid surface.

The physiological role of the cation channel of the sarcoplasmic reticulum membrane remains unclear. It is probably involved in an antiport system transferring monovalent cations in a direction opposite to that of the electrogenic calcium flux triggered by the action potential. Excitation in the form of an action potential propagates along the transverse tubule cylindric formations (which are part of the muscle plasma membrane) deeply into the myoplasm. In defined regions, the so-called triads, the transverse tubule membrane interacts with the sarcoplasmic reticulum membrane. The excitation of the transverse tubule membrane in these regions in some unknown way is coupled to a very large increase (100- to 1000-fold) in the permeability of the sarcoplasmic reticulum membrane for Ca^{2+}, which is released in the myoplasm and induces contraction of the actomyosin myofibrils. The monovalent cation channel of the sarcoplasmic reticulum may operate as a charge-compensating mechanism in this process of excitation–contraction coupling, countering the charge movements connected with the Ca^{2+} efflux.

Cukierman et al. (1985) obtained valuable information about the structure and properties of the SR K^+ channel in a study on Cs^+ block of the channel incorporated in BLM from rabbit skeletal muscle SR. An early observation about the Cs^+-blocking effect was revised on the basis of these surprising results, which showed that the permeability of Cs^+ through the channel is nearly the same as that of K^+ ($P_{Cs^+}/P_{K^+} = 0.7$), but the channel conductance in the presence of Cs^+ is 15–20 times lower than the K^+ conductance. The typical blocker of this channel, decamethonium (150 μM), exhibits inhibition of the Cs^+ current, showing that Cs^+ permeates through the same K^+ channel. A rectification is observed in symmetrical Cs^+ concentrations (18–600 μM), negative Cs^+ current passing more easily than positive, which indicates an asymmetric location of the Cs^+-binding site in the channel molecule. Cs^+ blocks the single-channel K^+ current in a voltage-dependent way. The results are consistent with the view that the Cs^+-binding site is located 40% of the way through the voltage gradient, from one of the membrane surfaces. The diameter of the selectivity filter of the channel is believed to be 0.4–0.5 nm, which permits free access of Cs^+ into the pore interior. The low Cs^+ conductance could therefore be explained by tight binding of the ion rather than by a steric hindrance effect. Indeed, the binding of Cs^+ is 20 times stronger than K^+, which may determine the Cs^+-induced block of the single-channel K^+ currents.

The structure of the channel can be envisioned as a tunnel, 0.8 nm in diameter, narrowing to 0.4–0.5 nm at a point 35–40% of the way down the voltage gradient. A model of the channel with a four barrier energy profile was considered. Bell (1985) envisioned the SR K^+ channel mouth to be located 5–10 Å from the membrane surface in a study of the influence of pH and of surface charge on the functioning of the channel. The protons were found to competitively inhibit the K^+ conductance with an inhibition constant of 0.5 μM. The K^+ conductance is lower in BLMs containing neutral phospholipids than in those composed of charged lipids due to the increased negative surface potential and the higher K^+ concentration near the membrane surface in the latter type of membrane.

The observed effects were attributed to changes in the membrane surface charge, which may influence the permeation of K^+ through the channel. The data are well described by the Gouy–Chapman–Stern theory. Bell (1985) suggested that the role of the surface charge should be considered more carefully when interpreting the rectifying behavior and other single-channel characteristics of different types of channel. Gray et al. (1985) found that the membrane surface charge influenced the blocking effect of the myorelaxant succinyl choline on the cardiac and skeletal muscle SR K^+ channel. Succinyl choline, as well as other positively charged blockers such as the quaternary ammonium compound bis Q11, exhibits a more pronounced block of the channel in BLMs containing negatively charged phospholipids. The inhibiting effect of succinyl choline is interestingly larger when added to the cis side of the membrane, while most of the other blockers (e.g., decamethonium) are more effective from the trans side. Due to the fact that the succinyl choline molecule is less hydrophobic than that of decamethonium, Gray *et al.* (1985) speculated that there are two regions for a blocker to bind on the two sides of the channel molecule and that the region to the cis side is less hydrophobic. The block is voltage dependent and is accompanied by "flickering" events when the channel resides in one of the two open states, α and β. The β state is more susceptible to inhibition by succinyl choline than the α state. They suggested that the single-channel conductance of the α state (100 pS in 75 mM K_2SO_4) is 60–70% of that of the β state (184 pS) and may be considered as a substate. One of the typical characteristics of the α state is the noisier open state current (Gray *et al.*, 1985). The peculiarity suggests that transient current fluctuations between the open and the closed state underlie the α-state opening events and that 30–40% of this time the channel spends in the closed state. Therefore, it may be considered as a gating rather than conducting state, with a characteristic time-averaged conductance. There is evidence that the channel has to be open in order to be blocked. Gray *et al.* (1985) observed that the channel resides more frequently in the open state in highly charged bilayers than in neutral bilayers. In neutral bilayers the number of the α-state opening events diminishes when increasing the K^+ activity. A difference between the SR K^+ channels in cardiac and in skeletal muscle was also revealed. The cardiac channel has a higher affinity for K^+ than the skeletal muscle SR channel (Coronado and Miller, 1980), although the maximal conductance of the channel is almost the same.

3.4. Chloride Channels

A Cl^- channel was characterized in BLM reconstituted with calf cardiac sarcolemma (Coronado and Latorre, 1982). The analysis of the channel kinetics and of the probability histogram revealed that the channel fluctuates between two closed states and three open states with conductances of 15, 39, and 61 pS. The current–voltage relationship showed that the current saturates at high positive potentials, while at negative potentials it does not saturate but increases nonlinearly. The reversal potential was almost the same as the expected diffusion potential in the absence of cations. The physiological role of this channel remains unclarified owing to the lack of sufficient information on similar channels in native membranes.

A new method for analyzing channel gating kinetics was proposed by Labarca et al. (1985) after analyzing the well-defined characteristics of the two-state cholinergic receptor channel and the three-state chloride channel from *Torpedo californica* electroplax membrane. Autocorrelation and autocovariance functions were used for obtaining information about the pathways connecting the channel's open and closed states.

A Cl^- channel is available in the cytoplasmic membrane of the electric organ of *Torpedo californica*. It is responsible for the control of the resistance of the electroplax cells serving as a source of electric current. This Cl^- channel has been reconstituted in BLM and its properties have been profoundly analyzed in several studies (Hanke and Miller, 1983; Labarca et al., 1985; Miller and White, 1984; White and Miller, 1979; Tank et al., 1982).

Several specific features of the channel were demonstrated:

1. Bursting activity with four conductance states including two open states, 10 and 20 pS, and two nonconducting states with largely differing lifetimes, 10 and 500 msec.
2. Voltage dependence of the gating process.
3. Specific block by several compounds: 4,4'-diisothiocyano-2, 2'-stilbenedisulfonate (DIDS), 4-acetamino-4'-isothiocyano-2,2'-disulfonic acid (SITS), and SCN.
4. Activation by protons.

The effect of low pH is observed only when the pH changes are provoked on the side of the BLM which was exposed to the added electroplax vesicles (Hanke and Miller, 1983).

DIDS also exhibits inhibitory activity only when added to the same side as the vesicles. About 5 sec after its addition, the higher conductance state disappears and only the 10 pS state resides. The channel activity disappears entirely about 25 sec later.

Miller and White (1984) interpreted these effects in favor of a model for the Cl^- channel based on a dimeric channel complex spanning the membrane. They proposed that the Cl^- channel operates as a dimeric complex consisting of two identical subunits, protochannels. Every protochannel opens and closes independently in the range of milliseconds, but both protochannels may enter and leave an inactivated state on a slower time scale.

It has been suggested that another anion-transport complex, the erythrocyte Cl^--exchange system, operates in a similar way as a functional dimer. The oligomeric structures of the transport protein assemblies obviously are quite important for their functional activities.

3.5. Reconstitution of Active Transport Mechanisms in Planar Bilayers

The sodium–potassium pump is a key enzyme complex involved in many physiological processes depending on the asymmetric ion distribution on both sides of the plasma membranes of different kinds of excitable and nonexcitable cells. The maintenance of the resting potential and the generation of action potentials in nerve and muscle cells as well as various processes in secretory and other cells cannot be performed normally without controlling Na^+ and K^+ concentrations on both membrane sides which is dependent mainly on the functioning of the Na, K pump. The enzyme responsible for this function (Na^+, K^+)-ATPase (Skou, 1957) operates by pumping out sodium ions and pumping potassium ions into the cells. It has been characterized as an enzyme complex spanning the plasma membrane and consisting of a catalytic α subunit with a molecular weight of 100,000 daltons and a glycoprotein β subunit with a molecular weight of 35,000–55,000 daltons. Some types of phospholipid and cholesterol are also specifically associated to the enzyme complex. The functional activity of the enzyme is regulated to a great extent by phosphorylation of the α subunit. The binding sites for the cations as well as for the specific inhibitor ouabain are also located on the α subunit. The functional importance of the β subunit is not well understood, but the enzyme is inactivated when the β subunit is separated from the α subunit.

The molecular mechanism of functioning of this enzyme complex in the membrane is not fully understood. A carrier type ion exchange mechanism is believed to be responsible for the pump operation, but recently some authors have claimed that channel activity might be related in some way to the functioning of the reconstituted (N^+,K^+)-ATPase in BLM (Reinhardt et al., 1984; Tien, 1988b; Tsong, 1983).

Last et al. (1983) isolated (Na^+,K^+)-ATPase from kidney medulla or eel electroplax and incorporated it into BLMs made from PE and PS (7:3) or from phosphatidylcholine (PC) and PS (4:1). Several specific features concerning the normal enzyme functioning in the membrane were observed:

1. Discrete conductance changes were conferred by the enzyme complex only in the presence of an ionic gradient and when applying a voltage with defined polarity.
2. The specific inhibitors of (Na^+,K^+)-ATPase activity, ouabain and vanadate, induced a quick and strong decrease of the membrane conductance. These inhibitory effects were found only when an ion gradient with a defined direction across the BLM was available.
3. In defined cases, ouabain and vanadate exerted their inhibitory effects only when added to opposite sides of BLMs, which correlates with the fact that

in the native systems vanadate inhibits from the cytoplasmic side and ouabain from the extracellular side of the plasma membrane.
4. Higher concentrations of KCl removed the inhibitory effect of ouabain. It is known that K^+ inhibits the binding of this agent to the enzyme in the native system.

Last et al. (1983) found two conductance states, but only the high conductance state was influenced by the inhibitors ouabain and vanadate. They reported that the value of the single-channel conductance was unusually large, 270 pS. The possibility that this was due to the simultaneous functioning of several pump units was tested later by Reinhardt et al. (1984). Lower unit conductances, 40–50 pS, were observed when liposomes containing only one intramembranous enzyme particle were incorporated into the BLM. In some cases subconductance states (<40 pS) were registered. No detailed analysis of the ionic selectivity of mean open times, probability of opening, and other characteristics of these channel events was provided. Their conductance is several times higher than that of the native Na^+ channels at similar Na^+ concentration, which shows that they cannot be identified as conventional sodium channels. However, the possibility that they represent some kind of K^+ channel cannot be excluded. The interpretation of the data from the studies on the properties of the (Na^+,K^+)-ATPase reconstituted in BLM meets some difficulties.

The results can be explained by the functioning of the enzyme as a disabled pump passing ions through a channel serving as a leakage pathway. The pump is not working actually as an active transporting system and it is not influenced by ATP. The enzyme may be inhibited by trypsin and by n-decane used as a solvent for preparing the BLM-forming solution.

Nevertheless, the information about the functioning even of the disabled and modified pump when incorporated into BLM may be of some significance. The finding that some components of the enzyme complex may operate through channel mechanisms might be useful for elucidating the structure–function relationships in this important membrane transport system.

4. RECONSTITUTION OF SYNAPTIC EVENTS

4.1. Acetylcholine-Receptor Mechanisms

The acetylcholine receptor (AR) is one of the best characterized membrane receptor–protein complexes, and experiments on its reconstitution in model membranes progressed relatively rapidly (Anholt et al., 1984; Boheim et al.,

1981; Labarca et al., 1984; Montal et al., 1984; Nelson et al., 1980; Schindler and Quast, 1980; Tank et al., 1983).

Most of the studies were performed by using AR isolated from the electric organ of T. electroplax where this amount is the highest known. After its isolation and purification from this source, its molecular structure has been well characterized (Lindstrom et al., 1979; Raftery et al., 1980). It was found that the AR is a pentamer; that is, it consists of five glycoprotein subunits, $\alpha 1$, $\alpha 2$, β, γ, δ. The molecular weight of the entire pentamer is 270,000 Dal. The binding sites for acetylcholine are on the α subunits, having a molecular weight of 40,000 Dal.

The AR has been reconstituted in BLM mainly by using the monolayer folding method (Labarca et al., 1984; Nelson et al., 1980) and the patch-clamp micropipette technique (Suarez-Isla et al., 1983; Tank et al., 1983). Several properties that are typical for the AR channels on native membranes (Hamill and Sakmann, 1981; Neher and Sakmann, 1976; Sakmann et al., 1980) have been demonstrated in the BLM experiments (Anholt et al., 1984; Boheim et al., 1981; Labarca et al., 1984; Montal et al., 1984; Nelson et al., 1980; Schindler and Quast, 1980; Suarez-Isla et al., 1983; Tank et al., 1983). The results may be summarized as follows:

1. The channels are activated by the cholinergic agonists acetylcholine (Ach), carbamoylcholine (Cch), and suberyldicholine (Sch). The channel-stimulating concentrations of these specific cholinergic ligands were similar to those used for activation of the native AR channels.
2. The channel activity is inhibited by specific inhibitors, α-bungarotoxin, β-tubocurarin, and hexamethonium.
3. Receptors are desensitized under defined conditions.
4. The channels exhibit characteristic unit channel conductance.
5. Concentration and voltage-dependent effect of the local anesthetic derivative of lidocaine, QX-222, which is known to block the permeation of ions through the open AR channel on native membranes, causes a reduction in the fast and slow components of the distribution of open times τ_{o1} and τ_{o2}, respectively.

Some differences between the actions of various agonists were found by Suarez-Isla et al. (1983). For instance, Sch provoked an increase of two lifetimes described by two exponential components characterizing the open time distribution.

At concentrations higher than 10 μM, Ach reduced typical bursting activity. The bursts were interrupted by relatively long quiescent periods. The unit conductance was not changed, 40 pS in 0.5 M NaCl, but the channel open times were longer at higher concentrations of the ligand than at lower concentrations. It is interesting that some agents, such as the disulfide reducing compound dithiothreitol, may alter the ligand affinity of the AR, but at the same time they do not influence the channel activity.

Concerning the ion selectivity, Schindler and Quast (1980) found that in the presence of Cch the membrane permeability was seven times higher for K^+ and Na^+ than for Cl^-. Labarca *et al.* (1984) determined the conductances for different ions at 0.3 M NaCl: 28 pS for Na^+, 30 pS for Rb^+, 38 pS for Cs^+, and 50 pS for NH^{4+}. When the Na^+ concentration is increased, the single-channel conductance reaches a maximal value of 95 pS. Two computer-simulated models were proposed:

1. A three-state model that consists of two closed states, one of them unliganded (C) and the other liganded (CL) and one open-liganded state (OL):

$$C \rightleftharpoons CL \rightleftharpoons OL \qquad (11)$$

2. A four-state model that includes an additional desensitized (liganded) state (DL):

$$\begin{array}{c} C \rightleftharpoons CL \rightleftharpoons OL \\ \updownarrow \\ DL \end{array} \qquad (12)$$

It was concluded that the second model describes quite well the AR channel kinetics.

Acetylcholine receptors from *Torpedo electroplax* membranes were reconstituted in liposomes and their single-channel properties were studied by Tank *et al.* (1983) using bilayer patches formed from the liposomal membranes on the tips of patch-clamp pipettes. They observed a main open state level with a slope conductance of 42 pS in 150 mM NaCl and a subconductance state. A double exponential distribution of the open times with very small time constants ($\tau_{fast} = 0.35$ msec and $\tau_{slow} = 2.8$ msec) was found. At lower temperatures the amplitude of the single-channel current decreased substantially. Bursting events and desensitization effects were observed. The frequency of the opening events decayed with a time constant of 92 sec in the presence of 1 μM acetylcholine. They observed a more rapid inactivation rate at higher concentrations of the agonist. Two inactivation processes were described: a fast (0.1–1 sec) and a slow (minutes) inactivation, which may be influenced in a different way by the agonists.

4.2. Dopamine Receptors

Murphy and Vodyanoy (1984) reconstituted another neurotransmitter-regulated channel, a dopamine-receptor channel, into BLMs and demonstrated the effects of some specific agonists and modifiers of its activities.

5. RECONSTITUTION OF SENSORY MECHANISMS

5.1. Visual Receptor Membranes

Hanke and Kaupp (1984) incorporated two types of channel from bovine rod outer segments into planar bilayers and patch-clamped bilayers. Neither of them was sensitive to light, but they did show several specific characteristics. One of the channels had relatively low unit conductance (20 pS in 150 mM NaCl) but was considerably more selective for Na$^+$ ($P_{Na+}/P_{K+} = 6 : 1$) and was considered a Na$^+$ channel. The other channel had a larger conductance (120 pS in 150 mM NaCl) but had low Na$^+$ selectivity ($P_{Na+}/P_{K+} = 1.6 : 1$). It was referred to as a cationic channel. One of its specific characteristics was its ability to activate in less than 10 msec and inactivate after applying a negative voltage step. The Na$^+$ channel was not influenced by Ca^{2+} concentrations in the range from 1 μM to 1 mM. Tetrodotoxin at concentrations up to 1 μM could not induce any block of the two channels. The channels may originate from disc or plasma membranes, but their physiological significance remains unclear.

5.2. Olfactory Receptors

BLM was used for investigation of the olfactory receptor function by Vodyanoy and Murphy (1983) (see Tien, 1974, 1988a,b). Homogenates of rat olfactory epithelium were incorporated into BLM made from egg PC by using a new technique (Vodyanoy et al., 1982) based on the Langmuir–Blodgett monolayer method (Tancrede et al., 1983; Tien, 1974). Single-channel events appeared about 30 sec after the addition of the homogenates. About 20 min later, the channel activity decreased and vanished. When adding subsequent nanomolar concentrations of odorous agents, diethylsulfide of (−)-carvone, the discrete fluctuations reappeared with changed characteristics. The odorous compounds did not significantly affect the unit channel conductance, which remained in the range of 60–65 pS, but changes in the mean open time and in the probability of opening were observed. Diethylsulfide induced increases in the probability of opening from 0.66 to 0.8 and in the mean open time. The lack of activation in the presence of NaCl and the block of the effect of both odorous compounds by the specific K^+-channel blocker 4-aminopyridine indicate that the channels may be specific for potassium. The authors suggested correlations with the Ach R channels. However, the very long open times and the lack of bursting activity show a quite different behavior.

6. PHYSIOLOGICAL PROCESSES IN NONEXCITABLE MEMBRANES

6.1. Erythrocyte Plasma Membranes

Different erythrocyte components such as sialoglycoproteins, band 3 protein, were found to induce large changes in the conductance of different types of BLMs (Benz et al., 1984; Bleuel et al., 1977; Schubert et al., 1977). A plant alkaloid, sanguinarine, may inhibit the erythrocyte sodium–potassium pump, but without any influence on the leak for Na^+ and K^+ (Cala et al., 1982). At the same time, the uncharged form of the compound induces an increased cation permeability in erythrocytes as well as in BLMs made from red cell lipids.

The major component of the erythrocyte membrane was defined to be the band 3 protein with a molecular weight of 97,000 D. This protein was found to be responsible for one of the main physiological processes in the red cell membrane, the anion exchange mechanism. The latter is related to the transport of nonelectrolytes such as glucose and water (Disalvo et al., 1988). After its reconstitution into BLM, Benz et al. (1984) found that the band 3 protein provoked an increase in the membrane conductance by several orders of magnitude. This effect was dependent on the fourth power of protein concentration, which was interpreted as an indication that formation of a tetrameric structure is involved in the mechanisms of the ion conduction through the BLM. The voltage did not influence the unit conductance, which was 50 pS. When the protein was pretreated with several different SM reagents, its action on the membrane conductance was inhibited. The same compounds are known to inhibit water transport through erythrocyte membranes. Other SM reagents that induce such an effect were not able to modify the influence of band 3 protein on BLM conductance. The subunits of this protein may form a tetramer that is involved in water transport across the erythrocyte membrane.

6.2. Epithelial Membranes

One of the most important physiological processes in epithelial cells is the transport of Na^+, which in some cells is stimulated by the hormones aldosterone and vasopressin, and inhibited by such diuretics as amiloride. Sodium-specific channels that could be modulated by amiloride were found in the membranes of such cells.

Epithelial membrane vesicles were recently inserted into BLMs from PE/PS, 7:3, and the properties of amiloride-sensitive single Na^+ channels were characterized (Olans *et al.*, 1984; Sariban-Sohraby *et al.*, 1984). A broad range of unit conductances were measured, from 4 to 80 pS at 0.2 M NaCl, which was explained by the presence of a variety of stable conductance states similar to the gramicidin A channel (Busath and Szabo, 1981). Changes in the applied potential did not influence the channel opening probabilities. There was one long-lived closed state with a mean dwell time of 5 sec and one fast closed state (< 200 msec). Amiloride provoked a dose-dependent decrease in the open state conductance when added to the cis side, and a flickering from the trans side. The influence of Na^+ concentration was studied separately by Olans *et al.* (1984) and at high [Na^+] a saturation of the amiloride-sensitive Na^+ channels was observed with Michaelis constants of 17 or 47 mM for maximal conductances, G^{max} of ~4 or 44 pS, respectively.

6.3. Kidney Plasma Membranes

Tosteson and Sapirstein (1981) extracted five main proteolipid fractions from the bovine kidney membrane vesicles and inserted them into BLMs made from diphytanoyl-PC by adding them to the aqueous phase on one side of the BLM. The addition of the proteolipid fractions provoked the appearance of discrete current fluctuations. When using protein concentrations lower than 0.6 μg/ml, the unit channel conductance was about 10 pS in 0.1 M KCl and 3 pS in 0.1 M NaCl. No voltage dependence was observed in this case. At higher protein concentration, 1μg/ml, a new channel population appeared with about 10 times higher conductance for both types of aqueous solution, 100 and 30 pS, respectively. In the presence of KCl, these channels were voltage dependent. Tosteson and Sapirstein (1981) obtained evidence showing that these channels appear owing to the formation of oligomer protein structures in the membrane. The permeability ratio P_{K+}/P_{Na+}, determined from the single-channel experiments, was only 2.5, but the steady-state conductance in the presence of 0.1 M KCl was 100–1000 times higher than that in 0.1 M NaCl. This was explained by assuming that the conformation of apoprotein molecules in KCl is more favorable for incorporation than in NaCl. When protein molecules integrate into the BLM, however, their conformation is changed in a way that always determines a defined permeability ratio.

The physiological role of the described channel activities is not clear. Although Na^+-specific channels such as the amiloride-sensitive channels are of particular interest in relation to the functions of the kidney epithelia, other transport mechanisms may also have some physiological implications.

A hormone influencing the transport of sodium and water across the kidney epithelial membranes, vasopressin, was found to induce a decrease in the resistance of BLMs (Bach and Miller, 1974; Fettiplace et al., 1972). The effect of this basic polypeptide was more pronounced when it was used at higher doses and at higher salt concentrations in negatively charged PS bilayers. Thus, polypeptide molecules may form channels by the aggregation of four or five polypeptide units.

6.4. Ciliary Membranes

The ciliary formations are responsible for one of the most important physiological functions, cell movement. The dynamic events accompanying this process depend to a great extent on the transport mechanisms across the ciliary membranes. The ciliary membranes of *Paramecium* are well characterized biochemically and electrophysiologically. Therefore, they have been used as a source for reconstituting ciliary membrane channel activities in BLMs (Ehrlich et al., 1984). Two types of Ca^{2+} channels were identified which were much more permeable to divalent than to monovalent ions:

1. A channel with relatively large unit conductance, 30 pS, was identified, with the following characteristics:
    (a) Low ionic selectivity, the permeability to Mg^{2+} being slightly higher than that to Ba^{2+}.
    (b) No voltage dependence in the range from -20 to 60 mV.
2. A channel with low unit conductance (1.5–2 pS) was found, with properties differing from those of the high conductance channel:
    (a) Good ionic selectivity with Ca^{2+} and Ba^{2+} permeability significantly exceeding that for Mg^{2+} (the permeability coefficient for Ba^{2+} is 10 times higher than that for Mg^{2+}).
    (b) Voltage dependence with opening probability increasing steeply with voltage. An e-fold increase in opening probability for a potential change of 9 ± 4 mV was observed, which is similar to that found for native Ca^{2+} currents.

It is worth noting that the applied potential apparently did not influence the mean open time, similar to the intact Ca^{2+} channels. When increasing $[Ba^{2+}]$ a saturating value of the single-channel conductance is reached at concentrations higher than 10 mM. This is another property that the reconstituted Ca^{2+} channels share with the native channels.

6.5. Outer Mitochondrial Membrane

The outer mitochondrial membrane is known to be highly permeable to different electrolytes and nonelectrolytes as well as for molecules with a molecular weight of several thousand daltons (Tien, 1974, 1988b). The physiological importance of this property is related to the high metabolic activity of these organelles involving the degradation and synthesis of many metabolites. The mechanism of this high permeability was not clear for a long time, but in the 1970s Sariban-Sohraby *et al.* (1984) and Colombini (1979) accumulated evidence showing that transmembrane channels with specific configurations are located in the outer mitochondrial membrane which may be responsible for this property. The functioning of outer mitochondrial membrane channels was demonstrated and characterized using the BLM system for the first time by Schein *et al.* (1976). Mitochondria from *Paramecium aureola* were found to produce voltage-dependent anion-selective channels (VDAC) in BLMs. The selectivity of the channel is quite high. Its permeability to Cl^- is about seven times higher than that to K^+. It is not permeable to Ca^{2+}. Parallel electron microscopy and BLM experiments by Mannela *et al.* (1983) demonstrated that porelike polymorphic crystalline arrays are located in outer *Neurospora crassa* mitochondrial membranes, which apparently are responsible for VDAC activity as shown in the BLM system reconstituted with VDAC. Rigid channel triplet structures were observed which can move in the plane of the membrane forming regularly arranged triplet aggregates. These data correlated with the finding that when inserted into BLM the mitochondria showed a tendency to induce changes in the BLM conductance in threefold multiple steps. Similar channels were also found in the outer walls of different kinds of bacteria.

7. CONCLUDING REMARKS AND PERSPECTIVE

Some of the recent studies on BLM were reviewed in an effort to point out the applicability of this model membrane system to the problems related to the structure, properties, and functions of biomembranes. The direct investigation of the cellular membranes meets with considerable difficulties due to their complex structure. The properties of the artificial bilayer lipid membranes (BLMs and liposomes) and the interactions between their components in relation to defined cellular functions can be studied easier than the more complicated intact biomembranes (Tien, 1988a,b). For example, investigations of the membrane transport mechanisms in BLM provide a framework of reference to which material transport in the native membranes can be compared. Further developments of BLM research can also facilitate the initial testing of working hypotheses, which may generate guidelines leading to a better choice of appropriate *in vivo* reconstituted membrane investigations.

Looking toward the future of research on BLM systems, it is expected that further achievements are likely in areas connected with membrane biology and biophysics such as active transport, oxidative phosphorylation, photosynthesis, vision, immunology, nerve conduction, and energy transduction. The BLM system should also be useful in understanding membrane biogenesis and various membrane-mediated processes.

ACKNOWLEDGMENTS. This work has been supported by the U.S. NIH grant GM-14971 and ONR grant N00014-85-0399. We thank Ms. Jan Reid and Theresa Hubbard for the preparation of the manuscript.

8. REFERENCES

Adams, P. R., Constanti, A., Brown, D. A., and Clark, R. B., 1982, Intracellular Ca^{2+} activates a fast voltage-sensitive K^+ current in vertebrate sympathetic neurons. *Nature (London)* **296**:746–479.

Affolter, H., and Coronado, R., 1986, Sidedness of reconstituted calcium channels from muscle transverse tubules as determined by D600 and D890 blockade. *Biophys. J.* **49**:767–771.

Andersen, O. S., 1978, Permeability properties of unmodified bilayer lipid membranes. In *Membrane Transport in Biology* (D. C. Tosteson, ed.), Springer-Verlag, New York.

Andersen, O. S., 1984, Gramicidin channels. *Annu. Rev. Physiol.* **46**:531–548.

Anholt, R., Lindstrom, J., and Montal, M., 1984, The molecular basis of neurotransmission: Structure and function of the nicotinic acetylcholine receptor. In *The Enzymes of Biological Membranes* (A Martonosi, ed.), Plenum Press, New York.

Bach, D., and Miller, I. R., 1974, Interaction of vasopressin with phosphatidylserine bilayers. *Biochim. Biophys. Acta* **339**:367–373.

Baer, M., Best, P. M., and Reuter, H., 1976, Voltage-dependent action of tetrodotoxin in mammalian cardiac muscle. *Nature (London)* **263**:344–345.

Barret, J. N., Magleby, K. L., and Pallota, B. S., 1982, Properties of single calcium-activated potassium channels in cultured rat muscle. *J. Physiol. (London)* **331**:211–230.

Bean, R. C., Shepherd, W. C., Chan, H., and Eichner, J., 1969, Discrete conductance fluctuations in lipid bilayer protein membranes. *J. Gen. Physiol.* **53**:741.

Bell, J., 1985, Protons decrease the single channel conductance of the sarcoplasmic reticulum K^+ channel in neutral and negatively charged bilayers. *Biophys. J.* **48**:349–353.

Bell, J. E., and Miller, C., 1984, Effects of phospholipid surface charge on ion conduction in the K^+ channel of sarcoplasmic reticulum. *Biophys. J.* **45**:279–287.

Benz, R., Tosteson, M. T., and Schubert, D., 1984, Formation and properties of tetramers of band 3 protein from human erythrocyte membranes in planar lipid bilayers. *Biochim. Biophys. Acta* **775**:347–355.

Bleuel, H., Wiedner, G., and Schubert, D., 1977, Variability of conductivity changes in black phosphatidylserine membranes induced by proteins from erythrocyte membranes. *Z. Naturforsch.* **32**:375–378.

Blumenthal, R., and Klausner, R. D., 1982, The interaction of proteins with black lipid membranes. In *Membrane Reconstitution* (G. Poste and G. L. Nicolson, eds.), pp. 43–82, Elsevier, Amsterdam.
Boheim, G., Hanke, W., Barrantes, F. J., Eibl, H., Sakmann, B., Fels, G., and Maelicke, A., 1981, Agonist-activated ionic channels in acetylcholine receptor reconstituted into planar lipid bilayers. *Proc. Natl. Acad. Sci. U.S.A.* **78:**3586–3590.
Busath, D., and Szabo, G., 1981, Gramicidin forms multistate rectifying channels. *Nature (London)* **294:**371–373.
Cala, P. M., Norby, J. G., and Tosteson, D. C., 1982, Effects of the plant alkaloid sanguinarine on cation transport by human red blood cells and lipid bilayer membranes. *J. Memb. Biol.* **64:**23–31.
Cohen, I. S., and Strichartz, G. R., 1977, On the voltage-dependent action of tetrodotoxin. *Biophys. J.* **17:**275–279.
Cohen, F. S., Zimmerberg, J., and Finkelstein, A., 1980, Fusion of phospholipid vesicles with planar phospholipid bilayer membranes. *J. Gen. Physiol.* **75:**251–270.
Cohen, C. J., Bean, B. P., Colatsky, T. J., and Tsien, R. W., 1981, Tetrodotoxin block of sodium channels in rabbit Purkinje fibers: Interactions between toxin binding and channel gating. *J. Gen. Physiol.* **78:**383–411.
Cohen, F. S., Akabas, M. M., and Finkelstein, A., 1982, Osmotic swelling of phsopholipid vesicles causes them to fuse with planar phospholipid bilayer membrane. *Science* **217:**458–460.
Colombini, M., 1979, A candidate for the permeability pathway of the outer mitrochondrial membrane. *Nature (London)* **279:**643–645.
Coronado, R., 1985, Effect of divalent cations on the assembly of neutral and charged phosopholipid bilayers in patch-recording pipettes. *Biophys. J.* **47:**851–857.
Coronado, R., and Latorre, R., 1983, Phospholipid bilayers made from monolayers on patch-clamp pipettes. *Biophys. J.* **43:**231–236.
Coronado, R., and Miller, C., 1979, Voltage-dependent caesium blockade of a cation channel from fragmented sarcoplasmic reticulum. *Nature (London)* **280:**807–810.
Coronado, R., and Miller, C., 1980, Decamethonium and hexamethonium block K^+ channels of sarcoplasmic reticulum. *Nature (London)* **288:**495–497.
Coronado, R., and Miller, C., 1982, Conduction and block by organic cations in a K^+-selective channel from sarcoplasmic reticulum incorporated into planar phospholipid bilayers. *J. Gen. Physiol.* **79:**529–547.
Coronado, R., Rosenberg, R. L., and Miller, C., 1980, Ionic activity, saturation, and block in a K^+-selective channel from sarcoplasmic reticulum. *J. Gen. Physiol.* **76:**425–446.
Coronado, R., Latorre, R., and Mautner, H. G., 1984, Single potassium channels with delayed rectifier behavior from lobster axon membranes. *Biophys. J.* **45:**289–299.
Cukierman, S., Yellen, G., and Miller, C., 1985, The K^+ channel of sarcoplasmic reticulum. A new look at Cs^+ block. *Biophys. J.* **48:**477–484.
Disalvo, E. A., Siddiqi, F. A., and Tien, H. T., 1988, Membrane transport with emphasis on water and non-electrolytes in experimental BLMs and biomembranes. In *Water Transport in Biological Membranes* (G. Benga, ed.), CRC Press, Boca Raton, FL.
Eisenman, G., and Horn, R., 1983, Ionic selectivity revisited: The role of kinetic and equilibrium processes in ion permeation through channels. *J. Membr. Biol.* **76:**197–225.
Ehrlich, B. E., Finkelstein, A., Forte, M., and Kung, C., 1984, Voltage-dependent calcium channels from *Paramecium* cilia incorporated into planar lipid bilayers. *Science* **225:**427–428.
Ehrlich, B. E., Schen, C. R., Garcia, M. L., and Kaczorowski, G. J., 1986, Incorporation of calcium channels from cardiac sarcolemmal membrane vesicles into planar lipid bilayers. *Proc. Natl. Acad. Sci. U.S.A.* **83:**193–197.

Etemadi, A.-H., 1983, Functional and orientational features of protein molecules in reconstituted lipid membranes. *Adv. Lipid Res.* **21**:281–428.

Feit, H., and Shay, J. W., 1980. The assembly of tubulin into membranes. *Biochem. Biophys. Res. Commun.* **94**:324–331.

Fettiplace, R., Haydon, D. A., and Knowles, C. D., 1972, The action of vasopressin on artificial lipid bilayers. *J. Physiol. (London)* **221**:18P–20P.

French, R. J., Worley, J. F. III, and Krueger, B. K., 1984, Voltage-dependent block by saxitoxin of sodium channels incorporated into planar lipid bilayers. *Biophys. J.* **45**:301–310.

Garcia, A. M., and Miller, C., 1984, Channel mediated monovalent cation-fluxes in isolated sarcoplasmic reticulum vesicles. *J. Gen. Physiol.* **83**:819–839.

Gordon, L. G. M., and Haydon, D. A., 1975, Potential dependent conductance in lipid membranes containing alamethicin. *Philos. Trans. R. Soc. (London) Sect. B* **270**:433.

Gray, M. A., Montgomery, R. A. P., and Williams, A. J., 1985, Asymmetric block of a monovalent cation-selective channel of rabbit cardiac sarcoplasmic reticulum by succinyl choline. *J. Membr. Biol.* **88**:85–95.

Hagiwara, S., and Byerly, L., 1981, Calcium channel. *Annu. Rev. Neurosci.* **4**:69–125.

Hamill, O. P., and Sakmann, B., 1981, Multiple conductance states of single acetylcholine receptor channels in embryonic muscle cells. *Nature (London)* **294**:462–464.

Hamill, O. P., Marty, A., Neher, E., Sakmann, B., and Sigworth, F. Y., 1981, Improved patch-clamp techniques for high-resolution current recording from cells and cell-free membrane patches. *Pflugers Arch.* **391**:85–100.

Hanke, W., and Kaupp, U. B., 1984, Incorporation of ion channels from bovine rod outer segments into planar lipid bilayers. *Biophys. J.* **46**:587–595.

Hanke, W., and Miller, C., 1983, Single chloride channels from *Torpedo electroplax*. Activation by protons. *J. Gen. Physiol.* **82**:25–45.

Hartshorne, R. P., and Caterall, W. A., 1981, Purification of the saxitoxin receptor of the sodium channel from rat brain. *Proc. Natl. Acad. Sci. U.S.A.* **78**:4620–4624.

Hartshorne, R. P., and Caterall, W. A., 1984, The sodium channel from rat brain. Purification and subunit composition. *J. Biol. Chem.* **259**:1667–1675.

Hartshorne, R. P., Keller, B. U., Talvenheimo, J. A., Caterall, W. A., and Montal, M., 1985, Functional reconstitution of the purified brain sodium channel in planar lipid bilayers. *Proc. Natl. Acad. Sci. U.S.A* **82**:240–244.

Heggeness, S. T., and Starkus, J. G., 1986, Saxitoxin and tetrodotoxin: Electrostatic effects on gating current in crayfish axons. *Biophys. J.* **49**:629–644.

Hescheler, J., Pelzer, D., Trube, G., and Trautwein, W., 1982, Does the organic calcium channel blocker D600 act from inside or outside on the cardiac cell membrane? *Pflugers Arch.* **393**:287–291.

Higashi, K-I., Suzuki, S., Fujii, H., and Kirino, Y., 1987, Preparation and some properties of giant liposomes and proteoliposomes. *J. Biochem.* **101**:433–440.

Hille, B., 1984, *Ionic Channels of Excitable Membranes*, Sinauer, Sunderland, MA.

Hodgkin, A. L., and Huxley, A. F., 1952, The dual effect of membrane potential on sodium conductance in the giant axon of Loligo. *J. Physiol. (London)* **116**:497–506.

Horn, R., Vandenberg, C. A., and Lange, K., 1984, Statistical analysis of single sodium channels: Effects of N-bromoacetamide. *Biophys. J.* **45**:323–336.

Krueger, B. K., Worley, J. F., and French, R. J., 1983, Single sodium channels from rat brain incorporated into planar lipid bilayer membranes. *Nature (London)* **303**:172–175.

Labarca, P., and Miller, C., 1981, A K^+-selective, three-state channel from fragmented sarcoplasmic reticulum of frog leg muscle. *J. Membr. Biol.* **61**:31–38.

Labarca, P., Coronado, R., and Miller, C., 1980, Thermodynamic and kinetic studies of the gating behavior of a K^+-selective channel from the sarcoplasmic reticulum membrane. *J. Gen. Physiol.* **76**:397–424.

Labarca, P., Lindstrom, J., and Montal, M., 1984, Acetylcholine receptor in planar lipid bilayers. Characterization of the channel properties of the purified nicotinic acetylcholine receptor from *Torpedo californica* reconstituted in planar lipid bilayers. *J. Gen. Physiol.* **83**:473–496.

Labarca, P., Rice, J. A., Fredkin, D. R., and Montal, M., 1985, Kinetic analysis of channel gating. Application to the cholinergic receptor channel and the chloride channel from *Torpedo californica*. *Biophys. J.* **47**:469–478.

Laeuger, R., 1982, Kinetic properties of ion carriers, channels, and pumps. In *Transport in Biomembranes: Model Systems and Reconstitution* (R. Antolini, A. Gliozzi, and A. Gorio, eds.), pp. 1–11, Raven Press, New York.

Last, T. A., Gantzer, M. L., and Tyler, C. D., 1983, Ion-gated channel induced in planar bilayers by incorporation of (Na^+, K^+)-ATPase. *J. Biol. Chem.* **258**:2399–2404.

Latorre, R., Vergara, C., and Hidalgo, C., 1982, Reconstitution in planar lipid bilayers of a Ca^{2+}-dependent K^+ channel from transverse tubule membranes isolated from rabbit skeletal muscle. *Proc. Natl. Acad. Sci. U.S.A.* **79**:805–809.

Latorre, R., Coronado, R., and Vergara, C., 1984, K^+ channels gated by voltage and ions. *Annu. Rev. Physiol.* **46**:485–495.

Liebovitch, L. S., and Fishbarg, J., 1985, Determining the kinetics of membrane pores from patch clamp data without measuring the open and closed times. *Biochim. Biophys. Acta* **813**:132–136.

Lindstrom, J., Merlie, J., and Yogesewaran, G., 1979, Biochemical properties of acetylcholine receptor subunits form *Torpedo californica*. *Biochemistry* **18**:4465–4470.

Mannela, C. A., Colombini, M., and Frank, J., 1983, Structural and functional evidence for multiple channel complexes in the outer membrane of *Neurospora crassa* mitochondria. *Proc. Natl. Acad. Sci. U.S.A.* **80**:2243–2247.

Margolis, R. L., 1983, Calcium and microtubules. In *Calcium and Cell Function*, Vol IV (W. Y. Cheung, ed.), p. 313, Academic Press, Orlando.

Marty, A., 1981, Ca^{2+}-dependent K^+ channels with large unitary conductance in chromaffin cell membranes. *Nature (London)* **291**:500.

Methfessel, C., and Boheim, C., 1982, The gating of single calcium-dependent potassium channels is described by an activation–blockage mechanism. *Biophys. Struct. Mech.* **9**:35–60.

Miller, C., 1978. Voltage gated cation conductance channel from fragmented sarcoplasmic reticulum: steady state electrical properties. *J. Membr. Biol.* **40**:1–23.

Miller, C. (ed.), 1986, *Ion Channel Reconstitution*, Plenum Press, New York.

Miller, C., 1982a, Bis-quarternary ammonium blockers as structural probes of the sarcoplasmic reticulum K^+ channel. *J. Gen. Physiol.* **79**:869–892.

Miller, C., 1982b, Feeling around inside a channel in the dark. In *Transport in Biomembranes: Model Systems and Reconstitution* (R. Antolini, A. Gliozzi, and A. Gorio, eds.), pp. 99–108, Raven Press, New York.

Miller, C., and Racker, E., 1976, Ca^{2+}-induced fusion of fragmented sarcoplasmic reticulum with artificial planar bilayers. *J. Membr. Biol.* **30**:283–300.

Miller, C., and Rosenberg, R. L., 1979, Modification of a voltage-gated K^+ channel from sarcoplasmic reticulum. *J. Gen. Physiol.* **74**:457–478.

Miller, C., and White, M. M., 1984, Dimeric structure of single chloride channels from *Torpedo electroplax*. *Proc. Natl. Acad. Sci. U.S.A.* **81**:2772–2775.

Miller, C., Bell, J. E., and Garcia, A. M., 1984, The K^+ channel of sarcoplasmic reticulum. *Curr. Top. Membr. Transport* **21**:99–132.

Moczydlowski, E., and Latorre, R., 1983, Gating kinetics of Ca^{2+}-activated K^+ channels from rat muscle incorporated into planar lipid bilayers. Evidence for two voltage-dependent Ca^{2+} binding reactions. *J. Gen. Physiol.* **82**:511–542.

Moczydlowski, E., Garber, S. S., and Miller, C., 1984a, Batrachotoxin-activated Na^+ channels in planar lipid bilayers. Competition of tetrodotoxin block by Na^+. *J. Gen. Physiol.* **84**:665–686.

Moczydlowski, E., Hall, S., Garber, S. S., Strichartz, G. S., and Miller, C., 1984b, Voltage-dependent blockade of muscle Na^+ channels by guanidinium toxins. Effect of toxin charge. *J. Gen. Physiol.* **84**:687–704.

Montal, M., and Mueller, P., 1972, Formation of bimolecular membranes from lipid monolayers and a study of their electrical properties. *Proc. Natl. Acad. Sci. U.S.A.* **69**:3561–3566.

Montal, M., Labarca, P., Fredkin, D. R., Suarez-Isla, B. A., and Lindstrom, J., 1984, Channel properties of the purified acetylcholine receptors from *Torpedo californica* reconstituted in planar lipid bilayer membranes. *Biophys. J.* **45**:165–174.

Mountz, J. D., and Tien, H. T., 1976, Protein–lipid interactions in bilayer lipid membranes (BLM). In *The Enzymes of Biological Membranes* (A. Martonosi, ed.), pp. 139–170, Plenum Press, New York.

Mueller, P., Rudin, D. O., Tien, H. T., and Wescott, W. C., 1962, Reconstitution of cell membrane in vitro and its transformation in an excitable system. *Nature (London)* **194**:979.

Mueller, P., and Rudin, D. O., 1968, Resting and action potentials in experimental bimolecular lipid membranes. *J. Theor. Biol.* **18**:222.

Muller, R. U., and Peskin, C. S., 1981, The kinetics of monazomycin-induced voltage-dependent conductance. II. Theory and a demonstration of a form of memory. *J. Gen. Physiol.* **78**:201.

Murphy, R. B., and Vodyanoy, V., 1984, Functional reconstitution of rat striatal dopamine agonist receptors into artificial lipid bimolecular membranes. *Biophys. J.* **45**:22–23.

Neher, E., and Sakmann, B., 1976, Single channel currents recorded from membrane of denervated frog muscle fibres. *Nature (London)* **260**:799–802.

Neher, E., Sakmann, B., and Steinbach, J. H., 1978, The extracellular patch clamp: A method for resolving currents through individual open channels in biological membranes. *Pflugers Arch.* **375**:219–228.

Nelson, M. T., 1986, Interactions of divalent cations with single calcium channels from rat brain synaptosomes. *J. Gen. Physiol.* **87**:201–222.

Nelson, M. T., and Reinhardt, R., 1984, Single ion-channel current measurements from rat brain synaptosomes in planar lipid bilayers. *Biophys. J.* **45**:60–62.

Nelson, N., Anholt, R., Lindstrom, J., and Montal, M., 1980, Reconstitution of purified acetylcholine receptors with functional ion channels in planar lipid bilayers. *Proc. Natl. Acad. Sci. U.S.A.* **77**:3057–3061.

Nelson, M. T., French, R. J., and Krueger, B. K., 1984, Voltage-dependent calcium channels from brain incorporated into planar lipid bilayers. *Nature (London)* **308**:77–80.

Nilius, B., Hess, P., Lansman, J. B., and Tsien, R. W., 1985, A novel type of cardiac calcium channel in ventricular cells. *Nature (London)* **316**:443–446.

Nowycky, M., Fox, A., and Tsien, R. W., 1985, Three types of neuronal calcium channels with different calcium agonist sensitivity. *Nature (London)* **316**:440–443.

Olans, L., Sariban-Sohraby, S., and Benos, D. J., 1984, Saturation behavior of single, amiloride-sensitive Na^+ channels in planar lipid bilayers. *Biophys. J.* **46**:831–835.

Oosting, P. H., 1979, Signal transmission in the nervous system. *Rep. Prog. Phys.* **42**:1479–1532.

Orozco, C. B., Suarez-Isla, B. A., Froelich, J. P., and Heller, P. F., 1985, Calcium channels in sarcoplasmic reticulum (SR) membranes. *Biophys. J.* **47**:57a.

Pallota, B. S., Magleby, K. L., and Barret, J. N., 1981, Single channel recordings of a Ca^{2+}-activated K^+ current in rat muscle cell culture. *Nature (London)* **293**:471–474.

Patlak, J., and Horn, R., 1982, Effect of *N*-bromoacetamide on single sodium channel currents in excised membrane patches. *J. Gen. Physiol.* **79**:333–351.

Raftery, M. A., Hunkapiller, M. W., Strader, C. D., and Hood, L. E., 1980, Acetylcholine receptor: Complex of homologous subunit. *Science* **208**:1454–1457.

Reinhardt, R., Lindemann, B., and Anner, B. M., 1984, Leakage-channel conductance of single (Na^+,K^+)-ATPase molecules incorporated into planar bilayers by fusion of liposomes. *Biochim. Biophys. Acta.* **779**:147–150.

Reuter, H., 1983, Calcium channel modulation by neurotransmitters, enzymes and drugs. *Nature (London).* **301**:569–574.

Rosenberg, R. L., Hess, P., Reeves, J. P., Smilowitz, M., and Tsien, R. W., 1986, Calcium channels in planar lipid bilayers: Insights into mechanisms of ion permeation and gating. *Science* **231**:1564–1566.

Sakmann, B., Patlak, J., and Neher, E., 1980, Single acetylcholine-activated channels show burst-kinetics in presence of desensitizing concentrations of agonist, *Nature (London)* **286**:71–73.

Sariban-Sohraby, S., Latorre, R., Burg, M., Olans, L., and Benos, D., 1984, Amiloride-sensitive epithelial Na^+ channels reconstituted into planar lipid bilayer membranes. *Nature (London)* **308**:80–82.

Schein, S. J., Colombini, M., and Finkelstein, A., 1976, Reconstitution in planar lipid bilayers of a voltage-dependent anion-selective channel obtained from *Paramecium* mitochondria. *J. Membr. Biol.* **30**:99–120.

Schindler, M., and Quast, U., 1980, Functional acetylcholine receptor from *Torpedo marmorata* in planar membranes. *Proc. Natl. Acad. Sci. U.S.A.* **77**:3052–3056.

Schubert, D., Bleuel, H., Domning, B., and Wiedner, G., 1977, Protein-induced conductivity changes in black lipid membranes and protein aggregation. *FEBS Lett.* **74**:47–49.

Schuerholz, T., and Schindler, H., 1983, Formation of lipid–protein bilayers by micropipette guided contact of two monolayers. *FEBS Lett.* **152**:187–190.

Schulman, H., 1984, Phosphorylation of microtubule-associated proteins by a Ca^{2+}/calmodulin-dependent protein kinase. *J. Cell Biol.* **99**:11–19.

Shamoo, A. E., and Tivol, W. F., 1980, Criteria for the reconstitution of ion transport systems. *Curr. Top. Membr. Transport* **14**:57.

Sigworth, F. J., and Neher, E., 1980, Single Na^+ channel currents observed in cultured rat muscle cells. *Nature (London)* **287**:447–449.

Skou, J. C., 1957, The influence of some cations on an adenosine triphosphatase from peripheral nerves. *Biochim. Biophys. Acta* **23**:394–401.

Smith, J. S., Coronado, R., and Meissner, G., 1985, Sarcoplasmic reticulum contains adenine nucleotide-activated calcium channels. *Nature (London)* **316**:446–449.

Soifer, D., and Czosnek, H., 1980, The possible origin of neuronal plasma membrane tubulin. In *Microtubules and Microtubule Inhibitors* (M. DeBrabander and J. DeMey, eds.), pp. 429–447, Janssen Research Foundation, Elsevier/North-Holland Biomedical Press, Amsterdam.

Suarez-Isla, A., Wan, K., Lindstrom, J., and Montal, M., 1983, Single-channel recordings from purified acetylcholine receptors reconstituted in bilayers formed at the tip of patch pipets. *Biochemistry* **22**:2319–2323.

Tancrede, P., Paquin, P., Houle, A., and Leblanc, R. M., 1983, Formation of asymmetrical planar BLM from characterized monolayers. *J. Biochem. Biophys. Methods* **7**:299–310.

Tank, D. W., Miller, C., and Webb, W. W., 1982, Isolated patch recording from liposomes containing functionally reconstituted chloride channels from *Torpedo electroplax*. *Proc. Natl. Acad. Sci. U.S.A.* **79**:7749–7753.

Tank, D. W., Huganir, R. L., Greengard, P., and Webb, W. W., 1983, Patch-recorded single-channel currents of the purified and reconstituted *Torpedo* acetylcholine receptor. *Proc. Natl. Acad. Sci. U.S.A.* **80:**5129–5133.

Tien, H. T., 1974, *Bilayer Lipid Membranes (BLM): Theory and Practice,* Marcel Dekker, New York.

Tien, H. T., 1988a, Bilayer lipid membrane in aqueous media. In *Thin Liquid Films* (I. Ivanov, ed.), Marcel Dekker, New Work.

Tien, H. T., 1988b, Membrane bioenergetics as viewed from reconstitution experiments. In *Redox Chemistry and Interfacial Behavior of Biological Molecules* (G. Dryhurst, ed.), Plenum Press, New York.

Tomlins, B., Williams, A. J., and Montgomery, R. A. P., 1984, The characterization of a monovalent cation-selective channel of mammalian cardiac muscle sarcoplasmic reticulum. *J. Membr. Biol.* **80:**191–199.

Tosteson, M. T., and Sapirstein, V. S., 1981, Protein interactions with lipid bilayers: The channels of kidney plasma membrane proteolipids. *J. Membr. Biol.* **63:**77–84.

Tsien, R. W., 1983, Calcium channels in excitable cell membranes. *Annu. Rev. Physiol.* **45:**341–358.

Tsong, T. Y., 1983, Voltage modulation of membrane permeability and energy utilization in cells. *Biosci. Rep.* **3:**487–505.

Vassilev, P. M., and Hume, J. R., 1986, Calcium channels from cardiac sarcoplasmic reticulum incorporated into bilayer membranes. *J. Gen. Physiol.* **88:**60a.

Vassilev, P. M., and Tien, H. T., 1985, Planar lipid bilayers in relation to biomembranes. In *Structure and Properties of Cell Membranes* (G. Benga, ed.), pp. 63–101, CRC Press, Boca Raton, FL.

Vassilev, P., Kanazirska, M., and Tien, H. T., 1985, Intermembrane linkage mediated by tubulin. *Biochem. Biophys. Res. Commun.* **126:**559–565.

Vassilev, P. M., Hadley, R. W., Lee, K. S., and Hume, J. R., 1986a, Voltage-dependent action of tetrodotoxin in mammalian cardiac myocytes. *Am. J. Physiol.* **251:**H475–H480.

Vassilev, P. M., Kanazirska, M., and Tien, H. T., 1986b, Calcium channels from intracellular membranes reconstituted in patch-clamped bilayers. *Biophys. J.* **49:**349a.

Vassilev, P., Kanazirska, M., and Tien, H. T., 1986c, Microtubule-dependent membrane interactions studied in two types of double bilayer membrane systems. *Bioelectrochem. Bioenerg.* **15:**395–406.

Vassilev, P. M., Kanazirska, M. P., and Tien, H. T., 1987, Ca^{2+} channels from brain microsomal membranes reconstituted in patch-clamped bilayers. *Biochim. Biophys. Acta* **897:**324–330.

Vergara, C., and Latorre, R., 1983, Kinetics of Ca^{2+}-activated K^+ channels from rabbit muscle incorporated into planar bilayers. Evidence for a Ca^{2+} and Ba^{2+} blockade. *J. Gen. Physiol.* **82:**543–568.

Vergara, C., Moczydlowski, E., and Latorre, R., 1984, Conduction, blockade and gating in a Ca^{2+}-activated K^+ channel incorporated into planar lipid bilayers. *Biophys. J.* **45:**73–76.

Vodyanoy, V., and Murphy, R. B., 1983, Single-channel fluctuations in bimolecular lipid membranes induced by rat olfactory epithelial homogenates. *Science* **220:**717.

Vodyanoy, V., Halverson, P., and Murphy, R. B., 1982, Hydrostatic stabilization of solvent-free lipid bimolecular membranes. *J. Colloid Interface Sci.* **88:**227.

White, M. M., and Miller, C., 1979, A voltage-gated anion channel from the electric organ of *Torpedo californica. J. Biol. Chem.* **254:**10161–10166.

Worley, J. F. III, French, R. J., and Krueger, B. K., 1986, Trimethyloxonium modification of single batrachotoxin-activated sodium channels in planar bilayers. Changes in unit conductance and in block by saxitoxin and calcium. *J. Gen. Physiol.* **87:**327–249.

Yamamoto, D., Yeh, J. Z., and Narahashi, T., 1984, Voltage-dependent calcium block of normal and tetramethrin-modified single sodium channels. *Biophys. J.* **45:**337–344.

Chapter 4

Fluorescence Studies of Membrane Dynamics and Heterogeneity

Lesley Davenport, Jay R. Knutson, and Ludwig Brand

1. INTRODUCTION

The cellular morphology of lipid bilayer membranes acting as dynamic boundaries is well established. Their biological function, however, is determined at the molecular level. Thus, techniques sensitive to molecular conformation are required to understand how membranes work.

A variety of biophysical methods have led to the current picture of the time-averaged structure of the bilayer (Singer and Nicolson, 1972). For example, translational and rotational diffusion studies have provided evidence for the "fluidity" of phospholipid bilayers (Cherry, 1979; Devaux and McConnell, 1972; Frye and Edidin, 1970; Jacobson *et al.*, 1982; McConnell *et al.*, 1972; Peters, 1981; Scandella *et al.*, 1972; Shinitzky *et al.*, 1971; Trauble and Sackmann, 1972; Webb *et al.*, 1981; Wolf and Edidin, 1981). This fluidity arises from the flexibility of fatty acyl chains, a property that has been extensively studied and documented both in biological membranes and model bilayer systems (Hubbel and McConnell, 1968; Jost *et al.*, 1971a; Levine and Wilkins, 1971; Luzzati, 1968; Phillips *et al.*, 1969; Stubbs *et al.*, 1981). Like their

---

Lesley Davenport   Department of Chemistry, Brooklyn College of the City University of New York, Brooklyn, New York 11210.   **Jay R. Knutson**   Laboratory of Technical Development, National Institutes of Health, Bethesda, Maryland 20892.   **Ludwig Brand**   Biology Department and McCollum–Pratt Institute, Johns Hopkins University, Baltimore, Maryland 21218.

parent alkanes, these acyl chains may exist in rigid (gel) or fluid (liquid-crystalline) phases. Calorimetry, EPR, NMR, and fluorescence spectroscopy have provided detailed information about the phase transitions in homogeneous liposomes as well as in bilayers containing mixed lipids or proteins.

Previous work from many laboratories (Bretscher, 1972; Casper and Kirshner, 1971; Giraud et al., 1981; Michaelson et al., 1973) indicates that the different lipids and proteins preferentially partition between the inner and outer leaflet of the membrane (Fong and Brown, 1978; Kates and Wassef, 1970; Michaelson et al., 1973; Rouser et al., 1968; Schroeder, 1985), thus rendering it asymmetric. Examples include red blood cell membranes (Bretscher, 1972; Morrot et al., 1986), viruses (Fong et al., 1976; Rothman et al., 1976), mitochondrial membranes (Getz, 1972), E. coli membranes (Cronan and Vagelos, 1972), and other sources (Rothman and Lenard, 1977). The asymmetric organization of phospholipids is of physiological importance (Haest et al., 1972; Op den Kamp et al., 1969, 1971; Steck and Fox, 1972; Tanaka and Ohnishi, 1976). It may provide the correct environment for enzymes that are asymmetrically located (Douce et al., 1973; Fressenden-Raden and Racker, 1971; Nicolson, 1972). In addition to inner/outer leaflet asymmetry, proteins often show a marked preference for activation by certain phospholipids (Kimelberg and Papahadojopoulos, 1972). This suggests a nonrandom (heterogeneous) organization of lipids within the plane of the membrane (Curatolo et al., 1977; Jain, 1983; Jain and White, 1977; Marsh and Barrantes, 1978). Such ordering might arise from favorable constraints of packing or specific interactions between membrane components. These interactions also provide phase separations (Karnovsky et al., 1982; Klausner and Wolf, 1980; Klausner et al., 1980a,b; Linden et al., 1973; Shimshick and McConnell, 1973a; Wolf et al., 1981a,b; Wu and McConnell, 1973; Yechiel and Edidin, 1987; Yechiel et al., 1986), patching (Betel and van den Berg, 1972; Bretscher, 1976; Edidin and Weiss, 1972; Taylor et al., 1971; Yahara and Edelman, 1972), phase boundaries (Jost et al., 1971b; Lee, 1977), multiple relaxations (Eck and Holzwarth, 1984; Gruenwald et al., 1981; Holzwarth et al., 1985; Kanehisa and Tsong, 1978; Mitaku and Date, 1982; Tsong, 1977), and anomalous temperature dependencies for membrane processes (Albrecht et al., 1978; Hawton and Boane, 1987; Mitaku et al., 1982, 1983; Nagle and Scott, 1978; Mitaku, 1981).

Lateral separations can even occur in a monolayer comprised of chemically identical molecules. Conformational differences (e.g., the number of kinks in acyl chains) lead to functional packing differences between molecules. These may provide energetic advantages for quiltlike segregations. Whether lateral separation occurs via these subtle packing differences or by more obvious chemical immiscibility, we refer to the separated regions as *domains*. Although the term has been used in many ways, some common features emerge. Implicit in the definition is the *identifiability* of the different phases; for example, some feature such as probe affinity, lipid chain length, or mixture composition must differ between the phases in a measurable fashion. The individual domains, however, should be effectively homogeneous by the same criteria. Moreover, the changes between domains as one moves around the monolayer should be relatively abrupt; that is, the composition (or other features described above) should change in a distance smaller than the domains themselves.

The very existence of domains in pure lipid systems is controversial, but the controversy usually involves differences in *definition*. For example, during vesicle melting, we refer to "gel" and "fluid" domains. A purist might suggest that "pure gel" (all trans) regions and "pure fluid" (highly kinked, as if at high temperatures) are not likely to coexist. "Gel" regions will probably contain a few kinked chains and "fluid" regions are usually populated with some moderately ordered chains. Nevertheless, a *functional* separation can occur, dependent on the measured parameter. To illustrate: a fluorescent probe might be effectively prevented from motion in any environment where at most one (near neighbor) lipid chain is kinked, and we would detect "gel." The same probe in any setting where two or more neighbors are multiply kinked would tumble more freely, and we would group all these environments under a single heading: "fluid." We are certainly aware that this categorization is incomplete. Inescapably, a definition of "domain" will be dependent on the type of measurement. (We stress this point again when considering the *duration* of domains.)

The potential for nonrandom lateral organization has been recognized for some time, especially for chemically heterogeneous systems. Domains are of important potential significance in biology. The aim of this chapter is to examine the concept of local lipid heterogeneity or "domain" formation within the phospholipid bilayer. We place special emphasis on the insights contributed to these questions by new associative methods in fluorescence spectroscopy. We ask: Do lipid domains exist? What is the lifetime and the size of domains? What induces domain formation and what are the biological implications of membrane heterogeneity?

2. DO DOMAINS EXIST?

Lipid heterogeneity (seen as lateral phase separations in binary mixtures of synthetic phosphatidylcholines) has previously been reported using X-ray diffraction (Engelman, 1972; Levine and Wilkins, 1971), electron microscopy (Hui, 1981; Hui and Parsons, 1975), differential scanning calorimetry (Mabrey and Sturtevant, 1976), EPR spectroscopy (Shimshick and McConnell, 1973b), and fluorescence spectroscopy (Klausner *et al.*, 1981a; Lentz *et al.* 1976a,b) techniques. Such lipid heterogeneity can be defined as a macroscopic separation of bilayer components and may be directly visualized. Indeed, electron microscopy provides the best evidence for these macroscopic lipid domains and the most accurate size, estimated to vary between 0.1 and 0.5 μm (Hui, 1981). Such lateral separations have been demonstrated for both liquid–liquid and liquid–gel mixed systems (Phillips et al., 1970; Shimshick and McConnell, 1973b).

These observations have been correlated with biological function (see Section 2.2.1). Domains have been implicated in fusion processes such as the acrosomal reaction (Bearer and Friend 1982; Wolf, 1986). The "immobile" lipid fraction observed by Wolf (1984) for sperm may result from gel domains. Large-size domains have been visualized by selective probes (Waggoner, 1986; Yechiel and Edidin, 1987). Lateral lipid phase segregations can effectively subdivide the membrane according to differential solubility of proteins and small molecules in different parts of the bilayer.

Another known cause of membrane subdivision is linkages to the cytoskeleton and exoskeleton. Such lipid immobilization has been reported for the red blood cell membrane (Cherry, 1979; Cherry et al., 1980; Golan and Veatch, 1980; Lux, 1979; Williamson et al., 1982; Bloom and Webb, 1983). The distribution of lipids into domains or clusters has been inferred from freeze–fracture electron microscopy of membranes treated with filipin, which interacts with cholesterol (Bittman, 1978; Elias et al., 1979), and polymyxin B, which interacts with anionic phospholipids (Bearer and Friend, 1982; Imai et al., 1975).

As mentioned before, the scale of domains depends on definition, which in turn relies on the type of measurement used. Large-scale differentiation of the lipid surface can be visualized directly, as well as by probes, but the converse need not hold true. For example, microscopic domains may provide environmental heterogeneity for fluorophores, spin probes, and other labels that do not extend over micron distances. Differential scanning calorimetry has also been useful in detecting the stepwise melting characteristic of microscopic heterogeneity, but it lacks specificity; for example, the *total* heat capacity (a mixed sum of the parts) is seen and dissection may be difficult. Using probe methods, acyl chains in erythrocytes also appear to undergo a phase change (Cullis and de Kruijff, 1979; Galla and Luisetti, 1980), suggesting a heterogeneous lipid distribution. Similarly, phase changes are believed to occur in *Tetrahymena pyriformis* (Wunderlich et al., 1975), in *E. coli* (Baldassare et al., 1976; Yang et al., 1979), in the sarcoplasmic reticulum (Davis et al., 1976), in *Bacillus stearothermophilus* (McElhaney and Souza, 1976), and in mitochondria (Raison et al., 1971). None of these transitions corresponds to the (smeared out) thermal transition observed by DSC, suggesting that these transitions are representative of selected regions only.

The state of the molecular packing in phospholipid bilayers can be affected by the lipid component(s), the temperature, ionic composition of the aqueous environment, and membrane modulators such as cholesterol. We group the transitions between these packing states into broad *thermal* and *chemical* categories.

2.1. Thermal Effects

Thermal transitions between coexisting (gel–fluid) phases in model membranes containing either single lipid types or mixtures have been studied in detail using calorimetry, fluorescence, and spin-label techniques. In the following, we focus on the contributions of fluorescence spectroscopy, although the body of knowledge from other methods is used to frame the problem. In particular, differential scanning calorimetry has stimulated a theoretical concept of bilayer lattices and the energy associated with chain packing. The endothermic, gel-to-fluid phase transition of pure (one-component) uncharged lipid bilayers appears as a broad continuous melt. Well below the transition, the (all-trans) bilayer is predominantly in a "gel" (dense, lateral packing) state with smaller patches of "fluid" (more loosely packed due to acyl kinks) scattered throughout. The number and size of fluid patches increase at the expense of gel, as temperature increases, with approximately equal portions present at transition temperature (T_c).

An obvious consequence of a lipid domain model is environmental heterogeneity within bilayers. Fluorescence spectroscopy has been used widely to investigate the induced membrane phases. The differential partitioning of probes into known "fluid" and "gel" regions (with altered fluorescence intensity) has been used to detect phase separation. Some fluorescent lipid probes selectively partition into different lipid domains. Membrane potential probes (whose intensities are known to be sensitive to environment) have been useful in this regard. Merocyanin 540 (Lelkes *et al.*, 1980; Sieber, 1987; Williamson *et al.*, 1982, 1983) is reported to partition preferentially into gel, whereas certain diacyl-3,3'-indocarbocyanine iodides ($C_n$diI, where $n = 10, 12, 20, 22$) (Klausner and Wolf, 1980; Wolf *et al.*, 1981b) prefer a more fluid environment. Most of these emission intensity-based studies used mixed lipid systems, but some of the probes have shown selective partitioning in single-component systems. For example, simple aromatic probes have often been seen to exhibit changes in yield through the phase transition.

2.1.1. Fluorescence Lifetime Studies

In addition to changes in intensity, probes used in these studies may express altered parameters such as fluorescence lifetime (τ) or rotational correlation time (ϕ) when located in different regions of the bilayer. The polyene probes such as 1,6-diphenyl-1,3,5-hexatriene (DPH) (Andrich and Vanderkooi, 1976; Heyn, 1979; Hildenbrand and Nicolau, 1979; Jahnig, 1979; Kawato *et al.*, 1977; Kinosita *et al.*, 1982; Klausner *et al.*, 1980a), 1,6-diphenyl-1,3,5,7-octatetraene (DPO), and isomers of octadecanoic acid (*cis-* and *trans*-parinaric acid) (Sklar *et al.*, 1975; Wolber and Hudson, 1981, 1982) have been widely used for membrane heterogeneity studies. Several groups (Florine-Casteel and Feigenson, 1988; Lentz *et al.*, 1976b) have reported equal partitioning of DPH into both fluid and gel regions of the lipid bilayer. The cis isomer of parinaric acid partitions nearly equally between the gel and fluid phases while the trans isomer partitions preferentially in the solid phase. In addition, the decay from the excited state depends on the type of environment in which the probe resides (Lakowicz, 1983). Some groups working with DPH in pure lipid systems have reported a monoexponential fluorescence decay (Kawato *et al.*, 1977; Klausner *et al.*, 1980a), whereas in mixed lipid systems, multiexponential decay is found. Furthermore, shorter lifetimes are reported for those DPH molecules imbedded in the fluid phase whereas in mixed lipid systems (e.g., vesicles composed of DLPC and DPPC at 25°C), it was found that the fluorescence decay was best fit using three exponential terms. The third short lifetime (2–3 nsec) was thought to represent the interface between domains, where DPH is quenched due to the binding of a small amount of water in this boundary. In disagreement with the above assignments, multiexponential decay was previously found by Dale *et al.* (1977), Chen *et al.* (1977), and Davenport *et al.* (1986b) in *single-component* systems. Recently, Parasassi *et al.* (1984) and Barrow and Lentz (1985) using newer multifrequency phase instruments, also found multiexponential decays in single-component systems. They reconcile the short lifetime as a photoproduct, possibly obtained via cis–trans isomerization. Distributed decay rates have also been suggested for DPH (Fiorini *et al.*, 1987; Parasassi *et al.*, 1987). Whatever the true nature of the short component, the heterogeneous lipid environment presents a complicated picture for the location of DPH within the bilayer matrix.

2.1.2. Fluorescence Polarization Studies

Time-dependent and steady-state fluorescence depolarization methods (Chuang and Eisenthal, 1972; Fleming *et al.*, 1977; Perrin, 1934, 1936; Zannoni, 1979) have become important tools in the investigation of anisotropic and/or restricted rotational diffusion of probes within membrane vesicles (Andrich and Vanderkooi, 1976; Kawato *et al.*, 1977; Lakowicz *et al.*, 1979; Shinitzky and Barenholz, 1974; Tao, 1969). Measurements have been performed on a variety of biological and model bilayer systems, often with the aim of studying the lipid organization and fluidity as well as the response to external agents, for example, cholesterol and anesthetics (Kawato *et al.*, 1978; Veatch and Stryer, 1977). The time-dependent anisotropy, $r(t)$, provides a direct measurement of reorientational motion on the 10^{-10} to 10^{-7} sec time scale. Anisotropic (e.g., nonspherical) motion is revealed as multiple exponential components in the decay of $r(t)$. Evidence for anisotropic rotational behavior of probes in bilayer membranes has been obtained by many workers (Brand et al., 1985; Dale et al., 1977; Davenport et al., 1986b,c; Glatz, 1978; Hildenbrand and Nicolau, 1979; Kawato et al., 1977; Kowalczyk et al., 1982; Knutson et al., 1981; Lakowicz and Knutson, 1980; Lakowicz et al., 1985; Lakowicz et al., 1979; Parola et al., 1979; Sene et al., 1978; Wolber and Hudson, 1982). Restriction of the range of reorientational motion is revealed as a constant asymptotic value of $r(t)$, that is, r_∞. The origins of r_∞ are not well understood and may arise from a homogeneous distribution of probes with restricted relaxation (Heyn, 1979; Jahnig, 1979; Kawato et al., 1977; Lakowicz et al., 1979; Lipari and Szabo, 1980; Zannoni et al., 1983) or, alternatively, from subensembles of probes located in gel or fluid domains (heterogeneous model; Chong and Weber, 1983; Dale et al., 1977; Davenport et al., 1986b,c; Knutson et al., 1986). Whatever its origins, the final anisotropy value (r_∞) furnishes information on the structural order of the acyl chain region of the phospholipid bilayers, while the fast correlation time (ϕ) provides information on kinetic properties such as microviscosity (Heyn, 1979; Jahnig, 1979; Kawato et al., 1977; Marcelja, 1974). However, this order is only relevant to the time scale over which the averaging occurs. Order parameters for lipid molecules in the bilayer membranes have been determined using a number of magnetic resonance methods, such as ESR (Hubbell and McConnell, 1971; Jost et al., 1971a), proton (^{1}H) (Seiter and Chan, 1973), and deuterium (^{2}H) NMR (Seelig and Seelig, 1974; Stockton et al., 1976). Corresponding order parameters (S_2) have been determined more recently via Raman spectroscopy (Mendelson et al., 1976) as well as polarized time-resolved fluorescence measurements (Chen et al., 1977; Heyn, 1979; Jahnig, 1979; Kawato et al., 1977; Wolber and Hudson, 1981). The traditional order parameter is defined by $S^2 = R_\infty/R_0$. Typically, the values of the order parameters determined by the various methods do not agree.

Much work has focused on studies using the polyene probes. Certain planar anisotropic probes (such as the anthroyloxy fatty acids; Thulborn, 1981; Thulborn and Sawyer, 1978; Vincent et al., 1982) with fluorescence moieties located at various points along the fatty acyl chain report "fluidity gradients." Knowledge of the location of the reporter group certainly provides an advantage over DPH, whose distribution across the bilayer is reported to alter with temperature (Davenport et al., 1985). In fact, trimethylamino-DPH (TMA-DPH), anchored by a polar moiety, has found favor for membrane studies, since it is restricted to the polar head group region (Cundall et al., 1979).

For years, a seemingly contradictory pair of concepts has coexisted. Patchwork domain structures strongly imply *heterogeneity*, while most membrane probe ordering models—wobbling within cones (Kawato et al., 1977) or other potentials—assume *homogeneity* of probe environment. Truly "model-free" approaches such as Lipari and Szabo (1980) or Zannoni *et al.* (1983) need not invoke homogeneous models, but the diffusion models that accompany them usually describe a "mean" (effectively homogeneous) environment. Remanent anisotropy (r_∞) was recognized by Dale *et al.* (1977) as a possible consequence of either type of ordering. Heterogeneity was considered, however, when interpreting the partitioning of anthroyloxy fatty acid probes (Thulborn and Sawyer, 1978). It was recognized that steady-state anisotropy can be considered a mixture of gel and fluid contributions, and that the partitioning and intensity characteristics of probes in the different phases will alter the *apparent* mean order (e.g., shifting melt curves) (de Kruijff *et al.*, 1974; Lentz *et al.*, 1976a,b; Sklar *et al.*, 1977; Thulborn, 1981). As will be discussed later, the *lifetime* of the probe can also play a role in selecting which portion of a melt profile is observed, so partitioning is not the sole answer for shifted apparent lipid phase transitions.

2.1.3. Anisotropy Decay-Associated Spectra

In this regard, it is pertinent to note that several different types of fluorescence information can be obtained from a particular system, such as excitation or emission spectra, quantum yields, lifetimes, and susceptibility to quenching (or other excited-state interactions). These different types of measurements can be *associated* (Knutson *et al.*, 1982) to provide more information about the system than is available from individual experiments (Figure 1). Complex fluorescence decay kinetics of tryptophan containing proteins have been analyzed using the association between lifetime and spectral envelope. These are called decay-associated spectra (DAS). In our studies of lipid heterogeneity, we found it appropriate to associate *rotation rates* with unique spectral distributions. These we have called anisotropy decay-associated spectra (ADAS) (Brand *et al.*, 1985; Davenport *et al.*, 1982, 1986b; Knutson *et al.*, 1986). We have used this associative technique to provide an explanation for the r_∞ observed for DPH in the context of microheterogeneity. While a more comprehensive description of the theory and methodology of ADAS has been presented elsewhere (Davenport *et al.*, 1986b; Knutson *et al.*, 1986), it is relevant to overview the technique here. The emission anisotropy, r, can be defined as

$$r(t) = \frac{\sum d_i(t) r_i(t) \alpha_i(\lambda)}{\sum \alpha_i(\lambda) d_i(t)} = \frac{y(\lambda, t)}{I_{\text{total}}(\lambda, t)} \qquad (1)$$

where $y = (I_{\text{parallel}} - I_{\text{perpendicular}})$ is the polarized difference decay and d_i is the total decay function, the α_i are spectra of each rotating species, and r_i are their corresponding anisotropy decays; for example,

$$r_i = r_{0,i} \exp\left(\frac{-t}{\phi_i}\right) \qquad (2)$$

Membrane Dynamics and Heterogeneity

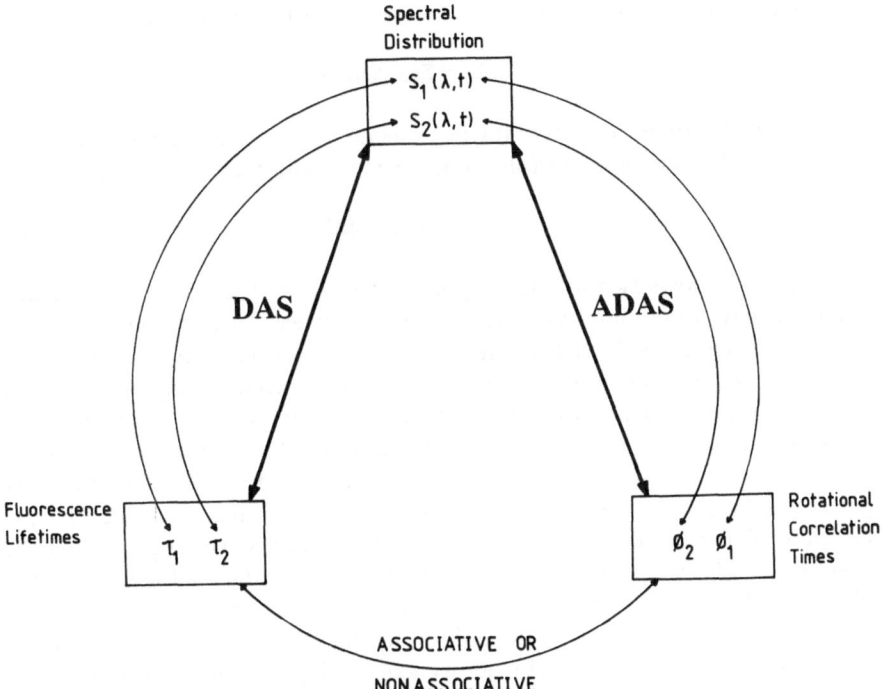

FIGURE 1. For a mixture of two fluorophores with different fluorescence lifetimes (τ), the mixture may be resolved into the component spectral distributions $S_n(\lambda/t)$ using decay-associated spectra (DAS) techniques. Similarly, heterogeneous mixtures may be resolved spectrally based on differences in their rotational characteristics (anisotropy decay-associated spectra—ADAS). For certain systems there may be a $\tau \rightarrow \varphi$ linkage (the associative model).

where $r_{0,i}$ and ϕ_i are the initial anisotropy and rotational correlation of the ith species, respectively. For the simple, nonassociative case,* where we have slowly rotating (large, bound, and immobile) and rapidly rotating (small, free, and mobile) probe fractions coexisting in the sample, the total intensity decay will be wavelength independent.

$$I_{\text{total}}(\lambda, t) = [\sum_i \alpha_i(\lambda)] d(t) \qquad (3)$$

---

*To avoid confusion, we should note that "associative" in this context refers to linkage between ϕ and τ components. Spectra associated with either τ or ϕ can be obtained whether or not ϕ_i and τ_i pair up, but for this section, we consider ϕ_i independent of τ_i; for example, the decay rates τ_i do not differ among the different rotating species ϕ_i.

The impulse decay of the *difference*, however, will be heterogeneous:

$$y(\lambda,t) = \Sigma y_i = \Sigma \alpha_i(\lambda) d(t) r_i(t) \qquad (4)$$

The observed difference decay, for instruments with a finite response time, is distorted by the convolution of the time response with the lamp function L:

$$Y_i(t) = \int L(t') y(\lambda, t - t') dt' \qquad (5)$$

Figure 2 shows two different Y_i decay functions. The division of the difference decay into "time windows" provides an easy way to obtain ADAS (spectra associated with each rotating species). Experimentally, an anisotropy decay analysis from one (or a number) of emission wavelengths provides r_i and d. This gives y_i and Y_i. Knowing Y_i functions, it is possible to predict how much of each species is present in a windowed TRES. For the simple case of a two-component system shown in Figure 2, almost no mobile contribution is present in the late window, so α_{slow} is just a multiple of that difference TRES. For pulse fluorometers, this is the easiest ADAS to observe. If ϕ_{fast} is small, even the *steady-state* difference spectrum (weighted by the total area under the difference decay curves) will provide a good approximation to α_{slow}.

We have used ADAS methodology to investigate rotationally homogeneous (wobbling-in-a-cone) and heterogeneous interpretations of anisotropy decay. A homogeneous model implies an anisotropic rotator that may also exhibit complex intensity decay. There is no association between anisotropy and inten-

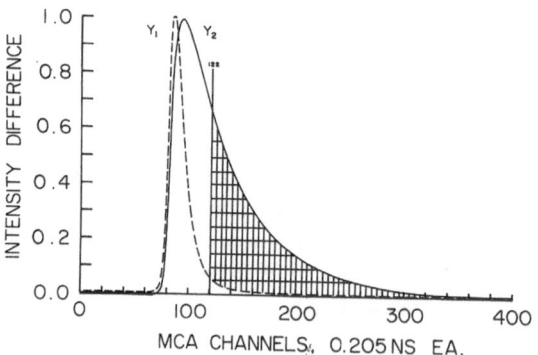

FIGURE 2. The convolved difference decays of a rapidly rotating Y_1 species and less mobile Y_2 species illustrate how a "late window" (shaded) difference spectrum will contain (predominantly) species 2 emission. The small amount of species 1 remaining can be removed from the spectrum by a matrix inversion method (see text). Reprinted with permission from the American Chemical Society (from Davenport *et al.*, 1986b) and the Royal Society of Chemistry (from Davenport *et al.*, 1986c).

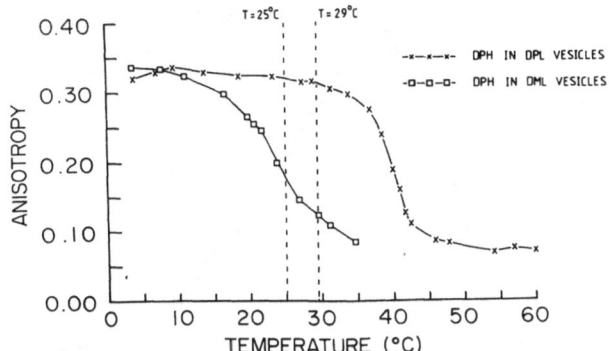

FIGURE 3. DPH steady-state emission anisotropies for DMPC and DPPC SUV bilayer melting curves, indicating temperatures where the mixture (29°C) would contain distinct gel and fluid components. If each melt *itself* represents an exchanging mixture, 25°C would similarly be a heterogeneous temperature for DMPC alone. Reprinted with permission from the American Chemical Society (from Davenport et al., 1986b).

sity decay terms. A heterogeneous model, on the other hand, does imply an association between anisotropy and intensity decay terms:

$$\text{heterogeneous model} \qquad \text{homogeneous model}$$
$$y = \Sigma \alpha_i r_i d_i \qquad y = \left(\Sigma \alpha_j d_j \right) \left(\Sigma \beta_i e^{-t/\phi_i} \right) \qquad (6)$$

Wavelength dependence of the β_i term necessitates the ADAS approach. A homogeneous distribution of probe rotations would not predict spectral ties to mobility.

As a test system, we first incorporated DPH into a 1:1 mixture of DMPC and DPPC vesicles. At 29°C, DMPC vesicles are in the fluid phase, whereas DPPC is below its phase transition (Figure 3). This creates a *known* rotationally heterogeneous environment for DPH. The DPPC-DMPC vesicle mixture provided a biphasic phospholipid phase transition [measuring the steady-state emission anisotropy ($<r>$) of DPH as a function of increasing temperature]. This suggested that minimal fusion between single bilayer vesicles occurred during the time course of the experiments. In contrast, a cosonicated phospholipid sample exhibited a very broad transition with a combined center at ~32°C.

Steady-state emission spectra for DPH in the two separate vesicle samples are shown in Figure 4. There are subtle vibrational shifts seen between these two vesicle samples. The DPPC-DMPC mixture, as expected, provides an emission spectrum that lies between the two component spectra.

The decay of the emission anisotropy for DPH incorporated into the mixed DMPC-DPPC vesicle sample was best expressed in terms of one correlation

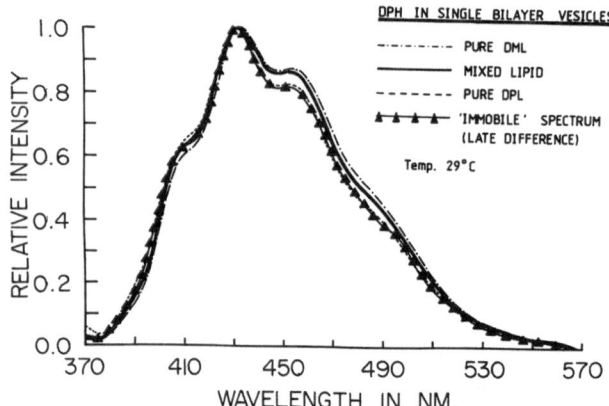

FIGURE 4. Anisotropy decay-associated spectrum for DPH-labeled 1 : 1 mixture of DMPC and DPPC vesicle preparations at 29°C. The steady-state emission spectra for the individual (- - -) DPPC and (-·-·-) DMPC vesicles and for the 1 : 1 mixture (——) are shown. The anisotropy decay-associated spectrum (▲·▲·▲) corresponding to the immobile probe fraction shows superposition with the DPPC steady-state spectrum. Excitation was at 355 nm, with excitation and emission bandwidths of 14 and 9.9 nm, respectively. Reprinted with permission from the American Chemical Society (from Davenport et al., 1986b) and the Royal Society of Chemistry (from Davenport et al., 1986c).

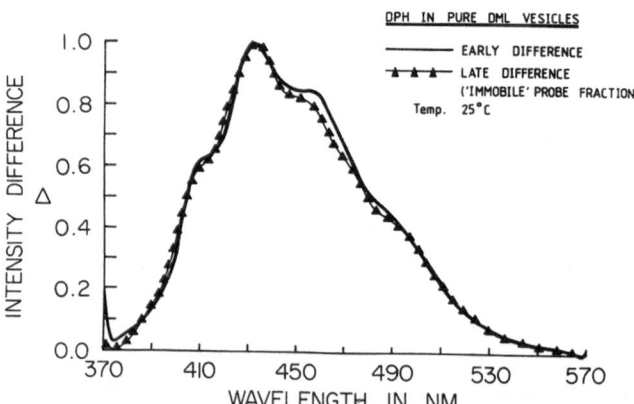

FIGURE 5. Anisotropy decay-associated spectrum (ADAS) for DPH-labeled DMPC (1 : 500 probe to lipid labeling ratio) SUVs at 25°C. the immobile ADAS corresponds to the "late" (-▲-▲-) difference spectra while the "early" (——) difference spectra contain both gel and fluid contributions. Excitation was at 355 nm, with excitation and emission bandwidths of 14 and 9.9 nm, respectively. Reprinted with permission from the American Chemical Society (from Davenport et al., 1986b) and the Royal Society of Chemistry (from Davenport et al., 1986c).

time plus a constant or remanent anisotropy (r_∞) term. A late window was chosen such that the difference decay (Y) associated with the more mobile fraction was easily excluded.

As is shown in Figure 4, the expected immobile (DPPC) signature was recovered. Having established a precedent for ADAS signatures differing between *known* gel and fluid environments, we examined an unknown system, namely, single-component (DMPC) vesicles near T_c. Again, the hallmarks of distinct mobile and immobile environments were apparent (Figure 5). We now find it difficult to reconcile models such as "wobbling-in-a-cone" with our observations, at least as those models are presently constituted. We recognize that the rotational heterogeneity may be due to more than two environments, or even a continuum or ordering forces. In any case, rotation models must now recognize the coexistence of more mobile versus less mobile phases represented by *different* spectra. The vibrational character of polyene probe emission has been shown to vary strongly with site polarity; hence, we expect that these probe environments correlate with different depth levels and/or water accessibilities. The changes in DPH depth distribution between gel and fluid were demonstrated previously (Davenport *et al.*, 1985). We may reasonably expect that any force redistributing DPH between phases will induce the same small spectral shifts. Hydrostatic pressure, for example, was recognized to do this by Chong and Weber (1983).

2.2. Chemically Induced Domain Formation

While we have been discussing microheterogeneity which invokes the concepts of *microclusters* of lipid molecules, most reports of and about domains involve *macroscopic* (>1 μm) structure and separations. As previously mentioned, these patches of lipid are readily viewed using electron microscopy. Such domain structure can be induced chemically, either by external cations or by the inclusion of cholesterol.

2.2.1. Extrinsic Membrane Effectors of Lipid Heterogeneity

The possibility of a chemically induced lipid phase separation has been clearly demonstrated using model membrane systems (Galla and Sackmann, 1975; Hartmann *et al.*, 1977; Hoekstra, 1982a; Hui *et al.*, 1983; Lai *et al.*, 1985). Extrinsic and intrinsic proteins are believed to be responsible in part for membrane lipid heterogeneity (Griffith *et al.*, 1973). Weakly associated extrinsic proteins, such as the positively charged lysozyme or cytochrome, show considerable effect on the bilayer packing, resulting in enhanced Na$^+$ permeability when associated with negatively charged phosphatidylserine vesicles (Kimelberg and Papahadjopoulos, 1971).

Dramatic effects in lipid distribution have been observed with the synthetic peptide polylysine and phosphatidylcholine bilayers containing negatively charged phosphatidic acid (Graham et al., 1985; Kouaouci et al., 1985; Liao and Prestegard, 1979, 1981). In addition, many studies have revealed that certain divalent cations and in particular Ca^{2+} have a profound effect on the induction of lipid phase separations (Florine and Feigenson, 1987a; Hoekstra, 1982a; Hui et al., 1983; Kouaouci et al., 1985). Indeed, Haverstick and Glaser (1987), using digital-image analysis, examined the size and distribution of Ca^{2+}-induced domains as a function of phospholipid composition in both vesicles and erythrocytes. This effect may have important biological relevance. Many membrane fusion mechanisms are dependent on the presence of Ca^{2+}. Such processes include release of neurotransmitters, hormones (Douglas, 1974), and macromolecular secretions (Lagunoff, 1973; Lawson et al., 1977).

The exact role of Ca^{2+} in promoting membrane fusion remains unclear (Volsky and Loyter, 1978); however, the effect of Ca^{2+} on the fusion of liposomes comprising acidic phospholipids such as phosphatidylglycerol, phosphatidylserine, cardiolipin, and phosphatidic acid or their mixtures with zwitterionic lipids is intimately related to the ability of this divalent cation to induce isothermal lipid phase changes (Duzgunes et al., 1984). This is proposed to result in a transient protein-free (Chi et al., 1976; Florine and Feigenson, 1987b; Kalderon and Gilula, 1979) destabilization of the membrane (Duzgunes et al., 1984; Hoekstra, 1982b; Ohki, 1984; Parente and Lentz, 1986; Wojcieszyn et al., 1983), local dehydration (Honda et al., 1981; Lucy and Ahkong, 1986), and ultimately fusion. Various models of the altered lipid packing of this destabilized state have been proposed (Hui et al., 1981; Lucy and Ahkong, 1986; Siegel, 1986). Reports of fusion through "point defects" in the bilayer may play a role in the susceptibility of two membranes to fuse (Hui et al., 1981; Papahadjopoulos et al., 1977). In addition, formation of dehydrated Ca^{2+}–phosphatidylserine complexes between apposed bilayers may be a key event in fusion initiation (Papahadjopoulos et al., 1978; Portis et al., 1979). Elegant experiments by Hoekstra (1982a), using labeled NBD and rhodamine-labeled fluorescent lipid analogues, have shown that macroscopic phase separations induced by Ca^{2+}, and to a lesser degree Mg^{2+}, may facilitate but do not induce fusion and thus are not directly involved in the actual fusion mechanism. While the mechanism of membrane fusion is still unresolved, point defects in the lipid packing may be critical initiators.

2.2.2. Cholesterol: An Intrinsic Effector of Lipid Heterogeneity

Many studies have been directed at understanding the precise nature of phase separations induced by cholesterol in lipid bilayers. The results of many of these studies appear contradictory. Apparent associations or "complexes" of cholesterol with phospholipid at ratios of 1:4, 1:2, or 1:1 have been suggested. Differential scanning calorimetry (DSC) (Mabrey *et al.*, 1978), freeze–fracture electron microscopy (Verkleij *et al.*, 1974), and ^{13}C-NMR (Opella *et al.*, 1976) experiments suggest a two-phase system, interpreted as a "cholesterol-free" phospholipid domain separated by a second phase of a 4:1 phospholipid–sterol complex (in bilayers containing less than 20 mol% cholesterol). A phase boundary at a cholesterol mole fraction of about 0.2 is detected. Interestingly, binary mixtures exhibit the characteristic phosphatidylcholine phase transition (only for cholesterol mole fractions less than 0.2).

In contrast with these studies, early calorimetric data (Ladbrooke *et al.*, 1968), X-ray diffraction experiments (Engleman and Rothman, 1972), and recent ^{13}C-NMR (Brulet and McConnell, 1977) studies report a phase boundary at the cholesterol mole fraction of 0.33. Above the boundary, a liquidlike mixed phase exists and below the boundary both an ordered gel phase of "pure" lecithin which completely disappears at around 33 mol% cholesterol and a mixed lecithin–cholesterol phase corresponding to almost 2:1 (from the model building calculations) exist. Haberkorn *et al.* (1977) have reported significant phase changes at both 0.2 and 0.33 mole fractions of cholesterol. Even at *very* low concentrations of cholesterol, heterogeneity within the bilayer has been reported (Melchior *et al.*, 1980). Self-quenching (a hallmark of self-association) of the fluorescence from the cholesterol analogue cholestatrienol in phosphatidylcholine bilayers is readily observed at sterol concentrations greater than 5% (Rogers *et al.*, 1979).

We became interested in the possibility of quantitating the extent of heterogeneity induced by the presence of cholesterol. Thus, our attention focused on lipid probes whose signals might provide us *more* than simple confirmations of domainlike heterogeneity. We have demonstrated that fluorescence anisotropy, through ADAS, can be used to detect such environmental heterogeneity within bilayers. Probes whose excited-state *lifetimes* can be directly altered by domain heterogeneity are potentially more useful, because they offer an easier route to the quantitation of domain structure. To this end, the use of excimer formation (via translational diffusion) seemed attractive, since the process of encounter will certainly be altered by sequestration of the probe, if any exists. The kinetics should reflect the size of any domain patches, as well as mobility within them. Previous studies using this kind of probe focused on complex *homogeneous* models and the mathematics of transient approach to diffusion equilibrium in two or three dimensions (Vanderkooi *et al.*, 1975). While we do not discount these effects, we are concerned with the possibility of much larger effects arising from *heterogeneity* and correlated immobilizations and/or "crowding."

To carry out sequestering studies, we desired a fluorescent probe that could mimic cholesterol in bilayers. To this aim, we employed the fluorescent cholesterol adduct, pyrenemethyl cholesterol (1-pyrene-methyl-3β-OH-22,23-bis-nor-5-cholenate; PMC) (Figure 6a). This analogue has many advantages over other cholesterol adducts. The fluorescent hydrophobic reporter group, a pyrene molecule, is attached through the 17 position on the cholesterol (rather than the usual 3' hydroxyl group). Modification of the 3' hydroxyl residue has been shown to alter positioning of the steroid within the bilayer (Shimshick and McConnell, 1973a; Worster and Franks, 1976).

Pyrene (Figure 6b) undergoes a well-studied, diffusion-controlled excimer formation (Birks, 1970). Excimer formation requires collision between two pyrene molecules, which in turn depends on membrane composition, temperature, and (especially) the effective concentration of the pyrene. Due to the relatively long lifetime of pyrene, probe molecules can diffuse large distances within the two-dimensional bilayer matrix while in the excited state. Thus, they may form excimers at extremely low probe concentrations. An inhomogeneous lateral distribution can enhance excimer formation rate (since some probes will then be closely spaced). Pyrene excimer fluorescence can easily be observed since the broad structureless blue-green excimer fluorescence (λ_{max} 480 nm) is readily distinguished from the highly structured monomer fluorescence (λ_{max} 397 nm).

Lateral diffusions of pyrene dispersed in phospholipid vesicles above the phase transition have been studied by numerous groups (Galla and Sackmann, 1974; Galla *et al.*, 1979; Zacharasse *et al.*, 1980). We have performed both steady-state and time-resolved fluorescence studies using pyrene and PMC, with the latter intended as a probe of cholesterol-induced heterogeneity. If "cholesterol-rich" regions are indeed formed within bilayers, then a fluorescent probe containing a cholesterol moiety might be expected to localize in such regions. Its local concentration would then be higher than that predicted by a random

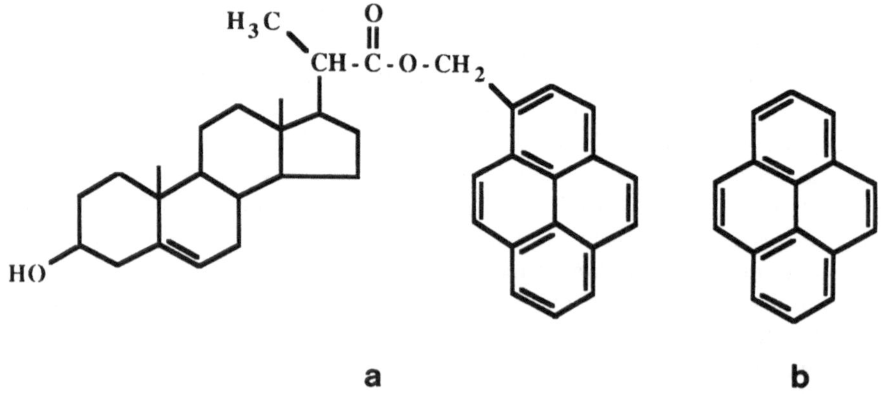

FIGURE 6. Structures of (a) pyrenemethylcholesterol adduct and (b) pyrene.

Membrane Dynamics and Heterogeneity

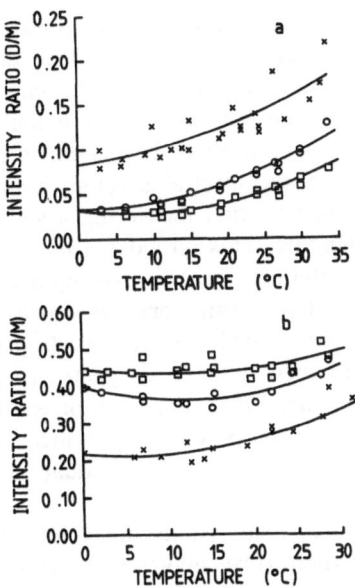

FIGURE 7. Temperature dependence of the ratio of the excimer fluorescence intensity to monomer fluorescence intensity (I_{480}/I_{397}) for (a) pyrene and (b) PMC incorporated into DMPC bilayer vesicles (molar labeling ratio of 1 : 50; probe to phospholipid) with increasing cholesterol concentrations: (+-+-+) 0 mol%; (O-O-O) 20 mol%; and (□-□-□) 33 mol%. The excitation wavelength was 340 nm with excitation and emission bandwidths of 4 and 3 nm, respectively.

distribution of the probe, and excimer formation might benefit. In our preliminary experiments, an interesting contrast between pyrene alone and PMC became apparent. We and others (Davenport and Brand, 1985; Galla and Sackmann, 1974) have shown that for pyrene-labeled DMPC vesicles, addition of cholesterol (20 mol%) results in a *reduction* in the ratio of excimer fluorescence intensity to monomer fluorescence intensity (I_{480}/I_{397}), that is, reduced excimer formation (Figure 7a). A further reduction in this intensity ratio is observed with increased cholesterol concentration (33 mol%). Apparently, the reduced overall fluidity of the acyl chain region on addition of cholesterol results in less facile excimer formation for pyrene. In contrast, for PMC-labeled DMPC vesicles, an *enhanced* excimer formation with increasing mole% cholesterol is observed (Figure 7b). This may reflect a localization (or phase separation) of probe molecules within cholesterol-rich regions. We have applied the DAS methodology to studies of both pyrene and PMC labeled DMPC samples, in the absence of cholesterol at 30°C (where the acyl chain region is in the fluid phase) in an attempt to resolve the experimentally observed complex fluorescence decay functions into different probe *subfractions*.

Decay-Associated Spectra. The technique of DAS and its ability to resolve heterogeneous fluorescence decays have been described in detail elsewhere (Knutson et al., 1982). Extraction of the component spectra can be achieved by a matrix inversion approach or by a multiparameter ("global") nonlinear least-squares fitting procedure (Beechem et al., 1985a; Knutson et al., 1983). In the latter case, the spectral envelope associated with a particular decay time (τ) is determined by first measuring a series of decay curves obtained at different wavelengths. This provides a three-dimensional decay data surface (intensity versus lifetime versus wavelength). A "global" nonlinear least-squares search, where the lifetimes are *linked* (i.e., τ_1 at $\lambda_1 = \tau_1$ at λ_2) but *not fixed* across the emission wavelength region, is performed. This analysis provides decay times and their associated spectra (preexponential terms for ground-state heterogeneity) across the emission or excitation band. The iterative multiparameter data analysis procedure provides "goodness-of-fit" criteria, for example, χ^2, autocorrelation, and residual functions for the overall surface.

The application of DAS to the study of excited-state reactions has been dealt with explicitly elsewhere (Davenport et al., 1986a). For a two-state excited-state reaction (e.g., excimer formation), the fluorescence decay is biexponential at any particular emission wavelength (Birks, 1970; Laws and Brand, 1979):

$$f_M(\lambda,t) = S_M(\lambda)(\alpha_1 e^{-t/\tau_1} + \alpha_2 e^{-t/\tau_2}) \tag{7}$$
$$f_D(\lambda,t) = S_D(\lambda)(\beta_1 e^{-t/\tau_1} + \beta_2 e^{-t/\tau_2}) \tag{8}$$

where $S_M(\lambda)$ and $S_D(\lambda)$ are the emission spectral envelopes for the monomer and dimer species at λ_i. The lifetimes τ_1 and τ_2 are identical for the decay of both species. The preexponential terms (amplitudes β_1 and β_2) are identical in magnitude, but opposite in sign (Birks, 1970; Laws and Brand, 1979), characteristic of an excited-state population that begins at zero, builds at the expense of the "parent" species, then decays away. The DAS (defined as the preexponential terms for a given lifetime, as a function of emission wavelength) for τ_1 and τ_2 are defined as follows:

$$S_1 = \alpha_1 S_M(\lambda) + \beta_1 S_D(\lambda) \tag{9}$$
$$S_2 = \alpha_2 S_M(\lambda) + \beta_2 S_D(\lambda) \tag{10}$$

For excited-state reactions, the preexponential terms obviously *do not* reflect the steady-state emission spectra or the individual species in the mixture. Rather, the DAS will contain contributions from both $S_M(\lambda)$ and $S_D(\lambda)$ (Davenport et al., 1986a; Knutson et al., 1982). The DAS are useful in identifying the kinetic relationships in the system, since the preexponential terms are related to the rates of depletion from M and D states:

$$\alpha_1 = \frac{k_{fM}(\lambda_2 - x)}{\lambda_2 - \lambda_1} \quad \alpha_2 = \frac{(x - \lambda_1)\alpha_1}{\lambda_2 - x} \quad |\beta| = \frac{k_{fD}k_1[M]}{\lambda_2 - \lambda_1} \quad (11)$$

where $\lambda_{1,2} = \bar{\tau}_{1,2}^{-1} = \frac{1}{2}[X + Y \pm \{(Y - X)^2 + 4k_{-1}k_1[M]\}^{1/2}]$, and

$$X = k_{NM} + k_{fM} + k_1[M] \quad (12)$$
$$Y = k_{ND} + k_{fD} + k_{-1} \quad (13)$$

where X and Y represent the combined rate constants for the depletion processes from the M^* and D^* states, respectively:

$$M + M^* \underset{k_{-1}}{\overset{k_1[M]}{\rightleftharpoons}} D^*$$
$$\downarrow k_{fM} \qquad\qquad \downarrow k_{fD}$$
$$M \qquad\qquad M + M$$

The expected DAS for molecules undergoing reversible two-state excimer formation are shown in Figure 8. Figure 8a shows the situation where the combined deactivation rates from the excimer species are faster than from the monomer ($Y > X$), whereas Figure 8b represents the typical DAS obtained when depletion from the excimer state is slower than from the monome state ($X > Y$). Note that the negative preexponential term is seen clearly in either case.

In some cases, one is also interested in obtaining SAS, the species-associated spectra (S_M and S_D above). They can be derived easily for the DAS in a remixing based on the knowledge of alpha and beta (Davenport et al., 1986a) and then compared with other known spectra (Figure 9). Alternatively, an eigenvector–eigenvalue model can be applied *directly* to the system (Beechem et al., 1985b). To obtain unique values for the rate constants and the species-associated spectra (SAS), it is necessary to change the approach to the excited-state equilibrium by changing the monomer concentration. Without such additional data, DAS are the last "assumption-free" step one can make in the analysis (Ameloot et al., 1986; Davenport et al., 1986a).

Our results (Davenport and Brand, 1985) reveal that for pyrene-labeled DMPC vesicles, a minimum of two probe ensembles are present (Figure 10). One class of molecules is involved in a classic excimer reaction. The second probe fraction appears to be prevented from formation of excimers, since its lone derived DAS is characteristic of a monomer species. A very short lifetime ($\tau = 8.4$ nsec) is also associated with this DAS. Since the lifetime is so quenched, we might expect this ensemble to be located in gel near the head groups, where transverse motion may be restricted while the fluorescence is quenched by the polar environment. A similar characterization about gel surroundings was made by Lianos and Georghiou (1979; Mukhopadhyay and Georghiou, 1980), based on vibronic shifts in the pyrene emission spectrum. We intend to obtain higher-resolution DAS to look for these more detailed shifts.

Using the relationship between alpha and beta signs and "mirror imaging" of DAS that vary among the kinetic schemes we considered, we were able to identify two PMC populations undergoing differing excited-state reactions (e.g.,

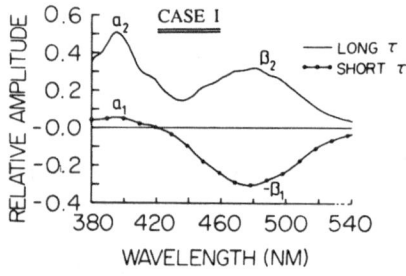

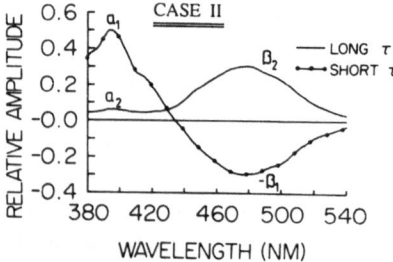

FIGURE 8. Simulated decay-associated spectra describing two contrasting schemes for depletion from the M^* and D^* states for reversible excited-state formation. The DAS are defined here as preexponential terms ($\alpha_1, \alpha_2, \beta_1, \beta_2$) for a given fluorescence lifetime (τ_1, τ_2) as a function of the emission wavelength. In case I, the combined rate constant Y, describing depletion from the D^* state, is greater in magnitude than X, the combined rate constant for depletion from the M^* state. This leads to the characteristic decay-associated spectra shown with a biexponential fluorescence decay function across both the M^* and D^* regions of the emission spectrum. In case II, $X > Y$. Under these conditions the longer lifetime component has a smaller absolute contribution in the M^* region of the spectrum. For both models the negative preexponential term is associated with the shorter lifetime component [(———) long τ, (---) short τ]. Reprinted with permission from the Royal Society of Chemistry (from Davenport *et al.*, 1986c).

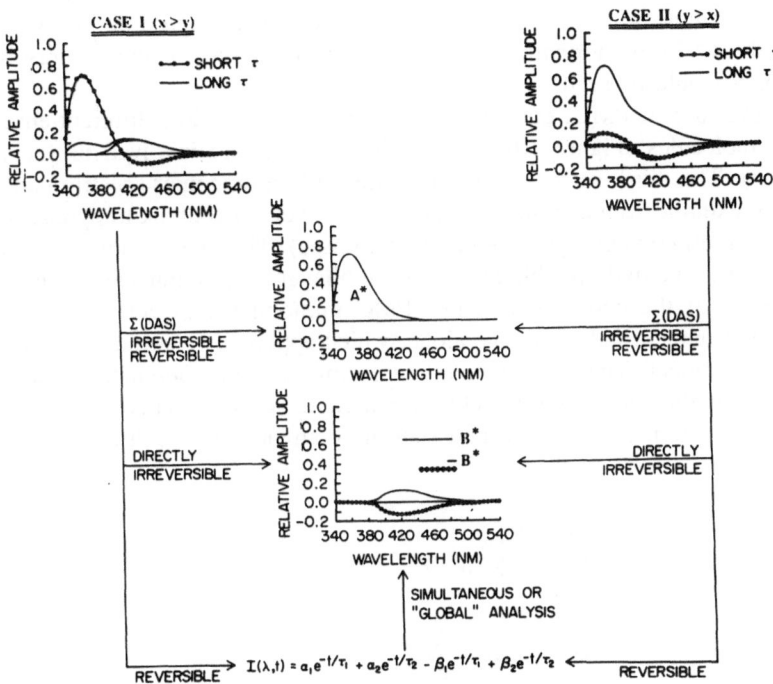

FIGURE 9. Simple combinations of DAS that yield SAS. The sum of DAS provides the profile of initially excited molecules (A^* alone if $B^*(0) = 0$). If the excited-state reaction is effectively irreversible, the B^* spectrum will appear in isolation as a DAS; otherwise, an analysis exploiting the mirror image nature of β (or other separate information) will be required.

with different rates). Figure 10 shows the DAS derived for pyrene and PMC-labeled systems. Fluorescence decay curves (4 min dwell at each wavelength) were collected at 2 nm intervals across the emission wavelength region. The resulting three-dimensional decay surface was then analyzed for a consistent set of decay times. Clearly, pyrene in DMPC is a heterogeneous population of probes, one fraction undergoing a reversible excited-state reaction (compare Figure 8a) with $Y>X$, and a second "noninteracting" subpopulation. Similarly, for PMC, the DAS reveal two fractions forming excimers (note two DAS with negative characteristics) and a third noninteracting probe fraction, with a long lifetime. One should expect to find mirroring of the preexponential term associated with the 8.7 nsec component. However, it is likely to be included in (not resolved from) the 82.2 nsec component. The Y values are too similar in the two active excimer-forming classes to permit resolution. The reader should imagine a simple summing of the amplitudes shown in Figure 8a ($Y>X$) and Figure 8b ($X>Y$), with identical Y creating decay constants too close to resolve: three DAS would be seen.

In the case of PMC, the four derived DAS suggest three different ensembles of molecules (Figure 10). Two probe fractions are involved in excimer formation. The third, as for pyrene, provides only a monomer spectrum and is therefore exempt from excimer formation. However, in contrast to pyrene, that fraction's lifetime is *not* quenched ($\tau = 120.5$ nsec). The pyrene moiety of PMC is confined to the hydrophobic region of the bilayer, due to linkage through the 17 position of the cholesterol. Thus, PMC immobilized in gel may retain a buried environment unlike pyrene alone, which apparently "surfaces" when in a gel environment. The DAS for the two excimer-forming populations indicate that their kinetics are different. One forms excimer by a relatively rapid process. The monomers in the ground state may already be proximate (i.e., via

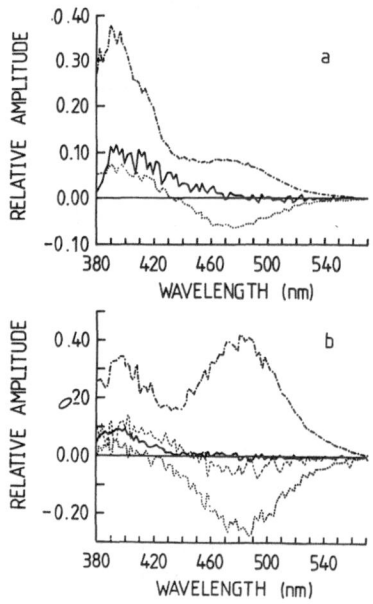

FIGURE 10. Decay-associated fluorescence emission spectra (DAS) for (a) pyrene and (b) PMC-labeled DMPC vesicles (1 : 50 probe to phospholipid molar labeling ratio) at 30°C. For pyrene the decay components were determined to be 128.2 nsec (-----), 33.9 nsec (···), and 8.4 nsec (——). For PMC, a minimum of four decay constants were required: 82.2 nsec (-----), 27.2 nsec (···), 8.7 nsec (---), and 120.5 nsec (——). For both molecules the negative prexponential terms are associated with the short decay components. Excitation was 340 nm, with excitation and emission bandwidths of 13 nm each, respectively. Global χ^2 values were 1.05 for pyrene and 1.05 for PMC. Reprinted with permission from the Royal Society of Chemistry (from Davenport et al., 1986c).

FIGURE 11. Representation of probe location within the lipid bilayer determined from the DAS for both pyrene and PMC. In the case of pyrene, evidence for two probe ensembles is obtained. One population forms excimer by a diffusion-controlled reaction and the second population experiences restricted motion and has a quenched fluorescence lifetime. It might be expected that these probes reside in a more gel environment toward the head group region of the bilayer. In the case of PMC, three populations of probe are proposed. Here the two ensembles undergo excimer formation but with altered rates of production as shown by the diagnostic DAS profiles. The probes that form excimer more readily are perhaps packed in clusters as a direct result of the cooperative packing constraints imposed by cholesterol. The third population of molecules is not involved in excimer formation but does not exhibit a quenched lifetime since the structure of the adduct forces the pyrene moiety into the center of the bilayer.

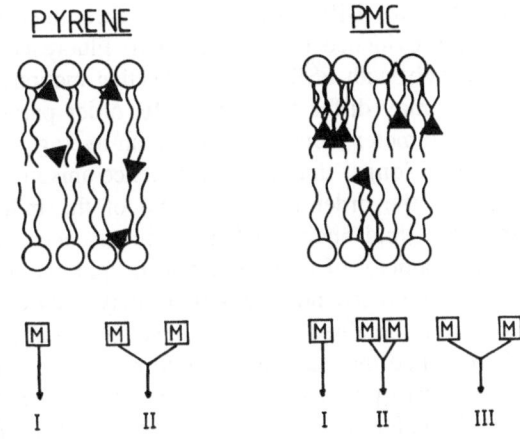

cholesterol-rich sequestering). The second population forms excited-state dimers by a slower process quite comparable with that obtained for pyrene alone. It is possible that, even at these low concentrations of probe, the cholesterol ($\cong 3$ mole %) associated with the PMC is sequestered from the bulk milieu (Figure 11).

The interpretation we have chosen is certainly not the only possible explanation for tetraexponential decay, and related models based on fluctuations (to be discussed later) may be more appropriate for this process. For now, it is instructive to note how widely separated the excimer formation rates are between fractions (fast, low, none). For future models, this will require the coexistence of very disparate fluctuation rates in the bilayer packing.

3. WHAT ARE THE RELAXATION TIMES FOR LIPID DOMAINS?

Differences between deuterium NMR, spin-label ESR, and emission anisotropy derived order parameters have been interpreted in terms of slow fatty acyl chain motions (Gaffney and McConnell, 1974; Jahnig, 1979; Peterson and Chan, 1977). The appropriate chain motions are assumed to proceed at a rate of 10^{-7}–10^{-8} sec. Angular motions in this frequency range will be averaged in a ^{2}H-NMR measurement but will appear as a static tilt in an ESR spin label measurement. Differences between these measurements may be taken as presumptive evidence for slow motion. Fluorescence depolarization results agree qualitatively with ESR results, reflecting motions over the same time scale (10^{-8} sec) but may differ from ^{2}H-NMR order parameters for two reasons: (1) the order parameter from ^{2}H-NMR is an average over a time scale of about 10^{-4} sec, much longer than the fluorescence anisotropy averaging time; (2) ^{2}H-NMR measures the order parameter (S_v) of the individual methylene segments (v) along the lipid chains, whereas fluorescence anisotropy measures an average order parameter of a probe between lipid chains.

Submicrosecond fluctuations may provide an alternative explanation for the r_∞ term obtained from time-resolved fluorescence anisotropy measurements. We reasoned that submicrosecond relaxation processes in the bilayer would be accessible to polarized fluorescence measurements, if probe lifetimes were extended. Under conditions where the averaging time (τ) is longer than the relaxation time (ϕ), equilibrium is established. The steady-state average, however, is limited by the fluorescence lifetime and relaxation processes with $\phi >> \tau$ are not detected. For instance, differences between ^{2}H-NMR and fluorescence anisotropy averaging times are irrelevant if intermediate relaxation processes (with $\phi >> \tau_{fl}$ but $<< \tau_{NMR}$) are absent. We have targeted "slow" reorientational motions by extending the available averaging regime. The fluorescence "time window" can be extended to span the critical submicrosecond region, using probes with fluorescence lifetimes on the order of several hundreds of nanoeconds. Both time-resolved and steady-state probe anisotropy studies should then detect submicrosecond acyl-chain motions.

To this aim, we have used *coronene*, a long-lived fluorescence probe, for investigation of membrane disordering (Brand et al., 1985; Davenport et al., 1983, 1987). Our results show that coronene, with a mean fluorescence lifetime greater than 200 nsec (not significantly quenched over the temperature range discussed here) is sensitive to "slow" lipid order changes within both small unilamellar vesicles (SUVs) and large unilamellar vesicles (LUVs). Previous studies using coronene and other long-lived fluorescence probes, for example, 1,12-benzperylene and derivatives of coronene (Clar and Schmidt, 1976; Katraro et al., 1979) were restricted to photophysical investigations from organic films and solvents (Lamotte et al., 1975; Lin and Topp, 1977; Vo Dinh et al., 1977; Zimmermann and Joop, 1961). Due to its disklike shape, coronene should be an anisotropic rotor. Fortunately, the polarized emission of this molecule, following the rules for D_{6h} planar symmetry ($r_0 = 0.1$), depends *only* on out-of-plane motions. Molecules with lower symmetry (e.g., perylene) depolarize via both fast in-plane and slow out-of-plane rotations (Barkley et al., 1981; Brand et al., 1985), while steady-state anisotropy values for coronene provide an exclusive detection of slow out-of-plane rotations ($<r>_{op}$).

Steady-state fluorescence emission anisotropies ($<r>_{op}$) as a function of

Membrane Dynamics and Heterogeneity

temperature ("melt" curves), for coronene imbedded in both DMPC and DPPC SUVs (Figure 12), show a much broader transition than reported for DPH (Figure 12). Coronene appears to be sensitive to slow depolarizing motions occurring within the bilayer well below (~10°C) the normal lipid melting temperatures ($T_c = 23$°C for DMPC and 41°C for DPPC). As mentioned previously, unequal partitioning of the probe between fluid or gel environments might help account for shifted phase transition curves. We have eliminated this alternative explanation via time-resolved measurements. Shifts due to slow partitioning alone will be unaltered by the time scale over which the experiment is performed, unlike the observations we present here.

Time-dependent fluorescence anisotropy data were obtained for coronene

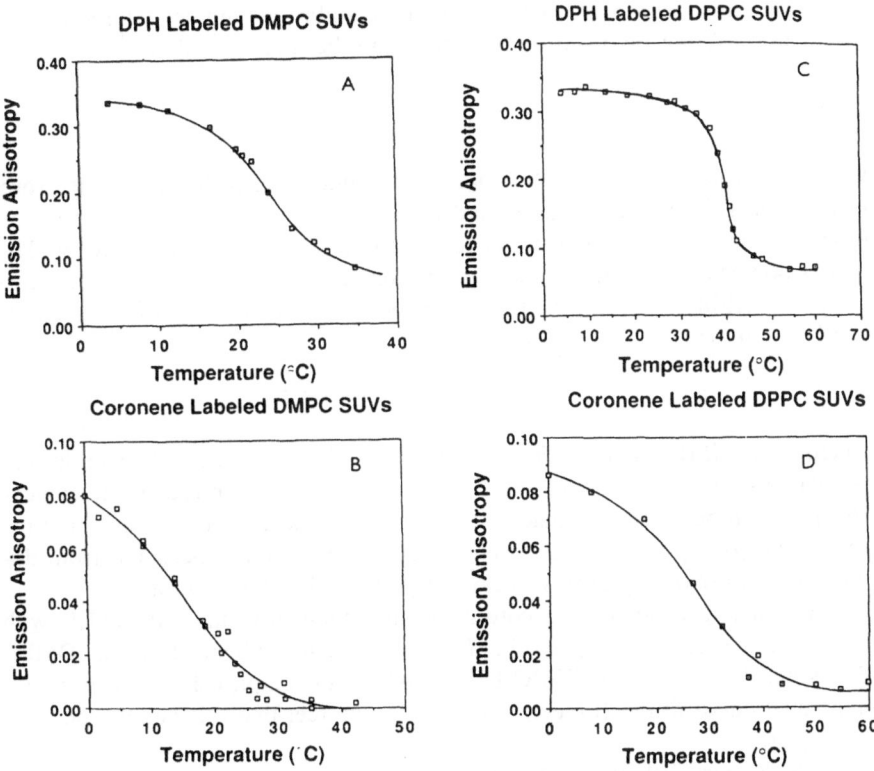

FIGURE 12. Steady-state fluorescence emission anisotropy ($\langle r \rangle$) for (A) DPH-labeled and (B) coronene-labeled DMPC SUVs and (C) DPH-labeled and (D) coronene-labeled DPPC SUVs in 0.01 M Tris, HCl containing 0.1 M NaCl (pH 8.5), as a function of temperature. Excitation emission wavelengths were 355, 430 nm and 337, 448 nm for DPH and coronene, respectively. Excitation and emission bandwidths were each 4 nm, respectively.

in DMPC (SUV and LUV) bilayer vesicles. The data could be resolved into three rotational correlation times. At 19.3°C, just below the phase transition (23°C), the three rotational correlation times are $\phi_1 = 2$ nsec, $\phi_2 = 50$ nsec, and $\phi_3 = 600$ nsec. The short decay, ϕ_1, may be assigned to the rotation of excited coronene in the fluid phase. Its relative proportion should increase with temperature through the phase transition. The long correlation time, ϕ_3, is on the order of 1000 nsec and can be assigned to whole vesicle rotation. What is the origin of ϕ_2? It is clearly too fast to represent whole-vesicle rotation. Moreover, it is the same for SUVs and LUVs. It is too slow to represent rotation of coronene in the fluid phase.

We suggest that motions represented by ϕ_2 are slow phospholipid packing fluctuations inherent to the acyl chain region of the bilayer.

One can begin by defining a compartmentalized model with a gel region, G, and a fluid region, F, which are in equilibrium and an additional gel subfraction, N, not involved in the equilibrium between F and G.

$$G \underset{}{\overset{k_1}{\rightleftharpoons}} F \tag{14}$$

where $k_1 = 1/\phi_{\text{eff}}$, $\phi_F \ll \tau \ll \phi_G$, τ is the fluorescence liftime of the probe in both the gel and fluid phases, and ϕ_G and ϕ_F are the rotational correlation times for the probes located in gel (G) and fluid (F) phases, respectively. The probes originally imbedded in these gel regions can only rotate by going through a "local melting" process (k_1). The decay of the emission anisotropy ($r(t)$), can be expressed as

$$r(t) = (\beta_1 e^{-t/\phi_F} + \beta_2 e^{-t/\phi_{\text{eff}}} + \beta_3) e^{-t/\phi_N} \tag{15}$$

where $\beta_1/r_0 = R/(G+F+N)$, $\beta_2/r_0 = G/(G+F+N)$, F and G are the equilibrium populations of fluid and gel localized probes, and N represents the nonexchangeable probe fraction, that is, those probes that do not melt on this time scale. The $(\beta_2 + \beta_3)$ term here is equivalent to the r_∞ term measured from fluorescence emission anisotropy decays of probes with short lifetimes.

We analyzed our data according to this compartmentalized model, with increasing temperatures up to the T_c, for coronene imbedded within DMPC SUVs and LUVs and for DPPC LUVs. While this empirical approach allows individual curve fitting (Figures 13 and 14), we feel that this simple model is inadequate.

The longest ϕ should correspond to whole-vesicle rotation. Even for SUVs, ϕ should then exceed 1 μsec (even near 20°C) and this is clearly not the case. ϕ_N should in fact obey a Perrin relationship with temperature and viscosity, unless another type of rotation can be ascribed to the shell. ϕ_N should also be

Membrane Dynamics and Heterogeneity

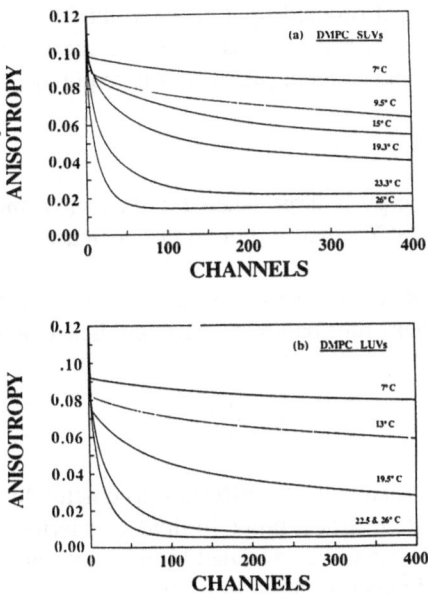

FIGURE 13. Impulse response functions for the decay of the emission anisotropy, $r(t)$, of coronene in (a) DMPC SUVs and (b) DMPC LUVs measured at the temperatures shown. The labeling ratio was 1 : 200 (probe to phospholipid molar ratio). (Time calibration is 0.372 nsec/channel.)

much longer for LUVs than for SUVs, in disagreement with our observations. Even if we could reconcile variations in ϕ_N, the compartmental model falls short.

The model is strictly empirical, and as such does not provide an expected relationship of rotational probe motions with temperature. An alternative choice is to predict a *distribution* of "melting rates." We have adapted the Landau model for describing gated packing fluctuations, based on the reviews of Hawton and Doane (1987), Jahnig (1981a,b), and Nagle and Scott (1978) and on the precise energy shapes derived by Mitaku *et al.* (1983) for DPPC.

In the framework of the Landau theory for phase transitions, the free energy (F) of a system can be expanded by the powers of an individual chain's order parameters, S:

$$F = -A_1 S + \tfrac{1}{2} A_2 S^2 - \tfrac{1}{3} A_3 S^3 + \tfrac{1}{4} A_4 S^4 \tag{16}$$

where A_2 is linearly related to temperature, $A_2 = a(T - T^*)$, and where T^* is the pseudocritical temperature (Jahnig, 1981a; Mitaku *et al.*, 1983).

For a small region of ordered lipid phase to fluidize, an activation energy, E, proportional to the difference between the free energy (per lipid molecule), between the fluid (F_f) and ordered phase (F_o) is required:

$$E = F_f - F_o \qquad (17)$$

This activation energy is large when T is far below the transition temperature (T_c) but vanishes at T_c; that is, lipid order decreases strongly with increasing temperature.

A normalized Boltzmann distribution defines the probability (P_i) of a molecule having a particular order parameter (S_i), and hence activation energy (E_i).

$$P_i = \frac{e^{-E_i/RT}}{\sum_i e^{-E_i/RT}} \qquad (18)$$

We expect that, in analogy to diffusion arguments by Jahnig (1981), the effective *rotational* rate, d_i, for a probe in lipid is determined by the activation energy (E_i) barrier that has to be exceeded by the chain in order for disordering (small value of S) and hence rotation to occur:

$$d_i = d_\infty \, e^{\gamma E_i/RT} \qquad (19)$$

Here γ accounts for the fact that some multiple of the activation barrier must be overcome (e.g., γ molecules must achieve "disorder" simultaneously to permit rotation).

When E_i is very small (or T is very large),

$$d_i \sim d_\infty \qquad (20)$$

d_∞ is a "frequency factor" that reflects the fastest rate of rotation of the probe, as if in pure "fluid" lipid:

$$d_\infty \doteq \frac{1}{\phi_F} \qquad (21)$$

We have used values of d_∞ near 0.2×10^9 sec^{-1}, which results in $\phi_F = 5 \times 10^{-9}$ sec, comparable with experimental observation. Thus, the distributed rates for the probe (from Equations (19) and (21)) may be expressed as

$$\frac{1}{\phi_i} = d_\infty e^{-\gamma E_i/RT} \qquad (22)$$

We can now predict the time-resolved emission anisotropy from the effective rotational rates [Equation (22)] and probabilities [Equation (18)]:

$$r(t) = r_0 \Sigma_i P_i e^{-d_i t} \qquad (23)$$

where

$$d_i = \frac{1}{\phi_i} \qquad (24)$$

Using Equations (18) and (22) describing P_i and ϕ_i, respectively, we have predicted the ϕ distributions and P_i distributions for initial order parameters $0 \to 1$ in increments of 0.05, at various temperatures below T_c, using parameters appropriate to coronene in DPPC. These are shown in Figure 15. The appearance of fluctuation-induced rotations is apparent *well* below the transition temperature $(T_c = 41°C)$. In fact, 8°C prior to the transition, a variety of medium to short correlation times (~50–200 nsec) are apparent [corresponding to the melting gel (G) in our simpler model]. This compares favorably with the time-resolved and steady-state data previously presented. It is interesting to observe that the preexponentials for long correlation times decrease with increasing temperature (i.e., N in our simpler model) (compare Figure 14).

A total of seven parameters (five in the free energy versus temperature expansion) (Jahnig, 1981a; Mitaku *et al.*, 1983), one (γ) estimating the gate requirement, and the limiting diffusion rate (frequency factor d_∞) suffice to predict complete anisotropy decay curves at several temperatures below T_c, although more detail can be added to the model (e.g., $F_f^{-d_\infty}$ distribution) as needed. The time-resolved curves appropriate to such simulations are shown in Figure 16. Although details of our multitemperature global model have not yet

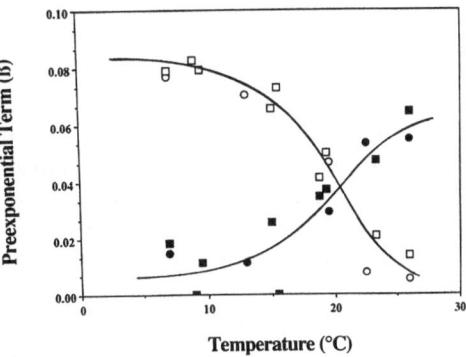

FIGURE 14. Plot of the preexponential terms (β), corresponding to the long (ϕ_N) or medium (ϕ_{eff}) rotational correlation times, as a function of temperature for both DMPC SUVs ($\square = \phi_N$; $\blacksquare = \phi_{eff}$) and DMPC LUVs ($\bigcirc = \phi_N$; $\bullet = \phi_{eff}$).

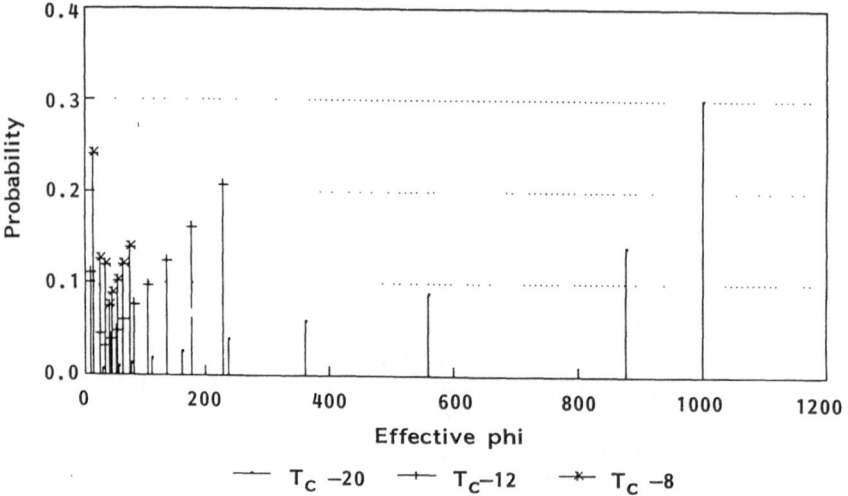

FIGURE 15. Examples of predicted correlation time (ϕ) distributions for coronene in DPPC vesicles. The probability P_i and effective rotational times (ϕ_i) were determined from equations X and Y for order parameters $0 \rightarrow 1$ (in steps of 0.05) according to the Landau model at various temperatures below T_c (precise energy shapes were derived from Mitaku *et al.*, 1983).

been rigorously fit to anisotropy data, the agreement seen between these preliminary curves and experimental coronene decays is remarkable. In summary, simple Landau modeling led to a distribution of fluctuation-induced rotation rates (at each T) with considerable correspondence to our observations.

4. FUTURE TRENDS

Using time-resolved fluorescence methods geared to microheterogeneity, we have sought to promote an alternative view of bilayer dynamics. Unlike homogeneous concepts for lipid probe order, using one potential and/or diffusion model, our view is one of a *heterogeneous* probe environment. In brief, we expect probes to inhabit an equilibrium between fluid and gel environments. Thus, some probes remain in each environment throughout their excited-state lifetime, while others reside in surroundings that change state during the probe lifetime. Since rotation in fluid surroundings (e.g., between phospholipid molecules containing several chain kinks) is generally rapid, while gel-state phos-

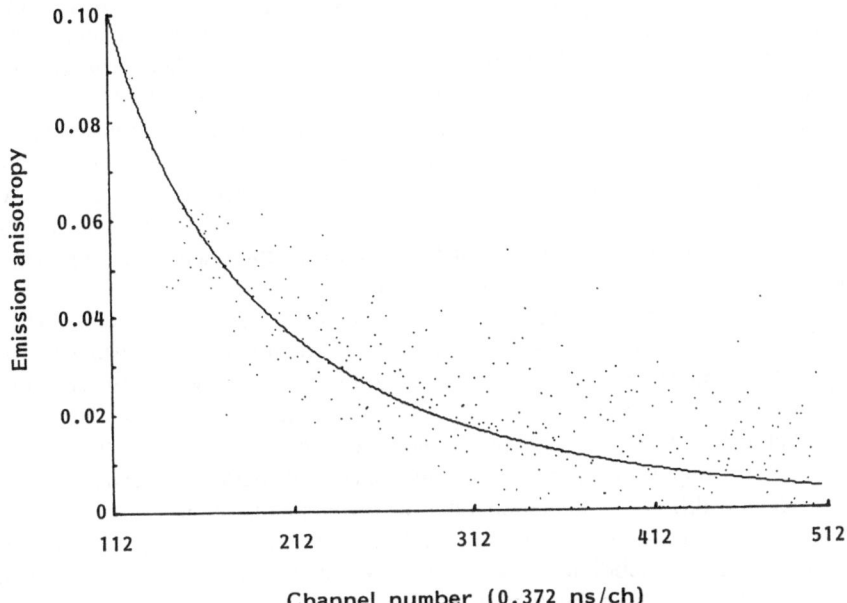

FIGURE 16. Empirical fit of Landau model simulation for packing fluctuations, to experimental time-resolved emission anisotropy data, obtained for coronene in DPPC LUVs at 35°C. Parameters varied to give the best "fit" were T_c, γ, and d_∞. Values used here were 41°C, 1, and 1×10^8 sec^{-1} ($1/d_\infty \doteq \phi_F = 10$ nsec), respectively.

pholipids immobilize their included probes, equilibrium must lead to a complex total depolarization. In addition to very fast ($\leq$nsec) and very slow ($\geq\mu$sec) rotational correlation times, one can observe intermediate (few to few hundred nsec) times characteristic of the equilibrium "melting" process. For short-lived probes, the latter are difficult to separate from the refractory gel, although it has long been recognized (Chen *et al.*, 1977; Dale *et al.*, 1977) that r_∞ for DPH in vesicles approaching transition could be replaced, without fitting error, by correlation times of a few hundred nsec. While homogeneous ordering models still provide a convenient shorthand for describing time-resolved data (e.g., narrow cone→large preexponential for $\phi \geq 100$), it is now unclear whether one should give literal meaning to their parameters.

We have examined the equilibrium rate(s) (gel–fluid) at several temperatures below the main thermal transition of pure phospholipid bilayers. In doing so, we discovered a need for more than one "melting" rate, especially a few degrees below the transition. This experimental constraint, along with our need to understand a wide range of conditions, led us to the Landau models for bilayer chains. The review of these concepts presented by Jahnig (1981a,b) was enlightening, especially when combined with the experimental studies of Mitaku et al. (1983). Interestingly, the foresight of Hudson and co-workers (1986) has led them to embrace fluctuation concepts, even though their probe (*trans*-parinaric acid) is effective over a much shorter lifetime than coronene.

In summary, these models can predict how a *homogeneous* ensemble of lipids will acquire heterogeneous structures that fluctuate between gel and fluid "domains" (more properly, between densely and loosely packed clusters or regions containing lipids in *similar* states). On the nanosecond time scale, the system *is* a heterogeneous quilt, but that pattern is continually shifting and is blurred to homogeneity within microseconds.

We shall soon apply these tools to "chemical" domain systems, where long-term heterogeneity is known to persist over relatively large distances. We expect fluorometry can again be extended to bridge differences between observations (previously in disagreement) made with different methods. We anticipate that a new flood of questions can now be addressed: Are slow fluctuations more or less important in membrane function? Will membrane-associated proteins change different fluctuation rates in different directions? What level of microscopic heterogeneity is present, and for how long? Are large macroscopic domains similarly subdivided, or do domains of different scale lead independent lives? Can simple fluctuation models predict complex biological effects such as anesthesia? How do fusion, permeability, and other membrane functions relate to the fluctuation gating process?

Abbreviations

DAS	decay-associated spectra
ADAS	anisotropy decay-associated spectra
TRES	time-resolved emission spectra
DPH	1,6-diphenyl-1,3,5-hexatriene
DPO	1,6-diphenyl-,3,5,7-octatetraene
DMPC	dimyristoyl phosphatidylcholine
DPPC	dipalmitoyl phosphatidylcholine
DOPC	dioleoyl phosphatidylcholine
DLPC	dilauroyl phosphatidylcholine
PMC	1-pyrene-methyl-3β-OH-22,23-bisnor-5-cholenate
SAS	species-associated spectra

ACKNOWLEDGMENTS. The authors would like to thank Dr. Robert E. Dale, Mr. David Markby, and Mr. Imtiaz Munshi and acknowledge expert technical assistance from Mr. Dana G. Walbridge. We thank Drs. Marcel Ameloot and Joseph Beechem for helpful discussions on compartmental systems. In addition, we would like to thank Ms. Karen Munday and Ms. Sion Kim for their help with the preparation of this manuscript and Dr. Michael Straher for graphic arts assistance. Acknowledgment is made to the donors of the Petroleum Research Fund, which is administered by the American Chemical Society (LD), The City University of New York PSC-CUNY Research Award Program (LD), and the National Institutes of Health (LB; GM 11632) for partial support of this research.

5. REFERENCES

Albrecht, O., Gruler, H., and Sackman, E., 1978, Polymorphism of phospholipid monolayers. *J. Physiol.* **39**:301–313.
Ameloot, M., Beechem, J. M., and Brand, L., 1986, Compartmental modeling of excited-state reactions: Identifiability of the rate constants from fluorescence decay surfaces. *Chem. Phys. Lett.* **129**:211–219.
Andrich, M. P., and Vanderkooi, J. M., 1976, Temperature dependence of 1,6-diphenyl-1,3,5-hexatriene fluorescence in phospholipid artificial membranes. *Biochemistry* **15**:1257–1261.
Baldassare, J. J., Rhinehart, K. B., and Silbert, D. F., 1976, Modification of membrane lipid: Physical properties in relation to fatty acid structure. *Biochemistry* **15**:2986–2994.
Barkley, M. D., Kowalczyk, A. A., and Brand, L., 1981, Fluorescence decay studies of anisotropic rotations of small molecules. *J. Chem. Phys.* **75**:3581–3593.
Barrow, D. A., and Lentz, B. R., 1985a, Membrane structural domains: Resolution limits using diphenylhexatriene fluorescence decay. *Biophys. J.* **48**:221–234.
Bearer, E. L., and Friend, D. S., 1982, Modifications of anionic-lipid domains preceding membrane fusion in guinea pig sperm. *J. Cell Biol.* **92**:604–615.
Beechem, J. M., Ameloot, M., and Brand, L., 1985a, Global analysis of fluorescence decay surfaces: Excited state reactions. *Chem. Phys. Lett.* **120**:466–472.
Beechem, J. M., Ameloot, M., and Brand, L., 1985b, Global and target analysis of complex decay phenomena. *Anal. Instrum.* **14**:379–402.
Betel, I., and van den Berg, K. J., 1972, Interactions of concanavalin A with rat lymphocytes. *Eur. J. Biochem.* **30**:571–578.
Birks, J. B., 1970, *Photophysics of Aromatic Molecules,* Wiley-Interscience, New York.
Bittman, R., 1978, Sterol–polyene antibiotic complexation: Probe of membrane structure. *Lipids* **13**:686–691.
Bloom, J. A., and Webb, W. W., 1983, Lipid diffusibility in the intact erythrocyte membrane. *Biophys. J.* **42**:295–305.
Brand, L., Knutson, J. R., Davenport, L., Beechem, J. M., Dale, R. E., Walbridge, D. G., and Kowalczyk, A. A., 1985, Time-resolved fluorescence spectroscopy: Some applications of associative behavior to studies of proteins and membranes. In *Spectroscopy and the Dynamics of Molecular Biological Systems* (P. M. Bayley and R. E. Dale, eds.), pp. 259–305, Academic Press, London.
Bretscher, M. S., 1972, Asymmetrical lipid bilayer structure for biological membranes, *Nature (London) New Biol.* **236**:11–12.
Bretscher, M. S., 1976, Directed lipid flow in cell membranes. *Nature (London)* **260**:21–23.
Brulet, P., and McConnell, H. M., 1977, Structural and dynamical aspects of membrane immunochemistry using model membranes. *Biochemistry* **16**:1209–1217.
Casper, D. L. D., and Kirshner, D. A., 1971, Myelin membrane—structure at 10Å resolution. *Nature (London) New Biol.* **231**:46–53.
Chen, L. A., Dale, R. E., Roth, S., and Brand, L., 1977, Nanosecond time-dependent fluorescence depolarization of diphenylhexatriene in dimyristoyllecithin vesicles and the determination of "microviscosity." *J. Biol. Chem.* **252**:2163–2169.

Cherry, R. J., 1979, Rotational and lateral diffusion of membrane proteins. *Biochim. Biophys. Acta* **559**:239–327.
Cherry, R. J., Nigg, E. A., and Beddard, G. S., 1980, Oligosaccharide motion in erythrocyte membranes investigated by picosecond fluorescence polarization and microsecond dichroism of an optical probe. *Proc. Natl. Acad. Sci. U.S.A.* **77**:5899–5903.
Chi, E. Y., Lagunoff, D., and Koehler, J. K., 1976, Freeze fracture study of mast cell secretion. *Proc. Natl. Acad. Sci. U.S.A.* **73**:2823–2827.
Chong, L. G., and Weber, G., 1983, Pressure dependence of 1,6-diphenyl-1,3,5-hexatriene fluorescence in single-component phosphatidylcholine liposomes. *Biochemistry* **22**:5544–5550.
Chuang, T. J., and Eisenthal, K. B., 1972, Theory of fluorescence depolarization by anisotropic rotational diffusion. *Chem. Phys.* **57**:5094–5097.
Clar, E., and Schmidt, W., 1976, Correlations between photoelectron and ultra-violet absorption spectra of polycyclic hydrocarbons and the number of aromatic sextets. *Tetrahedron* **32**:2263–2271.
Cronan, J. E., and Vagelos, P. R., 1972, Metabolism and function of the membrane phospholipid of *Escherichia coli*. *Biochim. Biophys. Acta* **265**:25–60.
Cullis, P. R., and de Kruijff, B., 1979, Lipid polymorphism and the functional roles of lipids in biological membranes. *Biochim. Biophys. Acta* **559**:399–420.
Cundall, R. B., Johnson, I. D., Jones, M. W., Thomes, E. W., and Munro, I. H., 1979, Photophysical properties of DPH derivatives. *Chem. Phys. Lett.* **64**:339–342.
Curatolo, W., Sakura, D. J., Small, D. M., and Schipley, D.G., 1977, Protein–lipid interactions: Recombinants of the proteolipid apoprotein of myelin with dimyristoyllecithin. *Biochemistry* **16**:2313–2319.
Dale, R. E., Chen, L. A., and Brand, L., 1977, Rotation and relaxation of the "microviscosity" probe DPH in paraffin oil and egg lecithin vesicles. *J. Biol. Chem.* **252**:7500–7510.
Davenport, L., and Brand, L., 1985, Fluorescence studies of excimer formation in single bilayer liposomes using a pyrene-methyl cholesterol adduct. *Biophys. J.* **47**:367a.
Davenport, L., Knutson, J. R., and Brand, L., 1982, Anisotropy-decay associated fluorescence spectra and the analysis of rotational heterogeneity. *Photochem. Photobiol.* **10**:69a.
Davenport, L., Markby, D. W., Knutson, J. R., and Brand, L., 1983, Restricted out-of-plane rotations of coronene as revealed by emission anisotropy. *Photochem. Photobiol.* **37**:S20.
Davenport, L., Dale, R. E., Bisby, R. H., and Cundall, R. B., 1985, Transverse location of the fluorescence probe 1,6-diphenyl-1,3,5-hexatriene in model lipid bilayer membrane systems by resonance excitation energy transfer. *Biochemistry* **24**:4097–4108.
Davenport, L., Knutson, J. R., and Brand, L., 1986a, Excited-state proton transfer of equilenin and dihydroequilenin: interaction with bilayer vesicles. *Biochemistry* **25**:1186–1195.
Davenport, L., Knutson, J. R., and Brand, L., 1986b, Anisotropy-decay associated fluorescence spectra and the analysis of rotational heterogeneity. 2. 1,6-diphenyl-1,3,5-hexatriene in lipid bilayers. *Biochemistry* **25**:1811–1816.
Davenport, L., Knutson, J. R., and Brand, L., 1986c, Studies of membrane heterogeneity using fluorescence associative techniques. *Faraday Discuss. Chem. Soc.* **81**:81–94.
Davenport, L., Knutson, J. R., and Brand, L., 1987, Coronene: A probe for structural fluctuations in phospholipid bilayers. *Biophys. J.* **51**:537a.
Davis, D. G., Inesi, G., and Gulik-Krzywicki, T., 1976, Lipid molecular motion and enzyme activity in sarcoplasmic reticulum membrane. *Biochemistry* **15**:1271–1276.
de Kruijff, B., Van Dijck, P. W. M., Demel, R. A., Schuijff, A., Brants, F., and Van Deenan, L. L. M., 1974, Non-random distribution of cholesterol in phosphatidylcholine bilayers. *Biochim. Biophys. Acta* **356**:1–7.
Devaux, P., and McConnell, H. M., 1972, Lateral diffusion in spin-labeled phosphatidyl choline vesicles. *J. Am. Chem. Soc.* **94**:4475–4481.

Douce, R., Mannella, C. A., and Bonner, W. D., 1973, External NADH dehydrogenase of intact plant mitochondria. *Biochim. Biophys. Acta* **292**:105-116.

Douglas, W. W., 1974, Mechanism of release of neurohypophyeal hormones: Stimulus secretion coupling. In *Handbook of Physiology. Section 7: Endocrinology* (R. O. Greep and E. B. Astwood, eds.), pp. 191-224, American Physiological Society (1972-1976), Washington D.C.

Duzgunes, N., Paiement, J., Freeman, K. B., Lopez, N. G., Wilschut, J., and Paphadojopoulos, D., 1984, Modulation of membrane fusion by ionotropic and thermotropic phase transitions. *Biochemistry* **23**:3486-3494.

Eck, V., Holzwarth, J. F., 1984, Fast dynamic phenomena in vesicles of phospholipids during phase transitions. In *Surfactants in Solution* (M. L. Mittal and B. Lindman, eds.), pp. 2059-2980, Plenum Press, New York.

Edidin, M., and Weiss, A., 1972, Antigen-cap formation in cultured fibroblasts: A reflection of membrane fluidity and of cell motility. *Proc. Natl. Acad. Sci. U.S.A.* **69**:2456-2459.

Elias, P. M., Friend, D. S., and Goerke, J., 1979, Membrane-stereo heterogeneity: Freeze-fracture detection with saponins and filipin. *J. Histochem. Cytochem.* **27**:1247-1260.

Engelman, D. M., 1972, The molecular structure of the membrane of Acholeplasma Laidlawii. *Chem. Phys. Lipids* **8**:298-302.

Engelman, D. M., and Rothman, J. E., 1972, The planar organization of lecithin-cholesterol bilayers. *J. Biol. Chem.* 247:3694-3697.

Fiorini, R., Valentino, M., Wang, S., Glaser, M., and Gratton, E., 1987, Fluorescence lifetime distributions of 1,6-diphenyl-1,3,5-hexatriene in phospholipid vesicles. *Biochemistry* **26**:3864-3869.

Fleming, G. R., Knight, A. E. W., Morris, J. M., Morrison, R. J. S., and Robinson, G. W., 1977, Picosecond fluorescence studies of xanthine dyes. *J. Am. Chem. Soc.* **99**:4306-4311.

Florine, K. I., and Feigenson, G. W., 1987a, Influence of calcium-induced gel phase on the behavior of small molecules in phosphatidylserine and phosphatidylserine-phosphatidylcholine multilamellar vesicles. *Biochemistry* **26**:1757-1768.

Florine, K. I., and Feigenson, G. W., 1987b, Protein redistribution in model membranes: Clearing of M13 coat protein from calcium-induced gel-phase regions in phosphatidylserine-phosphatidylcholine multilamellar vesicles. *Biochemistry* **26**:2978-2983.

Florine-Casteel, K. I., and Feigenson, G. W., 1988, On the use of partition coefficients to characterize the distribution of fluorescent membrane probes between coexisting gel and fluid phases. *Biochim. Biophys. Acta* **941**:102-106.

Fong, B. S., and Brown, J. C., 1978, Asymmetric distribution of phosphatidylethanolamine fatty acyl chains in the membrane of vesicular stomatitis virus. *Biochem. Biophys. Acta* **510**:230-241.

Fong, B. S., Hunt, R. C., and Brown, J. C., 1976, Asymmetric distribution of phosphatidylethanolamine in the membrane of vesicular stomatitis virus. *J. Virology* **20**:658-663.

Fressenden-Raden, J. M., and Racker, E., 1971, Structural and functional organization of mitochondrial membranes, In *Structure and Function of Biological Membranes* (L. Rothfield, ed.), pp. 401-438, Academic Press, Orlando.

Frye, L. D., and Edidin, M., 1970, The rapid intermixing of cell surface antigens after formation of mouse-human heterokaryon. *J. of Cell Sci.* **7**:319-335.

Gaffney, B. J., and McConnell, H. M., 1974, Paramagnetic resonance spectra of spin labels in phospholipid membranes. *J. Magn. Reson.* **16**:1-28.

Galla, H. J., and Luisetti, J., 1980, Lateral and transversal diffusion and phase transitions in erythrocyte membranes. An excimer fluorescence study. *Biochim. Biophys. Acta* **596**:108-117.

Galla, H. J., and Sackmann, E., 1974, Lateral diffusion in the hydrophobic region of membranes: Use of pyrene excimers as optical probes. *Biochim. Biophys. Acta* **339**:103-115.

Galla, H. J., and Sackmann, E., 1975, Chemically induced lipid phase separation in model membranes containing charged lipids: A spin label study. *Biochim. Biophys. Acta* **401:**509–529.

Galla, H. J., Theilen, U., and Hartmann, W., 1979, Transversal mobility in bilayer membrane vesicles: Use of pyrene lecithin as optical probe. *Chem. Phys. Lipids* **23:**239–251.

Getz, G. S., 1972, Organelle biogenesis. In *Membrane Molecular Biology* (C. F. Fox and A. Keith, eds.), pp 386–438, Sinauer, Sunderland, MA.

Giraud, F., Claret, M., Bruckdorfer, K. R., and Chailley, B., 1981, The effects of membrane lipid order and cholesterol on the internal and external cationic sites of the Na^+,K^+ pump in erythrocytes. *Biochim. Biophys. Acta* **647:**249–258.

Glatz, P., 1978, Limited rotational diffusion of DPH in human erythrocyte membranes. *Anal. Biochem.* **87:**187–194.

Golan, D. E., and Veatch, W., 1980, Lateral mobility of band 3 in human erythrocyte membrane studied by fluorescence photobleaching recovery: Evidence for control by cytoskeletal interaction. *Proc. Natl. Acad. Sci. U.S.A.* **77:**2537–2541.

Graham, I., Gagne, J., and Silvius, J. R., 1985, Kinetics and thermodynamics of calcium-induced lateral phase separations in phosphatidic acid containing bilayers. *Biochemistry* **24:**7123–7131.

Griffith, O. H., Jost, P., Capaldi, R. A., and Vanderkooi, G., 1973, Boundary lipid and fluid bilayer regions in cytochrome oxidase model membranes. *Ann. N.Y. Acad. Sci.* **222:**561–573.

Gruenwald, B., Frisch, W., and Holzwarth, J. F., 1981, The kinetics of the formation of rotational isomers in the hydrophobic tail region of phospholipid bilayers. *Biochim. Biophys. Acta* **641:**311–319.

Haberkorn, R. A., Griffin, R. G., Meadows, M. D., and Oldfield, E., 1977, Deuterium nuclear magnetic resonance investigation of the depalmitoyl lecithin–cholesterol–water system. *J. Am. Chem. Soc.* **99:**7353–7355.

Haest, C. W. M., Degier, J., Op den Kamp, J. A. F., Barter, P., and Van Deenen, L. L. M., 1972, Changes in permeability of *Staphylococcus aureus* and derived liposomes with varying lipid composition. *Biochim. Biophys. Acta* **255:**720–733.

Hartmann, W., Galla, H. J., and Sackmann, E., 1977, Direct evidence of charge-induced lipid domain structure in model membranes. *FEBS Lett.* **78:**169–172.

Haverstick, D. M., and Glaser, M., 1987, Visualization of Ca^{2+}-induced phospholipid domains. *Proc. Natl. Acad. Sci. U.S.A.* **84:**4475–4479.

Hawton, M. H., and Boane, J. W., 1987, Pretransitional phenomena in phospholipid/water multilayers. *Biophys. J.* **52:**401–404.

Heyn, M. P., 1979, Determination of lipid order parameters and rotational correlation times from fluorescence depolarization experiments. *FEBS Lett.* **108:**359–364.

Hildenbrand, K., and Nicolau, C., 1979, Nanosecond fluorescence anisotropy decays of 1,6-diphenyl-1,3,5-hexatriene in membranes. *Biochim. Biophys. Acta* **553:**365–377.

Hoekstra, D., 1982a, Fluorescence method for measuring the kinetics of Ca^{2+}-induced phase separations in phosphatidylserine-containing lipid vesicles. *Biochemistry* **21:**1055–1061.

Hoekstra, D., 1982b, Role of lipid phase separations and membrane hydration in phospholipid vesicle fusion. *Biochemistry* **21:**2833–2840.

Holzwarth, J. F., Eck, V., and Genz, A., 1985, Iodine laser temperature-jump: Relaxation processes in phospholipid bilayers on the picosecond to millisecond time-scale. In *Spectroscopy and the Dynamics of Molecular Biological Systems* (P. M., Bayley and R. E. Dale, eds.), pp. 351–377, Academic Press, London.

Honda, K., Maeda, Y., Sasakawa, S., Ohno, H., and Tsuchida, E., 1981, Activities of cell fusion and lysis of the hybrid type of chemical fusogens. I. Structure and function of the promotor of cell fusion. *Biochem. Biophys. Res. Commun.* **100:**442–448.

Hubbell, W. L., and McConnell, H. M., 1968, Spin-label studies of the excitable membranes of nerve and muscle. *Proc. Natl. Acad. Sci. U.S.A.* **61:**12–16.

Hubbell, W. L., and McConnell, H. M., 1971, Molecular motion in spin labeled phospholipids and membranes. *J. Am. Chem. Soc.* **93**:383–384.

Hudson, B. S., Harris, D. L., Ludescher, R. D., Ruggiero, A., Cooney-Freed, A., and Cavalier, S. A., 1986, Fluorescence probe studies of proteins and membranes. In *Applications of Fluorescence in the Biomedical Sciences* (D. L. Taylor, A. S. Waggoner, R. F., Murphy, and F. Lanni, eds.), pp. 159–202, Alan R. Liss, New York.

Hui, S. W., 1981, Geometry of phase-separated domains in phospholipid bilayers by diffraction-contrast electron microscopy. *Biophys. J.* **34**:383–395.

Hui, S. W., Parsons, D. F., 1975, Direct observation of domains in wet lipid bilayers. *Science* **190**:314–326.

Hui, S. W., Stewart, T. P., Boni, L. T., and Yeagle, P. L., 1981, Membrane fusion through point defects in bilayers. *Science* **212**:921–923.

Hui, S. W., Boni, L. T., Stewart, T. P., and Isac, T., 1983, Identification of phosphatidylserine and phosphatidylcholine in calcium-induced phase separated domains. *Biochemistry* **22**:3511–3516.

Imai, M., Inoue, K., and Nojima, S., 1975, Effect of polymyxin B on liposomal membranes derived from *Escherichia coli* lipids. *Biochim. Biophys. Acta* **375**:130–137.

Jacobson, K., Elson, E., Koppel, D., and Webb, W., 1982, Fluorescence photobleaching in cell biology. *Nature* **295**:283–284.

Jahnig, F., 1979, Structural order of lipids and proteins in membranes: Evaluation of fluorescence anisotropy data. *Proc. Natl. Acad. Sci. U.S.A.* **76**:6361–6365.

Jahnig, F., 1981a, Critical effects from lipid–protein interactions in membranes. I. Theoretical description. *Biophys. J.* **36**:329–345.

Jahnig, F., 1981b, Critical effects from lipid–protein interaction in membranes. II. Interpretation of experimental results. *Biophys. J.* **36**:347–357.

Jain, M. K., 1983, Non-random lateral organization in bilayers and biomembranes. In *Membrane Fluidity in Biology; Concepts of Membrane Structure*, Vol. 1 (R. C. Aloia, ed.), pp. 1–37, Academic Press, Orlando.

Jain, M. K., and White, H. B., 1977, Long range order in biomembranes. *Adv. Lipid Res.* **15**:1–60.

Jost, P., Waggoner, A. S., and Griffith, O. H., 1971a, Spin labeling and membrane structure. In *Structure and Function of Biological Membranes* (L. I. Rothfield, ed.), pp. 83–144, Academic Press, New York.

Jost, P. C., Griffith, O. H., Capaldi, R. A., and Vanderkooi, G., 1971b, Evidence for boundary lipid in membranes. *Proc. Natl. Acad. Sci. U.S.A.* **70**:480–484.

Kalderon, N., and Gilula, N. B., 1979, Membrane events involved in myoblast fusion. *J. Cell Biol.* **81**:411–425.

Kanehisa, K. I., and Tsong, T. Y., 1978, Cluster model of lipid phase transitions with application to passive permeation of molecules and structure relaxations in lipid bilayers. *J. Am. Chem. Soc.* **100**:424–432.

Karnovsky, M. J., Kleinfeld, A. M., Hoover, F. L., and Klausner, R. D., 1982, The concept of lipid domains in membranes. *Cell Biol.* **94**:1–6.

Kates, M., and Wassef, M. K., 1970, Lipid chemistry. *Annu. Rev. Biochem.* **39**:323–358.

Katraro, R., Ron, A., and Speiser, S., 1979, Photophysical studies of coronene and 1,12-benzperylene. Self-quenching, photo-quenching, temperature-dependent fluorescence, decay and temperature dependent electronic energy transfer to dye acceptors. *Chem. Phys.* **42**:121–132.

Kawato, S., Kinosita, J. R., and Ikegami, A., 1977, Dynamic structure of lipid bilayers studied by nanosecond fluorescence techniques. *Biochemistry* **16**:2319–2324.

Kawato, S., Kinosita, K., and Ikegami, A., 1978, Effect of cholesterol on the molecular motion in the hydrocarbon region of lecithin bilayer studied by nanosecond fluorescence techniques. *Biochemistry* **17**:5026–5031.

Kimelberg, H. K., and Papahadjopoulos, D., 1971, Interactions of basic proteins with phospholipid membranes: binding and changes in the sodium permeability of phosphatidyserine vesicles. *J. Biol. Chem.* **246**:1142–1148.

Kimelberg, H. K., and Papahadjopoulos, D., 1972, Phospholipid requirements for ($Na^+ + K^+$)-ATPase activity. Head group specificity and fatty acid fluidity. *Biochim. Biophys. Acta* **282**:277–292.

Kinosita, K., Ikegami, A., and Kawato, S., 1982, On the wobbling-in-cone analysis of fluorescence anisotropy decay. *Biophys. J.* **37**:461–464.

Klausner, R. D., and Wolf, D. E., 1980, Selectivity of fluorescent lipid analogues for lipid domains. *Biochemistry* **19**:6199–6203.

Klausner, R. D., Kleinfeld, A. M., Hoover, R. L., and Karnovsky, J. J., 1980a, Lipid domains in membranes: Evidence derived from structural perturbations induced by free fatty acids and lifetime heterogeneity analysis. *Biol. Chem.* **255**:1286–1295.

Klausner, R. D., Kleinfeld, A. M., Hoover, R. L., and Karnovsky, M. J., 1980b, Lipid domains in membranes: Evidence derived from structural perturbations induced by free fatty acids and lifetime heterogeneity analysis. *J. Biol. Chem.* **255**:1286–1295.

Knutson, J. R., Walbridge, D. G., and Brand, L., 1981, Resolution of fluorescent spectra in a mixture by means of nanosecond time resolved fluorescence spectroscopy. *Am. Soc. Photobiol.* **9**:62.

Knutson, J. R., Walbridge, D. G., and Brand, L., 1982, Decay-associated fluorescence spectra and the heterogeneous emission of alcohol dehydrogenase. *Biochemistry* **21**:4671–4679.

Knutson, J.R., Beechem, J. M., and Brand, L., 1983, Simultaneous analysis of multiple fluorescence decay curves: A global approach. *Chem. Phys. Lett.* **102**:501–507.

Knutson, J. R., Davenport, L., and Brand, L., 1986, Anisotropy decay associated fluorescence spectra and analysis of rotational heterogeneity. 1. Theory and application. *Biochemistry* **25**:5026–5031.

Kouaouci, R., Silvius, J. R., Graham, I., and Pezolet, M., 1985, Calcium induced lateral phase separations in phosphatidylcholine–phosphatidic-acid mixtures. A Raman spectroscopic study. *Biochemistry* **24**:7132–7140.

Kowalczyk, A. A., Knutson, J. R., Barkley, M. D., Christy, R., and Brand, L., 1982, Anisotropic rotations of perylene in liposomes. In *Conference Digest: Fourth Conference on Luminescence* (A. A. Kowalczyk, ed.), pp. 187–189, Debrecin Press, Budapest.

Ladbrooke, B. D., Williams, R. M., and Chapman, D., 1968, Studies of lecithin–cholesterol–water interactions by differential scanning calorimetry and X-ray diffraction. *Biochim. Biophys. Acta* **150**:333–340.

Lagunoff, D., 1973, Membrane fusion during mast cell secretion. *J. Cell Biol.* **57**:232–250.

Lai, M. Z., Vail, W. J., and Szoka, F. C., 1985, Acid- and calcium-induced structural changes in phosphatidylethanolamine membranes stabilized by cholesteryl hemisuccinate. *Biochemistry* **24**:1654–1661.

Lakowicz, J. R., 1983, *Principles of Fluorescence Spectroscopy*, Plenum Press, New York.

Lakowicz, J. R., and Knutson, J. R., 1980, Hindered depolarizing rotations of perylene in lipid bilayers. Detection by lifetime resolved fluorescence anisotropy measurements. *Biochemistry* **19**:905–911.

Lakowicz, J. R., Prendergast, F. G., and Hogen, D., 1979, Differential polarized phase fluorometric investigations of diphenylhexatriene in lipid bilayers. Quantitation of hindered depolarizing rotations. *Biochemistry* **18**:508–519.

Lakowicz, J., Cherek, H., and Maliwal, B., 1985, Time-resolved fluorescence anisotropies of diphenylhexatriene and perylene in solvents and lipid bilayers obtained from multifrequency phase-modulation fluorometry. *Biochemistry* **24**:376–384.

Lamotte, M., Lesclaux, R., Merle, A. N., and Joussot-Dubien, J., 1975, Spectroscopic studies of orientational interactions between straight-chain alkanes and aromatic hydrocarbons. *Faraday Discuss. Chem. Soç.* **58**:253–260.

Laws, W. R., and Brand, L., 1979, Analysis of two-state excited-state reactions: The fluorescence decay of 2-naphthol. *J. Phys. Chem.* **83**:795–802.

Lawson, D., Raff, M. C., Camperts, B., Fewtrell, C., and Gilula, N. B., 1977, Molecular events during membrane fusion. A study of exocytosis in rat peritoneal mast cells. *J. Cell. Biol.* **72**:242–259.

Lee, A. G., 1977, Annular events: Lipid–protein interactions. *Trends Biochem. Sci.* **2**:231–233.

Lelkes, P. I., Bach, D., and Miller, I. R., 1980, Perturbation of membrane structure by optical probes: II. Differential scanning calorimetry of dipalmitoyllecithin and its analogs interacting with merocyanine 540. *J. Membr. Biol.* **54**:141–148.

Lentz, B. R., Barenholz, Y., and Thompson, T., 1976a, Fluorescence depolarization studies of phase transitions and fluidity in phospholipid bilayers. 1. Single component phosphatidyl choline liposomes. *Biochemistry* **15**:4521–4528.

Lentz, B. W., Barenholz, Y., and Thompson, T., 1976b, Fluorescence depolarization studies of phase transitions and fluidity in phospholipid bilayers. 2. Two component phosphatidylcholine liposomes. *Biochemistry* **15**:4529–4536.

Levine, Y. K., and Wilkins, M. H. F., 1971, Structure of oriented lipid bilayers. *Nature (London) New Biol.* **230**:69–72.

Lianos, P., and Georghiou, S., 1979, Solute–solvent interaction and its effect on the vibronic and vibrational structure of pyrene spectra. *Photochem. Photobiol.* **30**:355–362.

Liao, M.-J., and Prestegard, J. H., 1979, Fusion of phosphatidic acid–phosphatidylcholine mixed lipid vesicles. *Biochim. Biophys. Acta* **550**:157–173.

Liao, M.-J., and Prestegard, J. H., 1981, Structural properties of a Ca^{2+} phosphatidic acid complex: Small angle X-ray scattering and calorimetric results. *Biochim. Biophys. Acta* **645**:147–156.

Lin, H. B., and Topp, M. R., 1977, Low quantum-yield molecular fluorescence. Aromatic hydrocarbons in solution at 300 K. *Chem. Phys. Lett.* **48**:251–255.

Linden, C., Wright, K. L., McConnell, H. M., and Fox, C. F., 1973, Lateral phase separations in membrane lipids and mechanisms of sugar transport in *Escherichia coli*. *Proc. Natl. Acad. Sci. U.S.A.* **70**:2271–2275.

Lipari, G., and Szabo, A., 1980, Effect of vibrational motion on fluorescence depolarization and nuclear magnetic resonance relaxation in macromolecules in membranes. *Biophys. J.* **30**:489–506.

Lucy, J. A., and Ahkong, Q. F., 1986, An osmotic model for the fusion of biological membranes. *FEBS Lett.* **199**:1–11.

Lux, S. E., 1979, Dissecting the red cell membrane skeleton. *Nature (London)* **281**:427–429.

Luzzati, V., 1968, X-ray diffraction studies of lipid–water systems. In *Biological Membranes: Physical Fact and Function* (D. Chapman, ed.), pp. 71–124, Academic Press, London.

Mabrey, S., and Sturtevant, J. M., 1976, Investigation of phase transitions of lipids and lipid mixtures by high sensitivity DSC. *Proc. Natl. Acad. Sci. U.S.A.* **73**:3862–3866.

Mabrey, S., Mateo, P. L., and Sturtevant, J. M., 1978, High sensitivity scanning calorimetric study of mixtures of cholesterol with dimyristoyl- and dipalmitoyl phosphatidylcholines. *Biochemistry* **17**:2464–2468.

Marcelja, S., 1974, Chain ordering in liquid crystals. 2. Structure of bilayer membranes. *Biochim. Biophys. Acta* **367**:165–176.

Marsh, D., and Barrantes, F. T., 1978, Immobilized liquid in acetylcholine receptor-rich membranes from *Torpedo marmorata. Proc. Natl. Acad. Sci. U.S.A.* **75**:4329–4333.
McConnell, H. M., Devaux, P., and Scandella, C. J., 1972, Lateral diffusion and phase separation in biological membranes. In *Membrane Research, ICN-UCLA Symposium on Molecular Biology, Proceedings 1st* (C. F. Fox, ed.), pp. 27–37, Academic Press, New York.
McElhaney, R. N., and Souza, K. A., 1976, The relationship between environmental temperature, cell growth and the fluidity and physical state of the membrane lipids in *Bacillus stearothermophilus. Biochim. Biophys. Acta* **443**:348–359.
Melchior, D. L., Scavitto, F. J., and Stein, J. M., 1980, Dilatometry of dipalmitoyllecithin–cholesterol bilayers. *Biochemistry* **19**:4828–4834.
Mendelson, R., Sunder, S., and Bernstein, H. J., 1976, The effect of sonication on the hydrocarbon chain conformation in model membrane systems: A Raman spectroscopic study. *Biochim. Biophys. Acta* **419**:563–569.
Michaelson, D. M., Horwitz, A. F., and Klein, M. P., 1973, Transbilayer asymmetry and surface homogeneity of mixed phospholipids in cosonicated vesicles. *Biochemistry* **12**:2637–2645.
Mitaku, S., 1981, Ultrasonic studies of lipid bilayer phase transition. *Mol. Cryst. Liq. Cryst.* **70**:1299–1306.
Mitaku, S., and Date, T., 1982, Anomalies of nanosecond ultrasonic relaxation in the lipid bilayer transition. *Biochim. Biophys. Acta* **688**:411–421.
Mitaku, S., Jippo, T., and Kataoka, R., 1983, Thermodynamic properties of the lipid bilayer transition. *Biophys. J.* **42**:137–144.
Morrot, G., Cribier, S., Devaux, P. F., Geldwerth, D., Davoust, J., Bureau, J. F., Fellmann, P., and Herve, P., 1986, Asymmetric lateral mobility of phospholipids in the human erythrocyte membrane. *Proc. Natl. Acad. Sci. U.S.A.* **83**:6863–6867.
Mukhopadhyay, A. K., and Georghiou, S., 1980, Solvent-induced enhancement of weakly allowed vibronic transitions of aromatic hydrocarbons. *Photochem. Photobiol.* **31**:407–411.
Nagle, J. F., and Scott, H. L. Jr., 1978, Lateral compressibility of lipid mono- and bilayers theory of membrane permeability. *Biochim. Biophys. Acta* **513**:236–243.
Nicolson, G. L., 1972, Concanavalin A: Modification of cell membrane site topography by proteolytic enzyme. *Nature (London) New Biol.* **239**:193–197.
Ohki, S., 1984, Effects of divalent cations, osmotic pressure gradient, and vesicle curvature on phosphatidylserine vesicle fusion. *J. Membr. Biol.* **77**:265–275.
Op den Kamp, J. A. F., Redai, I., and Van Deenen, L. L. M., 1969, Phospholipid composition of *Bacillus subtilis. J. Bacteriol.* **99**:298–303.
Op den Kamp, J. A. F., Verheij, H. M., and Van Deenen, L. L. M., 1971, Two isomers of glucosaminylphosphatidylglycerol. Their occurrence in *Bacillus megaterium,* structural analysis, and chemical synthesis. *Bioorg. Chem.* **1**:174–187.
Opella, S. J., Yesinowski, J. P., and Waugh, J. S., 1976, Nuclear magnetic resonance description of molecular motion and phase separation of cholesterol in lecithin dispersions. *Proc. Natl. Acad. Sci. U.S.A.* **73**:3812–3815.
Papahadjopoulos, D., Jacobson, K., Nir, S., and Isac, T., 1973, Phase transitions in phospholipid vesicles: Fluorescence polarization and permeability measurements concerning the effect of temperature and cholesterol. *Biochim. Biophys. Acta* **311**:330–348.
Papahadjopoulos, D., Vail, W. J., Newton, C., Nir, S., Jacobson, K., Poste, G., and Lazo, R., 1977, Studies of membrane fusion. III. The role of calcium-induced phase changes. *Biochim. Biophys. Acta* **465**:579–598.
Papahadjopoulos, D., Portis, A., and Pangborn, W., 1978, Calcium-induced lipid phase transitions and membrane fusion. *Ann. N.Y. Acad. Sci.* **308**:50–66.
Parasassi, T., Conti, F., Glaser, M., and Gratton, E., 1984, Detection of phospholipid phase

separation. A multifrequency phase fluorometry study of 1,6-diphenyl-1,3,5-hexatriene fluorescence. *J. Biol. Chem.* **259**:14011–14017.

Parasassi, T., De Stasio, G., Gratton, E., and Conti, F., 1987, Fluorescence lifetime distribution of parinaric acid isomers in isotropic solvents. *Biophys. J.* **51**:538a.

Parente, R. A., and Lentz, B. R., 1986, Rate and extent of poly(ethylene-glycol)-induced large vesicle fusion monitored by bilayer and internal contents mixing. *Biochemistry* **25**:6678–6688.

Parola, A. H., Robbins, P. W., and Blout, E. R., 1979, Membrane dynamic alterations associated with viral transformation and reversion: Decay of fluorescence emission and anisotropy studies of 3T3 cells. *Exp. Cell Res.* **118**:205–214.

Perrin, F., 1934, Mouvement Brownien D'un Ellipsoide: I. Dispersion Dielectrique des Molecules Ellipsoidales. *J. Phys. Radiat. Paris* **5**:497–511.

Perrin, F., 1936, Mouvement Brownien D'un Ellipsoide: II. Rotation libre et Depolarisation des Fluorescences Translation et Diffusion de Molecules Ellipsoidales. *J. Phys. Radiat. Paris* **7**:1–11.

Peters, R., 1981, Translation diffusion in the plasma membrane of single cells as studied by fluorescence microphotolysis. *Cell Biol. Int. Rep.* **5**:733–760.

Peterson, N. O., and Chan, S. I., 1977, More on the motional state of lipid bilayer membranes: Interpretation of order parameters obtained from nuclear magnetic resonance experiments. *Biochemistry* **16**:2657–2667.

Phillips, M. C., Williams, R. M., and Chapman, D., 1969, Hydrocarbon chain motions in lipid liquid crystals. *Chem. Phys. Lipids* **3**:234–244.

Phillips, M. C., Ladbrooke, B. D., and Chapman, D., 1970, Molecular interactions in mixed lecithin systems. *Biophys. Biochim. Acta* **196**:35–44.

Portis, A., Newton, C., Pangborn, W., and Papahadjopoulos, D., 1979; Studies of the mechanism of membrane fusion: evidence for an intermembrane Ca^{2+}–phospholipid complex, synergism with Mg^{2+}, and inhibition by spectrin. *Biochemistry* **18**:780–790.

Raison, J. K., Lyons, J. M., Melhorn, R. J., and Keith, A. D., 1971, Temperature-induced phase changes in mitochondrial membranes detected by spin labelling. *J. Biol. Chem.* **246**:4036–4040.

Rogers, J., Lee, A. G., and Wilton, D. C., 1979, The organization of cholesterol and ergosterol in lipid bilayers based on studies using non-perturbing fluorescent steroid probes. *Biochim. Biophys. Acta* **552**:23–37.

Rothman, J. E., and Leonard, J., 1977, Membrane asymmetry. The nature of membrane asymmetry provides clues to the puzzle of how membranes are assembled. *Science* **195**:743–753.

Rothman, J. E., Tsai, D. K., Davidowicz, E. A., and Lenar, J., 1976, Transbilayer phospholipid asymmetry and its maintenance in the membrane of influenza virus. *Biochemistry* **15**:2361–2370.

Rouser, G., Nelson, G. J., Fleischer, S., and Simon, G., 1968, Lipid composition of animal cell membranes, organelles, and organs. In *Biological Membranes: Physical Fact and Function* (D. Chapman, ed.), pp. 6–70, Academic Press, London.

Scandella, C. J., Devaux, P., and McConnell, H. M., 1972, Rapid lateral diffusion of phospholipids in rabbit sarcoplasmic reticulum. *Proc. Natl. Acad. Sci. U.S.A.* **69**:2056–2060.

Schroeder, F., 1985, Fluorescence probes unravel asymmetric structure of membranes. *Subcell. Biochem.* **11**:51–101.

Seelig, A., and Seelig, J., 1974, Deuterium magnetic resonance studies of phospholipid bilayers. *Biochem. Biophys. Res. Commun.* **57**:406–410.

Seiter, C. H. A., and Chan, S. I., 1973, Molecular motions in lipid bilayers. A nuclear magnetic resonance line with study. *J. Am. Chem. Soc.* **95**:7541–7553.

Sene, C., Genest, D., Obrenovitch, A., Wahl, P. H., and Monsigny, M., 1978, Pulse fluorimetry of 1,6-diphenyl-1,3,5-hexatriene incorporated into membranes of mouse leukemic L1210 cells. *FEBS Lett.* **88:**181–186.

Shimshick, E. J., and McConnell, H. M., 1973a, Lateral phase separations in binary mixtures of cholesterol and phospholipids. *Biochem. Biophys. Res. Commun.* **53:**446–451.

Shimshick, E. J., and McConnell, H. M., 1973b, Lateral phase separation in phospholipid membranes. *Biochemistry* **12:**2351–2360.

Shinitzky, M., and Barenholz, Y., 1974, Dynamics of the hydrocarbon layer in liposomes of lecithin sphingomyelin containing dicetylphosphate. *J. Biol. Chem.* **249:**2652–2657.

Shinitzky, M., Dianoux, A.-C., Gitler, C., and Weber, G., 1971, Microviscosity and order in the hydrocarbon region of micelles and membranes determined with fluorescent probes. 1. Synthetic micelles. *Biochemistry* **10:**2106–2113.

Sieber, F., 1987, Merocyanine 540. *Photochem. Photobiol.* **46:**1035–1042.

Siegel, D. P., 1986, Inverted micellar intermediates and the transitions between lamellar, cubic, and inverted hexagonal lipid phases. *Biophys. J.* **49:**1171–1183.

Singer, S. J., and Nicolson, G. L., 1972, The fluid mosaic model of the structure of cell membranes. *Science* **175:**720–731.

Sklar, L. A., Hudson, B. S., and Simoni, R. D., 1975, Conjugated polyene fatty acids as membrane probes: Preliminary characterization. *Proc. Natl. Acad. Sci. U.S.A.* **72:**1649–1653.

Sklar, L. A., Hudson, B. S., and Simoni, R. D., 1977, Conjugated polyene fatty acids as fluorescence probes: Synthetic phospholipid membrane studies. *Biochemistry* **16:**819–828.

Steck, T. L., and Fox, C. F., 1972, Membrane proteins. In *Membrane Molecular Biology* (C. F. Fox and A. D. Keith, eds.), pp. 27–75, Sinauer Associates, Stamford, Connecticut.

Stockton, G. W., Polnaszek, C. F., Tulloch, A. P., Hasa, F., and Smith, I. C. P., 1976, Molecular motion and order in single-bilayer vesicles and multilamellar dispersions of egg lecithin and lecithin–cholesterol mixtures. A deuterium nuclear magnetic resonance study of specifically labeled lipids. *Biochemistry* **15:**954–966.

Stubbs, C. D., Kouyama, T., Kinosita, K., and Ikegami, A., 1981, Effect of double bonds on the dynamic properties of the hydrocarbon region of lecithin bilayers. *Biochemistry* **20:**4257–4262.

Tanaka, K.-I., and Ohnishi, S.-I., 1976, Heterogeneity in the fluidity of intact erythrocyte membrane and its homogenization upon hemolysis. *Biochim. Biophys. Acta* **426:**218–231.

Tao, T., 1969, Time-dependent fluorescence depolarization and Brownian rotational diffusion coefficients of macromolecules. *Biopolymers* **8:**609–632.

Taylor, R. B., Duffus, W. P. H., Raff, M. C., and dePetris, S., 1971, Redistribution and pinocytosis of surface immunoglobulin molecules. *Nature (London) New Biol.* **233:**225–230.

Thulborn, K. R., 1981, The use of N-[9-anthroyloxy] fatty acids as fluorescence probes for biomembranes. In *Fluorescent Probes* (G. S. Beddard and M. A. West, eds.), pp. 113–141, Academic Press, London.

Thulborn, K. R., and Sawyer, T. W., 1978, Properties and the locations of a set of fluorescence probes sensitive to the fluidity gradient of the lipid bilayer. *Biochim. Biophys. Acta* **511:**125–140.

Trauble, H., and Sackmann, E., 1972, Studies of the crystalline–liquid crystalline phase transition of lipid model membranes. III. Structure of a steroid–lecithin system below and above the lipid-phase transition. *J. Am. Chem. Soc.* **94:**4499–4510.

Tsong, T. Y., 1977, Effect of phase transition on the kinetics of dye transport in phospholipid bilayer structures. *Biochemistry* **16:**2674–2684.

Vanderkooi, J. M., Fischkoff, S., Andrich, M., Podo, F., and Owen, C. S., 1975, Diffusion in two dimensions: Comparison between diffusional fluorescence quenching in phospholipid vesicles and in isostropic solution. *J. Chem. Phys.* **63:**3661–3666.

Veatch, W. R., and Stryer, L., 1977, Effect of cholesterol on the rotational mobility of DPH in liposomes: A nanosecond anisotropy study. *J. Mol. Biol.* **117**:1109–1113.

Verkleij, A. J., Ververgaert, P. H. J., de Kruijff, B., and Van Deenen, L. L. M., 1974, Distribution of cholesterol in bilayers of phosphatidylcholine as visualized by freeze–fracture. *Biochim. Biophys. Acta* **373**:495–501.

Vincent, M., deForester, B., Gallay, J., and Alfsen, A., 1982, Nanosecond fluorescence anisotropy decays of N-(9-anthroyloxy) fatty acids in dipalmitoylphosphatidylcholine vesicles with regard to isotropic solvents. *Biochemistry* **21**:708–716.

Vo Dinh, T., Leu Yen, E., and Winefordner, J. D., 1977, Room temperature phosphorescence of several polyaromatic hydrocarbons. *Talanta* **24**:146–148.

Volsky, D. J., and Loyter, A., 1978, Role of Ca^{2+} in virus-induced membrane fusion. Ca^{2+} accumulation and ultrastructural charges induced by Sendai virus in chicken erythrocytes. *J. Cell Biol.* **78**:465–479.

Waggoner, A. S., 1986, Fluorescent probes for the analysis of cell structure, function, and health by flow and imaging cytometry. In *Applications of Fluorescence in the Biomedical Sciences* (D. L. Taylor, R. F. Waggoner, R. F. Murphy, and F. Lanni, eds.), pp. 3–28, Alan R. Liss, New York.

Webb, W. W., Barak, L. S., Tank, D. W., and Wu, E.-S., 1981, Molecular mobility on the cell surface. *Biochem. Soc. Symp.* **46**:191–205.

Williamson, P., Bateman, J., Kozarshy, K., and Mattocks, K., 1982, Involvement of spectrin in the maintenance of phase-state asymmetry in the erythrocyte membrane. *Cell* **30**:725–733.

Williamson, P., Mattock, K., and Schlegel, R. A., 1983, Merocyanine 540, a fluorescent probe sensitive to lipid packing. *Biochim. Biophys. Acta* **732**:387–393.

Wojcieszyn, J. W., Schlegel, R. A., Lumley-Sapanski, K., and Jacobson, K. A., 1983, Studies on the mechanism of polyethylene glycol-mediated cell fusion using fluorescent membrane and cytoplasmic probes. *J. Cell Biol.* **96**:151–159.

Wolber, P. K., and Hudson, B. S., 1981, Fluorescence lifetime and time-resolved polarization anisotropy studies of acyl chain order and dynamics in lipid bilayers. *Biochemistry* **20**:2800–2808.

Wolber, P. K., and Hudson, B. S., 1982, Bilayer acyl chain dynamics and lipid–protein interaction: The effect of M13 bacteriophage coat protein on the decay of the fluorescence anisotropy of parinaric acid. *Biophys. J.* **37**:253–262.

Wolf, D. E., 1984, Overcoming random diffusion in polarized cells: Corralling the drunken beggar. *BioEssays* **6**:116–121.

Wolf, D. E., 1988, Probing the lateral organization and dynamics of membranes. In *Spectroscopic Membrane Probes (CRC Critical Reviews)* (L. Loew, ed.), CRC Press, Boca Raton, Florida, in press.

Wolf, D. E., and Edidin, M., 1981, Diffusion and mobility of molecules in surface membranes. In *Techniques in Cellular Physiology:* Part I (P. Baker, ed.), pp. 1–14, Elsevier/North Holland Scientific Publishers, Ireland.

Wolf, D. E., Edidin, M., and Handyside, A. M., 1981a, Changes in the organization of the mouse egg plasma membrane upon fertilization and first cleavage. Indications from the lateral diffusion rates of fluorescent lipid analogs. *Dev. Biol.* **85**:195–198.

Wolf, D. E., Kinsey, W., Lennarz, W., and Edidin, M., 1981b, Changes in the organization of the sea urchin egg plasma membrane upon fertilization: Indications from lateral diffusion rates of lipid-soluble fluorescent dyes. *Dev. Biol.* **81**:133–138.

Worster, D. L., and Franks, N. P., 1976, Structural analysis of hydrated egg lecithin and cholesterol bilayers. 2. Neutron diffraction. *J. Mol. Biol.* **100**:359–378.

Wu, S. H. W., and McConnell, H. M., 1973, Lateral phase separations and perpendicular transport in membranes, *Biochem. Biophys. Res. Commun.* **55** (2):484–491.

Wunderlich, F., Ronai, A., Speth, V., Seelig, J., and Blume, A., 1975, Thermotropic lipid clustering in tetrahymena membranes. *Biochemistry* **14**:3730–3734.

Yahara, I., and Edelman, G., 1972, Restriction of the mobility of lymphocyte immunoglobulin receptors by concanavalin. *Proc. Natl. Acad. Sci. U.S.A.* **69**:608–612.

Yang, R. D., Patel, K. M., Pownall, H. J., Knapp, R. D., Sklar, L. A., Crawford, R. B., and Morrisett, J. D., 1979, Biophysical properties of a major membrane phospholipid, dielaidoyl phosphatidylethanolamine, found in *Escherichia coli* fatty acid auxotroph. *J. Biol. Chem.* **254**:8256–8262.

Yechiel, E., and Edidin, M., 1987. Micrometer-scale domains in fibroblast plasma membranes. *J. Cell Biol.* **105**:755–760.

Yechiel, E., Barenholz, Y., and Hemis, Y. I., 1986, Lateral mobility and organization of phospholipids and proteins in rat myocyte membranes. *J. Biol. Chem.* **260**:9132–9136.

Zacharasse, K. A., Kuhnle, W., and Weller, A., 1980, Intramolecular excimer fluorescence as a probe of fluidity changes and phase transitions in phosphatidylcholine bilayers. *Chem. Phys. Lett.* **73**:6–11.

Zannoni, C., 1979, A theory of time-dependent fluorescence depolarization in liquid crystals. *Mol. Phys.* **38**:1813–1827.

Zannoni, C., Arcioni, A., and Cavatorta, P., 1983, Fluorescence depolarization in liquid crystals and membrane bilayers. *Chem. Phys. Lipids* **32**:179–250.

Zimmermann, H. V., and Joop, N., 1961, Polarization der Elektronenbanden von Aromaten. 5. Mitteilung: Benzol, Coronen, Triphenylen, Pyren, Perylen. *Z. Electrochem.* **65**:138–142.

Chapter 5

Membrane Fusion
Fusogenic Agents and Osmotic Forces

J. A. Lucy and Q. F. Ahkong

1. INTRODUCTION

Many of the dynamic features of the behavior of biological membranes, at cellular and subcellular levels, depend on the phenomenon of membrane fusion.

At the cellular level, membrane fusion occurs in the fusion of myoblasts into myotubes during the development of skeletal muscle. Cell fusion between a sperm and egg cell is also, of course, a central feature of fertilization. In addition, however, the developing egg invades and becomes implanted in the uterus following the destruction of uterine epithelium by the multinucleated, syncytial trophoblast, which is also formed by cell fusion. Multinucleated osteoclasts arise from the fusion of monocytes, although a consensus as to the identity of the precursor cell in this case has been reached only recently (Fawcett, 1986). The multinucleated cells that occur in pathological conditions, such as those seen in tuberculosis, arthritis, and malignant diseases may also arise from cell fusion. The majority of investigators, however, are probably best acquainted with hybridoma cells, produced in the laboratory by fusion of myeloma cells with lymphocytes with the aid of polyethylene glycol, as an illustration of membrane fusion occurring between cells.

J. A. Lucy and Q. F. Ahkong   Department of Biochemistry and Chemistry, Royal Free Hospital School of Medicine, University of London, London NW3 2PF, United Kingdom.

At the subcellular level, the most familiar and the most studied example of membrane fusion is exocytosis, which occurs in a wide variety of secretory cells. The phenomenon is no less important, however, in endocytosis. Several fusion reactions occur also in the flow of membranes, which links endocytosis with exocytosis, via endosomes, and which is responsible for the recycling of plasma membrane receptors. The compartmentalization and flow of newly synthesized proteins, via the rough endoplasmic reticulum and Golgi regions, also depend on membrane fusion occurring in the correct locations at the appropriate times, as does the coordinated functioning of the lysosomal degradative system. The very variety of the situations, both cellular and subcellular, in which membrane fusion occurs thus underlines the importance and potential benefits that may follow from a more complete understanding of the phenomenon than is available at present.

Studies on membrane fusion have concentrated on areas that are relatively accessible to experimentation, namely, the fusion of cells (particularly erythrocytes) induced by chemicals (Lucy, 1978) and by the reversible electrical breakdown of their membranes (Zimmermann, 1982), cell fusion induced by viruses (Spear 1987; White *et al.*, 1983), the fusion of myoblasts in tissue culture (Wakelam, 1985), exocytosis (Poste and Nicolson, 1978), and fusion in phospholipid vesicles. The last of these areas, which has been the most extensively studied, is well covered by recent reviews (Blumenthal, 1987; Düzgüneş, 1985; Ohki *et al.*, 1988; Sowers, 1987a; Wilschut and Hoekstra, 1984) and it is not considered in any detail here. The books edited by Poste and Nicolson (1978) and by Evered and Whelan (1984) provide comprehensive reviews of many aspects of earlier work on membrane fusion.

In this chapter an attempt is made to integrate findings that have been made over a number of years on the possible roles of osmotic forces in the artificially induced fusion of cells, in cell fusion induced by viruses, and in membrane fusion in exocytosis, and to relate them to the behavior of cytoskeletal structures in the fusion reactions.

2. FUSOGENIC LIPIDS

2.1. Osmotic Effects

In early work undertaken in the authors' laboratory, the hen erythrocyte was treated with a number of simple lipids in a study of their fusogenic properties. Fusogenic lipids were found to include saturated carboxylic acids containing 10–14 carbon atoms, longer but unsaturated carboxylic acids (e.g., oleic acid), and their monoesters of glycerol. Long-chain, saturated carboxylic acids (e.g., stearic acid) did not cause cell fusion. Erythrocytes that were treated with fusogenic lipids exhibited colloid osmotic swelling, and they became spherical within a few minutes. When EDTA was present, the swollen cells lysed more readily than they fused, but when two or more of the spherical cells came together in the presence of calcium ions, the cells aggregated and fused before they lysed (Ahkong et al., 1973a; Lucy, 1973). Significantly, fusion between oval, unswollen erythrocytes was never seen, and every chemical that was found to induce fusion also caused the cells to become spherical. We therefore concluded that cell swelling by colloidal osmosis plays an essential role in cell fusion (Ahkong et al., 1973b).

Kosower et al (1977, 1978) observed that 2-(2-methoxyethoxy) ethyl cis-8-(2-octylcyclopropyl)octanoate, designated A_2C and described as a membrane-mobility agent, promoted the fusion of erythrocytes from several strains of Leghorn hens with a high efficiency. They noted the following "typical" sequence of events: oval cells change to round cells, round cells approach and stick to other round cells, and then the cells fuse into binucleated cells that often participate in additional fusion steps to give multinucleated cells. Under the same conditions, erythrocytes from White Rock hens neither changed shape nor fused on treatment with A_2C. Maggio et al. (1978) similarly observed that hen erythrocytes fused by polysialogangliosides became round after 90 min and that binucleated and multinucleated cells then appeared after about 3 hr of incubation.

When the behavior of avian erythrocytes on a heating stage at 48–50°C was investigated in the absence of chemical fusogens, a similar relationship between cell swelling and cell fusion was seen (Ahkong et al., 1973b). Thus, it was noted that the cells rapidly changed in shape from oval to spherical. Some of the swollen, spherical cells then fused to produce multinucleated cells and, as with chemically induced fusion, lysis of both unfused and fused cells occurred.

The participation of cell swelling in the fusion process is not simply a vagary of the hen erythrocyte and its membrane since, during an investigation of the fusion of fibroblasts induced by the lipid fusogen, oleylamine, we found that fusion was again preceded by rounding of the cells (Bruckdorfer et al., 1974). The fibroblasts subsequently swelled to form large clear spaces immediately beneath the plasma membrane, which were attributed to a rapid influx of water. After 10–15 min, swollen cells that were in contact fused to form polykaryons. The dividing membranes between the fused cells "popped" with great speed, apparently having reached a critical point due to mechanical stress or chemical damage.

Since treating different cells in various ways induced them to swell and then to fuse, there seemed to be a causal relationship between swelling and fusion. If such a relationship exists, it might be anticipated that exposing cells to a hypotonic environment would induce them to fuse. However, this has not been observed in our laboratory, and it would also appear not to have been reported elsewhere. It could therefore be concluded that osmotic swelling did not cause cell fusion in the above work. Alternatively, the various fusogenic treatments may have induced several changes in the properties of the cells, including an increase in membrane permeability that resulted in cell swelling, and then caused the already modified cells to fuse. More recent studies have indicated that the latter interpretation appears to be the correct one. Thus, a likely reason for the failure of erythrocytes to fuse when they are subjected to osmotic swelling alone is that structural changes, which are induced in the plasma membrane by fusogenic treatments, do not occur. In relation to the heat-induced fusion of human erythrocytes, for example, it is relevant that the swelling of these cells at osmolarities between 200 and 700 mOsm is normally restricted by the membrane skeleton. After heating to 50°C for 60 min or exposing them to diamide, however, their swelling is approximately ideal and this is probably because these conditions are known to disrupt the spectrin–actin skeletal network (Heubusch et al., 1985). Coakley and Deeley (1980) showed that when human erythrocytes are heated to about 50°C in isotonic saline, they develop an unstable surface wave on the cell rim. Externalized vesicles then pinch off from the crests of the growing surface wave. Coakley and his colleagues also observed that cell fusion occurred at and above the denaturation temperature of the erythrocytes in the presence of 3–4 mM tetracaine (Coakley et al., 1983) and, in a recent article, they have reviewed the development of instability in cell membranes under a variety of conditions (Gallez and Coakley, 1986).

Structural damage to the membrane skeleton of erythrocytes appears to be the feature that is common to heating erythrocytes and exposing them to lipid fusogens, which allows the cells to fuse in response to subsequent osmotic swelling. Whereas heating human erythrocytes denatures their spectrin, treatment with lipid fusogens can activate endogenous proteinases.

2.2. Cellular Proteases

Membrane proteins in human erythrocytes were first shown to be degraded during cell fusion by Quirk et al. (1978). They found that bands 2.1 and 2.2 were lost and that band 2.3 was increased when the cells were induced to fuse by treatment with oleoylglycerol. Subsequently, band 2.3 disappeared almost completely, band 3 protein was decreased, and band 4.1 was lost. At the same time, a new band developed in the band 3 region which appeared to result from proteolysis of band 2.3. Bands 4.3 and 4.5 also became more prominent and a new component appeared that moved slightly ahead of band 6. Most, but not all, of the changes in the membrane proteins seemed to result from the entry of Ca^{2+} into the cells. The proteinase inhibitor, Tos-Phe-CH_2Cl partially inhibited both cell fusion and the loss of band 3. It was therefore proposed that the proteolytic breakdown of membrane proteins might be involved in cell fusion induced by oleoylglycerol, and that a consequent increase in the freedom of movement of other membrane proteins could be an important feature of membrane fusion. From work on the effects of chemical fusogens on the permeability of erythrocytes to Ca^{2+} and other ions, it was later also suggested that an increase in the concentration of intracellular Ca^{2+} initiates or facilitates events leading to the chemically induced fusion of erythrocytes (Blow et al., 1979). Interestingly, in the fusion of carrot protoplasts with *Xenopus* cells, protease treatment enhanced fusion by a high pH/high Ca^{2+} method (Ward et al., 1979).

A subsequent investigation on rat erythrocytes fused by benzyl alcohol indicated that a Ca^{2+}-activated thiol proteinase was important in the fusion process. In the treated cells, proteins corresponding to bands 2 and 3 in human erythrocytes were decreased, and a polypeptide with a slightly greater mobility than band 3 was produced. These changes were inhibited by EGTA, thiol reagents, and the proteinase inhibitor Tos-Lys-CH_2Cl. In addition, the intramembranous particles of the P-fracture face of the cells treated with benzyl alcohol were decreased in number and were susceptible to cold-induced aggregation; these phenomena were markedly inhibited by EGTA and partially inhibited by Tos-Lys-CH_2Cl and *N*-ethylmaleimide. A Ca^{2+}-activated thiol proteinase, which acts to degrade membrane proteins and to give freedom of lateral movement to intramembranous particles, therefore appeared to be an essential feature of membrane fusion in this system, and it was suggested that the proteinase degrades spectrin-binding proteins that attach band 3 protein to the membrane skeleton of the erythrocyte (Ahkong et al., 1980).

Thomas et al. (1983) found that treatment of chicken erythrocytes with the ionophore A23187 in the presence of Ca^{2+} resulted in a proteolytic breakdown of spectrin and microtubule-associated proteins to large well-defined fragments. A protein with an electrophoretic mobility similar to that of goblin in turkey erythrocytes was also degraded. Goblin has been shown by a variety of criteria to be the structural and functional analogue in avian erythrocytes of ankyrin (see below) in human erythrocytes (Nelson and Lazarides, 1984). A Ca^{2+}-sensitive protease again appeared to be responsible for these changes, which were inhibited by Tos-Lys-Ch_2Cl and by iodoacetamide, and it was suggested that the degradation of certain proteins associated with the membrane skeleton is needed for nucleus–plasma membrane fusion and for the release of spectrin-free vesicles into the extracellular medium.

From later work on the fusion of human erythrocytes by chlorpromazine, it was apparent that the increased lateral mobility of the intramembranous particles that accompanied fusion resulted from an activation by the fusogen of endogenous proteinases which act to degrade both band 2.1 protein (ankyrin) and band 3 protein (Lang et al., 1984). Since other cells have structures that are comparable with the spectrin–actin skeleton of the erythrocyte membrane, it was proposed that these observations may be relevant to the initiation of naturally occurring fusion reactions in a variety of biomembranes. In the fused human erythrocytes, proteolysis of ankyrin to band 2.3–2.6 proteins and to proteins that migrated in the band 3 region was apparently achieved by a serine proteinase that is inhibited by Tos-Lys-Ch$_2$Cl, whereas the breakdown of band 3 protein to band 4.5 was due to a Ca^{2+}-activated cysteine proteinase. Although the proteolysis of ankyrin was sufficient to permit fusion to occur, fusion was more rapid when band 3 protein was also degraded. We suggested that the proteolysis of band 3 might be due to the Ca^{2+}-dependent cysteine proteinase, termed calpain I (Murachi et al., 1981). As the Ca^{2+}-activated cysteine proteinase activity was much more important in the fusion of rat erythrocytes by benzyl alcohol (Ahkong et al., 1980) than with human erythrocytes fused by chlorpromazine, we also drew attention to the fact that the content of calpastatin (an endogenous inhibitor of calpain I) is very small in rat erythrocytes, whereas in human erythrocytes there is much more inhibitor than enzyme (Murachi et al., 1981).

Kosower et al. (1983) found that rat erythrocytes, but not human erythrocytes, were fused by the membrane-mobility agent A_2C. They related the difference between the fusible rat erythrocytes and the nonfusible human cells to Ca^{2+}-activated protease activity that was present in the rat cells but appeared to be deficient in the human erythrocytes. Again, the gel eletrophoresis patterns of membrane proteins derived from fusing cells or ghosts exhibited losses of proteins in the regions of bonds 1, 2, 2.1, and 3, accompanied by the production of lower molecular weight bands. Later, Glaser and Kosower (1986) showed that human erythrocyte ghosts can be induced to fuse on treatment with A_2C provided that they are pretreated with Ca^{2+} and calpain and are free from calpastatin. They therefore concluded that the human cells, unlike rat erythrocytes, are resistant to fusion by the membrane-mobility agent because the human erythrocyte has an excess of the endogenous inhibitor calpastatin.

Ca^{2+}-activated neutral proteinase activity appeared concomitantly with the fusion of rat myoblasts into myotubes, while other proteases such as cathepsin D and plasminogen activator did not show any change in their activities (Kaur and Sanwal, 1981). It may be relevant that Fulton et al. (1981) have also demonstrated that myoblasts have a well-developed and highly interconnected internal network structure that is extensively reorganized and destabilized immediately before fusion. Furthermore, the fusion of rat myoblasts requires the activity of a neutral metalloendoprotease at the time of fusion (Couch and Strittmatter, 1983). Various possible roles for metalloendoproteases in myoblast fusion, which appear to function after Ca^{2+} ions have entered the cells, have recently been reviewed by Strittmatter *et al.* (1987). Work undertaken by these scientists has also implicated metalloendoprotease activities in exocytosis in adrenal chromaffin cells and mast cells (Mundy and Strittmatter, 1985), in synaptic transmission in the presynaptic nerve terminal (Baxter *et al.*, 1983; Frederick *et al.*, 1984), and in the acrosome reaction in the sea urchin (Farach *et al.*, 1987).

Viral glycoproteins in certain enveloped viruses are responsible for mediating virion–cell fusion and cell–cell fusion. Attempts to relate the function of fusogenic viral proteins to their structure have focused on hydrophobic domains (other than membrane spanning regions) of the fusion proteins. This is because these domains seem to be more highly conserved than others among related viruses and also because, in some instances, they are at the new N-termini which are generated by the proteolytic cleavage of inactive precursor proteins (reviewed by Spear, 1987). It has been suggested that fusogenic viruses may have acquired the mechanism by which they induce fusion from a preexisting cellular mechanism (Helenius and Marsh, 1982). In view of the many associations between fusion and proteolysis that have been reported, it has also been proposed that endogenous proteinases may release hydrophobic fragments from integral membrane proteins or from membrane skeletal proteins, which act like fusogenic viral proteins and induce membrane fusion in the absence of viruses (Lucy, 1984).

2.3. Fusion Driven by Cell Swelling

Recently, it was found that osmotic swelling can drive the fusion of erythrocytes under conditions that are known to perturb the membrane skeleton. In this work, hen erythrocytes were preincubated with ionophore A23187 and Ca^{2+}, which induce both a proteolytic breakdown of spectrin and goblin (Thomas *et al.*, 1983) and a release of the intramembranous particles from the restraining effects of the membrane skeleton (Vos *et al.*, 1976). Prolonged incubation of hen erythrocytes treated in this way or incubation at 47°C results in cell fusion (Ahkong *et al.*, 1975c), which is probably facilitated by a Ca^{2+}-stimulated phosphodiesterase that yields a diacylglycerol (Allan and Michell, 1978) which may be fusogenic (Ahkong *et al.*, 1973a). However, under the experimental conditions employed, little fusion was apparent after 2 hr at 37°C in monolayers of the pre-treated cells, but only 5 min after adding water to dilute the buffered salt solution 2.5-fold, binucleate cells were clearly seen to have been formed by cell fusion. This finding was confirmed by pre-labeling a proportion of the erythrocytes with 6-carboxyfluorescein diacetate, which enters cells and is enzymatically hydrolyzed to 6-carboxyfluorescein (Goodall and Johnson, 1982).

Five minutes after the osmotic shock, fluorescent cells had fused with nonfluorescent cells to produce weakly fluorescent cells: they also fused with each other (Ahkong and Lucy, 1986).

Although human erythrocytes show greatly restricted volume changes at osmolarities between 200 and 700 mOsm, they swell like a simple osmometer in solutions that are more dilute than 150 mM (Heubusch et al., 1985). Experiments were therefore undertaken to see if a disruption of the membrane skeleton that could facilitate the fusion of human erythrocytes might be caused by cell swelling in very dilute media (Ahkong and Lucy, 1986). This proved to be the case when monolayers of the cells were incubated in an isotonic salt–buffer solution that was progressively diluted, first by 1.5-fold, then by a further 1.7-fold, and finally replaced by water alone. Dextran was present until the final stage to reduce the rate of swelling, and La^{3+} (0.5 mM) was present throughout to inhibit lysis. The procedure resulted in the formation of very large polycells of irregular shape that retained their hemoglobin, as well as giving rise to lysed single cells and polyghosts, thus confirming that osmotic swelling can induce the fusion of erythrocytes under appropriate conditions.

When La^{3+} was omitted or replaced by 2mM Ca^{2+} in this work, the swelling procedure caused lysis without fusion. Cations, including Ca^{2+}, are known to decrease the size of holes produced in human erythrocytes by osmotic lysis (Lieber and Steck, 1982), but La^{3+} (possibly because of its high charge) was found to be particularly effective in the above experiments. La^{3+} (1 mM) has also been reported to protect permeabilized protoplasts of *Saccharomyces cerevisiae* against rupture by hypotonic stress (Kovac et al., 1987). It should be noted, however, that as well as minimizing lysis of erythrocytes, 0.5 mM La^{3+} may additionally facilitate osmotically driven fusion by perturbing the polar groups (Hauser et al., 1975) of erythrocyte phospholipids. Thus, 0.1–0.3 mM La^{3+} disrupts phospholipid vesicles (Hammoudah et al., 1981) and, at much higher concentrations (10 mM), La^{3+} causes human erythrocytes to fuse in 325 mOsm saline (Majumdar et al., 1980). Very recently, the fusion of large lamellar liposomes of phosphatidylserine by lanthanum ions has also been reported (Bentz et al., 1988). Maximal fusion activity was observed in the concentration range 10–100 μM La^{3+}. Excess La^{3+} (1 mM at pH 7.4) abolished fusion, however, apparently because binding of excess La^{3+} ions reversed the electrostatic surface potential and thus abolished close apposition of the vesicles. Interestingly, unlike fusion of phosphatidylserine liposomes by Ca^{2+}, the La^{3+}-induced fusion was nonleaky even after extensive fusion had occurred. Bentz and his colleagues have suggested that, if fusion involves membrane rupture due to enhanced membrane tension, rupture can occur only at sites of interliposomal contact in the presence of La^{3+} (otherwise the vesicles would lyse) and that this consequently implies the existence of specific intermediate structures at the site of fusion.

3. FUSION INDUCED BY POLYETHYLENE GLYCOL

3.1. Purity of the Polymer

In early work on cell fusion, commercial preparations of polyethylene glycol (PEG) were used to fuse plant protoplasts (Kao and Michayluk, 1974; Wallin *et al.*, 1974), hen erythrocytes (Ahkong *et al.*, 1975a), mammalian cells (Pontecorvo, 1975), and yeast protoplasts with hen erythrocytes (Ahkong *et al.*, 1975b). Subsequently, they have been extensively employed to obtain hybridoma cells, by fusing B-lymphocytes with myeloma cells, in the preparation of monoclonal antibodies. However, Honda *et al.* (1981) claimed that the fusogenic properties of commercial preparations of polyethylene glycol are due to impurities and that, although the polymer aggregates human erythrocytes, it has no intrinsic ability to fuse them. These investigators also suggested that diethyl ether-soluble and water-soluble contaminants in the commercial polymers promote cell fusion, and that the contaminants may be antioxidants in view of the well-known autoxidation of polyethylene glycol. By contrast, Smith *et al.* (1982) found that the fusogenic behavior of four commercial preparations of polyethylene glycol were unaffected by purification, although a preparation like that used by Honda and his colleagues contained contaminating substances that enhanced the fusion of erythrocytes. The addition to polyethylene glycol of small quantities of fusogenic lipid-soluble compounds, for example monoleoyl glycerol and alpha-tocopherol (Ahkong *et al.*, 1973a), enhanced the fusion of hen erythrocytes by up to 50%. Butylated hydroxyanisole and butylated hydroxytoluene acted similarly. It appeared, however, that the enhancement of cell fusion by the latter compounds, and by alpha-tocopherol, was not related to their antioxidant properties since the water-soluble antioxidants N-propyl gallate, methyl hydroquinone, and ethoxyquin did not enhance fusion at comparable concentrations.

In related work, Hui *et al.* (1985) observed that purified polyethylene glycol induced the fusion of erythrocyte ghosts. In addition, not only has recrystallized polyethylene glycol been found to fuse carrot protoplasts as efficiently as the unpurified polymer, but it has also been reported that a sample of PEG 8000, which was inefficient in fusing erythrocytes with fibroblasts, was fusogenic for carrot protoplasts (Boss, 1983). Boni *et al.* (1984) also found that alterations in phospholipid polymorphism were caused by both purified and commercial preparations of polyethylene glycol, while Parente and Lentz (1986) observed that impurities in commercial polyethylene glycol had no effect on the rate or extent of the fusion of phospholipid vesicles. Roos and Choppin (1985) similarly found no substantial differences in the fusogenic properties of polyethylene glycol before and after purification in experiments with mouse fibroblast cell lines.

Despite these various observations, Wojcieszyn et al. (1983) nevertheless found that polyethylene glycol did not fuse erythrocytes with fibroblasts after it had been purified by recrystallization from chloroform/ether. Although the fusogenic action of polyethylene glycol on mouse erythrocytes apparently involves the activity of a Ca^{2+}-activated cysteine proteinase (Nakornchai et al., 1983) (cf. the lipid-soluble fusogens discussed above), polyethylene glycol seems not to activate endogenous proteinases when it fuses human erythrocytes (Schindler et al., 1980; Smith and Palek, 1982) perhaps because of their high content of calpastatin (see above). Instead, the polymer may physically disrupt the membrane skeleton by its dehydrating action. It thus seems possible that polyethylene glycol may not be able to fuse human erythrocytes with fibroblasts (Wojcieszyn et al., 1983) unless a lipid-soluble fusogen is also present, as an impurity, to activate proteinases which degrade the membrane skeletons of one or both cells sufficiently to allow them to fuse. Interestingly, in this connection, it was found in early work on homokaryon formation that monolayers of avian and mammalian erythrocytes fused to a very high degree (60–95% multinucleated cells) when they were exposed to a protease preparation prior to treatment with polyethylene glycol in Ca^{2+}-free media (Hartmann et al., 1976). Cells that were not treated with proteolytic enzymes were observed to agglutinate in the presence of polyethylene glycol but did not fuse to a significant extent (0.01%) under the conditions used. It is also relevant that Huang and Hui (1986) have shown that 0.05% spermidine, 100 mM trichloracetic acid, and 4% ethanol each lower the threshold concentration of polyethylene glycol required for the fusion of human erythrocytes to 25%, apparently by creating zones in the plasma membranes that are free from intramembranous particles. They proposed that the additional polyethylene glycol needed for most cell fusion work (up to 45%) is likely to be associated with the property of polyethylene glycol that aggregates and precipitates membrane proteins to expose intramembranous particle-free areas of membrane.

In recent observations made in the authors' laboratory, pretreatment of polyethylene glycol with a Chelex-100 resin has been found to abolish its fusogenic properties toward human erythrocytes, provided that the isotonic saline used for suspending the cells is similarly treated. This finding once again raises the possibility that polyethylene glycol is not intrinsically fusogenic. However, rather than antioxidants being involved, it seems that the fusogenic properties of commercial preparations of polyethylene glycol may be mediated by trace quantities of certain metal ions.

Work by Kao and Saleem (1986) has indicated that, as deionization of polyethylene glycol with a mixed-bed exchange resin to remove acidic impurities promotes the fusion and viability of plant protoplasts, its fusogenic properties are decreased if acidic products are formed by autoxidation. Commercial preparations of polyethylene glycol that have a high content of carbonyl compounds are similarly unsuitable for use in the preparation of hybridomas, and autoclaving solutions of the polymer results in an increased oxidative decomposition and formation of carbonyl products (Kadish and Wenc, 1983). Deleterious effects of carbonyl compounds have also been observed in the fusion of plant protoplasts, and an improved procedure for protoplast fusion has been developed by Chand et al. (1988) that is based on the use of preparations of the polymer having a low carbonyl content.

3.2. Dehydration

Cell fusion induced by polyethylene glycol comprises two experimental steps. Different laboratories employ slightly different procedures but, in essence, the cells are first briefly exposed to a concentrated solution of the polymer, and they are then gradually returned to an isotonic environment that is free from polyethylene glycol. In early studies on the fusion of fibroblasts by polyethylene glycol, no unbound water was observed in aqueous preparations of the polymer that were sufficiently concentrated (30–50%) to induce cell fusion. It was therefore proposed that polyethylene glycol induces cell fusion by altering the physical state of bulk water adjacent to the cell surface and/or the water of hydration of the phospholipid polar groups in the cell membrane (Blow et al., 1978). Tilcock and Fisher (1979) concluded that each water molecule is either bound to polyethylene glycol or is at least "structured" in solutions containing more than 13% of the solute. It was also found that concentrated solutions of polyethylene glycol, in common with lipid-soluble fusogens, allowed Ca^{2+} ions to enter the cytoplasm of erythrocytes before fusion occurred (Blow et al., 1979).

A first requirement for the fusion of two cells is that the repulsive forces between them must be overcome. For membranes that are 2–3 nm apart, the repulsive forces between the phospholipid bilayers are essentially electrostatic. These forces are sensitive to the identity and charge of the lipid polar groups and also to the ionic composition of the medium. At closer distances, hydration repulsion, arising from solvation of the polar head groups of the phospholipids, dominates the interactions of two approaching membranes (reviewed by Parsegian et al., 1984, and by Blumenthal, 1987). Hydration repulsion is negligibly dependent on the ionic composition of the medium and the electrical charge of the bilayers. Both the electrostatic and hydration forces of repulsion may be overcome, for example, by adding Ca^{2+}. Consequently, although bilayers of phosphatidylserine repel each other strongly, they approach so closely in the presence of Ca^{2+} that there is virtually no water between them. This is because the binding energy of Ca^{2+} ions to the polar groups of the phospholipid molecules apparently exceeds their initial energy of hydration. Phospholipid vesicles, erythrocytes, and even myoblasts have been reported to fuse in the absence of calcium (reviewed by Lucy, 1984). Calcium is nevertheless very important in many fusion phenomena, and it is likely to be multifunctional since it can have an extracellular role (in overcoming membrane repulsion) and at least one intracellular role (stimulation of proteinase activity). Ca^{2+} ions are also much more effective than other cations in inducing fusion, both in physiological and in simple model systems. For example, they are more effective than Mg^{2+} ions in inducing the fusion of vesicles of phosphatidylserine. This is probably a reflection of the fact that the interbilayer space is dehydrated when a Ca^{2+}-phosphatidylserine is formed, whereas the corresponding complex with Mg^{2+} maintains its layer of hydration (reviewed by Düzgüneş, 1985).

High concentrations of polyethylene glycol (30–60%) remove at least 90% of the water from phospholipid bilayers, again enabling membranes to come extremely close together and ultimately to fuse. In the past 10 years, much work has been undertaken on the effects of polyethylene glycol on model phospholipid systems with the aim of elucidating its fusogenic mechanism of action (reviewed by Boni and Hui, 1987). The studies pursued include investigations on transition temperature and transition entropies in saturated phospholipids (Tilcock and Fisher, 1979), motions of the choline methyl group (Ohno et al., 1981), phospholipid exchange (Tilcock and Fisher, 1982), surface potentials (Maggio et al., 1976), polarity (Herrmann et al., 1983) and dielectric constant (Arnold et al., 1985) of the aqueous phase, phospholipid polymorphism (Boni et al., 1984), and a comparison of the fusogenic effects of polyethylene glycol, dextran, and sucrose on vesicles of phosphatidylcholine (MacDonald, 1985). Arnold et al. (1988) recently concluded that there is no evidence for a direct interaction between phospholipid molecules and polyethylene glycol, and they suggest that the aggregation of membranes induced by the polymer is caused by the high osmotic pressure of its solutions together with a reduced solubility of the polymer in the water layers near to the surface of membranes.

These physicochemical studies on the behavior of phospholipid systems in concentrated solutions of polyethylene glycol, however, have not shown clearly how the dehydrating action of the polymer causes cell fusion, rather than simply inducing the very close apposition of the plasma membranes of the treated cells. In this connection, it is important to note that the fusion of phospholipid vesicles by polyethylene glycol occurs in concentrated solutions of the polymer and the dilution step, which is important for cell fusion, is not required (Aldwinckle et al., 1982; Boni et al., 1981; Parente and Lentz, 1986). The mechanism by which polyethylene glycol fuses cells may therefore differ, at least in some respects, from its fusogenic effect on cells. Nevertheless, it has been suggested that, as the surface concentration of Pr^{3+} and of Mn^{2+} ions was found to be increased about threefold when a 30% solution of polyethylene glycol was added to vesicles of phosphatidylcholine (Gawrisch, 1986), ion binding may be a general feature of the action of polyethylene glycol on neutral phospholipid bilayers. Arnold *et al.* (1988) propose that this may provide an explanation for the enhanced permeabilities to Ca^{2+} of liposomes and erythrocytes in the presence of high concentrations of the polymer (Aldwinckle *et al.*, 1982; Blow *et al.*, 1979).

3.3. Rehydration

A number of assays, based on the use of fluorescent probes, have been devised for determining the occurrence of membrane fusion. Some show the mixing of membrane components, while others demonstrate the mixing of labeled material that was previously trapped in separate membrane compartments. Both types of assay have been applied primarily to the fusion of phospholipid vesicles (reviewed by Düzgüneş, 1985, and by Blumenthal, 1987). Most of the assays for the mixing of membrane components in phospholipid vesicles are based on fluorescence energy transfer between donor and acceptor groups that are initially in separate vesicular membranes and are attached to lipid molecules that are free to diffuse in the membranes. In using such assays for the mixing of membrane components, it is necessary to exclude the possibility of lipid exchange between the membranes. Lipid exchange may be eliminated by a suitable choice of probe (Struck *et al.*, 1981). However, the mixing of the membrane components in fusing phospholipid vesicles does not necessarily mean that their contents are also mixing, and the most rigorous criterion for membrane fusion in phospholipid vesicles is the coalescence of the internal aqueous contents with concomitant stoichiometric intermixing of membrane constituents (Düzgüneş, 1985; Düzgüneş *et al.*, 1987; Rosenberg *et al.*, 1983).

An alternative way of using fluorescent probes for investigating the mixing of membrane components is to follow the movement of a probe from labeled to unlabeled membranes. This technique was employed by Wojcieszyn *et al.* (1983), who found that a carbocyanine membrane–probe spread from labeled erythrocytes to unlabeled fibroblasts within 1–2 min in a concentrated solution of polyethylene glycol, whether or not the polymer had been purified by recrystallization from chloroform/ether. However, because purified polyethylene glycol did not induce fusion of the two cell types, they concluded that the polymer promoted a unique, but unspecified, type of membrane apposition that permits a nonfusogenic mechanism of lipid–probe exchange between erythrocytes and cultured cells, and that the cell fusion event was caused by unknown impurities in commercial polyethylene glycol (Schlegel and Lieber, 1987; Wojcieszyn *et al.*, 1983).

Wojcieszyn *et al.* (1983) also found that water-soluble fluorescent proteins, introduced into human erythrocytes by hypotonic hemolysis, did not diffuse into fibroblasts on fusion induced by impure polyethylene glycol until after the polymer had been removed. Since fluorescent serum albumin was immobile in 44% polyethylene glycol but diffused freely (with a diffusion coefficient of about 10^{-8} cm^2/sec) after removal of the polymer, this finding was attributed to immobilization of cytoplasmic proteins by the concentrated polyethylene glycol. In subsequent work, Herrmann et al. (1985) similarly observed that the rotational correlation time of a small spin label molecule was increased threefold with human erythrocytes in 30% polyethylene glycol. Several investigators, however, have reported that cytoplasmic bridges between cells treated with polyethylene glycol are not seen by electron microscopy until the polymer solution has been diluted and the cells resuspended in isotonic buffer (Ahkong and Lucy, 1986; Knutton, 1979; Krähling, 1981; Robinson et al., 1979). It therefore seems possible that the failure of fluorescent albumin to diffuse into unlabeled fibroblasts from labeled erythrocytes until the polyethylene glycol had been removed (Wojcieszyn et al., 1983) was due to the cytoplasms of the fusing cells being separated by a membranous boundary until the polyethylene glycol was replaced by isotonic buffer.

Recent observations with a low-light video camera on the movements of fluorescent probes in erythrocytes treated with polyethylene glycol are consistent with this alternative interpretation (Ahkong et al., 1987). In this work, purified polyethylene glycol 6000 (recrystallized/dialyzed) induced both aggregation and fusion of human erythrocytes. A proportion of the cells were doubly labeled with carboxyfluorescein and with octadecyl rhodamine B chloride. The latter probe does not dissociate from membranes either by spontaneous transfer of free monomers through the aqueous phase or by a collision-mediated transfer process, possibly as a consequence of interactions between the rhodamine head group and the phospholipid head groups of membranes (Hoekstra et al., 1984, 1985). La^{3+} (50 μM) was used to prevent the leakage of cytoplasmic carboxyfluorescein that otherwise resulted when human erythrocytes were treated with 40% polyethylene glycol (Ahkong and Lucy, 1986). No movement of the lipid probe from labeled to unlabeled cells occurred in the absence of polyethylene glycol, or with cells that were aggregated by treatment with concanavalin A, protamine, or spermine, but in 40% polyethylene glycol it started to diffuse into the membranes of the unlabeled cells at least within 30 sec. By contrast, transfer of the cytoplasmic carboxyfluorescein from labeled to unlabeled cells did not occur until after the polyethylene glycol had been replaced by isotonic buffer. Equilibrium was then achieved within 320 msec in some cells, and this was considered to be more consistent with the rupture of membranous barriers between the rehydrating cells than with the return of their cytoplasmic viscosity to normal.

It was also suggested by Ahkong et al. (1987) that the membranous barriers between the cytoplasms of labeled and unlabeled cells might consist of regions of shared bilayer that were not broken until replacement of the polyethylene glycol by an isotonic buffer caused water to enter the cells (cf. Figures 1e and 1f). Osmotic swelling would then impart a stretching force that might

Membrane Fusion and Osmotic Forces

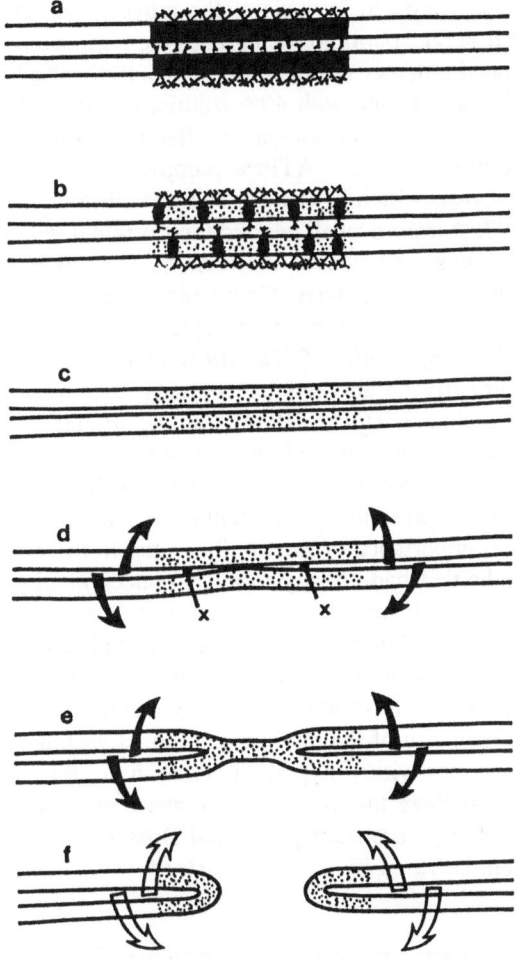

FIGURE 1. A schematic diagram that illustrates six proposed events in the fusion–fission reactions of biological membranes. (a) The initial close approach of two membranes. (b) The phospholipid bilayers of the membranes are perturbed at the potential fusion site. This is illustrated diagrammatically by the change from solid to hatched shading. (c) The membrane proteins are perturbed. The absence of integral membrane proteins, membrane-skeleton, and cytoskeleton proteins from the potential fusion site is intended to indicate that the perturbed lipid bilayers are now relatively unrestrained by membrane proteins, and it does not imply that the bilayers are totally free from all polypeptides at the developing site of fusion. (d) Solid arrows represent the entry of water into one, or both (as illustrated), of the two membrane-bound compartments, which causes swelling either because osmotically active particles have accumulated in one or both compartments or because functionally impermeable membranes that previously prevented swelling now have an increased permeability. The regions marked × are indicative of structures or mechanisms that hold the membranes together adjacent to the fusion site as swelling proceeds. (e) Under the influence of the pressure generated by the osmotic gradient, the two phospholipid bilayers are forced together at the fusion site and they form an intermediate, shared bilayer that is common to both of the fusing membranes. (f) Osmotic swelling imparts a stretching force that ruptures the intermediate, shared bilayer at one or both of the points where three bilayers meet. The original bilayers then reseal; continued entry of water (open arrows) will cause further swelling and may cause lysis. Reproduced with permission from Lucy and Ahkong (1986).

rupture the shared bilayer at one or both of the points where three bilayers meet. It is important to note in this connection that polyethylene glycol-treated cells exhibit extensive colloid osmotic lysis on rehydration, and that cell fusion (monitored by the transfer of carboxyfluorescein) was frequently and immediately followed by lysis of the rehydrating cells. This presumably occurred because phospholipid membranes are permeabilized by treatment with the concentrated polymer (Aldwinckle et al., 1982). Na^+ and Ca^{2+} thus enter erythrocytes that are treated with 40% solutions of polyethylene glycol (Blow et al., 1979). Since Na^+ is no longer an effectively impermeable cation (and Ca^{2+} ions inhibit the Na^+/K^+ ATPase pump), water will also enter the treated cells. Erythrocytes, which have been dehydrated in concentrated solutions of the polymer, will therefore swell beyond their original volume when they are rehydrated and, as suggested above, the osmotic force may be responsible for cell fusion as well as for cell lysis. Comparable considerations apply to the permeabilization, fusion, and lysis of erythrocytes that are incubated with lipid-soluble fusogens (Ahkong et al., 1973a; Blow et al., 1979), and phospholipid vesicles may similarly be stressed by applying a gradient of osmotic pressure across their membranes (Ohki, 1984). Hui et al. (1985) reported that osmotic swelling, achieved by using a hypotonic incubation medium after treatment with polyethylene glycol, is necessary for the fusion of sealed, "pink" erythrocyte ghosts (which presumably still contained some hemoglobin). However, they attributed the importance of the swelling step to the opening of connecting sites between cells that had already fused in concentrated polyethylene glycol, rather than to the establishment of cytoplasmic continuity.

It is interesting that Kao and Saleem (1986) found that the frequency of protoplast fusion was enhanced by eluting the polyethylene glycol with a hypotonic salt solution, whereas fusion was greatly reduced by increasing the osmolarity of the eluting solution. It was therefore suggested that elution of the polymer with a hypotonic solution, in the presence of sufficient calcium ions to stabilize the treated membranes, might lead to better fusion without loss of viability. Kanchanapoom and Boss (1986) also reported that plant protoplasts can be fused by an osmotic stress.

4. ELECTRICALLY INDUCED CELL FUSION

When cells are exposed to a short high-voltage DC pulse, their membranes are permeabilized. This treatment causes colloid osmotic swelling and lysis of the treated cells, which may be prevented by the inclusion of impermeant substances in the medium (Kinosita and Tsong, 1977; Schwister and Deuticke, 1985; Serpersu et al., 1985; Zimmermann et al., 1976). Cells that are brought into close contact and aligned into chains by the application of an AC field (dielectrophoresis) are fused when a high-voltage pulse is subsequently applied (Zimmermann, 1982), and they remain viable provided that the electrically induced membrane breakdown is reversible. Reversibility is achieved by appropriate limitations of the magnitude and of the duration of the high-voltage pulse.

Much attention has been paid in recent years to electrically induced cell fusion because of its intrinsic interest and because of its potential applications in, for example, the formation of hybridoma cells and the production of monoclonal antibodies (for reviews see Arnold and Zimmermann, 1984; Bates et al., 1987; and Zimmermann, 1986).

The transfer of fluorescein isothiocyanate-dextran from labeled to individual, unlabeled erythrocyte ghosts in 30 mM phosphate buffer (Sowers, 1984), and of carboxyfluorescein from a labeled erythrocyte into a chain of unlabeled erythrocytes in a hypotonic solution of 150 mM erythritol (Ahkong and Lucy, 1986), was found to occur virtually instantaneously with the breakdown pulse. Movement of carboxyfluorescein from labeled to unlabeled NS1 mouse myeloma cells in 200 mM inositol also occurred very rapidly when these cells were exposed to a breakdown pulse, although several seconds were needed for the fluorophore in a labeled cell to be distributed equally throughout a chain of 11–17 fused myeloma cells (Brown et al., 1986). On the other hand, Ohno-Shosaku and Okada (1985) observed that populations of mouse lymphoma cells, which were labeled with Janus Green and Neutral Red, exhibited an intermixed color several minutes after exposure to the breakdown pulse. This apparently slower rate of fusion, however, may have been due to fusion being less readily detectable with vital stains than with fluorescent labels. Alternatively, as discussed below, the cells may not have fused until several minutes after the breakdown pulse had been applied.

Since exposure of cells to a high-voltage pulse causes colloidal osmotic swelling that can lead to cell lysis under appropriate conditions, the possibility that osmotic forces are involved in electrically induced cell fusion has been investigated. By contrast to the long chains of fluorescent cells that were formed by transfer of carboxyfluorescein from a labeled erythrocyte to adjacent unlabeled cells in 150 mM erythritol, the fluorophore was found to transfer only to single cells when less swollen cells in 200 mM erythritol were exposed to the breakdown pulse (Ahkong and Lucy, 1986). In 400 mM erythritol, the instantaneous transfer of fluorophore was negligible, and no fused cells were produced. The degree to which the cells were swollen in a hypotonic medium when the electrical breakdown pulse was applied thus affected the extent to which the cells were fused. These observations have some resemblance to those made by Zimmerberg et al. (1980b), who found that phospholipid vesicles that were preswollen fused more readily with a planar phospholipid bilayer than more flaccid vesicles under the same osmotic gradient. It is known that the breakdown potential of the cell membrane decreases with increasing turgor pressure in *Valonia utricularis* (Zimmermann et al., 1977) and with increasing hydrostatic pressure in erythrocytes (Zimmermann et al., 1980). Consequently, it seems probable that the extent of instantaneous electrically induced fusion in erythrocytes is governed by a combination of the electrical-compressive force due to the breakdown pulse and the osmotic force applied prior to electrical breakdown (Ahkong and Lucy, 1986).

As with erythrocytes, transfer of carboxyfluorescein from labeled to unlabeled cells occurred more extensively when mixed populations of myeloma cells and lymphocytes were exposed to breakdown voltages of 1.5–3.5 kV/cm in 200 mM inositol than when the cells were in 300 mM inositol (Brown et al., 1986). NS1 myeloma cells in a 280 mM solution of inositol, containing 0.1 mM Ca^{2+} and 0.5 mM Mg^{2+} ions, also fused much more readily in the presence than in the absence of pronase, as judged by movement of the fluorophore. The pronase-free inositol medium had previously been found to be particularly useful for the preparation of hybridoma cells by electrofusion by Vienken and Zimmermann (1985), who observed that (contrary to their earlier studies) pronase was not needed. As these investigators were, however, monitoring the production of hybridoma cells rather than cell fusion, it seems possible that hybridoma cells are not derived from cells that are fused immediately by the breakdown pulse. Instead, they may result from fusion that occurs some time afterward. The feasibility of this is indicated by the fact that an intercellular transfer of fluorophore was observed with erythrocytes during the colloid osmotic swelling that occurred several minutes after permeabilization by a breakdown pulse (Lucy, 1986). This indicates that, depending on the experimental conditions used, an electrical breakdown pulse may induce both an instantaneous fusion (governed by the electrical and osmotic pressures) and an osmotically driven fusion of the permeabilized cells which occurs subsequently.

From a study of the kinetics of the ultrastructural changes that accompany the electrically induced fusion of human erythrocytes, Stenger and Hui (1986) have concluded that fusion sites arise where electrically ruptured regions of bilayer adjacent cells make contact (Figure 2a). Membrane continuities (Figure 2b) develop within 2 sec from such regions, which, after 10 sec, yield expanding cytoplasmic bridges between the cells (Figure 2c). The regions of membrane continuity (Figure 2b) in their reconstructed model for electrically induced cell fusion are similar to the regions of shared bilayer (Figure 1e) that feature in the general model that we have proposed for membrane fusion processes (see Section 5) (Lucy and Ahkong, 1986).

Erythrocytes, which are unfused after exposure to a breakdown pulse, can nevertheless be permanently attached to their neighbors as a consequence of being exposed to the voltage pulse (Lucy and Ahkong, 1988). This was shown by the fact that, when the AC field was disconnected, some pulsed erythrocytes did not separate but remained in chains that were not dispersed by thermal or mechanical motion. Furthermore, when some of the erythrocytes were prelabeled with octadecyl rhodamine B, the lipid fluorophore diffused into the membranes of the unlabeled cells to which they became attached (but not fused) on exposure to the breakdown pulse. It has therefore been suggested that a break-

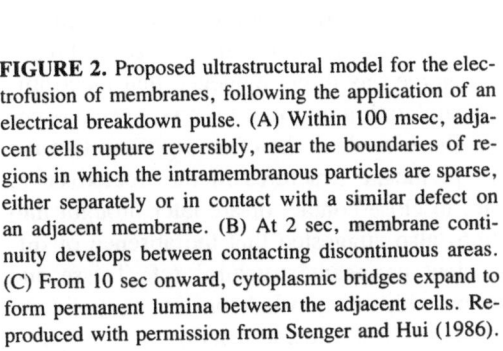
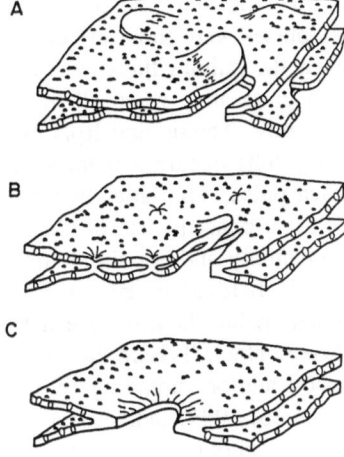

FIGURE 2. Proposed ultrastructural model for the electrofusion of membranes, following the application of an electrical breakdown pulse. (A) Within 100 msec, adjacent cells rupture reversibly, near the boundaries of regions in which the intramembranous particles are sparse, either separately or in contact with a similar defect on an adjacent membrane. (B) At 2 sec, membrane continuity develops between contacting discontinuous areas. (C) From 10 sec onward, cytoplasmic bridges expand to form permanent lumina between the adjacent cells. Reproduced with permission from Stenger and Hui (1986).

down pulse, which is just below the threshold that is necessary to induce sufficient membrane thinning for complete fusion, may nevertheless have enough energy to compress the membranes of aligned cells into shared bilayers (cf. Figure 1e) at points of very close contact. The linked cells, which are also permeabilized by the pulse, may then fuse completely if the shared bilayer between the cells is subsequently ruptured osmotically (Lucy, 1986).

Suspensions of cells in random positions that are initially permeabilized by exposure to a DC breakdown pulse, and are subsequently aligned in an AC field to bring about membrane–membrane contact, might also be expected to fuse if the DC pulse initiates a (delayed) osmotically driven fusion. This phenomenon has in fact been observed with erythrocyte ghosts that were loaded with fluorescein isothiocyanate-dextran, although it was not interpreted in this way (Sowers, 1984, 1986). It was also shown that ghosts brought into contact at places where the transmembrane DC field strength had been highest (i.e., at the poles) fused more readily than ghosts brought into contact where the field had been at or near zero. Areas on membranes which became fusogenic on exposure to the breakdown pulse therefore appeared to be highly localized, and related experiments showed that the fusogenic sites were laterally immobile. Sowers (1987b) has suggested that the latter finding indicates that fusogenic alterations in membrane structure, which are induced by the breakdown pulse, may somehow be associated with the cytoskeleton. Since the membrane skeleton of the treated cells is likely to be damaged preferentially at the poles of the treated cells by the breakdown pulse, these observations may alternatively be interpreted in terms of osmotically driven, delayed fusion occurring preferentially at the electrically damaged sites.

Teissie and Blangero (1984) found in earlier experiments that repetitive DC pulses were more effective in inducing the fusion of cells cultured *in vitro* when the culture flasks were rotated through 90° after half of the pulses had been given than when all the pulses were applied in the same direction. Subsequently, Teissie and Rols (1986) showed that the fusion of Chinese hamster ovary cells occurred if the cells were centrifuged after exposure to the breakdown pulse. The latter finding was regarded as direct evidence that electrofusion is a consequence of electropermeabilization and that the two events occur sequentially and not simultaneously. These workers therefore proposed that the electric field gives rise to transient permeant structures in the treated cells, which enable them to fuse if they are in close contact or are later brought into close contact. Teissie and Rols (1986) also suggested that the absence of diffusion and/or spreading of the transient permeant structures (cf. Teisse and Blangero, 1984) indicated a connection with the cytoskeleton, but they did not invoke osmotic swelling as a driving force in the fusion reaction.

Data relevant to the behavior of the cytoskeleton/membrane skeleton in electrically fused cells has been obtained by Donath and Arndt (1984), who fused untreated human erythrocytes with erythrocytes in which the surface charge density had been decreased by treatment with neuraminidase. Fusion of an untreated cell with an enzyme-treated cell gave a "doublet" that had a permanent dipole moment which decreased with time as a consequence of lateral diffusion of surface proteins. A surprisingly high lateral diffusion constant of about 3×10^{-9} cm^2/sec was obtained in this way for the fused cells, and it was concluded that this was due to a diminished interaction of the cytoskeleton with the cell membrane. Donath and Arndt (1984) suggested that this, in turn, may have been a consequence of the very low ionic strength used, or of the fusion process itself, or both. The importance of the membrane skeleton was further demonstrated by Glaser and Donath (1987), who recently observed the frequency of electrofusion to increase sharply when human erythrocytes were heated above the denaturation temperature (50°C) of the cytoskeleton before exposure to the breakdown pulse.

5. MOLECULAR MODELS FOR MEMBRANE FUSION

Any agent, either chemical or physical, that interacts with phospholipids to yield a nonbilayer configuration is likely to be capable of inducing membranes to fuse because of its perturbing action. It has therefore been suggested that micellar structures (Lucy, 1970), a hexagonal phase (Cullis and Hope, 1978), and inverted micelles (Verkleij *et al.*, 1980) may be involved in membrane fusion reactions. Such structures need not necessarily be intermediates in membrane fusion reactions, however, if perturbation of phospholipid bilayers is of primary importance in the fusion process rather than formation of structures that normally arise as end products under equilibrium conditions when phospholipid bilayers are perturbed in model systems. Early studies on the fusion of phospholipid vesicles also indicated a close correlation between fusion induced by divalent cations and their ability to induce a phase separation of acidic phospholipids from zwitterionic lipids in the membrane. It was therefore proposed that such a phase separation was a key event in membrane fusion (Papahadjopoulos, 1978). Later work showed that large unilamellar vesicles composed of phosphatidylserine and dimyristoylphosphatidylethanolamine (1:1) were however, fused in the presence of Mg^{2+} without any observable phase separation, while vesicles composed of phosphatidylserine and dipalmitoylphosphatidylcholine (1:1) also fused in the presence of Ca^{2+} without exhibiting any massive separation of the lipid species (Düzgüneş et al., 1984).

Another potential difficulty with the concept that nonbilayer intermediates participate in fusion reactions is that, if such an intermediate forms spontaneously on treating membranes with a fusogenic agent, what causes it to rupture in the subsequent membrane fission reaction? However, a kinetic model proposed by Siegel (1984, 1986, 1987) for the rates of formation of inverted micellar intermediates predicts that they should be short-lived structures that are not only in dynamic equilibrium with the lamellar bilayer phase from which they are formed, but can also rapidly give rise to structures (referred to as interlamellar attachments) by a process that corresponds to membrane fission.

Alternatively, some kind of bilayer may persist throughout the membrane fusion–fission reactions. Palade (1975) proposed a model for exocytosis that featured a shared bilayer as an intermediate structure, on the basis that ultrastructural studies indicated that thinning of the closely apposed granule and plasma membranes preceded their rupture at the site of exocytosis. Thus, in a study of vascular endothelia, Palade and Bruns (1968) reported a single membrane (20 nm in length) that was common to both the plasma membrane and a vesicle membrane, and which separated the contents of the vesicle from the extracellular fluid prior to their discharge by exocytosis. A single bilayer intermediate has also been described in the secretion of zoospores in *Phytophthora palmivora* (Pinto da Silva and Nogueira, 1977), in the fusion of myoblasts (Kalderon and Gilula, 1979), and in the fusion of plant protoplasts by polyethylene glycol (Kanchanapoom and Boss, 1986). Nevertheless, relatively few such observations have been made, and it is not clear whether a single bilayer intermediate is absent from other fusing systems or whether its formation is extremely rapid and/or labile, thus making it difficult to capture (Kalderon, 1980).

Electron microscopy of membrane fusion in phospholipid vesicles has additionally given rise to the suggestion that a shared bilayer may occur as an intermediate in the fusion process. Thus, from a study of fusion between bilayers of mixed egg phosphatidylcholine and soybean phosphatidylethanolamine that was induced by freezing and thawing, Hui et al. (1981) proposed that a single bilayer may be formed via point defects during fusion.

We have pointed out (Lucy and Ahkong, 1986) that several workers have shown that a new bilayer is formed when two phospholipid bilayers are pushed together with a sufficient force to overcome their mutual repulsion (Fisher and Parker, 1984; Horn, 1984; Melikyan et al., 1983). In the light of these observations, we have proposed a new, general model for membrane fusion processes and have suggested that application of the model to, for example, the fusion of myoblasts into myotubes may clarify some existing observations on fusion phenomena in biological systems (Lucy and Ahkong, 1986). A central feature of the model is that fusion–fission reactions depend on two perturbed phospholipid bilayers being merged into a single, shared bilayer, which is subsequently ruptured by osmotic forces (Figure 1). Similar ideas have more recently been developed by Chernomordik et al. (1987).

6. VIRALLY INDUCED MEMBRANE FUSION

Since thin sections of Sendai virions reveal tightly folded, parallel nucleocapsid strands just beneath the viral membrane, it has been proposed that the viral M protein and nucleocapsids may together be regarded as a "viroskeleton" (Kim et al., 1979). Although Sendai virions normally exhibit a spherical and unconvoluted appearance, they develop a grooved and convoluted surface after attachment to erythrocytes at 37°C, which is essential for fusion with the cells (Knutton, 1977). From work with three-dimensional models, it has been suggested that this convoluted appearance is consistent with viral swelling and that virus–cell fusion may occur when a region of viral membrane swells against the plasma membrane of a cell to which it is already firmly attached (Lucy and Ahkong, 1986). Recently, the abilities of young and aged Sendai virions to undergo virion-to-virion fusion induced by polyethylene glycol were compared by Kim and Okada (1987). Aged virions fused readily to form large virion vesicles, but young virions were resistant to fusion. In addition, aged virions fused on incubation at 37°C, even without the fusogen. These differences were attributed to differing interactions of the M protein with the envelope membrane, particularly as the asymmetrical distribution of spikes was also lost in aged virions, but not in young virions, on treatment with polyethylene glycol.

Citovsky et al. (1987) reported that osmotic swelling allows fusion of Sendai virions with membranes of desialized erythrocytes and with chromaffin granules. Since virus–liposome fusion occurred independently of the osmolarity of the medium, it was apparently the structure of the recipient membrane that was affected by the osmotic swelling and not the biological activity of the virus. The findings of other investigators that enveloped virions, including Sendai virus, fuse extensively with phospholipid vesicles that lack receptors for the virus have demonstrated that the eventual target of the virus is the phosopholipid bilayer of the recipient cell plasma membrane. Rather than osmotic swelling driving the fusion reaction, Citovsky *et al.* have therefore suggested that the swelling of erythrocyte vesicles and chromaffin granule vesicles in a medium of low osmolarity exposes their phospholipid bilayers, which can then interact with glycoproteins of the viral envelope. By contrast, Blumenthal *et al.* (1987) have observed that at low pH the rate of fusion between membranes of vesicular stomatitis virus and Vero cells is about the same in hypo-osmotic solutions (in which the cells are disrupted) as in iso- or hyperosmotic solutions. Osmotic swelling therefore appeared not to play a role in this membrane fusion system.

The participation of cell swelling in the formation of giant erythrocytes, or other polykaryons, by hemolytic Sendai virus is well known. Pasternak and his colleagues have shown, however, that, at least for fusion that involves a virion acting as a bridge between two cells, swelling is responsible only for the rounding of already fused cells, rather than for the fusion reaction itself (Pasternak, 1984). The evidence for this is twofold. First, the formation of giant cells is prevented when osmotic swelling is inhibited by hypertonic media (Impraim *et al.*, 1980; Knutton and Pasternak, 1979). Second, if early harvested (nonhemolytic and nonleaky) Sendai virions are used instead of the usual hemolytic virions, giant cells are not formed (Knutton and Pasternak, 1979; Wyke *et al.*, 1980). The treated cells can nevertheless be shown to have fused, since they expand into giant cells when they are subsequently exposed to a hypotonic medium (Knutton and Bachi, 1980).

Direct fusion between cells treated with hemolytic Sendai virus, rather than via a virion that fuses simultaneously with two cells and acts as a bridge, has also been reported (Bächi *et al.*, 1973; Toister and Loyter, 1973), and it has been pointed out in this connection that the fusion of erythrocytes by hemolytic Sendai virus is similar in many ways to the fusion of these cells by lipid-soluble fusogens (Lucy and Ahkong, 1986). In particular, (1) since the viral membrane is permeable, its incorporation into the erythrocyte membrane (by virus–cell fusion) causes colloid-osmotic swelling and lysis of the cells (Hosaka and Shimizu, 1977); (2) the time that precedes the increased permeability of cells treated with the virus is not due to a lag in virus–cell fusion but to the development of a threshold level of membrane damage, that is, the introduction of a sufficient number and/or size of permeability pores (Micklem *et al.*, 1985); (3) cell lysis, rather than cell fusion, occurs when erythrocytes are treated with large numbers of Sendai virions in the absence of Ca^{2+} or other divalent cations (Hart, et al., 1976; Peretz *et al.*, 1974), and erythrocytes similarly lyse more readily than they fuse when they are treated with lipid fusogens in the absence of Ca^{2+} (Lucy, 1973); (4) as with chemically induced cell fusion, either polycells or polyghosts are formed depending on whether or not lysis follows fusion (Peretz et al., 1974); and (5) proteolytic activity that is associated with the virus may be an essential feature of its ability to fuse cells (Israel et al., 1983). It has therefore been suggested that osmotic swelling drives the fusion of erythrocytes with one another after they have been permeabilized by the virus (Lucy and Ahkong, 1986).

7. EXOCYTOSIS

7.1. Model Systems

Early work demonstrated that, under appropriate experimental conditions, an osmotic gradient will induce the fusion of phospholipid vesicles with one another. Miller et al. (1976) showed that unilamellar proteoliposomes containing phosphatidylserine, which ceased to fuse in the presence of calcium ions after they had reached diameters of about 100 nm, could be induced to fuse further by an osmotic gradient (internal osmotic pressure higher than external) to yield single-walled liposomes with diameters that exceeded 1 μm, as well as forming multilayered vesicles. Vesicles of phosphatidylserine, which were aggregated by Mg^{2+} ions and then suspended in a hypotonic solution, were also observed to fuse (Ohki, 1984). The larger the diameter of the initial vesicles, the smaller was the osmotic pressure gradient (outside more hypotonic than inside) needed for fusion. For vesicles that were 1 μm in diameter, the threshold gradient required for fusion was 0.104 M sucrose; whereas for vesicles that were only 0.1 μm in diameter, there was no appreciable fusion even with 0.138 M sucrose.

Osmotically induced fusion of phospholipid vesicles with one another may not be especially relevant to exocytosis, but Finkelstein and his colleagues have published an important series of papers which demonstrate that phospholipid vesicles can also be induced to fuse with a planar phospholipid bilayer by osmotic means. They found that an essential requirement for fusion was an osmotic gradient across the planar membrane, with the cis side (the side containing the vesicles) being hyperosmotic to the opposite (trans) side; a divalent cation was also required on the cis side. The applied osmotic gradient led to discharge of the contents of multilamellar vesicles across (decane-containing) planar phospholipid membranes. It was proposed that fusion occurred by the osmotic swelling of vesicles in contact with the planar membrane followed by rupture of the vesicular and planar membranes in the region of contact (Cohen et al., 1980, 1982; Zimmerberg et al., 1980a,b). Later, it was shown that unilamellar vesicles fuse with "hydrocarbon-free" planar bilayers under the same conditions. It was also found that vesicles made from uncharged lipids readily fused with planar membranes of phosphatidylethanolamine in the near absence of divalent cations, that is, with just an osmotic gradient (Cohen et al., 1984). Two experimentally distinguishable steps in the fusion of phospholipid vesicles with planar bilayers have been demonstrated. In the first, the vesicles form a stable, tightly bound, prefusion state with the planar membrane, for which divalent cations are necessary if either membrane contains negatively charged lipids. Osmotic swelling of the bound vesicle and its fusion with the planar bilayer occur subsequently in a second step (Akabas et al., 1984). These studies have been reviewed by Finkelstein et al. (1986). Very recently, video fluorescence microscopy has been used to study the adsorption and fusion of unilamellar phospholipid vesicles to solvent-free planar bilayer membranes (Niles and Cohen, 1987). In this work it was demonstrated that, when Ca^{2+} ions promote adhesion of the vesicle to the bilayer in the first step of the fusion process, the vesicular and planar membranes remain as individual phospholipid bilayers rather than melding to form a shared bilayer at the point of contact. This finding does not, however, exclude the possibility of a shared bilayer acting as a transient intermediate in the ensuing fusion–fission reaction (cf. Figure 1).

Extrapolations from the fusion of phospholipid vesicles with a planar bilayer to biological exocytosis have assumed that, if an analogous osmotic gradient drives the naturally occurring phenomenon, secretory vesicles will swell as a whole immediately prior to their fusion with the plasma membrane of a secreting cell. Several observations have now been made, however (see below), which show that fusion actually precedes gross swelling of the secretory vesicles (at least in mast cells of beige mice), indicating that the fusion–fission reaction of exocytosis is not driven by an osmotic gradient. Recent findings on the fusion of erythrocytes induced by polyethylene glycol are relevant to this problem because the hemoglobin of erythrocytes that are dehydrated in a 50% solution of the polymer may be regarded as a model for the tightly packed, dehydrated contents of the secretory granules in many types of secretory cell. When the dehydrated erythrocytes (labeled with carboxyfluorescein) were rehydrated, it was observed that entering water initially accumulated in microdroplets, and then in a thin layer, between the hemoglobin and the plasma membrane. Electron micrographs showed that cell fusion occurred locally in association with the accumulating aqueous droplets and preceded cell swelling (Figure 3). It has therefore been proposed that, if water enters secretory granules and accumulates adjacent to the granule membrane early in exocytosis as a consequence of an increase in membrane permeability (cf. Ornberg and Reese, 1981), fusion of the granule membrane with the plasma membrane may similarly precede overall granule swelling (Ahkong and Lucy, 1988). On this basis, the

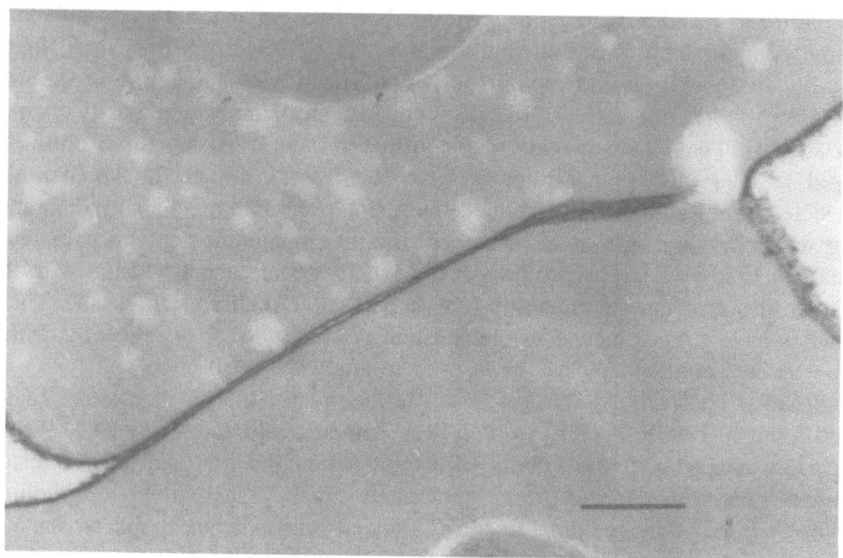

FIGURE 3. An electron micrograph of a thin section of hen erythrocytes that were exposed to a 40% solution of polyethylene glycol 6000 containing 5 mM EGTA for 5 min at 37°C. The cells were fixed for electron microscopy by the addition of 9 vol of a 0.5% solution of glutaraldehyde in buffered isotonic saline at pH 7.4, after which they were allowed to stand for 1 hr at 37°C. In this way, the cells were fixed at an early stage in their rehydration. The initial hydration was accompanied by the development of cytoplasmic, aqueous microdroplets at differing rates in adjacent cells. A cytoplasmic bridge between two closely apposed cells is seen at the site of an aqueous microdroplet; cell fusion induced in this way preceded cell swelling.

contents of the granule will not swell and disperse until larger quantities of external water subsequently enter through the enlarging fusion pore.

7.2. Biological Systems

Early investigations showed that isolated chromaffin granules released their contents under isotonic conditions at 37°C when they were exposed to Mg^{2+}-ATP and large quantities of Cl^- or other permeant anions. In addition, the release was suppressed by raising the osmotic strength of the medium with salt or sucrose. From these studies it became apparent that the release reaction is the result of osmotic lysis, which permits the entire contents of the granules to escape (reviewed by Pollard *et al.*, 1979). Several workers in this field considered that lysis of the chromaffin granule might provide a general model for the fission step by which its contents are released in the exocytosis reaction, and a chemiosmotic model for exocytosis was proposed by Pollard *et al.* (1979).

In their model, it was envisaged that close juxtaposition of the granule membrane to the cytoplasmic surface of the plasma membrane of a chromaffin cell is mediated by Ca^{2+} ions and synexin. A region of shared bilayer (cf. Palade 1975; Palade and Bruns, 1968) is then formed by fusion of the two membranes. Anions subsequently enter the granule via transport sites in the shared bilayer, along with H^+ ions pumped in by the ATPase of the granule membrane, thus increasing the osmotic strength of the granule interior. This leads finally to osmotic rupture of the shared bilayer and release of the granule contents by exocytosis.

Grinstein et al. (1982) suggested an alternative model in which a H^+-cation antiport becomes operative across the granule membrane during stimulation by secretagogues so that intragranular H^+ is exchanged for extragranular cations. Geisow and Burgoyne (1982a), who observed that a cation-dependent, osmotically driven lysis of isolated chromaffin granules occurred with the ionophore monensin under iso-osmotic conditions in the absence of a permeant anion, similarly proposed that increased permeability of the granule membrane to sodium or potassium ions might provide a mechanism for osmotically driven exocytosis. A higher osmotic activity of the matrix would then result from dissociation of impermeant, weak acid species consequent on an efflux of H^+, coupled to an influx of K^+ (Geisow and Burgoyne, 1982b). More recently, Stanley and Ehrenstein (1985) have alternatively proposed that the role of Ca^{2+} in Ca^{2+}-induced exocytosis is to open Ca^{2+}-activated K^+ channels in the vesicle membrane which, coupled with anion transport across the membrane, would result in an influx of K^+ and anions that increases the osmotic pressure of the vesicle.

An elevation of osmolarity strength with sucrose does, in fact, inhibit catecholamine secretion by exocytosis in chromaffin cells (Baker and Knight, 1981; Hampton and Holz, 1983; Pollard et al., 1984). Also, Hampton and Holz (1983) showed that hyperosmotic solutions inhibit secretion at a step after the entry of Ca^{2+} ions, while Pollard et al. (1984) demonstrated that an approximately exponential relationship exists between osmotic strength and suppression of epinephrine release regardless of whether the osmotic strength is raised by NaCl or by sucrose. However, Pollard et al. (1984) additionally observed a poor quantitative correlation between the anion dependence of chemiosmotic granule lysis and anion dependence of cell secretion. Furthermore, stilbene disulfonates inhibited granule lysis but not secretion. These observations therefore showed that granule lysis and secretion are not identical, even though they have strong qualitative similarities. In a review article, Baker and Knight (1984) subsequently concluded that there is no clear evidence for an essential involvement of either anions or monovalent cations in the exocytotic release of catecholamines, and that the ATP-dependent proton pump in the membranes of secretory vesicles is very unlikely to play a part in exocytosis. More recently, Holz

(1986) proposed that an increase in the permeability of the granule membrane to extracellular components at the fusion site (or to cytosolic components at nonfusion sites), without a permeability increase to the contents of the granules, could provide a mechanism for producing the necessary increase in osmotic pressure. Because granules would never reach osmotic equilibrium, they would lyse. Finkelstein et al. (1986) suggested that this could occur if an elevated concentration of cytosolic Ca^{2+} were to induce the formation of a gap-junction-like structure between the vesicle and the plasma membrane that would expose the granular contents to the external medium. In this case, however, as has recently been observed (see below), the change in membrane capacitance that accompanies formation of the initial fusion pore would precede vesicular swelling, rather than follow it.

Many investigations have indicated that, despite ambiguities regarding the origin of the osmotic gradient in chromaffin granules, osmotic forces appear to have an important role at least in the following exocytotic systems: mucocyst secretion in Tetrahymena (Satir et al., 1973), the release of parathyroid hormone (Brown et al., 1978), endotoxin-induced degranulation in amebocyte blood cells of the horseshoe crab, Limulus polyphemus (Ornberg and Reese, 1981), the discharge of nematocysts in sea anemones (Lubbock et al., 1981), exocytosis induced in the toad urinary bladder by antidiuretic hormone (Kachadorian et al., 1981), the release of insulin from the pancreas (Hermans and Henquin, 1986; Orci and Malaisse, 1980), the secretion of serotonin from platelets (Pollard et al., 1977), and the cortical granule reaction in sea urchin eggs (Zimmerberg and Whitaker, 1985; Zimmerberg et al., 1985).

In general, the contents of the intracellular storage granules of secretory cells are very closely packed. For example, the majority of the secretory granules in human pulmonary mast cells have been reported to contain crystalline structures (Caulfield et al., 1980); the contents of mucocysts in Tetrahymena are also crystalline (Satir et al., 1973). The insolubility of neurophysin within pituitary secretory granules allows the hormone to be stored at very high concentrations without increasing the intragranular osmotic pressure to the point of lysis (Breslow, 1984) and, in most species, insulin is stored in the secretory vesicles of pancreatic B cells as a central, insoluble aggregate, usually with a crystalline structure (Gold and Grodsky, 1984). It might be expected therefore that hydration of the stored material will play an important part in its dispersal, and Whitaker and Zimmerberg (1987) have demonstrated that movements of water are in fact essential for exocytosis of the contents of cortical granules in sea urchin eggs. Thus, a variety of polymers prevented exocytosis by preventing dispersal of the granule contents after fusion of the secretory granule membrane with the plasma membrane had occurred.

In order for osmotic forces to be responsible for establishing continuity between the contents of secretory granules and the extracellular medium (as well as for subsequently expelling the contents of the granules), osmotically driven changes must clearly precede the formation of the initial fusion pore. This point has been investigated in mast cells from beige mice by utilizing changes in membrane capacitance to detect the moment at which the granule membrane and the plasma membrane become continuous in exocytosis. Exocytosis was found to begin with the sudden formation of a pore that had a conductance of about 230 pS, which was thus similar to a single gap-junction channel (conductance 160 pS). The conductance then increased to much larger values within milliseconds, as if the pore dilated soon after opening (Breckenridge and Almers, 1987a). However, the change in capacitance was found to precede swelling of the granules (Zimmerberg, 1987; Zimmerberg et al., 1987). To rule out the possibility that fusion was induced by a mechanical stress, imparted by the internal pressure of a taut secretory granule, experiments were also done with cells in which the vesicles were shrunken with hyperosmotic solutions. With flaccid granules, the rise in capacitance again preceded granule swelling, and it was concluded that swelling cannot be the driving force for membrane fusion in this system. Breckenridge and Almers (1987b) likewise found that individual granules swelled after stepwise changes in capacitance and that the capacitance steps were followed by the stepwise loss of a fluorescent dye that had been loaded into the vesicles. They similarly concluded that the formation of an electrical connection between the cell exterior and the inside of a secretory vesicle in mast cells of beige mice cannot be driven by granule swelling. Work by Holz and Senter (1986) has additionally indicated that shrunken chromaffin granules can undergo exocytosis and that elevated ionic strength (rather than shrinkage of the granules) contributes to the inhibition of secretion observed in intact cells with hyperosmotic solutions. It is relevant that fusion of the trichocyst membrane with the plasma membrane in *Paramecium* cells can also occur before decondensation or expansion of the contents of the trichocysts (Gilligan and Satir, 1983; Matt and Plattner, 1983).

Nevertheless, all is not lost for osmotic hypotheses in exocytosis because a localized entry of water, which is not readily seen, may be responsible for the fusion of secretory granule membranes with the plasma membrane. It is interesting therefore that, in a rapid-freezing study of the release of blood-clotting proteins from amebocyte blood cells of horseshoe crabs, the earliest change observed was a separation between the granule core and its membrane that left a clear crescent beneath the granule membrane (Ornberg and Reese, 1981). It was considered that this could be due to an influx of water resulting from an osmotic gradient that was associated with the initial formation of small fusion pores. Similar clear spaces in isolated chromaffin granules, aggregated by Ca^{2+} ions, were also attributed in an earlier study to an osmotically driven influx of water (Edwards et al., 1974). As discussed in Section 7.1, it is also relevant that the fusion of human erythrocytes induced by polyethylene glycol is associated with the localized accumulation of interstitial water between the dehydrated hemoglobin and the plasma membrane, which precedes cell swelling on rehydrating the cells (Figure 3) (Ahkong and Lucy, 1988).

By applying ultrastructural and metachromatic staining techniques to adjacent thin and semithin sections, Schmauder-Chock and Chock (1987) demonstrated that the perigranular membrane can indeed enlarge prior to fusion with the plasma membrane. This may reflect a surge in intragranular osmotic pressure which triggers a rapid influx of water into the granule. However, the membrane enlargement observed was so extensive that it could not be explained by membrane stretching alone, and they postulated that insertion of additional membrane occurred during activation of the secretory granules.

As with the fusion of erythrocytes induced by chemicals or by an electrical discharge (see Sections 2.3 and 4, respectively), modifications apparently need to be made to cytoskeletal structures in order for exocytosis to proceed (reviewed by Pearl and Taylor, 1985; Linstedt and Kelly, 1987). A considerable number of observations have been reported which indicate that Ca^{2+}-activated proteases in sea urchin eggs function in the discharge of their cortical granules by exocytosis during fertilization. Ca^{2+}-activated proteases and serine proteases may indeed participate in exocytosis generally (Schuel, 1978). For example, it has been proposed that the degradation of fodrin by a Ca^{2+}-activated protease in chromaffin cells could enable secretory granules to have access to exocytotic sites on the plasma membrane (Burgoyne and Cheek, 1987a). Phosphorylation of cytoskeletal components (e.g., of synapsin I), as well as changes in the organization of the network that are mediated by proteolytic activities, additionally appears to be important in exocytosis (Linstedt and Kelly, 1987).

Connecting filamentous structures between secretory granules and the plasma membrane have been observed in actively secreting chromaffin cells. Consequently, some cytoskeletal elements may conversely be responsible for interactions between granules and plasma membranes. In this connection, Burgoyne and Cheek (1987b) have recently considered the possibility that the membrane-binding proteins—caldesmon, p70, and p36—may participate in such interactions immediately prior to exocytosis.

8. REFERENCES

Ahkong, Q. F., and Lucy, J. A., 1986, Osmotic forces in artificially induced cell fusion. *Biochim. Biophys. Acta* **858**:206–216.

Ahkong, Q. F., and Lucy, J. A., 1989, Localised osmotic forces and membrane fusion. *Biochem. Soc. Trans.*, in press.

Ahkong, Q. F., Fisher, D., Tampion, W., and Lucy, J. A., 1973a, The fusion of erythrocytes by fatty acids, esters, retinol and alpha-tocopherol. *Biochem. J.* **136**:147–155.

Ahkong, Q. F., Cramp. F. C., Fisher, D., Howell, J. I., Tampion, W., Verrinder, M., and Lucy,

J. A., 1973b, Chemically-induced and thermally-induced cell fusion: Lipid–lipid interactions. *Nature New Biol.* **242:**215–217.

Ahkong, Q. F., Fisher, D., Tampion, W., and Lucy, J. A., 1975a, Mechanisms of cell fusion. *Nature (London)* **253:**194–195.

Ahkong, Q. F., Howell, J. I., Lucy, J. A., Safwat, F., Davey, M. R., and Cocking, E. C., 1975b, Fusion of hen erythrocytes with yeast protoplasts induced by polyethylene glycol. *Nature (London)* **255:**66–67.

Ahkong, Q. F., Tampion, W., and Lucy, J. A., 1975c, Promotion of cell fusion by divalent cation ionophores. *Nature (London)* **256:**208–209.

Ahkong, Q. F., Botham, G. M., Woodward, A. W., and Lucy, J. A., 1980, Calcium-activated thiol-proteinase activity in the fusion of rat erythrocytes induced by benzyl alcohol. *Biochem. J.* **192:**829–836.

Ahkong, Q. F., Desmazes, J.-P., Georgescauld, D., and Lucy, J. A., 1987, Movements of fluorescent probes in the mechanism of cell fusion induced by poly(ethylene glycol). *J. Cell Sci.* **88:**389–398.

Akabas, M. H., Cohen, F. S., and Finkelstein, A., 1984, Separation of the osmotically driven fusion event from vesicle-planar membrane attachment in a model system for exocytosis. *J. Cell Biol.* **98:**1063–1071.

Aldwinckle, T. J., Ahkong, Q. F., Bangham, A. D., Fisher, D., and Lucy, J. A., 1982, Effects of poly(ethylene glycol) on liposomes and erythrocyte permeability changes and fusion. *Biochim. Biophys. Acta* **689:**548–560.

Allan, D., and Michell, R. H., 1978, A calcium-activated polyphosphoinositide phosphodiesterase in the plasma membrane of human and rabbit erythrocytes. *Biochim. Biophys. Acta* **508:**277–286.

Arnold, K., Herrmann, A., Pratsch, L., and Gawrisch, K., 1985, The dielectric properties of aqueous solutions of poly(ethylene glycol) and their influence on membrane structure. *Biochim. Biophys. Acta* **815:**515–518.

Arnold, K., Herrmann, A., Gawrisch, K., and Pratsch, L., 1988. Water-mediated effects of PEG on membrane properties and fusion, in *Molecular Mechanisms of Membrane Fusion* (S. Ohki, D. Doyle, T. D. Flanagan, S. W. Hui, and E. Mayhew, eds.), Plenum Press, New York.

Arnold, W. M., and Zimmermann, U., 1984, Electric field-induced fusion and rotation of cells, in *Biological Membranes,* Vol. 5 (D. Chapman, ed.), pp. 389–454, Academic Press, London.

Bächi, M., Aguet, M., and Howe, C., 1973, Fusion of erythrocytes by Sendai virus studied by immuno-freeze etching. *J. Virol.* **11:**1004–1012.

Baker, P. F., and Knight, D. E., 1981, Calcium control of exocytosis and endocytosis in bovine adrenal medullary cells. *Philos. Trans. R. Soc. London (Biol.)* **296:**83–103.

Baker, P. F., and Knight, D. E., 1984, Chemiosmotic hypotheses of exocytosis: A critique. *Biosci. Rep.* **4:**285–298.

Bates, G. W., Saunders, J. A., and Sowers, A. E., 1987, Electrofusion. Principles and applications, in *Cell Fusion* (A. E. Sowers, ed.), pp. 367–395, Plenum Press, New York.

Baxter, D. A., Johnston, D., and Strittmatter, W. J., 1983, Protease inhibitors implicate metalloendoprotease in synaptic transmission at the mammalian neuro-muscular junction. *Proc. Natl. Acad. Sci. U.S.A.* **80:**4174–4178.

Bentz, J., Alford, D., Cohen, J., and Düzgüneş, N., 1989, La^{3+} induced fusion of phosphatidylserine liposomes: Close approach, intermembrane intermediates and the electrostatic surface potential. *Biophys. J.,* in press.

Blow, A. M. J., Botham, G. M., Fisher, D., Goodall, A. H., Tilcok, C. P. S., and Lucy, J. A., 1978, Water and calcium ions in cell fusion induced by poly(ethylene glycol). *FEBS Lett.* **94:**305–310.

Blow, A. M. J., Botham, G. M., and Lucy, J. A., 1979, Calcium ions and cell fusion. Effects of

chemical fusogens on the permeability of erythrocytes to calcium and other ions. *Biochem. J.* **182**:555–563.
Blumenthal, R., 1987, Membrane fusion, in *Current Topics in Membranes and Transport,* Vol. 29 (R. D. Klausner, C. Kempf, and J. van Renswoude, eds.), pp. 203–254, Academic Press, New York.
Blumenthal, R., Bali-Puri, A., Walter, A., Covell, D., and Eidelman, O., 1987, pH-dependent fusion of vesicular stomatitis virus with Vero cells. *J. Biol. Chem.* **262**:13614–13619.
Boni, L. T., and Hui, S. W., 1987, The mechanism of polyethylene glycol-induced fusion in model membranes, in *Cell Fusion* (A. E. Sowers, ed.), pp 301–330, Plenum Press, New York.
Boni, L. T., Stewart, T. P., Alderfer, J. L., and Hui, S. W., 1981, Lipid–polyethylene glycol interactions: I. Induction of fusion between liposomes. *J. Membr. Biol.* **62**:65–70.
Boni, L. T., Stewart, T. P., and Hui, S. W., 1984, Alterations in phospholipid polymorphism by polyethylene glycol. *J. Membr. Biol.* **80**:91–104.
Boss, W. F., 1983, Poly(ethylene glycol)-induced fusion of plant protoplasts. A spin-label study. *Biochim. Biophys. Acta* **730**:111–118.
Breckenridge, L. J., and Almers, W., 1987a, Currents through the fusion pore that forms during exocytosis of a secretory vesicle. *Nature (London)* **328**:814–817.
Breckenridge, L. F., and Almers, W., 1987b, Final steps in exocytosis observed in a cell with giant secretory granules. *Proc. Natl. Acad. Sci. U.S.A.* **84**:1945–1949.
Breslow, E., 1984, Neurophysin: Biology and chemistry of its interactions, in *Cell Biology of the Secretory Process* (M. Cantin, ed.), pp. 276–308, Karger, Basel.
Brown, S. M., Ahkong, Q. F., and Lucy, J. A., 1986, Osmotic pressure and the electrofusion of myeloma cells. *Biochem. Soc. Trans.* **14**:1129–1130.
Brown, E. M., Pazoles, C. J., Creutz, C. E., Aurbach, G. D., and Pollard, H. B., 1978, Role of anions in parathyroid hormone release from dispersed bovine parathyroid cells. *Proc. Natl. Acad. Sci. U.S.A.* **75**:876–880.
Bruckdorfer, K. R., Cramp, F. C., Goodall, A. H., Verrinder, M., and Lucy, J. A., 1974, Fusion of mouse fibroblasts with oleylamine. *J. Cell Sci.* **15**:185–199.
Burgoyne, R. D., and Cheek, T. R., 1987a, Role of fodrin in secretion. *Nature (London)* **326**:448.
Burgoyne, R. D., and Cheek, T. R., 1987b, Reorganisation of peripheral actin filaments as a prelude to exocytosis. *Biosci. Rep.* **7**:281–288.
Caulfield, J. P., Lewis, R. A., Hein, A., and Austen, K. F., 1980, Secretion in dissociated human pulmonary mast cells. *J. Cell Biol* **85**:299–311.
Chand, P., Davey, M. R., Power, J. B., and Cocking, E. C., 1988, An improved procedure for plant protoplast fusion using polyethylene glycol. *J. Plant Physiol.* in press.
Chernomordik, L. V., Melikyan, G. B., and Chizmadzhev, Y. A., 1987, Biomembrane fusion: A new concept derived from model studies using two interacting planar lipid bilayers. *Biochim. Biophys. Acta* **906**:309–352.
Citovsky, V., Laster, Y., Schuldiner, S., and Loyter, A., 1987, Osmotic swelling allows fusion of Sendai virions with membranes of desialized erythrocytes and chromaffin granules. *Biochemistry* **26**:3856–3864.
Coakley, W. T., and Deeley, J. O. T., 1980, Effects of ionic strength, serum protein and surface charge on membrane movements and vesicle production in heated erythrocytes. *Biochim. Biophys. Acta* **602**:355–375.
Coakley, W. T., Nwafor, A., and Deeley, J. O. T., 1983, Tetracaine modifies the fragmentation mode of heated human erythrocytes and can induce heated cell fusion. *Biochim. Biophys. Acta* **727**:303–312.
Cohen, F. S., Zimmerberg, J., and Finkelstein, A., 1980, Fusion of phospholipid vesicles with

planar phospholipid bilayer membranes. II. Incorporation of a vesicular membrane marker into the planar membrane. *J. Gen. Physiol* **75**:251–270.

Cohen, F. S., Akabas, M. H., and Finkelstein, A., 1982, Osmotic swelling of phospholipid vesicles causes them to fuse with a planar phospholipid bilayer membrane. *Science* **217**:458–460.

Cohen, F. S., Akabas, M. H., Zimmerberg, J., and Finkelstein, A., 1984, Parameters affecting the fusion of unilamellar phospholipid vesicles with planar bilayer membranes. *J. Cell Biol.* **98**:1054–1062.

Couch, C. B., and Strittmatter, W. J., 1983, Rat myoblast fusion requires metalloendoprotease activity. *Cell* **32**:257–265.

Cullis, P. R., and Hope, M. J., 1978. Effects of fusogenic agent on membrane structure of erythrocyte ghosts and the mechanism of membrane fusion. *Nature (London)* **271**:672–674.

Donath, E., and Arndt, R., 1984, Electric-field-induced fusion of enzyme-treated human red cells: Kinetics of intermembrane protein exchange. *Gen. Physiol. Biophys.* **3**:239–249.

Düzgüneş, N., 1985, Membrane fusion, in *Subcellular Biochemistry*, Vol. 11 (D. B. Roodyn, ed.), pp. 195–286, Plenum Press, New York.

Düzgüneş, N., Paiement, J., Freeman, K. B., Lopez, N. G., Wilschut, J., and Papahadjopoulos, D., 1984, Modulation of membrane fusion by ionotropic and thermotropic phase transitions. *Biochemistry* **23**:3486–3494.

Düzgüneş, N., Allen, T. M., Fedor, J., and Papahadjopoulos, D., 1987, Lipid mixing during membrane aggregation and fusion: Why fusion assays disagree. *Biochemistry* **26**:8435–8442.

Edwards, W., Phillips, J. H., and Morris, S. J., 1974, Structural changes in chromaffin granules induced by divalent cations. *Biochim. Biophys. Acta* **356**:164–173.

Evered, D., and Whelan, J., 1984, *Cell Fusion*, Ciba Foundation Symposium 103, Pitman, London.

Farach, H. A., Mundy, D. I., Strittmatter, W. J., and Lennarz, W. J., 1987, Evidence for the involvement of metalloendoproteases in the acrosome reaction in sea urchin sperm. *J. Biol. Chem.* **262**:5483–5487.

Fawcett, D. W., 1986, *A Textbook of Histology*, 11th ed., W. B. Saunders, Philadelphia.

Finkelstein, A., Zimmerberg, J., and Cohen, F. S., 1986, Osmotic swelling of vesicles: Its role in the fusion of vesicles with planar phospholipid bilayer membranes and its possible role in exocytosis. *Annu. Rev. Physiol.* **48**:163–174.

Fisher, L. R., and Parker, N. S., 1984, Osmotic control of bilayer fusion. *Biophys. J.* **46**:253–258.

Frederick, J. M., Hollyfield, J. G., and Strittmatter, W. J., 1984, Inhibitors of metalloendoprotease activity prevent K^+-stimulated neurotransmitter release from the retina of *Xenopus laevis*. *J. Neurosci.* **4**:3112–3119.

Fulton, A. B., Prives, J., Farmer, S. R., and Penman, S., 1981, Developmental reorganization of the skeletal framework and its surface lamina in fusing muscle cells. *J. Cell Biol.* **91**:103–112.

Gallez, D., and Coakley, W. T., 1986, Interfacial instability at cell membranes. *Prog. Biophys. Mol. Biol.* **48**:155–199.

Gawrisch, K., 1986, Molekulare mechanismen und membranveranderungen bei der durch polyethylenglykol induzierten zellfusion, Thesis B, Karl Marx Universitat, Leipzig.

Geisow, M., and Burgoyne, R. D., 1982a, Cation-dependent lysis of chromaffin granules—an alternative hypothesis for osmotically-driven exocytosis. *Cell Biol. Int. Rep.* **6**:353–359.

Geisow, M., and Burgoyne, R. D., 1982b, Effect of monensin on chromaffin cells and the mechanism of organelle swelling. *Cell Biol. Int. Rep.* **6**:933–939.

Gilligan, D. M., and Satir, B. H., 1983, Stimulation and inhibition of secretion in *Paramecium:* Role of divalent cations. *J. Cell Biol.* **97**:224–234.

Glaser, R. W., and Donath, E., 1987, Hindrance of red cell electrofusion by the cytoskeleton. *Studia Biophys.* **121**:37–43.

Glaser, T., and Kosower, N. S., 1986, Calpain–calpastatin and fusion. Fusibility of erythrocytes is determined by a protease–protease inhibitor [calpain–calpastatin] balance. *FEBS Lett.* **206**:115–120.

Gold, G., and Grodsky, G. M., 1984, The secretory process in B cells of the pancreas, in *Cell Biology of the Secretory Process* (M. Cantin, ed.), pp. 359–388, Karger, Basel.

Goodall, H., and Johnson, M. H., 1982, Use of carboxyfluorescein diacetate to study formation of permeable channels between mouse blastomers. *Nature (London)* **295**:524–526.

Grinstein, S., Meulen, J. V., and Furuya, W., 1982, Possible role of H^+-alkali cation counter transport in secretory granule swelling during exocytosis. *FEBS Lett.* **148**:1–4.

Hammoudah, M. M., Nir, S., Bentz, J., Mayhew, E., Stewart, T. P., Hui, S. W., and Kurland, R. J., 1981, Interactions of La^{3+} with phosphatidylserine vesicles. Binding, phase transition, leakage, ^{31}P-NMR and fusion. *Biochim. Biophys. Acta* **645**:102–114.

Hampton, R. Y., and Holz, R. W., 1983, Effects of changes in osmolality on the stability and function of cultured chromaffin cells and the possible role of osmotic forces in exocytosis. *J. Cell Biol.* **96**:1082–1088.

Hart, C. A., Fisher, D., Hallinan, T., and Lucy, J. A., 1976, Effects of calcium ions and the bivalent cation ionophore A23187 on the agglutination and fusion of chicken erythrocytes by Sendai virus. *Biochem. J.* **158**:141–145.

Hartmann, J. X., Galla, J. D., Emma, D. A., Kao, K. N., and Gamborg, O. L., 1976, The fusion of erythrocytes by treatment with proteolytic enzymes and polyethylene glycol. *Can. J. Genet. Cytol.* **18**:503–512.

Hauser, H., Phillips, M. C., Levine, B. A., and Williams, R. J. P., 1975, Ion-binding to phospholipids. Interaction of calcium and lanthanide ions with phosphatidylcholine (lecithin). *Eur. J. Biochem.* **58**:133–144.

Helenius, A., and Marsh, M., 1982, Endocytosis of enveloped animal viruses, in *Membrane Recycling* (D. Evered and G. M. Collins, eds.), pp. 59–76, Pitman, London.

Hermans, M. P., and Henquin, J. C., 1986, Is there a role for osmotic events in exocytotic release of insulin? *Endocrinology* **119**:105–111.

Herrmann, A., Pratsch, L. Arnold, K., and Lassmann, G., 1983, Effect of poly(ethylene glycol) on the polarity of aqueous solutions and the structure of vesicle membranes. *Biochim. Biophys. Acts* **733**:87–94.

Herrmann, A., Arnold, K., and Pratsch, L., 1985, The effect of osmotic pressure of aqueous PEG solutions on red blood cells. *Biosci. Rep.* **5**:689–696.

Heubusch, P., Jung, C. Y., and Green, F. A., 1985, The osmotic response of human erythrocytes and the membrane skeleton. *J. Cell Physiol.* **122**:266–272.

Hoekstra, D., de Boer, T., Klappe, K., and Wilschut, J., 1984, Fluorescence method for measuring the kinetics of fusion between biological membranes. *Biochemistry* **23**:5675–5681.

Hoekstra, D., Klappe, K., de Boer, T., and Wilschut, J., 1985, Characterisation of the fusogenic properties of Sendai virus: Kinetics of fusion with erythrocyte membranes. *Biochemistry* **24**:4739–4745.

Holz, R. W., 1986, The role of osmotic forces in exocytosis from adrenal chromaffin cells. *Annu. Rev. Physiol.* **48**:175–189.

Holz, R. W., and Senter, R. A., 1986, Effects of osmolality and ionic strength on secretion from adrenal chromaffin cells permeabilized with digitonin. *J. Neurochem.* **46**:1835–1842.

Honda, K., Maeda, Y., Sasakawa, S., Ohno, H., and Tsuchida, E., 1981, The components contained in polyethylene glycol of commercial grade (PEG-6000) as cell fusogen. *Biochim. Biophys. Acta* **101**:165–171.

Horn, R. G., 1984, Direct measurement of the force between two lipid bilayers and observation of their fusion. *Biochim. Biophys. Acta* **778**:224–228.

Hosaka, Y., and Shimizu, K., 1977, Cell fusion by Sendai virus, in *Virus Infection and the Cell Surface* (G. Poste and G. L. Nicolson, eds.), pp. 129–155, North-Holland, Amsterdam.

Huang, S. K., and Hui, S. W., 1986, Chemical co-treatments and intramembrane particle patching in the poly(ethylene glycol)-induced fusion of turkey and human erythrocytes. *Biochim. Biophys. Acta* **860**:539–548.

Hui, S. W., Stewart, T. P., and Yeagle, P. L., 1981, Membrane fusion through point defects in bilayers. *Science* **212**:921–923.

Hui, S. W., Isac, T., Boni, L. T., and Sen, A., 1985, Action of polyethylene glycol on the fusion of human erythrocyte membranes. *J. Membr. Biol.* **84**:137–146.

Impraim, C. C., Foster, K. A., Micklem, K. J., and Pasternak, C. A., 1980, Nature of virally mediated changes in membrane permeability to small molecules. *Biochem. J.* **186**:847–860.

Israel, S., Ginsberg, D., Laster, Y., Zakai, N., Milner, Y., and Loyter, A., 1983, A possible involvement of virus-associated protease in the fusion of Sendai virus envelopes with human erythrocytes. *Biochim. Biophys. Acta* **732**:337–346.

Kachadorian, W. A., Muller, J., and Finlekstein, A., 1981, Role of osmotic forces in exocytosis: Studies of ADH-induced fusion in toad urinary bladder. *J. Cell Biol.* **91**:584–588.

Kadish, J. L., and Wenc, K. M., 1983, Contamination of polyethylene glycol with aldehydes: Implications for hybridoma fusion. *Hybridoma* **2**:87–89.

Kalderon, N., 1980. Muscle cell fusion, in *Membrane–Membrane Interactions* (N. B. Gilula, ed.), pp. 99–118, Raven Press, New York.

Kalderon, N., and Gilula, N. B., 1979, Membrane events involved in myoblast fusion. *J. Cell Biol.* **81**:411–425.

Kanchanapoom, K., and Boss, W. F., 1986, Osmoregulation of fusogenic protoplast fusion. *Biochim. Biophys. Acta* **861**:429–439.

Kao, K. N., and Michayluk, M. R., 1974, A method for high-frequency intergeneric fusion of plant protoplasts. *Planta (Berlin)* **115**:355–367.

Kao, K. N., and Saleem, M., 1986, Improved fusion of mesophyll and cotyledon protoplasts with PEG and high pH-Ca^{2+} solution. *J. Plant Physiol.* **122**:217–225.

Kaur, H., and Sanwal, B. D., 1981, Regulation of the activity of a calcium-activated neutral protease during differentiation of skeletal myoblasts. *Can.J. Biochem.* **59**:743–747.

Kim, J., and Okada, Y., 1987, Differences in capacities for virion-to-virion fusion of young and aged HJV (Sendai virus): A model of membrane fusion. *J. Membr. Biol* **97**:241–249.

Kim, J., Hama, K., Miyake, Y., and Okada, Y., 1979, Transformation of intramembrane particles of HJV (Sendai virus) envelopes from an invisible to visible form on aging of virions. *Virology* **95**:523–535.

Kinosita, K., and Tsong, T. Y., 1977, Formation and resealing of pores of controlled sizes in human erythrocytes by electrical breakdown. *Nature (London)* **268**:438–441.

Knutton, S., 1977. Studies of membrane fusion. II. Fusion of human erythrocytes by Sendai virus. *J. Cell Sci.* **28**:189–219.

Knutton, S., 1979, Studies of membrane fusion. III. Fusion of erythrocytes with polyethylene glycol. *J. Cell Sci.* **36**:61–72.

Knutton, S., and Bachi, T., 1980, The role of cell swelling and hemolysis in Sendai virus-induced cell fusion and in the diffusion of incorporated viral antigens. *J. Cell Sci.* **42**:153–167.

Knutton, S., and Pasternak, C. A., 1979, The mechanisms of cell–cell fusion. *Trends Biochem. Sci.* **4**:220–223.

Kosower, E. M., Kosower, N. S., and Wegman, P., 1977, Membrane mobility agents. IV. The mechanism of particle–cell and cell–cell fusion. *Biochim. Biophys. Acta* **471**:311–329.

Kosower, N. S., Wegmaan, P. O., Neiman, T., and Kosower, E. M., 1978, Membrane mobility agents. V. Genetic variability in the fusibility of hen red cells. *Exp. Cell Res.* **116**:454–456.

Kowoser, N. S., Glaser, T., and Kosower, E. M., 1983, Membrane-mobility agent-promoted fusion of eryhrocytes: Fusibility is correlated with attack by calcium-activated cytoplasmic proteases on membrane proteins. *Proc. Natl. Acad. Sci. U.S.A.* **80**:7542–7546.

Kovac, L., Bohmerova, E., and Necas, O., 1987, The plasma membrane of yeast protoplasts exposed to hypotonicity becomes porous but does not disintegrate in the presence of protons or polyvalent cations. *Biochim. Biophys. Acta* **899**:265–275.

Krähling, H., 1981, Investigations on polyethylene glycol-induced cell fusion: Freeze–fracture observations. *Acta Histochem. Suppl.* **23**:S219–S223.

Lang, R. D. A., Wickenden, C., Wynne, J., and Lucy, J. A., 1984, Proteolysis of ankyrin and of band 3 protein in chemically induced cell fusion. *Biochem. J.* **218**:295–305.

Lieber, M. R., and Steck, T. L., 1982, Dynamics of the holes in human erythrocyte membrane ghosts. *J. Biol. Chem* **257**:11660–11666.

Linstedt, A. D., and Kelly, R. B., 1987, Overcoming barriers to exocytosis. *Trends Neurochem. Sci.* **10**:446–448.

Lubbock, R., Gupta, B. L., and Hall, T. A., 1981, Novel role of calcium in exocytosis: Mechanism of nematocyst discharge as shown by X-ray microanalysis. *Proc. Natl. Acad. Sci. U.S.A.* **78**:3624–3628.

Lucy, J. A., 1970. The fusion of biological membranes. *Nature (London)* **227**:814–817.

Lucy, J. A., 1973, The chemically-induced fusion of cells, in *Membrane-Mediated Information* (P. W. Kent, ed.), Vol. 2, pp. 117–128, Medical and Technical Publishing, Lancaster, United Kingdom.

Lucy, J. A., 1978, Mechanisms of chemically induced cell fusion, in *Membrane Fusion* (G. Poste and G. L. Nicolson, eds.), pp. 267–304, North-Holland, Amsterdam.

Lucy, J. A., 1984, Do hydrophobic sequences cleaved from cellular polypeptides induce membrane fusion reactions in vivo? *FEBS Lett.* **166**:223–231.

Lucy, J. A., 1986, Salient features of artificially induced cell fusion. *Biochem. Soc. Trans.* **14**:250–251.

Lucy, J. A., and Ahkong, Q. F., 1986, An osmotic model for the fusion of biological membranes. *FEBS lett.* **199**:1–11.

Lucy, J. A., and Ahkong, Q. F., 1988, Osmotic forces and the fusion of biomembranes, in *Molecular Mechanisms of Membrane Fusion* (S. Ohki, D. Doyle, T. D. Flanagan, S. W. Hui, and E. Mayhew, eds.), pp. 163–179, Plenum Press, New York.

MacDonald, R. I., 1985, Membrane fusion due to dehydration by polyethylene glycol, dextran, or sucrose. *Biochemistry* **24**:4058–4066.

Maggio, B., Ahkong, Q. F., and Lucy, J. A., 1976, Poly(ethylene glycol), surface potential and cell fusion. *Biochem. J.* **158**:647–650.

Maggio, B., Cumar, F. A., and Caputto, R., 1978, Induction of membrane fusion by polysialogangliosides. *FEBS Lett.* **90**:149–152.

Majumdar, S., Baker, R. F., and Kalra, V. K., 1980, Fusion of human erythrocytes induced by uranyl acetate and rare earth metals. *Biochim. Biophys. Acta* **598**:411–416.

Matt, H., and Plattner, H., 1983, Decoupling of exocytotic membrane fusion from protein discharge in *Paramecium* cells. *Cell Biol. Int. Rep.* **7**:1025–1031.

Melikyan, G. B., Abidor, I. G., Chernomordik, L. V., and Chailakhyan, L. M., 1983, Electrostimulated fusion and fission of bilayer lipid membranes. *Biochim. Biophys. Acta* **730**:395–398.

Micklem, K. J., Nyaruwe, A., and Pasternak, C. A., 1985, Permeability changes resulting from virus–cell fusion: Temperature dependence of the contributing processes. *Mol. Cell. Biochem.* **66**:163–173.

Miller, C., Arvan, P., Telford, J. N., and Racker, E., 1976, Ca^{2+}-induced fusion of proteoliposomes: Dependence on transmembrane osmotic gradient. *J. Membr. Biol.* **30**:271–282.

Mundy, D. I., and Strittmatter, W. J., 1985, Requirement for metalloendoprotease in exocytosis: Evidence in mast cells and adrenal chromaffin cells. *Cell* **41**:645–656.

Murachi, T., Tanaka, K., Hatanaka, M., and Murakami, Y., 1981, Intracellular Ca^{2+}-dependent protease (calpain) and its high-molecular-weight endogenous inhibitor (calpastatin). *Adv. Enzyme Regul.* **19**:407–424.

Nakornchai, S., Sathitudsahakron, C., Chongchirasiri, S., and Yuthavong, Y., 1983, Mechanism of enhanced fusion capacity of mouse red cells infected with *Plasmodium berghei. J. Cell Sci.* **63**:147–154.

Nelson, W. J., and Lazarides, E., 1984, Goblin (ankyrin) in striated muscle: Identification of the potential membrane receptor for erythroid species in muscle cells. *Proc. Natl. Acad. Sci. U.S.A.* **81**:3292–3296.

Niles, W. D., and Cohen, F. S., 1987, Video fluorescence microscopy studies of phospholipid vesicle fusion with a planar phospholipid membrane. *J. Gen. Physiol.* **90**:703–735.

Ohki, S., 1984, Effects of divalent cations, temperature, osmotic pressure gradient, and vesicle curvature on phosphatidylserine vesicle fusion. *J. Membr. Biol.* **77**:265–275.

Ohki, S., Doyle, D., Flanagan, T. D., Hui, S. E., and Mayhew, E., 1988, *Molecular Mechanisms of Membrane Fusion,* Plenum Press, New York.

Ohno, H., Maeda, Y., and Tsuchida, E., 1981, ^{1}H-NMR study of the effect of synthetic polymers on the fluidity, transition temperature and fusion of dipalmitoyl phosphatidylcholine small vesicles. *Biochim. Biophys. Acta* **642**:27–36.

Ohno-Shosaku, T., and Okada, T., 1985, Electric pulse-induced fusion of mouse lymphoma cells: Roles of divalent cations and membrane lipid domains. *J. Membr. Biol.* **85**:269–280.

Orci, L., and Malaisse, W., 1980, Single and chain release of insulin secretory granules is related to anionic transport at exocytotic sites. *Diabetes* **29**:943–944.

Ornberg, R. L., and Reese, T. S., 1981, Beginning of exocytosis captured by rapid-freezing of *Limulus* amebocytes. *J. Cell Biol.* **90**:40–54.

Palade, G. E., 1975, Intracellular aspects of protein synthesis. *Science* **189**:347–358.

Palade, G. E., and Bruns, R. R., 1968, Structural modifications of plasmalemmal vesicles. *J. Cell Biol.* **37**:633–649.

Papahadjopoulos, D., 1978. Calcium-induced phase changes and fusion in natural and model membranes. In *Membrane Fusion* (G. Poste and G. L. Nicolson, eds.), pp. 765–790, North-Holland, Amsterdam.

Parente, R. A., and Lentz, B. R., 1986, Rate and extent of poly(ethylene glycol)-induced large vesicle fusion monitored by bilayer and internal contents mixing. *Biochemistry* **25**:6678–6688.

Parsegian, V. A., Rand, R. P., and Gingell, D., 1984, Lessons for the study of membrane fusion from membrane interactions in phospholipid systems, in *Cell Fusion* (D. Evered and J. Whelan, eds.), pp. 9–27, Pitman, London.

Pasternak, C. A., 1984, Virally mediated changes in cellular permeability, in *Membrane Processes* (G. Benga, H. Baum, and F. A. Kummerow, eds.). pp. 140–166, Springer-Verlag, New York.

Pearl, M., and Taylor, A., 1985, Role of the cystoskeleton in the control of transcellular water flow by vasopressin in amphibian urinary bladder. *Biol. Cell* **55**:163–172.

Peretz, H., Toister, Z., Laster, Y., and Loyter, A., 1974, Fusion of intact human erythrocytes and erythrocyte ghosts. *J. Cell Biol.* **63**:1–11.

Pinto da Silva, P., and Nogueira, M. L., 1977, Membrane fusion during secretion. A hypothesis based on electron microscope observation of *Phytophthora palmivora* zoospores during encystment. *J. Cell Biol.* **73**:171–181.

Pollard, H. B., Tack-Goldman, K., Pazoles, C. J., Creutz, C. E., and Shulman, N. R., 1977, Evidence for control of serotonin secretion from human platelets by hydroxyl ion transport and osmotic lysis. *Proc. Natl. Acad. Sci. U.S.A.* **74**:5295–5299.

Pollard, H. B., Pazoles, C. J., Creutz, C. E., and Zinder, O., 1979, The chromaffin granule and possible mechanisms of exocytosis. *Int. Rev. Cytol.* **58**:159–197.

Pollard, H. B., Pazoles, C. J., Creutz, C. E., Scott, J. H., Zinder, O., and Hotchkiss, A., 1984, An osmotic mechanism for exocytosis from dissociated chromaffin cells. *J. Biol. Chem.* **259**:1114–1121.

Pontecorvo, G., 1975, Production of mammalian somatic hybrids by means of polyethylene glycol treatment. *Somat. Cell Genet.* **1**:397–400.

Poste, G., and Nicolson, G. L. (eds.), 1978, *Membrane Fusion*, North-Holland, Amsterdam.

Quirk, S. J., Ahkong, Q. F., Botham, G. M., Vos, J., and Lucy, J. A., 1978, Membrane proteins in human erythrocytes during cell fusion induced by oleoylglycerol. *Biochem. J.* **176**:159–167.

Robinson, J. M., Roos, D. S., Davidson, R. L., and Karnovsky, M. J., 1979, Membrane alterations and other morphological features associated with polyethylene glycol-induced cell fusion. *J. Cell Sci.* **40**:63–75.

Roos, D. S., and Choppin, P. W., 1985, Biochemical studies on cell fusion. II. Control of fusion response by lipid alteration. *J. Cell Biol.* **101**:1591–1598.

Rosenberg, J., Düzgüneş, N., and Kayalar, C., 1983, Comparison of two liposome fusion assays monitoring the intermixing of aqueous contents and of membrane components. *Biochim. Biophys. Acta* **735**:173–180.

Satir, B., Shooley, C., and Satir, P., 1973, Membrane fusion in a model system: Mucocyst secretion in *Tetrahymena*. *J. Cell Biol.* **56**:153–176.

Schindler, M., Koppel, D. E., and Sheetz, M. P., 1980, Modulation of membrane protein lateral mobility by polyphosphates and polyamines. *Proc. Natl. Acad. Sci. U.S.A.* **77**:1457–1461.

Schlegel, R. A., and Lieber, M. R., 1987, Microinjection of cultured cells via fusion with loaded erythrocytes. In *Cell Fusion* (A. E. Sowers, ed.), pp. 457–478, Plenum Press, New York.

Schmauder-Chock, E. A., and Chock, S. P., 1987, Mechanism of secretory granule exocytosis: Can granule enlargement precede pore formation? *Histochem. J.* **19**:413–418.

Schuel, H., 1978, Secretory functions of egg cortical granules in fertilization and development. *Gamete Res.* **1**:299–382.

Schwister, K., and Deuticke, B., 1985, Formation and properties of aqueous leaks induced in human erythrocytes by electrical breakdown. *Biochim. Biophys. Acta* **816**:332–348.

Serpersu, E. H., Kinosita, K., and Tsong, T. Y., 1985, Reversible and irreversible modification of erythrocyte membrane permeability by electric field. *Biochim. Biophys. Acta* **812**:779–785.

Siegel, D. P., 1984, Inverted micellar structures in bilayer membranes. *Biophys. J.* **45**:399–420.

Siegel, D. P., 1986, Inverted micellar intermediates and the transitions between lamellar, cubic, and inverted hexagonal lipid phases. II. Implications for membrane–membrane interactions and membrane fusion. *Biophys. J.* **49**:1171–1183.

Siegel, D. P., 1987, Membrane–membrane interactions via intermediates in lamellar-to-inverted hexagonal phase transitions, in *Cell Fusion* (A. E. Sowers, ed.), pp. 181–207, Plenum Press, New York.

Smith, C. L., Ahkong, Q. F., Fisher, D., and Lucy, J. A., 1982, Is purified poly(ethylene glycol) able to induce cell fusion? *Biochim. Biophys. Acta* **692**:109–114.

Smith, D. K., and Palek, J., 1982, Modulation of lateral mobility of band 3 in the red cell membrane by oxidative cross-linking of spectrin. *Nature (London)* **297**:424–425.

Sowers, A. E., 1984, Characterization of electric field-induced fusion of erythrocyte ghost membranes. *J. Cell Biol.* **99**:1989–1996.

Sowers, A. E., 1986, A long-lived fusogenic state is induced in erythrocyte ghosts by electric pulses. *J. Cell Biol.* **102**:1358–1362.

Sowers, A. E. (ed.), 1987a, *Cell Fusion*, Plenum Press, New York.

Sowers, A. E., 1987b, The long-lived fusogenic state induced in erythrocyte ghosts by electric pulses is not laterally mobile. *Biophys. J.* **52**:1015:1020.

Spear, P. G., 1987, Virus-induced cell fusion, in *Cell Fusion* (A. E. Sowers, ed.), pp. 3–32, Plenum Press, New York.

Stanley, E. F., and Ehrenstein, G., 1985, A model for exocytosis based on the opening of calcium-activated potassium channels in vesicles. *Life Sci.* **37**:1985–1995.

Stenger, D. A., and Hui, S. W., 1986, Kinetics of ultrastructural changes during electrically induced fusion of human erythrocytes. *J. Membr. Biol.* **93**:43–53.

Strittmatter, W. J., Couch, C. B., and Mundy, D. I., 1987, Role of metalloendoprotease in the fusion of biological membranes, in *Cell Fusion* (A. E. Sowers, ed.), pp. 99–121, Plenum Press, New York.

Struck, D. K., Hoekstra, D., and Pagano, R. E., 1981, Use of resonance energy transfer to monitor membrane fusion. *Biochemistry* **20**:4093–4099.

Teissie, J., and Blangero, C., 1984, Direct experimental evidence of the vectorial character of the interaction between electric pulses and cells in cell electrofusion. *Biochim. Biophys. Acta* **775**:446–448.

Teissie, J., and Rols, M. P., 1986, Fusion of mammalian cells in culture is obtained by creating the contact between cells after their electropermeabilization. *Biochem. Biophys. Res. Commun.* **140**:258–266.

Thomas, P., Limbrick, A. R., and Allan, D., 1983, Limited breakdown of cytoskeletal proteins by an endogenous protease controls Ca^{2+}-induced membrane fusion events in chicken erythrocytes. *Biochim. Biophys. Acta* **730**:351–358.

Tilcock, C. P. S., and Fisher, D., 1979, Interaction of phospholipid membranes with poly(ethylene glycol)s. *Biochim. Biophys. Acta* **577**:53–61.

Tilcock, C. P. S., and Fisher, D., 1982, The interaction of phospholipid membranes with poly(ethylene glycol). Vesicle aggregation and lipid exchange. *Biochim. Biophys. Acta* **688**:645–652.

Toister, Z., and Loyter, A., 1973, The mechanism of cell fusion. II. Formation of chicken erythrocyte polykaryons. *J. Biol. Chem.* **248**:422–432.

Verkleij, A. J., van Echteld, C. J. A., Gerritsen, W. J., Cullis, P. R., and de Kruijff, B., 1980, The lipidic particle as an intermediate structure in membrane fusion processes and bilayer to hexagonal hII transitions. *Biochim. Biophys. Acta* **600**:620–624.

Vienken, J., and Zimmermann, U., 1985, An improved electrofusion technique for production of mouse hybridoma cells. *FEBS Lett.* **182**:278–280.

Vos, J., Ahkong, Q. F., Botham, G. M., Quirk, S. J., and Lucy, J. A., 1976, Changes in the distribution of intramembranous particles in hen erythrocytes during cell fusion induced by the bivalent-cation ionophore A23187. *Biochem. J.* **158**:651–653.

Wakelam, M. J. O., 1985, The fusion of myoblasts. *Biochem. J.* **228**:1–12.

Wallin, A., Glimelius K., and Erikkson, T., 1974, The induction of aggregation and fusion of *Daucus carota* protoplasts by polyethylene glycol. *Z. Pflanzenphysiol.* **74**:64–80.

Ward, M., Davey, M. R., Mathias, R. J., Cocking, E. C., Clothier, R. H., Balls, M., and Lucy, J. A., 1979, Effects of pH, Ca^{2+}, temperature, and protease pretreatment of interkingdom fusion. *Somat. Cell Genet.* **5**:529–536.

Whitaker, M., and Zimmerberg, J., 1987, Inhibition of secretory granule discharge during exocytosis in sea urchin eggs by polymer solutions. *J. Physiol.* **389**:527–539.

White, J., Kielian, M., and Helenius, A., 1983, Membrane fusion proteins of enveloped animal viruses. *Q. Rev. Biophys.* **16**:151–195.

Wilschut, J., and Hoekstra, D., 1984, Membrane fusion: From liposomes to biological membranes. *Trends Biochem. Sci.* **9**:479–483.

Wojcieszyn, J. W., Schlegel, R. A., Lumley-Sapanski, K., and Jacobson, K. A., 1983, Studies on the mechanism of polyethylene glycol-mediated cell fusion using fluorescent membrane and cytoplasmic probes. *J. Cell Biol.* **96**:151–159.

Wyke, A. M., Impraim, C. C., Knutton, S., and Pasternak, C. A., 1980, Components involved in virally medicated membrane fusion and permeability changes. *Biochem. J.* **190**:625–638.

Zimmerberg, J., 1987, Molecular mechanisms of membrane fusion: Steps during phospholipid and exocytotic membrane fusion. *Biosci. Rep.* **7**:251–268.

Zimmerberg, J., and Whitaker, M. 1985, Irreversible swelling of secretory granules during exocytosis caused by calcium. *Nature (London)* **315**:581–584.

Zimmerberg, J., Cohen, F. S., and Finkelstein, A., 1980a, Fusion of phospholipid vesicles with planar phospholipid bilayer membranes. I. Discharge of vesicular contents across the planar membrane. *J. Gen. Physiol.* **75**:241:250.

Zimmerberg, J., Cohen, F. S., and Finkelstein, A., 1980b, Micromolar Ca^{2+} stimulates fusion of lipid vesicles with planar bilayers containing a calcium-binding protein. *Science* **210**:906–908.

Zimmerberg, J., Sardet, C., and Epel, D., 1985, Exocytosis of sea urchin egg cortical vesicles *in vitro* is retarded by hyperosmotic sucrose: Kinetics of fusion monitored by quantitative light-scattering microscopy. *J. Cell Biol.* **101**:2398–2410.

Zimmerberg, J., Curran, M., Cohen, F. S., and Brodwick, M., 1987, Simultaneous electrical and optical measurements show that membrane fusion precedes secretory granule swelling during exocytosis of beige mouse mast cells. *Proc. Natl. Acad. Sci. U.S.A.* **84**:1585–1589.

Zimmermann, U., 1982, Electric field-mediated fusion and related electrical phenomena. *Biochim. Biophys. Acta* **694**:227–277.

Zimmermann, U., 1986, Electrical breakdown, electropermeabilization and electrofusion. *Rev. Physiol. Biochem. Pharmacol.* **105**:175–256.

Zimmermann, U., Pilwat, G., Holzapfel, C., and Rosenheck, K., 1976, Electrical hemolysis of human and bovine red blood cells. *J. Membr. Biol.* **30**:135–152.

Zimmermann, U., Beckers F., and Coster, H. G. L., 1977, The effect of pressure on the electrical breakdown in the membranes of *Valonia utricularis*. *Biochim. Biophys. Acta* **464**:399–416.

Zimmermann, U., Pilwat, G., Pequeux, A., and Gilles, R., 1980, Electromechanical properties of human erythrocyte membranes: The pressure-dependence of potassium permeability. *J. Membr. Biol.* **54**:103–113.

Chapter 6

Lectin–Carbohydrate Interactions in Model and Biological Membrane Systems

Dick Hoekstra and Nejat Düzgüneş

1. INTRODUCTION

Many of the diverse activities of cells are manifested at the cell surface; that is, the start of many intracellular events find their origin in a signal triggered at the cell surface. Such a response may arise from the specific attachment of macromolecules to carbohydrate-bearing molecules (glycolipids and glycoproteins), which are abundantly present at the outer surface of the plasma membrane. Typical examples include the operation of hormones and toxins (Kelly *et al.*, 1979; Neville and Hudson, 1986; Ross and Gilman, 1980). Glycoconjugates, covalently linked to either lipids or proteins, also mediate the attachment of certain viruses, representing the initial event in the viral entry mechanism that eventually may lead to infection (Bächi *et al.*, 1977; Hoekstra *et al.*, 1988; White *et al.*, 1983). Secretory processes (Ling *et al.*, 1985), mitogenic effects (Rosoff *et al.*, 1987), myotube production from myoblasts (Cates *et al.*, 1984; Knudsen, 1985), and the formation of multinucleate macrophages during the inflammatory response (Papadimitriou, 1978) further illus-

---

Dick Hoekstra   Laboratory of Physiological Chemistry, University of Groningen, 9712 KZ Groningen, The Netherlands.   **Nejat Düzgüneş**   Cancer Research Institute and Department of Pharmaceutical Chemistry, University of California, San Francisco, California 94143-0128.

trate the variety of cell surface mediated responses. The origin of these events can be traced back to a specific association between external macromolecules and cell surface glycolipids and/or glycoproteins that act as a specific receptor and transmitting unit for these responses. The diversity, specificity, and frequency by which these responses occur throughout the biological life of the cell also dictate that the macromolecule itself, which recognizes the specific code provided by the carbohydrate part of the receptor, displays a high degree of specificity. Under defined conditions, it is in principle possible to detect the biological results of such macromolecule–receptor interactions. Yet, due to the complexity of the cell-surface structure, the molecular mechanisms underlying this interaction, as well as its immediate consequences, are still rather obscure. Therefore, many investigators have focused on studying the dynamic properties of glycolipids and/or glycoproteins as mobile membrane receptors by incorporating them into phospholipid membranes. With such model systems, parameters such as carbohydrate-specific recognition, the density of the receptor, and the physical properties of the lipid matrix surrounding the receptor, all of which may affect the interaction with a specific macromolecule, can be studied in a relatively convenient way. The interaction of multidentate carbohydrate binding molecules such as lectins with glycoreceptors has become a valuable model system for studying macromolecule–receptor interactions.

In this chapter we first describe some general properties of lectins and of the molecules that serve as their receptors, that is, glycoproteins and glycolipids. It is not our intention to discuss these aspects at length but rather to introduce nonspecialist readers to the field. Subsequently, we discuss detailed structural studies concerning lectin–receptor interactions, as carried out with artificial membrane systems. Finally, we present a summary of work carried out with biological systems, paying particular attention to cellular responses upon addition of *exogenous* (plant) lectins. In addition, we review and discuss the biological role of *endogenous* lectins.

2. LECTINS: ABUNDANCE AND PROPERTIES

Lectins were first detected in and isolated from plants, but it has been well established now that these sugar-specific binding proteins are ubiquitous in nature (Barondes, 1984, 1986; Goldstein and Hayes, 1978; Koch and Uhlenbruck, 1982; Kokourek, 1986; Lis and Sharon, 1986). Thus, they can also be found in microorganisms, including bacteria, invertebrates, and mammalian cells. More than 100 different lectins have been isolated thus far, but those from plants are still the best characterized group. Table I lists a number of lectins that are frequently used in lectin–receptor studies and some of their molecular properties. Recently, their properties, functions, and applications have been extensively reviewed (Damjanov, 1987; Liener et al., 1986; Lis and Sharon, 1986). Therefore, we only briefly summarize these various aspects of lectins. For further details the reader is referred to the aforementioned reviews.

2.1. Structural Features of Lectins

Lectins are proteins of molecular weights that may vary between approximately 40 and 400 kDa. They do not display enzyme activity. Rather, their primary property is their ability to recognize and bind to specific carbohydrate residues, the biological significance of which is still largely obscure (see Section 5). Two types of lectins can be defined: those that are integral membrane proteins that can be isolated only upon extraction from the membranes with nonionic detergents, and those that are soluble and present intracellularly and/or extracellularly. Many lectins themselves contain carbohydrates; however, there is no evidence that the lectin-bound carbohydrate is necessary for expression of their biological properties. Usually, lectins consist of two or more subunits, which may be cross-linked via disulfide bridges (Giga et al., 1985, 1987). With few exceptions (e.g., see Schnell and Etzler, 1987) the subunits contain one sugar binding site per subunit. These binding sites are equivalent and display the same sugar specificity. By virtue of having more than one subunit, each binding a carbohydrate residue, lectins cause agglutination of cells or receptor-containing artificial bilayers (see Sections 3 and 4). This activity can be used to establish the carbohydrate binding specificity of lectins. Monovalent lectins, like ricin (Utsumi et al., 1987) and RCA_{II} (Baenziger and Fiete, 1979) from castor bean, can bind to glycoreceptors but do not induce the cross-linking reaction.

Several lectins can be classified as metalloproteins; that is, they require metal ions for binding to a saccharide (Liener et al., 1986; Shimizu and Hatano, 1983; Young, 1983). The metal ions *per se* are not directly involved in sugar binding. Rather, they appear to affect the conformation of the lectin. Thus far this has been most clearly described in the case of the binding of a sugar residue to the plant lectin concanavalin A (ConA) (Brewer et al., 1983; Brown et al., 1977). Sequential binding of certain metal ions (Ca^{2+}, Mn^{2+}, Mg^{2+}) by the lectin transforms the conformation of the protein, or part thereof, from an "unlocked" conformation, displaying weak saccharide binding, to a "locked" conformation, representing the lectin structure that actively binds the saccharide. The position of the metal ions in various lectins appears to be very similar and hence the metal ion binding sites of various lectins appear to display a very close homology (Bhattacharyya et al., 1987). Interestingly, when Ca^{2+} and Mn^{2+}, present in both lentil and pea lectins, were substituted for CD^{2+}, the physicochemical properties of the modified lectins appeared to be the same as those of the corresponding native proteins (Bhattacharyya et al., 1987).

Table I
Molecular Properties of Some Frequently Used Lectins[a]

Lectin	M_r (kDa)	Subunit structure	Carbohydrate specificity[b]	Metal requirement	Primary structure	Three-dimensional structure
Concanavalin A (ConA) (*Canavalia ensiformis*)	106	α4	-GlcNAc -Manα1-2Manα1-2Man -Manα1-2Man -αMan -αGlc -αGlcNAc	Mn^{2+}, Ca^{2+}	Wang et al., 1975; Cunningham et al., 1975	Edelman et al., 1972; Hardman and Ainsworth, 1972
Dolichos biflorus lectin (DBL)	110–120	α2β2	-GalNAcα1-3GalNAc -αGalNAc	Ca^{2+}, Mg^{2+}, Mn^{2+}, Zn^{2+}		
Fava bean lectin (Favin) (*Vicia faba*)	52.5	α2β2	-αMan -αGlc -αGlcNAc	Ca^{2+}, Mg^{2+}	Cunningham et al., 1979	
Helix pomatia agglutinin (HPA)	79	α6	-GalNAcα1-3GalNAc -αGalNAc -αGlcNAc			
Lentil lectin (LL) (*Lens culinaris*)	46	α2β2	-αMan -αGlc -αGlcNAc	Mn^{2+}, Ca^{2+}	Foriers et al., 1981	

Lectin	MW (kDa)	Subunit	Specificity[b]	Metal requirement	Reference
Peanut lectin (PL) (*Arachis hypogaea*)	98–111	α4	-Galβ1-3GalNAc -GalNH₂ -αand βGal	Ca²⁺, Mg²⁺	
Phytohemagglutinin (PHA) (*Phaseolus vulgaris*)	126	α4	-Galβ1-4GlcNAcβ1-2Man	Mn²⁺, Ca²⁺	Hoffman et al., 1982
Pisum sativum lectin (PSL)	50	α2β2	-αMan -αGlc -αGlcNAc	Mn²⁺, Ca²⁺	Higgens et al., 1983
Ricinus communis agglutinin I (RCA₁)	120	α2β2	-βGal -αGal -GalNAc	None	Goldstein and Poretz, 1986
Sophora japonica agglutinin (SJA)	132	α2β2	-αGalNAc, βGalNAc -αGal, βGal	Ca²⁺, Mn²⁺	Meehan et al., 1982
Soy bean agglutinin (SBA) (*Glycine max*)	120	α4	-αGalNAc, βGalNAc -αGal	Ca²⁺, Mn²⁺	Vodkin et al., 1983
Wheat germ agglutinin (WGA) (*Triticum vulgare*)	43	α2	-[GlcNAcβ1-4]₂GlcNAc -GlcNAcβ1-4GlcNAc -βGlcNAc		Wright et al., 1984 Wright, 1980, 1981

[a] Based on data reviewed by Goldstein and Poretz (1986) and Strosberg et al. (1986).
[b] The carbohydrate specificity is listed in decreasing order.

The primary structures of a number of lectins have been determined, among others those of ConA (Cunningham *et al.*, 1975), soy bean agglutinin (SBA; Vodkin *et al.*, 1983), pea lectin (Higgens *et al.*, 1983), favin (Zeeuws and Strosberg, 1978), lentil lectin (Foriers *et al.*, 1981), *Ricinus communis* agglutinin, type I (RCA_I) and type II (RCA_{II}) (Goldstein and Poretz, 1986), and *Dolichos biflorus* lectin (Schnell and Etzler, 1987). These lectins are all isolated from leguminous seed and comparison of their primary structures indicates that there exists substantial homology in amino acid sequence. The homology does not extend to the regions that are thought to be involved in carbohydrate binding, which, in the light of the different specificities, can be expected.

Work concerning the biosynthesis of lectins is progressing (Carrington *et al.*, 1985; Schnell and Etzler, 1987). *Dolichos biflorus* lectin, an *N*-acetylgalactosamine (GalNAc)-recognizing lectin, is a tetramer, consisting of equal amounts of two different subunits of which only one type exhibits sugar binding properties. Yet both subunits are encoded by a single mRNA. It appears that the difference between the subunit types arises by a post-translational modification, involving a proteolytic removal of a small decapeptide from the COOH terminus of one of the subunits (Schnell and Etzler, 1987). An analogous phenomenon has been reported to occur during the biosynthesis of ConA (Carrington *et al.*, 1985). This removal apparently conveys saccharide binding properties to the lectin as the unmodified subunit does not bind carbohydrate residues. Indeed, the binding activity is located at the carboxyl-terminal domain of the subunit (Borrebaeck and Etzler, 1981; Schnell and Etzler, 1987). The significance of a COOH-terminal domain as the carbohydrate binding domain is also apparent in a variety of mammalian lectins (Giga *et al.*, 1987). In the case of a sialic acid binding lectin from frog, ϵ-amino groups of lysyl residues have been proposed to play a specific role in carbohydrate binding (Titani *et al.*, 1987). Finally, as far as the secondary and tertiary structure of lectins is concerned, it appears that many lectins contain mainly β-sheet and β-turn and little α-helix (Jirgensons, 1980; Giga *et al.*, 1987). This might be important for stabilizing their tertiary structure, as it would favor the formation of interpeptide hydrogen bonding.

2.2. Carbohydrate Binding Properties

Many lectins recognize and bind to terminal, nonreducing saccharides. Near-terminal residues may also be involved in or affect to a certain extent the binding of a lectin to a carbohydrate. In addition, in model systems it has been shown that the local environment of the carbohydrate receptor influences its binding properties (see Section 3). Several other parameters should also be taken into account. One of these, for example, is the stereochemical orientation of the terminal sugar moiety. It has been reported that some lectins display anomeric specificity while others do not, or to a lesser extent (Goldstein and Hayes, 1978; Hammarström et al., 1977; Kahn et al., 1986). Also, apart from the terminal sugar, peripheral branching patterns of the carbohydrate structure can modulate lectin interactions (Green et al., 1987). When determining the ability of a certain lectin to bind to a carbohydrate residue, it is also relevant to take into account the nature of the substrate. Differences in binding properties may emerge when determining the interaction of a lectin with a lectin-specific terminal sugar residue, being part of either one of the following structures: carbohydrate-containing biopolymers, (soluble) protein conjugates, or lectin receptors such as a glycolipid incorporated in a supporting lipid matrix (cf. Hoekstra et al., 1985). Furthermore, the carbohydrate specificity exhibited by a lectin may very much depend on the assay systems used (Molin et al., 1986). Recently, lectin agarose affinity HPLC was introduced to define lectin specificity (Green et al., 1987). This technique appears superior to previous methods and allows high sensitivity analysis of subtle aspects of oligosaccharide–lectin interactions to be performed.

The binding of a lectin, which is likely of a positive cooperative nature, is a relatively slow process and proceeds for several minutes (Grant and Peters, 1984). Half-times on the order of many minutes have been reported for the binding of wheat germ agglutinin (WGA) and ConA to glycophorin and band 3, respectively, both receptors being reconstituted into artificial bilayers (Chicken and Sharom, 1983; Ketis and Grant, 1983). The binding rate constants of lectin–sugar interactions are quite low (Harrington et al., 1981; Kahn et al., 1986; Neurohr et al., 1982), that is, approximately 10^5 slower than found for a diffusion-controlled reaction. Furthermore, the binding is usually quite weak and reversible. Affinity constants on the order of 10^3–10^6 M^{-1} have been reported (Grant and Peters, 1984)—too low for irreversible binding. Hence, covalent bonds appear not to be involved in lectin–carbohydrate interactions. Rather, it has been proposed that the binding between a lectin and a sugar residue is mediated by hydrogen bonding and presumably hydrophobic interactions (Biswas et al., 1987; Hammarström et al., 1977; Kahn et al., 1986; Sekharudu et al., 1986; Spohr et al., 1985). Studies in which the binding of the basic lectin from winged bean with fluorescently labeled galactosamine was examined showed an enhancement of the fluorescence quantum yield which was accompanied by a blue shift of the emission maximum (Kahn et al., 1986). This would suggest that after binding the fluorophore senses a hydrophobic environment. In principle, this environment can correspond to the binding state. That the derivatized sugar did bind to this site was convincingly shown by its specific displacement with haptenic sugars.

The binding sites of a lectin can be composed of an extended region (Clegg et al., 1983), such that more than one monosaccharide unit of an oligosaccharide can interact simultaneously at each site. Peanut agglutinin (Momoi et al., 1982; Pereira et al., 1976) and *Croton tiglium* lectin (Banerjee and Sen, 1983) show a specific affinity for recognizing and binding disaccharide structures. Also RCA_I, which interacts specifically with galactosyl residues, displays an enhanced binding to lactosyl residues relative to the binding to monosaccharides (Kabat, 1978). Other lectins, such as *Lotus tetragonolobus*, can accommodate as many as four or five residues in their binding sites (Rao and Biswas, 1982).

The binding mechanism and the nature of forces controlling lectin–sac-

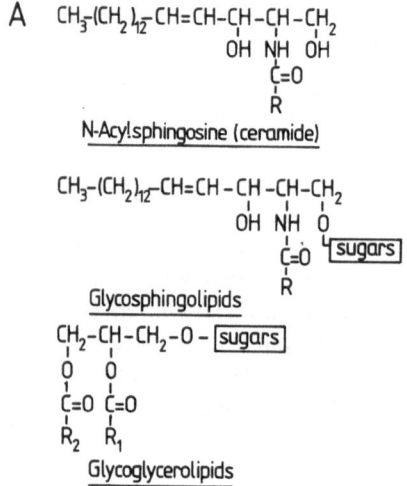

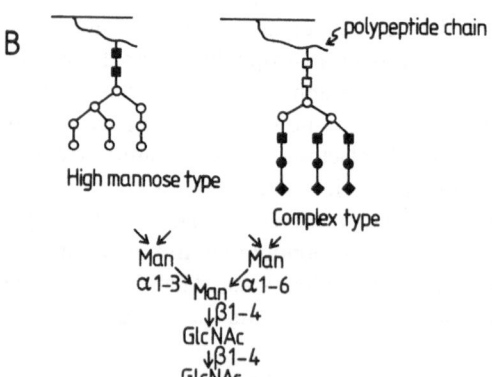

FIGURE 1. (A) Molecular structures of glycosphingo- and glycoglycerolipids. (B) Oligosaccharides on glycoproteins: high mannose and complex type. ■, GlcNAc; ○, Man; ●, Gal; △, Fuc; ◆, NeuAc; $(Man)_3(GlcNAc)_2$ represents the asparagine-linked core structure that is found in many glycoproteins.

charide interaction have been most extensively studied in the case of conA. Both experimental work (Baenziger and Fiete, 1979; Brewer and Brown, 1979; Goldstein et al., 1973) and theoretical predictions, based on computer modeling studies (Sekharudu et al., 1986), indicate that binding is accomplished by interaction of the lectin with particular mannosidic hydroxyl groups that allow formation of hydrogen bonds between the terminal mannose residue and the lectin binding site. In addition, hydrophobic interaction forces contribute to the stability of the binding. Although the sugar binding site of ConA only allows for a single mannose residue, the interaction extends slightly beyond the sugar residue that is placed in the binding site. This indicates that lectins may possess the ability to recognize fine differences in complex carbohydrate structures, affected by the sugar residue positioned next to the specifically recognized terminal residue. This ability is presumably dependent on the possible shapes that the oligosaccharide chain can assume. However, very little is known about the possible conformations of the oligosaccharides free in solution and when bound to a protein or lipid. Model studies have been carried out (Biswas et al., 1986) to predict the possible conformations of some oligosaccharides which have the trimannoside core that interact with ConA. ConA displays a high binding affinity for the $Man_3GlcNAc_2$ structure (Baenziger and Fiete, 1979), commonly found in a wide variety of glycoproteins (Figure 1B). A distortion in the three-dimensional structure of the oligosaccharide seriously weakens the oligosaccharide–ConA interaction (Schachter et al., 1982).

3. GLYCOLIPIDS AND GLYCOPROTEINS: BIOLOGICAL RECEPTORS FOR LECTINS

As outlined in Section 2, lectins display a selective binding for carbohydrate residues. Such residues are found both intracellularly and at the cell surface where they are usually covalently attached to lipids and proteins. Various distinct classes of glycolipids can be distinguished (Makita and Taniguchi, 1985; Wiegandt, 1985): those based on the long-chain fatty base, sphingosine, and those based on glycerol (Figure 1). In the first class, sphingosine has an amide-linked fatty acid; N-acyl-sphingosine is called a ceramide which constitutes the backbone of the neutral and acidic glycosphingolipids. The neutral glycosphingolipids have one or more neutral sugars attached to the ceramide backbone. Acidic glycosphingolipids contain, in addition to the neutral sugar portion, one or more N-acetyl-neuraminic acid (NANA) or, in short, sialic acid residues. Glycolipids based on glycerol have a structure analogous to phospholipids with the sugar attached glycosidically to the 3-position of the glycerol backbone and the fatty acids at the 1- and 2-positions. This latter class of glycolipids is mainly found in plants and microorganisms, whereas the ceramide-based glycolipids are most relevant in mammalian systems. For convenience we will refer to these glyco*sphingo*lipids as glycolipids. Although the structural knowledge of glycolipids has rapidly advanced in recent years (Wiegandt, 1985), the exact functional roles and modes of action of glycolipids remain largely unknown. The prevailing view has been that glycolipids are mainly, if not exclusively, localized in the outer leaflet of the plasma membrane (Barbosa and Pinto da Silva, 1983). Yet, evidence is accumulating that these lipids can also be found in intracellular membranes (Symington *et al.*, 1987). Apart from their localization, they usually constitute a minimal fraction of the total membrane lipid. Usually, some 5% of the plasma membrane lipids are contributed by glycolipids. However, in myelin membranes, a unique multilamellar compaction of the plasma membrane of oligodendrocytes (in the central nervous system) or Schwann cells (in the peripheral nervous system), the glycolipid content can be as high as 30% of the total lipid (Morell, 1984).

Glycolipids have been related to many biological functions, although the mechanisms through which these lipids exert their effects are largely unclear. They are thought to be involved in intercellular communication, the immune response, the regulation of the cell growth and differentiation, and the pathogenesis of cancer (Hakomori, 1981, 1986; Nudelman *et al.*, 1983; Wiegandt, 1985). Metastatic properties of tumor cell lines have been related to the level of gangliosides; that is, relatively high levels of these acidic glycolipids convey an enhanced metastatic capacity to the tumor cells (Laferte *et al.*, 1987), while certain glycolipids become specifically expressed during malignancy (Hersey *et al.*, 1986; Junqua *et al.*, 1987). More recently, gangliosides have attracted considerable attention because of their ability to modulate transmembrane signaling via a specific ganglioside-mediated stimulation of endogenous protein phosphorylation (Chan, 1987; Tsuji *et al.*, 1985).

As far as the complexity of the carbohydrate content and composition are concerned, the glycolipids are relatively simple structures compared to those of the other carbohydrate-containing molecules, the glycoproteins. Glycoproteins may contain as many as 100 carbohydrate residues or more per molecule. Despite the diversity of glycoproteins (Rauvala, 1983; Sharon, 1984a), their carbohydrate components have many common structural features. The oligosaccharide chains are N-glycosidically linked to an asparagine residue of the protein backbone. Usually, they possess an almost invariant trimannosidic core structure ($Man_3GlcNAc_2$, Figure 1) although the precursor structure, the high mannose oligosaccharide, may be an end product as well (Kornfeld, 1982; Kornfeld and Kornfeld, 1985).

Lactosaminoglycan proteins are a special group of plasma membrane glycoproteins of which the carbohydrates are composed of a trimannosyl di-*N*-acetylchitobiosyl core structure and side chains made of Galβ1-4GlcNAcβ1-3 repeats. These glycoproteins appear to be present in many cell types from human and animal origin (Cummings and Kornfeld, 1984; Fukuda et al., 1987; Järnefelt et al., 1978; Li et al., 1980). The physiological significance of their presence is unclear. Another family of glycoproteins that appears to be abundantly distributed in mammalian cells are those that resemble the properties of glycoproteins GPIIb and GPIIIa, which are the major surface glycoproteins in blood platelets (Charo et al., 1987; Fitzgerald et al., 1985, 1987; Pytela et al., 1986). The function that these proteins express in platelets involves their ability to bind adhesive proteins, such as fibrinogen, fibronectin, and von Willebrand factor (Bennett et al., 1983; Parise and Phillips, 1986; Ruggeri et al., 1983). In most other cells, the function(s) of these glycoproteins is unknown but their presence on the endothelial cell surface has been related to binding circulating cells (Fitzgerald et al., 1985). From a point of view of lectin–carbohydrate interactions, it appears attractive to imply such interactions in any type of extracellular communication, not in the least because of specificity. However, the observation that the interaction of GPIIb and GPIIIa with adhesive proteins is inhibited by the tripeptide Arg-Gly-Asp (Parise and Phillips, 1986) illustrates that carbohydrate-mediated binding is not necessarily a general phenomenon. The question thus arises, what is the purpose of these proteins being glycosylated? Most proteins that enter the secretory pathway are covalently modified by the addition of carbohydrates in the Golgi apparatus. After N-glycosylation, the nascent peptide chain(s) folds (Rothman and Lodish, 1977). It turns out that the mechanism of protein folding does not depend on the presence of the carbohydrate residues, nor does the carbohydrate moiety interfere with the correct folding of the polypeptide chain (Grafl et al., 1987). By using drugs that inhibit N-glycosylation, it was concluded that generalizations concerning the role of protein-bound carbohydrates could not be made (Schwartz and Datema, 1982). Yet, two common features became apparent: (1) the presence of carbohydrate may reduce the protein's susceptibility to proteolytic degradation by local shielding of protein surface regions (Bernard et al., 1983) and (2) carbohydrates may provide a specific code to the proteins that may determine their intracellular transport. From the latter it follows that, analogous to a "key-and-lock" mechanism, specific carbohydrate-recognizing molecules, that is, lectins or lectinlike molecules, should be involved in guiding certain glycoproteins to their appropriate cellular sites.

Although the potential biological significance of lectin–carbohydrate interactions in mammalian systems becomes increasingly clear, the molecular mechanisms of such interactions are, for obvious reasons, quite difficult to tackle within a biological system. For this reason a substantial amount of work has been carried out with simple model systems. We therefore first discuss the work carried out with artificial membranes and subsequently focus on the interaction of lectins with biological membranes and the biological significance of lectin and lectinlike activities, both at the cell surface and intracellularly.

4. LECTIN–CARBOHYDRATE INTERACTIONS IN MODEL SYSTEMS

There are a number of obvious advantages related to studies of carbohydrate–lectin interactions in simplified model systems. Carbohydrates are abundantly present at the cell surface, as outlined in Section 3. This fact makes a detailed molecular study of the interaction of a certain lectin with a particular glycolipid and/or protein virtually impossible. In a model system, all parameters can, in principle, be carefully controlled. Furthermore, certain lectins, considered identical in terms of monosaccharide specificity, possess the ability to recognize fine differences in complex structures (Debray *et al.*, 1981). Obviously, such properties are best revealed when the carbohydrate receptor can be modified in a systematic manner. This is also true for studies of the effect of carbohydrate conformation on lectin binding. Such model studies can be carried out with oligosaccharides free in solution or with glycolipid and glycoproteins, reconstituted in a supporting bilayer matrix of phospholipids. The emphasis of our discussion is on the latter systems, and where appropriate, we refer to studies with solubilized oligosaccharides.

4.1. Properties of Glycolipid-Containing Membranes

The hydroxyl (from the sugar residue and sphingosine) and the amide groups (from sphingosine) of glycosphingolipids can readily act as hydrogen bond vectors which allow specific intra- and intermolecular contacts (Boggs, 1980). Therefore, glycosphingolipids tend to aggregate in water, forming high molecular weight micelles, presumably mediated by the formation of a hydrogen bond network. Hence, to study their molecular properties, in particular those related to their functioning as mobile membrane receptors, the glycolipid of interest is usually incorporated in phospholipid model membranes, such as liposomes. By doing so, additional parameters become an inherent part of studies that were initially primarily concerned with glycolipid–ligand interaction only. These parameters include the physical and structural properties of the lipid matrix surrounding the glycolipid that may affect, directly or indirectly, the properties of the glycolipid *per se*. This implies that if a certain parameter affects the interaction of a lectin with a glycolipid the immediate question arises whether the observed effect originates from a direct alteration in the structural and/or physical properties of the glycolipid and/or from a lipid matrix-mediated effect on the glycolipid. Another commonly observed phenomenon is that the incorporation of glycolipids reduces the fluidity of the membrane in a concentration and glycolipid species-dependent manner as there is an increased ordering of the lipids surrounding the glycolipid receptor (Bertoli *et al.*, 1981; Correa-Freire *et al.*, 1979; Curatolo *et al.*, 1978; Maggio *et al.*, 1985; Sharom *et al.*, 1976). One important parameter includes the glycolipid distribution in the lipid matrix. When incorporated at relatively high concentrations (above approximately 10–15 mol%, depending on the glycolipid), glycolipids display the tendency to phase separate and cluster in undispersed patches (Bertoli *et al.*, 1981; Bunow and Bunow, 1979; Utsumi *et al.*, 1984). The degree of clustering will in turn affect the stability of the bilayer, as the coexistence of (phospholipid) bilayer structure and (glycolipid) micellar structures will lead to a perturbation or disruption of the membrane, the extent of which depends on the mole ratio of a glycolipid (Bertoli *et al.*, 1981; Delmelle *et al.*, 1980). Neutral glycolipids seem to display a higher tendency to phase separate from gangliosides (Curtain *et al.*, 1980; Maggio, 1985; Maggio *et al.*, 1985; Rintoul *et al.*, 1986; Sharom *et al.*, 1976; Tillack *et al.*, 1983). For example, the ganglioside GM_1 can be incorporated into phosphatidylcholine (PC) bilayers up to approximately 25 mol% before phase separation occurs and gross membrane disruption can then be detected (Bertoli *et al.*, 1981). Extensive clustering of neutral glycolipids can occur at substantially lower concentration, particularly at temperatures below the chain melting transition temperature of the supporting lipid membrane. When galactosylceramide (GalCer) is incorporated in dipalmitoylphosphatidylcholine (DPPC) bilayers at a concentration of 10 mol%, more than 85% of the glycolipid is clustered at 15°C, where the host lattice is in the solid state (Utsumi *et al.*, 1984). Upon raising the temperature, GalCer becomes randomly dispersed in the bilayer; that is, fluidity appears to regulate the lateral distribution of GalCer. The experimental evidence in this case was based on the use of a nitroxide spin probe, attached to the acyl chain of the glycolipid. In principle, these bulky probes could affect mutual interactions between the glycolipids of interest by interfering with hydrogen bond formation, involving hydroxyl groups on the sphingosine chain, which very much determines the interactions of these lipids in the plane of the bilayer (Boggs, 1980; Rintoul *et al.*, 1986). With head group-labeled GalCer, a clustering has been reported in fluid PC liposomes (Sharom *et al.*, 1976). On the other hand, it has also been shown (Maggio *et al.*, 1985), by high sensitivity differential scanning calorimetry (DSC), that neutral natural glycolipids with a relatively short hydrocarbon chain (mono- and dihexosylceramides) are readily miscible with DPPC. No indication of phase-separated pure cerebroside domains could be found. It has also been reported that GluCer is randomly distributed in DPPC bilayers, provided that its concentration does not exceed 15 mol% (Correa-Freire *et al.*, 1979). For gangliosides the picture seems to be more straightforward: in fluid phospholipid membranes, gangliosides, when present at low concentrations, form almost ideal mixtures with phosphatidylcholines (Brown and Thompson, 1987; Maggio *et al.*, 1985; Ollmann *et al.*, 1987; Rintoul *et al.*, 1986; Uchida *et al.*, 1981). Presumably, the presence of the negative charge in gangliosides, provided by sialic acid, can be related to this phenomenon as it may cause intermolecular electrostatic repulsions. Support for this explanation can be derived from the work of Brown and Thompson (1987), who studied the spontaneous transfer of GM_1 and asialo-GM_1 between phospholipid vesicles. These authors demonstrated that GM_1 transfers to acceptor vesicles with a half-time of approximately 40 hr, whereas the majority of the asialo-GM_1 (approximately 84%) transfers with an extreme slow rate ($t_{1/2} > 500$ hr), thus reflecting differences in the lateral distribution of the lipid, which is presumably attributable to a clustering of the "neutral" asialo-GM_1 in the lateral plane of the bilayer. In this respect, Tillack et al. (1983) have demonstrated, using DSC and freeze-etch electron microscopy, that asialo-GM_1 exists in small clusters in both fluid and solid PC model membranes. Moreover, the extent of clustering also appears to be dependent on the phospholipid species. Hence, the relationship between the physical state of the bilayer and the lateral distribution of glycolipids is rather complex.

It seems likely that the simple presence of sialic acid is not the only factor in determining whether or not a neutral versus acidic glycolipid tends to aggregate. In a study in which the molecular behavior of GT_{1b} was compared to that of GD_{1a}, it was shown that in spite of an enhanced charge density in the GT_{1b} species, this acidic glycolipid displayed a higher intrinsic tendency than GD_{1a} to phase separate into ganglioside-enriched domains (Myers et al., 1984). This is in line with observations that vesicle destabilization occurs when the amount of ganglioside incorporated into fluid PC vesicles exceeds a certain range. Thus, approximately 25% GM_1, 15% GD_{1a}, or 10% GT_{1b} can be incorporated in a PC bilayer structure before membrane disruption occurs (Bertoli et al., 1981). Presumably, the relevance of intermolecular hydrogen bond interactions that may occur between the gangliosides and the possibility for acidic sialic acid residues to provide potential negative charges for these interactions can explain the inverse relationship between charge density and membrane stability. It has become apparent, however, that the lateral distribution of glycolipids in the plane of the bilayer is an important parameter as far as carbohydrate head group accessibility is concerned. When clustered, sialic acid residues are no longer accessible to cleavage by neuraminidase (Myers et al., 1984), while the antigenic reactivity of the head group in the presence of specific antibodies is abolished (Utsumi et al., 1984). It is obvious that these structural aspects of the glycolipids per se also bear relevance to their functioning as receptors for lectins.

4.2. Lectin–Glycolipid Interactions

Studies concerning the interaction of lectins with glycolipid-containing bilayers may serve a dual purpose. First, they may serve as a model system for ligand–receptor interactions and second, in such studies the lectins can also be a valuable tool for analyzing the structural and physical features of the glycolipids (e.g., in studies such as those described in Section 4.1). In addition, lectin–glycolipid interactions may constitute a useful model to study membrane–membrane interactions.

4.2.1. Parameters Affecting Lectin–Glycolipid Interaction

Upon their interaction with specific carbohydrate residues, multivalent lectins can cross-link adjacent glycolipid-containing vesicles, a process that can be conveniently monitored by turbidity measurements (Curatola *et al.*, 1978; Hoekstra *et al.*, 1980; Surolia *et al.*, 1975). To agglutinate glycolipid-containing vesicles, a certain threshold glycolipid concentration is required (approximately 5 mol%), that is, a certain minimum density of receptors is necessary to support lectin-mediated agglutination (e.g., see Curatola *et al.*, 1978). Vesicle agglutination is further modulated by the physical properties of the lipid matrix surrounding the glycolipid receptor. Obviously, such effects become most prominent when the lectin-specific carbohydrate residue of the receptor is in the vicinity of the bilayer–water interface. The extent to which a carbohydrate head group protrudes beyond the vesicle surface is therefore an additional parameter of paramount significance that codetermines the interaction between lectin and glycolipid receptor. Furthermore, different lectins display different sensitivities to changes in parameters such as the lipid environment of the receptor.

The availability of synthetic "glycolipids" (Williams *et al.*, 1979) has greatly facilitated the efforts to characterize at the molecular level the relationship between head group exposure, receptor lipid environment, and the ability of a lectin to interact with that receptor. Different classes of synthetic "glycolipids" have been used (Hampton *et al.*, 1980; Rando *et al.*, 1980; Sundler, 1984; Sundler and Wijkander, 1983; Williams *et al.*, 1979). As can be seen (Figure 2), their chemical structure is quite different from that of natural gly-

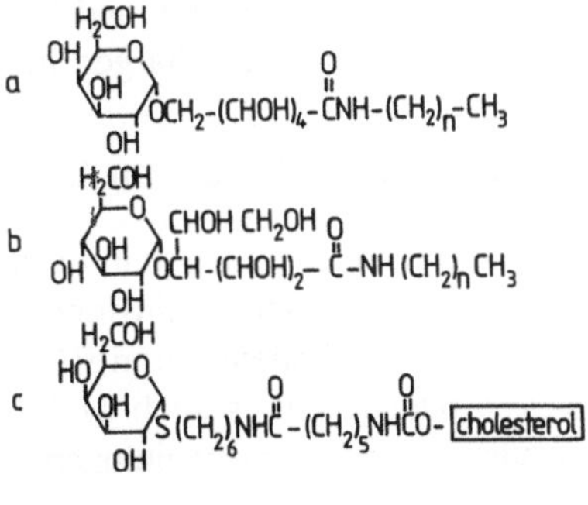

FIGURE 2. Chemical structures of synthetic glycolipids. (a) *N*-alkylmelibionamide; (b) *N*-alkylmaltobionamide; (c) a cholesterol-anchored, synthetic glycolipid as synthesized by Slama and Rando (1980). For synthesis and chemical characteristics of (a) and (b) the reader is referred to Williams *et al.* (1979).

colipids but in all studies reported specific interactions between a lectin and the carbohydrate structure, linked to a hydrophobic core via a spacer, were convincingly demonstrated by showing that competing haptenic sugars abolished the interaction. The successful use of synthetic glycolipids is dictated by the necessity that the carbohydrate residue is linked to the hydrophobic membrane anchoring structure (be it cholesterol or an alkyl chain, in the case of bionamides) via a hydrophilic spacer (Rando et al., 1980). With a hydrophobic spacer reorientations in the carbohydrate head group may take place, such that the terminal sugar is no longer available for interaction. As the *spacer length* can be varied in a controllable manner, a detailed picture can be obtained on the effect of the parameters that affect lectin binding, as outlined above. An interesting illustration of how the binding of a lectin can be influenced by both effective exposure of carbohydrate receptor and the lipid environment of the receptor is revealed by studies of ConA interaction with synthetic "glycolipid"-containing phospholipid vesicles. With a relatively short spacer length (4–6 atoms) and when incorporated into zwitterionic vesicles consisting of PC or PC/phosphatidylethanolamine (PE), no vesicle agglutination is observed (Hampton et al., 1980). However, when part of the PC is replaced by an anionic lipid, such as phosphatidylserine (PS), phosphatidic acid (PA), or cardiolipin (CL) agglutination will occur. In this case agglutination is reversed by elevated ionic strength as well as upon addition of the haptenic sugar α-methylmannoside. In the absence of the glycolipid receptor, no agglutination is apparent. Thus, it appears that both a carbohydrate-specific interaction and electrostatic forces are involved in ConA-mediated agglutination of the vesicles. This phenomenon is observed in systems where the carbohydrate receptor is in the direct vicinity (i.e., within a few angstroms) of the bilayer–water interface. Thus, PC vesicles containing the thiomannoside-linked cholesterol as synthetic lipid can be agglutinated by ConA (Orr et al., 1979). Here, the sugar residue is linked to either a 6-aminohexyl or a 6-(6-aminohexanamido) hexyl spacer, which, when fully extended, are approximately 10 and 16 Å long, respectively. From these studies, the conclusion seems justified that ConA does not require charged lipids for interactions with a carbohydrate receptor, provided that the distance between receptor and bilayer–water interface is at least 10 Å. At shorter distances, negatively charged lipid seems to facilitate the binding of ConA.

4.2.2. Effect of Ca^{2+} on Lectin-Induced Agglutination

Other lectins are equally sensitive to the lipid environment surrounding the glycolipid, and acidic phospholipids play a particularly prominent role in that respect. The divalent lectin RCA_I agglutinates PC and PC vesicles that contain acidic phospholipids such as PS, PA, or CL. As the glycolipid receptor, the synthetic *N*-tetradecylmelibionamide was incorporated, which has a 6 atom spacer arm (Sundler, 1984). The increase in spacer arm by only 2 atoms (i.e., from a 4- to a 6-atom spacer) suffices to allow interaction between the lectin and the carbohydrate head group. In accordance with the observations on ConA (see above), RCA_I is unable to agglutinate neutral vesicles containing glycolipids with only a 4 atom spacer (Sundler, 1984). RCA_I, however, displays another feature: in contrast to glycolipid-containing PC/PA and PC/CL vesicles, PC/PS vesicles are only agglutinated upon addition of Ca^{2+}, while agglutination is reversed upon the addition of EDTA which chelates the Ca^{2+} ions (Hampton et al., 1980). Evidently, charge per se is not the regulating factor in this case, as PA- and CL-containing vesicles are agglutinated by RCA_I. As the glycolipid used in this study is the same in each vesicle system, the possibility that Ca^{2+} would be required for binding of the lectin to the glycolipid can also be excluded. More likely, the effective exposure of the head group of the synthetic glycolipid could be affected in an indirect manner, presumably mediated by a Ca^{2+}-modifiable association of PS with the synthetic glycolipid. Possibly, this distinct behavior of an acidic phospholipid could reflect the stronger ability of PS, as opposed to CL and PA, to hydrogen bond with adjacent glycolipid molecules (cf. Boggs, 1980). Although this point remains a matter of speculation, it is clear that the species of the charged lipid may, in addition, play a role in glycolipid-lectin interaction, particularly when the receptor is situated at a relatively short distance from the membrane surface. As shown by Sundler (1984), when synthetic lipids with spacer arms of 10–12 atoms (which would correspond to a distance between terminal sugar and interface of approximately 22 Å) are incorporated in lipid vesicles, influences of the properties of the lipid environment surrounding the receptors are completely eliminated. Even the physical state of the supporting lipid matrix exerts very little effect on the efficiency of agglutination in this case (Rando and Bangerter, 1979).

Soy bean agglutinin-induced agglutination of vesicles composed of lipids derived from erythrocyte membranes is also inhibited when the erythrocyte lipids are supplemented with PS (Rendi et al., 1979). Addition of Ca^{2+} to such vesicles stimulates the agglutination to an extent that largely exceeds that of SBA-induced agglutination of vesicles not containing PS. When Ca^{2+} is added to the latter vesicles, however, an identical kinetics of agglutination is observed. N-acetylgalactosamine, a haptenic sugar, completely reverses aggregation, indicating that the agglutination is due to a carbohydrate-specific interaction of the lectin and that the influence of Ca^{2+} relies on a secondary effect. Mg^{2+} displays an identical effect (see also below), bbt elevated ionic strength is ineffective, suggesting that cation-mediated enhancement of the SBA-induced agglutination is more specific than simply a neutralization of charges. In this regard, the possibility should be considered whether divalent cations may somehow directly interact with a lectin. Croton tiglium lectin (CTL), which binds to disaccharide units of galactose or their N-acetylated derivatives, is also affected by Ca^{2+} in its ability to agglutinate glycolipid-containing vesicles (Banerjee and Sen, 1983). With forssman-antigen-containing lipid vesicles, CTL can induce agglutination with essentially the same kinetics and to the same extent, independent of whether the supporting matrix of the glycolipid consists of PC, PC/PS, or PC/PE, provided that Ca^{2+} is present in all cases. The long-chain Forssman antigen (GalNAc-GalNAc-Gal-Gal-Glc-Cer) should protrude beyond the vesicle surface to such an extent that it it reasonable to expect that the bilayer composition does not substantially affect its effective exposure, analogous to observations with synthetic "glycolipids" (Sundler, 1984). Furthermore, pure PC bilayers do not bind significant amounts of Ca^{2+} (Düzgüneş and Papahadjopoulos, 1983). These considerations would therefore suggest that Ca^{2+} acts at the level of the lectin. Although binding of the lectin to the membrane appears to be carbohydrate-dependent, agglutination is reversed upon addition of EDTA. That Ca^{2+} may directly interact with a lectin in mediating the cross-linking of glycolipid-containing vesicles was suggested in a study (Hoekstra et al., 1985) in which the monovalent lectin RCA_{II} was added to globoside-containing PE/PA vesicles. due to its monovalency (Baenziger and Fiete, 1979), no agglutination takes place, as expected. However, in the presence of Ca^{2+}, at a concentration that does not cause the aggregation of negatively charged vesicles *per se*, agglutination does take place. The process of agglutination is initiated at a fairly high lectin concentration (approximately 50 μ/ml), which indicates that a threshold number of liposomal receptor sites have to be occupied before agglutination is triggered. These receptor sites were the globoside molecules, since Ca^{2+}/RCA_{II}-induced agglutination was prevented upon prior addition of the haptenic sugar GalNAc. No dissociation of aggregated vesicles is observed when the sugar is added during the course of agglutination. Rather, the process is arrested, and dissociation only occurs upon addition of EDTA. An additional feature of the combined action of RCA_{II} and Ca^{2+} is that vesicle agglutination is only observed when Ca^{2+} is added *after* the lectin. The conclusion of this work was that RCA_{II} apparently contains Ca^{2+} binding sites and that the cation, by its binding to vesicle-bound lectin, established intervesicular cross-links via adjacent RCA_{II} molecules, thus causing vesicle aggregate formation (Hoekstra et al., 1985).

4.2.3. Role of Dehydration of Phospholipid Head Groups

From the previous paragraph it is clear that divalent cations enhance lectin-induced vesicle agglutination, particularly when such vesicles contain acidic phospholipids (Figure 3; Hampton et al., 1980; Hoekstra and Düzgüneş, 1986; Hoekstra et al., 1985; Rendi et al., 1979; Sundler, 1982, 1984). Various mechanisms could explain this phenomenon. In acidic phospholipid-containing bilayers, Ca^{2+} is capable of inducing lipid phase separations (Düzgüneş, 1985;

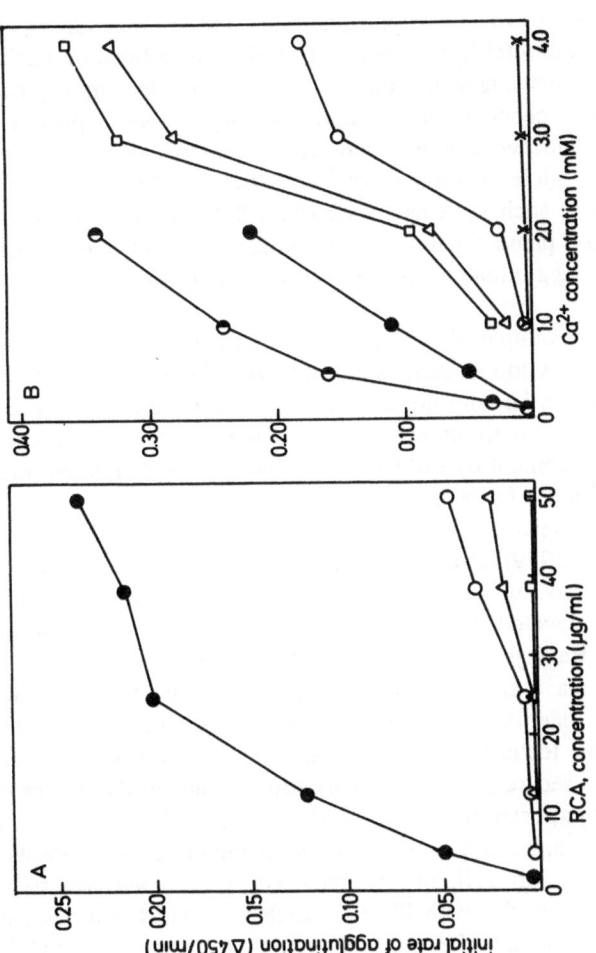

FIGURE 3. Effect of glycolipid head group structure on Ca^{2+}, lectin, and Ca^{2+}/lectin-induced aggregation of lipid vesicles. (A) PE/PA vesicles containing GalCer (□), LacCer (△), or Gb_3 (○) were agglutinated in the presence of various amounts of RCA_I and the initial rates were plotted as a function of the lectin concentration. When the medium contains, in addition, 1 mM Ca^{2+}, a considerable increase in the initial rate of agglutination occurs, as illustrated for Gb_3-containing vesicles (●). (B) Ca^{2+}-induced aggregation of the same glycolipid-containing vesicles as in (A) but in the absence of lectin (open symbols). Note that the larger the size of the carbohydrate head group, the lower the tendency of the vesicles to aggregate. (x) Gb_4-containing vesicles. The combined action of RCA_I and Ca^{2+} (◐, 12.5 and ●, 50 μg/ml RCA_I) reveals that vesicle aggregation already occurs under conditions where separate addition (of Ca^{2+} and RCA_I) does not induce aggregation. Data from Hoekstra and Düzgüneş (1986).

Hoekstra and Wilschut, 1988; Wilschut and Hoekstra, 1986). Consequently, when such bilayers also contain glycolipids, glycolipid-enriched patches may be formed. The consequent increase in the local density of lectin receptors would thus facilitate lectin-induced agglutination. However, the rate of phase separation is very much dependent on the relative concentration of acidic phospholipids to total lipid (Hoekstra, 1982). In phospholipid vesicles composed either of pure acidic lipids or of mixed phospholipids where the acidic lipids form a minor fraction of the total lipid, the effect of divalent cations on lectin-mediated agglutination is instantaneous, that is, much faster than lipid phase separation if the latter can occur at all. Furthermore, Mg^{2+} also displays the ability to enhance lectin-induced agglutination in, for example, PS-containing glycolipid/phospholipid vesicles (Rendi *et al.*, e1979). Yet, Mg^{2+} does not induce phase separation in such systems (Hoekstra, 1982). It would therefore appear that cation-induced phase separations constitute an unlikely mechanism to explain the enhanced agglutination in the presence of lectins and divalent cations.

As illustrated by the example of RCA_{II} and Ca^{2+}, as described above, the possibility of a *direct* interaction of cations and lectins, which somehow leads to agglutination, should also be considered. Yet, Ca^{2+}–lectin binding is not a general phenomenon and in many cases it could be demonstrated or inferred that the initiation or enhancement by Ca^{2+} of lectin-induced agglutination originates from a different source. In the case of Ca^{2+}-stimulated increase in SBA-induced agglutination of vesicles prepared from total lipids of erythrocytic membranes (Rendi *et al.*, 1979), it was suggested that a Ca^{2+}, by virtue of its binding to negative charges, would diminish intervesicular repulsive electrostatic forces. However, elevated ionic strength does not always mimic the effect of divalent cations (Banerjee and Sen, 1983; Rendi *et al.*, 1979), suggesting that the origin of the Ca^{2+} enhancement of agglutination may lie elsewhere. In model systems it has been demonstrated that the stimulating effect of the divalent cation to facilitate lectin-induced agglutination is very much dependent on the particular cation used (e.g., Ca^{2+} versus Mg^{2+}) and on the extent of carbohydrate head group exposure beyond the vesicle surface. Thus, it has been shown that RCA_I-induced agglutination of vesicles containing acidic phospholipids and various synthetic "glycolipids" is enhanced by Ca^{2+} and less so by Mg^{2+} (Sundler, 1984). Similarly, with PE/PA vesicles, in which various natural glycolipids are incorporated, the same distinction in efficiency of Ca^{2+} over Mg^{2+} to enhance RCA_I-induced agglutination is observed (D. Hoekstra and N. Düzgüneş, unpublished observation). On the other hand, the combination of the lectin with the polycation spermine did not produce a synergistic effect, suggesting that charge neutralization could be excluded as the cause of enhanced agglutination (Hoekstra and Düzgüneş, 1986).

As an alternative explanation, it has been proposed that divalent cation binding to the acidic phospholipid head groups affects the state of hydration of these head groups and hence that dehydration of phospholipid environment of the receptor facilitates the latter's interaction with the lectin (Düzgüneş and Hoekstra, 1986; Hoekstra and Düzgüneş, 1986; Sundler, 1984). From studies of the binding of Ca^{2+} and Mg^{2+} to pure acidic phospholipids, it is known that Ca^{2+} dehydrates the membranes more efficiently than Mg^{2+} (Düzgüneş, 1985; Hoekstra and Wilschut, 1988; Newton et al., 1978; Portis et al., 1979). Accordingly, when the divalent cation effect is to be attributed to a cation-meditated modification of the supporting matrix, one would also anticipate that the closer the lectin receptor site to the phospholipid–water interface the more pronounced the effect of Ca^{2+} should be. Indeed, this distinction has been reported with both synthetic "glycolipids" as well as "natural" glycosphingolipids (Hoekstra and Düzgüneş, 1986; Sundler, 1984). A possible clue as to how dehydration of surrounding phospholipid may indirectly affect the molecular behavior of the glycolipid receptor, and thereby the latter's efficiency to act as a receptor, was provided by studies in which the fusogenic behavior of glycolipid-containing vesicles was investigated (Hoekstra and Düzgüneş, 1986; Hoekstra et al., 1985).

4.2.4. Lectins and Glycolipids as Modulators of Intermembrane Interactions

Several reports have been published that indicate that lectins can induce fusion between lipid vesicles in the absence of Ca^{2+}. As early as 19775, Van den Bosch and McConnell described the ability of ConA to induce fusion between glycolipid-devoid DPPC vesicles. Furthermore, WGA appears to induce fusion of ganglioside/PC vesicles (Redwood and Polefka, 1976). More recently, it was observed that SBA-induced agglutination of vesicles consisting of PE, PA, and globoside Gb_4 was accompanied by a limited fusion event (Hoekstra et al., 1985). In this case, fusion was only seen when the PE/PA vesicles contained the glycolipid receptor. Hence, prior binding of the lectin to Gb_4 was a prerequisite for fusion. It was proposed that the binding process causes changes in the tertiary structure of the lectin (a phenomenon reported to occur upon sugar binding; Bhattacharyya et al., 1987) which in turn allows for nonspecific interactions, possibly hydrophobic in nature, causing membrane perturbations that can result in merging of closely apposed bilayers. Again, however, the phenomenon of nonspecific lectin-induced fusion of vesicles is not a general one and depends on the lectin and the lipid composition of the membrane vesicles. For example, in the same system where SBA induces fusion, SJA does not, although it agglutinates the vesicles (Hoekstra et al., 1985).

How do glycolipids *per se* affect close approach of bilayers and thus membrane fusion? It can readily be envisaged that a carbohydrate protrusion on a membrane surface can act as a steric barrier (see Figure 3). This can be demonstrated when glycosphingolipids with carbohydrate head groups of increasing sugar length are incorporated into acidic phospholipid-containing bilayers that, in the absence of glycolipids, rapidly fuse in the presence of Ca^{2+} (Hoekstra and Düzgüneş, 1986). GalCer-containing vesicles fuse upon addition of Ca^{2+} as rapidly as vesicles that do not contain the cerebroside. This observation indicates that the monosaccharide is more or less positioned at the level of the phospholipid head group. However, Ca^{2+}-induced fusion was virtually abolished when the carbohydrate head group consisted of more than three carbohydrate units; that is, fusion is still seen for vesicles containing trihexosylceramide (Gb_3) but is almost abolished when Gb_3 is replaced by the globoside Gb_4 (Hoekstra and Düzgüneş, 1986). In studies in which the spontaneous transfer of glycolipids between membranes was investigated, it has been noted that trisialogangliosides interfere with fusion. Below the gel–liquid-crystalline phase transition temperature, small unilamellar DPPC vesicles fuse spontaneously. However, when such vesicles also contain trisialogangliosides, fusion is impeded, possibly due to electrostatic (Felgner et al., 1981) as well as steric repulsion. In one study, however, it has been shown that polysialogangliosides, when added at fairly high concentrations (100 μg/ml) to chicken erythrocytes, can induce cell–cell fusion (Maggio et al., 1978). The mechanism is unclear, although Ca^{2+} appears to be required for ganglioside-induced fusion. In light of the relatively low affinity of gangliosides for Ca^{2+} (Ollmann et al., 1987), it remains to be seen whether Ca^{2+}–ganglioside interaction is necessary for triggering the fusion event.

4.2.5. Effect of Glycolipid Carbohydrate Head Group on Membrane Fusion

In studies specifically dealing with cation-induced fusion of membrane vesicles, it has become apparent that a key feature of the fusion mechanism involves a local dehydration of the apposed bilayers (Düzgüneş, 1985; Hoekstra and Wilschut, 1988; Wilschut and Hoekstra, 1986). Lipid-bound water molecules constitute the ultimate barrier for fusion and any fusogen requires the intrinsic ability to dehydrate the (local) lipid environment, thereby perturbing the strong interbilayer repulsive hydration forces (Rand and Parsegian, 1984). It is therefore not too surprising that bulky and strongly hydrated carbohydrate head groups (Pasher, 1976) can interfere with the close approach and (local) dehydration of adjacent membranes. When studying the overall kinetics of fusion of various glycolipid-containing vesicles in the presence of lectins and Ca^{2+}, we observed (Düzgüneş and Hoekstra, 1986; Hoekstra and Düzgüneş, 1986; Hoekstra et al., 1985) that when the vesicles were in the process of lectin-induced agglutination the rate of Ca^{2+}-induced fusion increased dramat-

FIGURE 4. Effect of lectin-induced vesicle agglutination on the threshold Ca^{2+} concentration for Ca^{2+}-induced vesicle fusion. Vesicles, consisting of PE, PA, and 10 mol% of the synthetic glycolipid PE-N-lactobionamide, were fused in the presence of various Ca^{2+} concentrations, and the initial fusion rates were calculated and plotted versus the cation concentration (open symbols). The threshold Ca^{2+} concentration for fusion decreases approximately 10-fold when the vesicles were incubated (for 1 min) with RCA_I, prior to the addition of Ca^{2+} (filled symbols). Data from Sundler and Wijkander (1983).

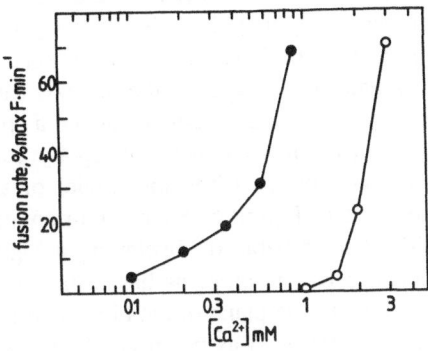

ically. Furthermore, the threshold Ca^{2+} concentration, that is, the minimal cation concentration necessary to initiate fusion, also decreased (see Figure 4). More specifically, the smaller the size of the carbohydrate head group, the lower the Ca^{2+} concentration that was required to induce fusion. This trend is expected, since a limited carbohydrate head group protrusion will allow for a smaller interbilayer distance between apposed bilayers. The latter is essential in cation-induced fusion of acidic phospholipid vesicles, as it enables the formation of an anhydrous cation–acidic phospholipid complex of specific stoichiometry that presumably serves as the trigger for membrane fusion (Ekerdt and Papahadjopoulos, 1982). The formation of this complex is apparently impaired when the order of lectin and Ca^{2+} addition is reversed, since fusion is not facilitated when Ca^{2+} is added to the glycolipid-containing vesicles *prior to* the lectin. Yet, the stimulation of lectin-induced agglutination by Ca^{2+} is *independent* of the order of addition (Hoekstra and Düzgüneş, 1986). Furthermore, when monitoring continuously the kinetics of fusion of lactosylceramide (LacCer)-containing vesicles (in which case the carbohydrate head group barely protrudes beyond the vesicle surface) versus those of GB_3- or GB_4-containing vesicles, a remarkable difference became apparent (Figure 5). LacCer vesicles undergo multiple rounds of fusion, whereas the fusion of Gb_3- or GB_4-containing vesicles halts abruptly, shortly after the addition of Ca^{2+} to vesicles, agglutinating in the presence of appropriate lectins. In a parallel experiment it could be shown that the enhanced rate of agglutination, resulting from the combined action of cation and lectin, proceeded normally (Figure 5). Apparently, the addition of Ca^{2+} leads to an enhanced exposure of the glycolipid head group, in such a way that the distance of the terminal sugar residue relative to the bilayer–water interface increases. As a consequence, the accessibility for its interaction with a lectin will increase. On the other hand, the protrusion of the carbohydrate head group will also cause an increase in interbilayer distance within the lectin-induced vesicle aggregates or between vesicles that are in the process of aggregation, thus causing mitigation of the fusion reaction. In summary, the combined action of cations and lectins in their effects on agglutination and fusion suggests that cations capable of dehydrating membrane surfaces will, as a secondary effect, induce a spatial reorientation of the strongly hydrophilic carbohydrate head group by forcing it to tilt toward a more aqueous environment. A schematic model of the events outlined in this paragraph is depicted in Figure 6. Such a reorientation requires that glycolipid head groups should be capable of displaying a high degree of motional freedom. This appears to be the case, as has been reviewed elsewhere (Maggio et al., 1981). Relative to the plane of the bilayer, oligosaccharide chains in the aqueous phase can be displaced as much as 80° from their perpendicular position, relative to the plane of the bilayer. The observation that SBA, bound to Gb_4, induces vesicle fusion will presumably require such a strong folding that the carbohydrate chain is almost in a parallel position relative to the plane of the bilayer, in order to allow for lectin-mediated intervesicular interactions (Hoekstra et al., 1985, see above). It appears that the head groups of both gangliosides (Maggio et al., 1981) and neutral glycolipids such as globoside (Peters et al., 1982) display a relatively unrestricted motion, when the glycolipids are more or less randomly distributed within the lateral plane of the bilayer (see Section 4.1). The motion is limited to the aqueous region of the interface between lipid and surrounding medium. One would conclude therefore that interface properties are likely to affect the carbohydrate head group mobility. Indeed, the work carried out on the effects of divalent cations that dehydrate the interface on lectin-induced agglutination points to a restricted motional freedom of the head groups; that is, the head groups tend to adopt a more perpendicular orientation relative to the plane of the bilayer.

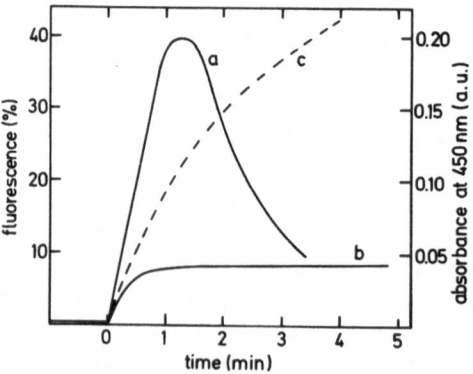

FIGURE 5. Ca^{2+}-induced fusion of lectin-mediated, agglutinating glycolipid-containing vesicles. Effect of the size of the glycolipid head group. (a) LacCer- and (b) Gb_3-containing PE/PA vesicles were agglutinated by RCA_I. After 2 min, Ca^{2+} was added and fusion was monitored with an aqueous contents mixing assay. Note that fusion, as reflected by the increase of fluorescence, is very fast for the LacCer vesicles (a), whereas it halts rather abruptly in the case of Gb_3-containing vesicles (b), in spite of the fact that vesicle agglutination proceeds (c). For details, see Hoekstra and Düzgüneş (1986).

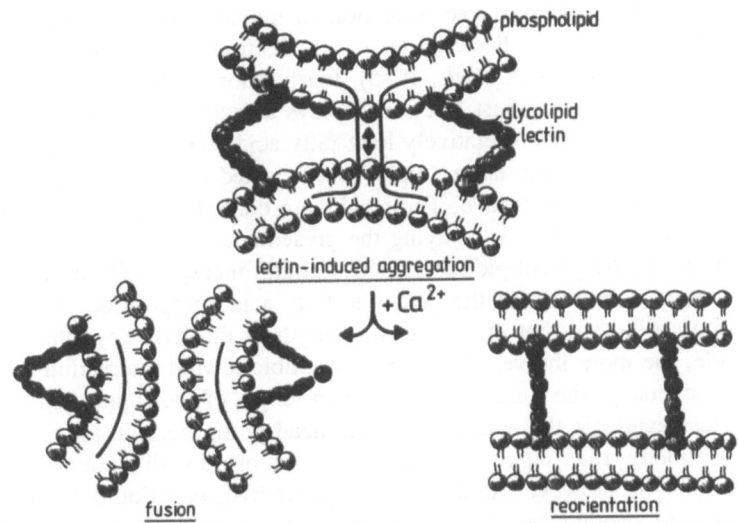

FIGURE 6. Schematic model of Ca^{2+}-induced modulation of intermembrane distances of lectin-agglutinated, glycolipid-containing phospholipid bilayers. Vesicles, containing glycolipids with a carbohydrate head group size of three or more sugar residues, are agglutinated in the presence of a specific lectin (not on scale). Depending on the interbilayer distance in the agglutinated state (arrow), addition of Ca^{2+} induces either fusion (left) or a spatial reorientation of the carbohydrate head group, causing an interbilayer separation to such extent that fusion is inhibited (see text for details).

4.2.6. Effect of Ceramide Fatty Acid Composition on Head Group Exposure

Apart from exogenous factors such as divalent cations, the steric orientation and thereby the effective exposure of the carbohydrate head group can also be affected by the fatty acid composition of the ceramide moiety. Kannagi *et al.* (1982) noted that there appears to exist a close correlation between the identity of the terminal carbohydrate structure and the fatty acid, contained in the ceramide portion of the "lacto series" of glycosphingolipids, isolated from human erythrocytes. In conjunction with the observation that the ceramide composition of various murine tumor cell lines affects their antigenicity, it was proposed that the ceramide structure may modulate the organization and orientation of the carbohydrate chain in the plasma membrane. In this context, cell surface carbohydrates have been implicated as receptors in the process of bacterial infections (see Section 5) as inferred from observations that haptenic sugars compete with bacterial adhesion to cell surfaces (Jones and Isaacson, 1983). With glycolipids, adhered to a solid phase, it has been found that *Escherichia coli* strains, causing urinary tract infection in humans, bound more strongly (presumably mediated by a bacterial lectin, cf. Section 5) to glycolipids having α-hydroxy fatty acids (Bock *et al.*, 1985). Furthermore, it has been shown that a monoclonal antibody against LacCer displays a much higher reactivity with LacCer species that contain relatively long fatty acid chains (C_{20}–C_{24}; Symington *et al.*, 1984). Also the latter studies were carried out with isolated glycolipids, coated on a solid surface. The results are consistent, however, with the work of Crook *et al.* (1986), studying the presentation of an antibody-binding epitope of the acidic glycolipid cerebroside sulfate, incorporated into liposomal membranes, as a function of the ceramide fatty acid composition. The results of this study indicated that the longer the length of the fatty acyl chain in the cerebroside, the more the vesicles were susceptible toward interaction with the antibody. Inversely, the longer the acyl chain length of the supporting PC matrix, the less available the acidic glycolipid head group became for interaction with the antibody, thus suggesting a decreased exposure with an increase in PC acyl chain length. However, in this model system hydroxylation of the cerebroside fatty acid residue caused an inhibition of the binding of the antibody, suggesting a modified surface exposure. The discrepancy with the preference of bacterial lectins, as noted above, may arise from differences in the specificity of the different systems used. However, with whole cells the reactivity of the antibody is also opposite to the effect of hydroxylation observed in a model system (Hakamori and Kannagi, 1983). Evidently, the molecular environment of receptors at the cell surface differs from that at the liposomal surface. As such, this provides further clues as to traditional factors regulating the organization and orientation of glycolipid receptors. On the other hand, these distinct results also warrant the notion that enhanced or decreased specificity of ligand–receptor interactions may also depend on the assay system (cf. Section 2.2). In this respect, a distinction was also reported concerning the interaction of certain lectins with soluble glycoproteins or synthetic substrates versus their interaction with membrane-associated glycolipids (Hammarström *et al.*, 1977; Hoekstra *et al.*, 1985; Mansson and Olofsson, 1983). In this context it is interesting to note that the nature of the ceramide-linked fatty acid governs the degree of hydration of the carbohydrate head group. The more unsaturated the fatty acid, the higher the state of hydration (Bach *et al.*, 1982b; Ruocco and Shipley, 1983). Hence, if hydration forces interfere with lectin–carbohydrate head group interaction (Sundler, 1984; see above), it can be envisaged that certain lectins display a preference for a particular glycolipid only, when it contains the appropriate fatty acids, in spite of the presence of the appropriate carbohydrate receptor.

4.2.7. Lateral Clustering of Glycolipid–Lectin Complexes

Besides intervesicular cross-linking, lateral cross-linking may obviously also occur upon lectin–glycolipid interactions. This will eventually result in a clustering of glycolipid–lectin complexes within the lateral plane of the bilayer. With spin-probe techniques such a cross-linking can be demonstrated directly (Suzuki et al., 1983; Utsumi et al., 1984). Due to spin–spin interactions, the spectrum will become broadened. The occurrence of lectin-mediated lateral phase separations can also be inferred from studies involving the fusion of glycolipid-containing vesicles with pure phospholipid vesicles. WGA facilitates the Ca^{2+}-induced fusion of PE/PA/GD_{1a} vesicles with PE/PA vesicles, that is, vesicles devoid of glycolipid, by a factor of approximately 40 (Düzgüneş et al., 1984). This indicates that in the absence of the lectin the GD_{1a} head group interferes strongly with the close approach of the surfaces of the vesicles. Since WGA does not interact with the vesicles devoid of glycolipid, and thus does not cause nonspecific cross-linking, it would appear that the lectin abolishes the steric interference of the carbohydrate head group by phase separating the glycolipid molecules and allowing the phospholipids of the two membranes to establish fusion-susceptible contact. Indeed, the longer the preincubation with the lectin, prior to the addition of Ca^{2+}, the more efficient the fusion reaction (Düzgüneş et al., 1984).

Lateral phase separation or local clustering of glycolipids induced by glycolipid–lectin interaction can result in a substantial reorganization of membrane components. Does such a reorganization induce a destabilization of perturbation of the membrane specific? SBA can, in a concentration-dependent manner, perturb the membrane of vesicles composed of PE, PA, and Gb_4 (Hoekstra et al., 1985). SJA does not, although it agglutinates such vesicles. Similarly, RCA_1-induced agglutination of vesicles composed of the same phospholipids, but containing LacCer or Gb_3 rather than Gb_4, was not accompanied by a release of the aqueous contents of the vesicles. The distinction between SBA and the other lectins when interacting with glycolipid-containing PE/PA vesicles is that SBA is capable of interacting hydrophobically with the phospholipid bilayer, whereas the other lectins are not. It would therefore appear that lateral clustering or phase separation of lectin–glycolipid complexes is not necessarily accompanied by destabilization of the membrane structure. It should be noted, however, that in these systems the molar ratio of glycolipid to other lipids was less than 0.1. It cannot be excluded that when the relative amount of glycolipid increases, membrane stability may weaken upon lectin-induced clustering. In this context, when brain gangliosides are incorporated into DPPC bilayers, "spontaneous" phase separation of the glycolipid can be induced when the vesicles are incubated at temperatures below the gel–liquid-crystalline phase transition of the supporting matrix (Delmelle et al., 1980). Over a time interval of 30 min, only approximately 10% of the aqueous vesicle contents are released, reflecting a fairly modest perturbation of the bilayer. The perturbing effect may become more pronounced, however, when lipids are included in the bilayer that display the capacity to adopt nonbilayer structures (see Section 4.3). For example, a massive disruption of the membrane permeability barrier is triggered when glycophorin, incorporated in a bilayer consisting of unsaturated PEs, is clustered upon addition of WGA, which allows the PEs to undergo the bilayer-to-nonbilayer, that is, hexagonal H_{II} phase transition (Taraschi *et al.*, 1982b).

4.3. Interactions of Lectins with Glycoprotein-Containing Membranes

Most studies that have dealt with the interaction of lectins with glycoproteins focus on glycophorin and band 3, isolated from erythrocytes and reconstituted in phospholipid vesicles, as lectin receptors. Glycophorin, the major integral sialoglycoprotein of the human erythrocyte membrane, is one of the best characterized membrane glycoproteins and has a high content of GlcNAc and sialic acid (Verpoorte, 1975). Presumably, the fairly limited application of defined membrane glycoproteins as lectin receptors originates from the complexity of isolating and purifying such proteins (e.g., see Campbell *et al.*, 1983) and to reconstitute them properly into artificial bilayers. By contrast, glycophorin can readily be isolated and reconstituted into phospholipid vesicles (MacDonald and MacDonald, 1975). In this section, we only briefly summarize work carried out with glycoprotein-containing bilayers as the recent literature is fairly scant while data obtained before 1983 have been excellently reviewed elsewhere (Grant and Peters, 1984).

The dimensions of the extracellular part of a glycoprotein are obviously quite distinct from those of a glycosphingolipid (Grant and Peters, 1984). With liposomes containing either a glycoprotein or glycolipid in their bilayer, this distinction is nicely illustrated by the work of Redwood and co-workers (1975, 1976). These authors demonstrated that WGA induces agglutination of both GM_1-containing PC vesicles and glycophorin-containing PC vesicles but only the former can engage in lectin-induced fusion. PS vesicles, which readily fuse upon addition of Ca^{2+}, display no Ca^{2+}-induced fusion activity when glycophorin is included in the bilayer (De Kroon *et al.*, 1985). Furthermore, prior addition of WGA, although substantially agglutinating such vesicles, only leads to a limited extent of fusion upon addition of Ca^{2+}. Under these conditions, however, the threshold Ca^{2+} concentration required to induce fusion is lower than the concentration of Ca^{2+} necessary to induce fusion of glycophorin-free PS vesicles. This implies that to some extent, the WGA–glycophorin interaction establishes a close interbilayer contact which facilitates the immediate formation of a (dehydrated) *trans*-Ca^{2+}–PS complex (Ekerdt and Papahadjopoulos, 1982), when the cation is added to the system (cf. Section 4.2.5). As observed for glycolipid-containing vesicles, the fusion reaction is limited in this case and is impeded shortly after addition of the cation (De Kroon *et al.*, 1985). The latter observation suggests that glycophorin displays a tendency to reorient itself sterically, thereby increasing the interbilayer distance and eliminating the ability of *trans*-metal–phospholipid complex formation. This situation would be analogous to what has been proposed for Ca^{2+}-induced fusion of glycolipid-containing vesicles, provided that, in the latter case, the carbohydrate head group contains at least three sugar residues (cf. Figure 6; see Section 4.2.5). By taking into account the large size of the extravesicular glycosylated head group of glycophorin, it is evident that the latter will interfere with the close approach of bilayers. Hence, a complete inhibition of Ca^{2+}-induced fusion of such vesicles in the absence of WGA is anticipated. Removal of sialic acid does allow Ca^{2+}-induced fusion to an extent that is similar to the case of fusion mediated by the combined action of WGA and Ca^{2+} (De Kroon et al., 1985). It is known that removal of sialic acid from glycophorin by neuraminidase does not induce changes in the mobility of the extravesicular glycopeptide portion of the protein (Lee and Grant, 1980). Furthermore, trypsinization of the glycophorin-containing vesicles, which removes almost the entire extravesicular part of glycophorin, further enhances the Ca^{2+}-induced fusion activity, although remaining lower than that seen for pure PS vesicles. It would thus appear that in this system lectin–carbohydrate interaction as such does not significantly affect the fusion event *per se*, except for establishing close intermembrane approach which lowers the threshold Ca^{2+} concentration (cf. Figure 4; Sundler, 1984). The inhibiting effect of glycophorin on Ca^{2+}-induced fusion probably originates from a combination of steric interference, attributable to the bulky glycosylated head group, and to some extent from electrostatic repulsion, as reflected by the enhancement of fusion upon removal of sialic acid. Finally, the WGA–glycophorin interaction in this system (1 molecule of glycophorin per 400–500 molecules of PS) does not result in an overall perturbation of the bilayer structure; upon addition of WGA only 4–5% of the contents leak instantaneously. No subsequent leakage is seen (De Kroon et al., 1985), except upon addition of Ca^{2+}, which is analogous to observations reported for Ca^{2+}/lectin–glycolipid interactions (Hoekstra and Düzgüneş, 1986; Hoekstra et al., 1985). Similarly, when RCA_I was added to glycophorin/egg PC vesicles, less than a factor of 2 enhancement in leakage was seen (Juliano and Stamp, 1976); that is, also in this case, substantial membrane perturbation was not induced.

When reconstituted with phospholipids that, in isolation, adopt the bilayer structure, glycophorin-containing bilayers display no structural reorganization of the membrane phospholipid bilayer upon addition of WGA (Taraschi et al., 1982a). However, when reconstituted with phospholipids such as unsaturated PEs, which in isolation can adopt nonbilayer structures, two effects become apparent (Taraschi et al., 1982b). First, depending on the molar ratio of glycophorin/phospholipid (i.e., when the ratio exceeds 1 : 200), the protein can stabilize the bilayer structure of such lipids. Second, upon addition of WGA, the bilayer becomes destabilized, resulting in a drastic reorganization of the phospholipid from a bilayer to a nonbilayer, hexagonal H_{II} phase. In the latter

FIGURE 7. Freeze–fracture image of DOPE–glycophorin vesicle next to DOPE in the hexagonal (H_{II}) phase. The hexagonal phase in the glycophorin-containing vesicles is triggered upon addition of WGA. As a result, the bilayers become highly permeabilized. The photograph was kindly provided by Dr. Arie Verkleij. For details, see Taraschi *et al.* (1982a,b).

case, WGA cross-links adjacent glycophorin molecules (Ketis and Grant, 1982), which leads to the formation of protein-free lipid patches in which the lipid molecules are no longer "stabilized" by protein molecules. The lipids then revert to the hexagonal H_{II} phase, as would occur in isolation (Figure 7).

In light of the substantial distance between the membrane surface and the glycoprotein carbohydrates that act as lectin receptors, it is reasonable to assume that the surface properties of the lipid matrix barely, if at all, influence lectin–protein interaction. Thus far, no such systematic study, as in the case of glycolipid-containing bilayers, has been carried out. It has been noted, however, that the lateral distribution of glycophorin does affect the rate of lectin-induced agglutination of glycophorin-containing lipid vesicles. Thus, rigid vesicles display a 6–20 times faster agglutination rate than fluid ones (Goodwin *et al.*, 1982). This effect may be due to an enhanced receptor density in the solid lipid bilayers, caused by protein–lipid phae separation. Apparently, the (partial) exclusion of glycophorin from the lipid phase does not necessarily lead to tight protein–protein interactions as far as the (extravesicular) glycosylated portion of the molecule is concerned. This can be inferred from the observation that the dynamic state of the glycoprotein–carbohydrate structure is not affected by the physical state of the bilayer (Lee and Grant, 1980). Furthermore, as noted by Grant and Peters (1984), the evidence for the necessity of a minimal concentration of receptors is less clear than the well-defined threshold concentration of glycolipids necessary to induce lectin-induced agglutination.

From the literature published thus far one tends to conclude that due to the extensive extracellular part of a glycosylated protein, and hence to the considerable distance between the lectin–receptor and the bilayer surface, the ability of a glycoprotein to recognize and bind lectins is far less dramatically affected by the surface properties of the lipid bilayer than appears to be the case with glycolipids. As such, one would then anticipate from these model studies that the "specificity" of glycoprotein–lipid interaction is less pronounced than the interaction of lectins with glycolipids. Besides sugar specificity, the latter interactions can, in addition, be modulated by the molecular *environment* of the glycolipid. Whether these speculations imply that in biological systems (soluble) lectin–glycoprotein interactions are also less specific than lectin–glycolipid interactions, in terms of a biological response, will be discussed in Section 5.

5. LECTIN–CARBOHYDRATE INTERACTIONS IN BIOLOGICAL SYSTEMS

To describe the significance and role of both exogenous and endogenous lectins in biological systems, it appears useful to make a distinction between aspects that deal particularly with the application of a lectin as a tool and those focusing on the biological relevance. In studies concerning the application of lectins, lectins of plant origin are commonly used to probe for specific carbohydrate sites, both intracellularly and extracellularly. After interacting with a cellular carbohydrate receptor, the same exogenous lectins can be capable of triggering molecular mechanisms that might bear potential biological significance. Such information may then lead to knowledge concerning the presence and functioning of endogenous lectins in eukaryotic cells. These various aspects of lectins are briefly discussed below, particularly in the light of the work carried out with model systems.

5.1. Exogenous Lectins: Applications

For several decades, plant lectins have been extensively applied as tools for numerous purposes. Since lectin binding is selective, they obviously provide a valuable tool for mapping carbohydrate residues both at the cell surface and intracellularly. Lectins also serve a valuable purpose in cancer research as they allow one to distinguish "normal" cells, where lectin receptors are randomly distributed, from tumor cells, where the receptors readily aggregate into clusters. In addition, a dramatic change in the synthesis of glycoconjugates at the cell surface is a typical feature of neoplastic transformation (Hakomori, 1981, 1986). As a result, lectins that do not bind to the normal parent cell will do so to the transformed counterpart (Damjanov, 1987). As a consequence, transformed cells tend to be more agglutinable than their normal counterparts. Based on studies with model systems (Section 4), this observation can be rationalized by taking into account the strong dependence of lectin–receptor interaction on receptor density. It is tempting to relate changes in receptor density in transformed cells with membrane fluidity changes (Van Blitterswijk, 1983) and, accordingly, with an enhanced receptor mobility in tumor cell membranes, relative to that in normal cells. Yet, in model systems there is no clear picture as to the relationship of lateral distribution of lectin receptors and membrane fluidity. As outlined in Section 4, a variety of parameters (i.e., charge, hydration, etc.) affect the interaction of a lectin with its receptor. In this context, the extent to which the carbohydrate head group of a glycolipid protrudes beyond the cell surface may represent a parameter of major importance. It has been reported (Rando et al., 1980) that ConA, which does not agglutinate bovine erythrocytes, can induce such agglutination after insertion of a (synthetic) glycolipid in the membrane, provided that the distance between the terminal ConA-binding sugar residue and the bilayer–water interface amounts at least 26 Å. This would imply, analogous to observations with artificial systems (Section 4), that the cell surface *per se* and/or the presence of adjacent cell surface molecules modulate the glycolipid receptor reactivity. It has been noted (Kannagi et al., 1982) in this respect, that glycolipid reactivity may be greatly influenced by the organization and dynamic state of coexisting glycoproteins on the surface of the cell membrane. It is clear, however, that many details concerning the extent to which each of these parameters modulates the interaction of lectins with biological membranes remain to be elucidated.

Lectins have found a wide application in cytochemistry and histochemistry (Damjanov, 1987). Particularly in this field of research it has become apparent that although many lectins may display the same monosaccharide specificity, they show different affinities for the same cell or tissue structure in histochemical preparations. Hence, these results fit with the picture obtained from model studies that lectin–carbohydrate interactions can be modulated by the molecular nature of the receptor site *per se* and presumably by receptor-environment factors as well.

In research on oncogenesis, it has not been possible thus far to relate the development of malignancy to the presence or appearance of specific lectin receptors at the cell surface of all tumor cells (Damjanov, 1987; Hakomori, 1981). The extensive heterogeneity in cells comprising the tumor explains this inability. The diagnostic relevance of lectins in this case is therefore limited to a direct comparison of lectin binding sites present or absent on normal cells versus their malignant counterparts.

Of the numerous applications of lectins described in the literature (for a recent review, see Lis and Sharon, 1986) we finally want to emphasize their particular fruitful use in affinity chromatography. A selective use of lectins, coupled to chromatographic carriers, allows one to isolate and purify a variety of membrane glycoproteins which can be subsequently reconstituted into artificial membranes for further examination of structural and functional properties. For example, the membrane glycoprotein receptors for epidermal growth factor (Cohen et al., 1982) and insulin (Hedo et al., 1981) have been purified by such a procedure, as well as various viral membrane glycoproteins involved in binding and fusion of enveloped virions (Gething et al., 1978; Hayman et al., 1973).

5.2. Exogenous Lectins: Biological Effects

Due to their multivalent binding character, lectins, when added exogenously, will cross-link membrane surface receptors, leading to lateral molecular reorganizations in the plane of the bilayer. In several instances, the interaction between a lectin and cellular receptors has been shown to result in a cellular response, implying that the interaction leads to transmission of a transmembrane signal.

Lectin activation of lymphocytes serves as a model to study the biochemical mechanisms underlying cellular growth and immunoregulation. Resting T-lymphocytes can be stimulated by mitogens, which cause the activation of several membrane signal transduction systems, leading to changes in the cytosolic Ca^{2+} concentration, pH, and turnover of phosphatidylinositol. ConA, PHA, and WGA (under certain conditions; Clevers et al., 1986) are such mitogens and trigger this cascade of events (Clevers et al., 1986; Hersey et al., 1986; Hesketh et al., 1985; Metcalfe et al., 1985; Rosoff et al., 1987). These effects are not specifically linked to lectins, however, as phorbol esters (e.g., see Boon et al., 1985; Metcalfe et al., 1985) give rise to similar effects. The lectins presumably bind to high-affinity oligosaccharides present on the T-cell receptor as well as to other membrane glycoproteins (Rosoff et al., 1987). That the T-cell receptor is activated directly in this manner, thereby initiating the pathway of proliferation via the breakdown of phosphatidylinositol(4,5)-biphosphate to generate a Ca^{2+} signal and to activate protein kinase C, seems unlikely (Metcalfe et al., 1985). Rather a complex mechanism, subsequent to lectin binding and possibly involving a secondary role of gangliosides, appears to participate in the proliferation response of the cells. It has been noted, in this regard, that the interaction of PHA with lymphocytes results in an indirect stimulation of sialyltransferase activities preceding the maximal proliferative response (Basu et al., 1986). This activation causes the predominant synthesis of GM_3, and it was suggested that sialylated glycolipids are probably linked to cell activation. Such a conclusion is tempting, particularly in light of the observation that the level of GM_3 is also enhanced in leukemic cells (Westrick et al., 1983). Furthermore, lymphocytes express low levels of GD_3 and the expression (Hersey et al., 1986) substantially increases upon mitogenic response to PHA. However, an enhanced mitogenic response is also seen when PHA and monoclonal antibodies against GD_3 are co-cultured with the lymphocytes. This observation thus precludes a definite conclusion as to the role of ganglioside levels and synthesis in lectin-induced cell proliferation, although a cell-type-dependent modulating effect of gangliosides in this process is apparent (Basu et al., 1986; Hersey et al., 1986; Marcus et al., 1987).

Finally, exogenous lectins are also capable of activating macrophages in vitro, that is, rendering macrophages capable of expressing tumor cytotoxic activity (Nakajima et al., 1986). The mechanism by which lectins bring about this tumoricidal mechanism is not known. However, also in this case, the activation mechanism is not an exclusive property of lectins, as bacterial lipopolysaccharides and synthetic muramyl dipeptide derivatives can induce the same effect (Fidler, 1985).

5.3. Endogenous Lectins

It is now well established that lectins are also active in eukaryotic cells. They can be found both at the cell surface, as integral glycosylated membrane proteins, and in the intra- and extracellular environment of the cell, as water-soluble proteins. Ultimately, many of the soluble lectins appear to be secreted into the extracellular space (Barondes, 1984; Roberson et al., 1985). There exists no clear picture yet, as far as their biological functions are concerned. However, from work carried out thus far, a picture emerges that indicates that cellular lectins *may* play important roles in processes such as intercellular communication via specific lectin–cell surface carbohydrate recognition processes, cell–cell fusion, as occurs during fertilization and myotube formation, specific intracellular transport of glycoproteins and perhaps in vesicular traffic between various organelles, and endocytosis. We discuss these various aspects by summarizing the properties of a few mammalian lectins that have been fairly well characterized.

The best-characterized integral membrane lectin is a membrane glycoprotein, usually referred to as asialoglycoprotein receptor, found in the plasma membrane of hepatocytes (Ashwell and Harford, 1982). The lectin specifically interacts with galactose or, depending on the species, N-acetylgalactosamine residues. These residues become exposed upon desialylation of sialoglycoproteins. Upon intravenous injection of such asialoglycoproteins into animals, the proteins are rapidly cleared from the bloodstream by specific binding to the receptor and the complex is subsequently processed via the endocytic pathway. The asialoglycoprotein receptor has been studied extensively, serving as a model for lectin–carbohydrate recognition and receptor-mediated endocytosis. Recently, it has become apparent that in rat major and minor receptor forms exist, both displaying galactose but not GalNAc activity, and that the major and minor polypeptide species each form homo-oligomers (Drickamer et al., 1984; Halberg et al., 1987). It has been proposed that the various receptor forms may be active at different stages during the endocytic process. The receptor in human liver membranes is composed of two different polypeptide species that form an oligomeric complex, perhaps a hexamer, thereby providing multiple binding sites for galactose and GalNAc (Bischoff and Lodish, 1987). In rat, the lectin contains a carbohydrate binding site near its carboxyl terminal, and binding strongly depends on Ca^{2+} (Halberg et al., 1987).

Studies concerning the role of protein-bound carbohydrate have revealed that the carbohydrate structure may also provide a specific code for intracellular transport (Schwarz and Datema, 1982). By inference, it is obvious then to assume that lectins or lectinlike molecules must be functioning intracellularly as well. The best-known example in this case is a receptor specific for phosphomannosyl residues which plays an essential role in the segregation and targeting of lysosomal enzymes to the lysosomes (Von Figura and Hasilik, 1986), after post-translational modification in the Golgi apparatus. This modification involves the generation of a mannose-6-phosphate (Man-6-P) residue that can bind specifically to the Man-6-P receptors in the Golgi, after which the complex is translocated to a prelysosomal compartment where the low pH causes dissociation. There appear to be two distinct Man-6-P receptors, one that is cation-dependent and one that is cation-independent in binding its ligand (Hoflack and Kornfeld, 1985; Von Figura and Hasilik, 1986). At present, it is not known whether the different receptors, both containing multiple carbohydrate binding sites, also display distinct functions.

The core-specific lectin, probably identical to rat mannan binding protein (Drickamer et al., 1986), is a soluble mammalian lectin, synthesized and secreted by rat hepatocytes (Colley and Baenziger, 1987). The lectin has a capacity for hydrophobic interactions and displays specificity for mannose and N-acetylglucosamine residues. For exit from the cell, which is an active process, hydroxylation and/or glycosylation are required. The physiological relevance of the lectin has not yet been established.

In light of the abundant presence of complex carbohydrates at the surfaces of cells in conjunction with the presence of membrane-associated or soluble lectins, there has been much speculation on the involvement of lectin–carbohydrate interactions in mediating intercellular communication. Yet, there is very little solid evidence that supports this possibility. Discoidin I, a lectin of the cellular slime mold *Dictyostelium discoideum*, promotes cell attachment and spreading as well as ordered cellular migrators during morphogenesis (Springer et al., 1984). Furthermore, experiments involving the inhibition of N-glycosylation of glycoproteins in *D. discoideum* with tunicamycin have suggested that the carbohydrate group of specific glycoproteins may participate in cell adhesion (Bozzaro, 1985).

The pathogenesis of several infectious diseases can be related to a lectin–carbohydrate interaction, serving as the first step in the overall process. Glycoproteins carrying N-glycosyl units of the oligomannose type have been suggested as cell surface receptors for mannose-specific lectins that can be detected on the surface of bacterial membranes (Sharon, 1984b). This binding would thus provide an initial and essential step in bacterial infections. Observations, showing that the bacterium *Actinomyces naeslundii* WVU45 binds to glycolipids that contain terminal Galβ1-3GalNAc (e.g., GM_1) or GalNAcβ1-3Gal (e.g., GB_4) residues, have led to the suggestion that glycolipids may also serve as receptors for bacteria on mammalian cells (Brennan et al., 1987). The binding is mediated by a bacterial membrane-associated lectin (see also Section 4.2.6).

Trypanosoma cruzi, a unicellular parasite causing Chagas' disease, penetrates host cells via a mannose-containing receptor (Villalta and Kierszenbaum, 1983). This conclusion was inferred from observations that the mannose-specific lectin ConA blocks infection while the same occurs upon revamol of host-cell surface mannose residues by treatment with alpha-mannosidase. Whether a lectin, located in the parasite membrane, is mediating this infectious entry is unclear.

Enveloped viruses, that is, viruses in which the nucleocapsid is surrounded by a membrane, enter the cell by a mechanism in which the first essential step represents a binding to the cell surface, followed by either fusion of the viral membrane with the plasma membrane, or by fusion between the viral membrane and the endosomal membrane after endocytosis of the viral particle (Hoekstra et al., 1988; White et al., 1983). In general, viral binding with cell surface receptors involves the interaction of the viral binding protein, a membrane glycoprotein, with sialic acid-containing target cell membrane receptors. In essence, the nature of this interaction is electrostatic (Bächi et al., 1977); however, the conformation of the viral binding protein is an important parameter in optimizing the binding between virus and receptor (Hoekstra and Klappe, 1986), at least in the case of Sendai virus. This is also true for the stereochemical orientation of the carbohydrate receptor. For example, different strains of influenza virus, which infect cells via the endocytic pathway, recognize different sialic acid linkages while the binding is also influenced by sugars present in addition to sialic acid (Rogers and Paulson, 1983). Sendai virus, a virus that binds and fuses with the plasma membrane at neutral pH, displays a particular preference of binding to terminal sialic acid in an β2,3-ketosidic linkage to the adjacent carbohydrate residue, which is usually a galactose moiety (Holmgren et al., 1980; Umeda et al., 1984). The Sendai virus binding protein, HN, is present as dimers and tetramers in the membrane (Markwell and Fox, 1980).

In order to meet the criteria for being classified as a lectin (Goldstein and Hayes, 1978), HN should bear at least two sugar binding sites and display no enzymatic activity. To the best of our knowledge, the sugar binding sites have not been characterized yet but since the HN molecules are composed of identical polypeptide chains, it seems reasonable to assume that they display as a dimer or tetramer a multivalent binding character. However, since the same protein also contains neuraminidase activity, it does not strictly fit the definition of a lectin. In the case of influenza virus, the protein (HA) that mediates the binding also displays a low pH-induced fusion activity. It is a trimer (White et al., 1983) but does not display neuraminidase activity, which is located on a separate protein. By the same token, that is, based on assumptions as described above for Sendai virus, one tends to conclude then that the HA protein of influenza virus can, in principle, be designated as a fusogenic lectin. Both viral sialic acid-binding proteins, as described in the foregoing, can also interact in a nonspecific manner with artificial membranes devoid of sialic acid, provided that the membranes contain negatively charged phospholipids (Hoekstra et al., 1988). In that respect, their behavior as a lectin is quite distinct from plant lectins, which show no tendency to associate with vesicles composed of only pure negatively charged phospholipids (see Section 4).

Sialic acid-binding lectins have recently been demonstrated in frog eggs and it has been suggested that the lectin might be involved in fertilization and/or play a role in the development of the frog embryo (Titani et al., 1987). In the sperm of marine animals, adhesive proteins (bindins) have been found that mediate the adhesion of sperm to the egg surface (Glabe and Lennarz, 1981; Hong et al., 1987; Vacquier and Moy, 1977). Proteoglycan-like glycoconjugates containing sulfated fucose heteropolysaccharide chains appear to function as receptors on the egg surface (Glabe and Lennarz, 1981; Kinsey and Lennarz, 1981; Rossignol et al., 1984). Both the charged sulfate esters and the fucan structure appear critical to the interaction with bindin (Glabe et al., 1982). Hence, it appears possible that sperm–egg fusion may be initiated by a specific lectinlike interaction, which moreover could also explain why fusion in this case is restricted to specific membrane domains.

Fusion of myoblasts, which leads to the formation of multinucleated myotubes, is a characteristic and early event in the process of muscle differentiation. Studies of this cell system in particular have focused on the potential involvement of lectin–carbohydrate interactions in cell differentiation and membrane fusion (Gartner and Podleski, 1976; Nowak et al., 1976). How membrane fusion between myoblasts is accomplished is still unknown, although it has been proposed that the breakdown of inositol phospholipids is involved in fusion (Wakelam and Pette, 1984). Whatever the precise mechanism of fusion, myoblast adhesion has to occur prior to fusion. Mannosylated glycoproteins seem to be involved in this process as inferred from the observations that differentiation and fusion proceed normally with myoblast mutants lacking the ability to synthesize glycoproteins that lack terminal sialic acid and galactose (Gilfix and Sanwal, 1982). ConA-resistant mutants are unable to differentiate or to fuse, while binding of ConA to wild-type myoblasts blocks their capacity to fuse (Cates et al., 1984; Den et al., 1975; Parfett et al., 1983). This suggests that high-mannose-type glycoproteins are likely to participate in myoblast differentiation. Furthermore, the requirement for Ca^{2+} to fuse myoblasts in culture can at least partly be attributed to a Ca^{2+}-induced conformational change of these proteins, which appears necessary for rendering them adhesion competent (Knudsen, 1985). There appears to be no consensus as to the identity of the glycoprotein(s) involved (Cates et al., 1984; Pauw and David, 1979; Walsh and Phillips, 1981). It has been shown that 46-kilodalton protein was conspicuously absent in a mutant defective in differentiation and fusion, as revealed by ConA-binding experiments. Somatic hybrids, in which the synthesis of this protein was restored, differentiated and fused normally (Cates et al., 1984). Whether lectins play a role in cell–cell adhesion or fusion is unclear. In chick embryo myoblasts two endogenous lectins are produced, just prior to fusion, with binding specificities to lactose and GalNAc (Mir-Lechaire and Barondes, 1978). The involvement of the lactose lectin in fusion has been ruled out (Den and Chin, 1981) but a possible function of the GalNAc-specific lectin in the early developmental stages of muscle cells remain to be established.

Abbreviations

CL	cardiolipin
ConA	concanavalin A
CTL	*Croton tiglium* lectin
DPPC	dipalmitoylphosphatidylcholine
DSC	differential scanning calorimetry
EDTA	ethylenediaminetetraacetic acid
Fuc	fucose
Gal	galactose
GalCer	galactosylceramide, Galβ1-1Cer
GalNAc	*N*-acetylgalactosamine
Gb$_3$	trihexosylceramide, Galα1-4Galβ1-4Glcβ1-1Cer
Gb$_4$	globoside, GalNAcβ1-3Galα1-4Galβ1-4Glcβ1-1Cer
GD$_{1a}$	NeuAcα2-3Galβ1-3GalNAcβ1-4[NeuAcα2-3]Galβ1-4Glcβ1-1Cer
GD$_3$	NeuAcα2-8NeuAcα2-3Galβ1-4Glcβ1-1Cer
Glc	glucose
GlcNAc	*N*-acetylglucosamine
GluCer	glucosylceramide, Gluβ1-1Cer

GM_1	Galβ1-3GalNAcβ1-4[NeuAcα2-3]Galβ1-4Glcβ1-1Cer
GT_{1b}	NeuAcα2-3Galβ1-3GalNAcβ1-4[NeuAcα2-3]Galβ1-4Glcβ1-1Cer
LacCer	lactosylceramide, Galβ1-4Gluβ1-1Cer
Man	mannose
Man-6-P	mannose-6-phosphate
NANA or NeuAc	N-acetylneuraminic acid
PA	phosphatidic acid
PC	phosphatidylcholine
PHA	phytohemagglutinin
PS	phosphatidylserine
RCA	*Ricinus communis* agglutinin
SBA	soy bean agglutinin
WGA	wheat germ agglutinin

ACKNOWLEDGMENTS. The work cited in this chapter and carried out in the authors' laboratories was supported by NIH Grants GM 28117 and AI 25534 and by a NATO Research Grant RG 151.81. We are especially thankful to Mrs. Lineke Klap and Mrs. Rinske Kuperus for their help in preparing the manuscript.

6. REFERENCES

Ashwell, G., and Harford, J., 1982, Carbohydrate-specific receptors of the liver. *Annu. Rev. Biochem.* **51**:531–554.

Bach, D., Miller, I. R., and Sela, B.-A., 1982a, Calorimetric studies on various gangliosides and ganglioside–lipid interactions. *Biochim. Biophys. Acta* **686**:233–239.

Bach, D., Sela, B., and Miller, I. R., 1982b, Compositional aspects of lipid hydration. *Chem. Phys. Lipids* **31**:381–394.

Bächi, T., Daes, J. E., and Howe, C., 1977, Virus–erythrocyte membrane interactions, in *Virus Infection and the Cell Surface* (G. Poste and G. L. Nicholson, eds.), pp. 83–127, Elsevier Biomedical Press, New York.

Baenziger, J. V., and Fiete, D., 1979, Structural determinants of *Ricinus communis* agglutinin and toxin specificity for oligosaccharides. *J. Biol. Chem.* **254**:9795–9799.

Banerjee, K. K., and Sen, A., 1983, Glycolipid-dependent agglutination of liposomes by *Croton tiglium* lectin. *FEBS Lett.* **162**:248–251.

Barbosa, M. L. F., and Pinto da Silva, P., 1983, Restriction of glycolipids to the outer half of a plasma membrane: Concanavalin A labeling of membrane halves in *Acanthamoeba castellanii*. *Cell* **33**:959–966.

Barondes, S. H., 1984, Soluble lectins: A new class of extracellular proteins. *Science* **223**:1259–1264.

Barondes, S. H., 1986, Vertebrate lectins: Properties and functions, in *The Lectins. Properties, Functions and Applications in Biology and Medicine* (I. E. Liener, N. Sharon, and I. J. Goldstein, eds.), pp. 437–466, Academic Press, Orlando, FL.

Basu, S. K., Whisler, R. L., and Yates, A. J., 1987, Effects of lectin activation on sialyltransferase activities in human lymphocytes. *Biochemistry* **25:**2577–2581.
Bennett, J. S., Hoxie, J. A., Leitman, S. F., Vilaire, G., and Cines, D. B., 1983, Inhibition of fibrinogen binding to stimulated human platelets by a monoclonal antibody. *Proc. Natl. Acad. Sci. U.S.A.* **80:**2417–2421.
Bernard, B. A., Newton, S. A., and Olden, K., 1983, Effect of size and location of the oligosaccharide chain on protease degradation of bovine pancreatic ribonuclease. *J. Biol. Chem.* **258:**12198–12202.
Bertoli, E., Masserini, M., Sonnino, S., Ghidoni, R., Cestaro, B., and Tettamanti, G., 1981, Electron paramagnetic resonance studies on the fluidity and surface dynamics of egg phosphatidylcholine vesicles containing gangliosides. *Biochim. Biophys. Acta* **647:**196–202.
Bhattacharyya, L., Marchetti, P. S., Ellis, P. D., and Brewer, C. F., 1987, Nuclear magnetic resonance investigation of cadmium 113 substituted pea and lentil lectins. *J. Biol. Chem.* **262:**5616–5621.
Bischoff, J., and Lodish, H. F., 1987, Two asialoglycoprotein receptor polypeptides in human hepatoma cells, *J. Biol. Chem.* **262:**11825-11832
Biswas, M., Sekharudu, Y. C., and Rao, V. S. R., 1986, Complex carbohydrates: 1. Conformational studies on some oligosaccharides related to N-glycosylproteins which interact with concanavalin A. *Int. J. Biol. Macromol.* **8:**2–8.
Biswas, M., Sekharudu, Y. C., and Rao, V. S. R., 1987, The conformation of glycans of the oligo-*D*-mannosidic type, and their interaction with concanavalin A: A computer-modelling study, *Carbohydr. Res.* **160:**151–170.
Bock, K., Breimer, M. E., Brignole, A., Hansson, G. C., Karlsson, K.-A., Larson, G., Leffler, H., Samuelsson, B. E., Strömberg, N., Eden, C. S., and Thurin, J., 1985, Specificity of binding of a strain of uropathogenic *Escherichia coli* to Galα1-4Gal-containing glycosphingolipids. *J. Biol. Chem.* **260:**8545–8551.
Boggs, J. M., 1980, Intermolecular hydrogen bonding between lipids: Influence on organization and function of lipids in membranes. *Can. J. Biochem.* **58:**755–770.
Boon, A. M., Beresford, B. J., and Mellors, A., 1985, A tumor promoter enhances the phosphorylation of polyphosphoinositides while decreasing phosphatidylinositol labelling in lymphocytes. *Biochem. Biophys. Res. Commun.* **129:**431–438.
Borrebaeck, C. A., and Etzler, M. E., 1981, Production and characterization of a monoclonal antibody against the seed lectin of the *Dolichos biflorus* plant. *J. Biol. Chem.* **256:**4723–4725.
Bozzaro, S., 1985, Cell surface carbohydrates and cell recognition in *Dictyostelium*. *Cell Differ.* **17:**67–82.
Brennan, M. J., Joralmon, R. A., Cisar, J. O., and Sandberg, A. L., 1987, Binding of *Actinomyces naeslundii* to glycosphingolipids. *Infect. Immun.* **55:**487–489.
Brewer, C. F., and Brown, R. D., 1979, Mechanism of binding of mono- and oligosaccharides to Concanavalin A: A solvent proton magnetic relaxation dispersion study. *Biochemistry* **18:**2555–2562.
Brewer, C. F., Brown, R. D., and Koenig, S. H., 1983, Kinetics of transitions of demetalized concanavalin A. *Biochem. Biophys. Res. Commun.* **112:**595–601.
Brown, R. D., Brewer, C. F., and Koenig, S. H., 1977, Conformation states of Concanavalin A: kinetics of transitions induced by interaction with Mn^{2+} and Ca^{2+} ions. *Biochemistry* **16:**3883–3896.
Brown, R. E., and Thompson, T. E., 1987, Spontaneous transfer of ganglioside GM1 between phospholipid vesicles. *Biochemistry* **26:**5454–5460.
Bunow, M. R., and Bunow, B., 1979, Phase behavior of ganglioside–lecithin mixtures. Relation to dispersion of gangliosides in membranes. *Biophys. J.* **27:**325–337.
Campbell, C. D., Ross, T. E., and Sharom, F. J., 1983, Functional reassembly of lymphocyte

lentil lectin receptor glycoproteins into lipid bilayer vesicles. *Biochim. Biophys. Acta* **730**:95–103.
Carrington, D. M., Auffret, A., and Hanke, D. E., 1985, Polypeptide ligation occurs during post-translational modification of Concanavalin A. *Nature* **313**:64–67.
Cates, G. A., Brickenden, A. M., and Sanwal, B. D., 1984, Possible involvement of a cell surface glycoprotein in the differentiation of skeletal myoblasts. *J. Biol. Membr.* **259**:2646–2650.
Chan, K. F. J., 1987, Ganglioside-modulated protein phosphorylation. *J. Biol. Chem.* **262**:5248–5255.
Charo, I. F., Bekaert, L. S., and Phillips, D. R., 1987, Platelet glycoprotein IIb-IIIa-like proteins mediate endothelial cell attachment to adhesive proteins and the extracellular matrix. *J. Biol. Chem.* **262**:9935–9938.
Chicken, C. A., and Sharom, F. J., 1983, The Concanavalin A receptor from human erythrocytes in lipid bilayer membranes. *Biochim. Biophys. Acta* **729**:200–208.
Clegg, R. M., Lootiens, F. G., Sharon, N., and Jovin, T. M., 1983. Dynamic evidence for extended structure of the ligand combining site on wheat germ agglutinin: Temperature-jump relaxation with fluorescence detection. *Biochemistry* **22**:4797–4804.
Clevers, H. C., De Bresser, A., Kleinveld, H., Gmelig-Myeling, F. H., and Ballieux, R. E., 1986, Wheat germ agglutinin activates human T lymphocytes by stimulation of phosphoinositide hydrolysis. *J. Immunol.* **136**:3180–3183.
Cohen, S., Fava, R. A., and Sawyer, S. T., 1982, Purification and characterization of epidermal growth factor receptor/protein kinase from normal mouse liver. *Proc. Natl. Acad. Sci. U.S.A.* **79**:6237–6241.
Colley, K. J., and Baenziger, J. U., 1987, Identification of the post-translational modifications of the core-specific lectin. *J. Biol. Chem.* **262**:10290–10295.
Correa-Freire, M. C., Freire, E., Barenholz, Y., Biltonen, R. L., and Thompson, T. E., 1979, Thermotropic behavior of monoglucocerebroside–dipalmitoylphosphatidylcholine multilamellar liposomes. *Biochemistry* **18**:442–445.
Crook, S. J., Boggs, J. M., Vistines, A. I., and Koshy, K. M., 1986, Factors affecting surface expression of glycolipids: Influence of lipid environment and ceramide composition on antibody recognition of cerebroside sulfate in liposomes. *Biochemistry* **25**:7488–7494.
Cummings, R. D., and Kornfeld, S., 1984, The distribution of repeating (Galβ1,4GlcNAcβ1,3) sequences in asparagine-linked oligosaccharide of the mouse lymphoma cell lines BW 5147 and PHA^{R}2.1. *J Biol. Chem.* **259**:6253–6260.
Cunningham, B. A., Wang, J. L., Waxdal, M. J., and Edelman, G. M., 1975, The covalent and three-dimensional structure of Concanavalin A. *J. Biol. Chem.* **250**:1503–1512.
Cunningham, B. A., Hamperly, J. J., Hopp, T. P., and Edelman, G. M., 1979, Favin versus concanavalin A: Circularly permute amino acid sequences. *Proc. Natl. Acad. Sci. U.S.A.* **76**:3218–3222.
Curatolo, W., Yau, A. O., Small, D. M., and Sears, B., 1978, Lectin-induced agglutination of phospholipid/glycolipid vesicles. *Biochemistry* **17**:5740–5744.
Curtain, C. C., Looney, F. D., and Smelstorius, J. A., 1980, Lipid domain formation and ligand-induced lymphocyte membrane changes. *Biochim. Biophys. Acta* **596**:43–56.
Damjanov, I., 1987, Biology of disease: Lectin cytochemistry and histochemistry. *Lab. Invest.* **57**:5–20.
Debray, H., Decout, D., Strecker, G., Spik, G., and Montreuil, J., 1981, Specificity of twelve lectins towards oligosaccharides and glycopeptides related to N-glycosylproteins. *Eur. J. Biochem.* **117**:41–55.
De Kroon, A. I. P. M., Van Hoogenvest, P., Geurts, W. S. M., and De Kruijff, B., 1985, Influence of glycophorin and Ca^{2+}-induced fusion of phosphatidylserine vesicles. *Biochemistry* **24**:6382–6389.

Delmelle, M., Dufrane, S. P., Brasseur, R., and Ruysschaert, J. M., 1980, Clustering of gangliosides in phospholipid bilayers. *FEBS Lett.* **121**:11–14.
Den, H., and Chin, J. H., 1981, Endogenous lectin from chick embryo skeletal muscle is not involved in myotube formation in vitro. *J. Biol. Chem.* **256**:8069–8073.
Den, H., Malinzak, D. A., Keating, H. J., and Rosenberg, A., 1975, Influence of Concanavalin A, wheat germ agglutinin, and soy bean agglutinin on the fusion of myoblasts in vitro. *J. Cell Biol.* **67**:826–834.
Drickamer, K., Mamon, J. F., Binns, G., and Leung, J. O., 1984, Primary structure of the rat liver asialoglyprotein receptor. *J. Biol. Chem.* **259**:770–778.
Drickamer, K., Dordal, M. S., and Reynolds, L., 1986, Mannose-binding proteins isolated from rat liver contain carbohydrate-recognition domains linked to collagenous tails. *J. Biol. Chem.* **261**:6878–6887.
Düzgüneş, N., 1985, Membrane fusion, in *Subcellular Biochemistry* (D. B. Roodyn, ed.), Vol. 11, pp. 195–286, Plenum Press, New York.
Düzgüneş, N., and Hoekstra, D., 1986, Agglutination and fusion of glycolipid–phospholipid vesicles mediated by lectins and calcium ions. *Studia Biophys.* **111**:5–10.
Düzgüneş, N., and Papahadjopoulos, D., 1983, Ionotropic effects of phospholipid membranes: Calcium/magnesium specificity in binding, fluidity and fusion, in *Membrane Fluidity in Biology*, Vol. 2, *General Principles* (R. C. Aloia, ed.), pp. 187–216, Academic Press, Orlando, FL.
Düzgüneş, N., Hoekstra, D., Hong, K., and Paphadjopoulos, D., 1984, Lectins facilitate calcium-induced fusion of phospholipid vesicles containing glycosphingolipids. *FEBS Lett.* **173**:80–84.
Edelman, G. M., Cunningham, B. A., Reeke, G. N., Jr., Becker, J. W., Waxdal, M. J., and Wang, J. L., 1972, The covalent and three-dimensional structure of concanavalin A. *Proc. Natl. Acad. Sci. U.S.A.* **69**:2580–2585.
Ekerdt, R., and Papahadjopoulos, D., 1982, Intermembrane contact affects calcium binding to phospholipid vesicles. *Proc. Natl. Acad. Sci. U.S.A.* **79**:2273–2277.
Felgner, P. L., Barenholz, Y., and Thompson, T. E., 1981, Asymmetric incorporation of trisialoganglioside into dipalmitoylphosphatidylcholine vesicles. *Biochemistry* **20**:2168–2172.
Fidler, I. J., 1985, Macrophages and metastases. A biological approach to cancer therapy: Presidential address. *Cancer Res.* **45**:4714–4726.
Fitzgerald, L. A., Charo, I. F., and Phillips, D. R., 1985, Human and bovine endothelial cells synthesize membrane proteins similar to human platelet glycoproteins IIb and IIIa. *J. Biol. Chem.* **260**:10893–10896.
Fitzgerald, L. A., Steiner, B., Rall, S. C., Lo, S. S., and Phillips, D. R., 1987, Protein sequence of endothelial glycoprotein IIIa derived from a cDNA clone. *J. Biol. Chem.* **262**:3936–3939.
Foriers, A., Lebrun, E., Van Rapenbusch, R., De Neve, R., and Strosberg, A. D., 1981, The structure of the lentil *(Lens culinaris)* lectin, *J. Biol. Chem.* **256**:5550–5560.
Fukuda, M. N., Dell, A., and Scartezzini, P., 1987, Primary defect of congenital dyserythropoietic anemia type II. *J. Biol. Chem.* **262**:7195–7206.
Gartner, T. K., and Podleski, T. R., 1976, Evidence that a membrane bound lectin mediates fusion of L_6 myoblasts. *Biochem. Biophys. Res. Commun.* **67**:972–978.
Gething, M. J., White, J. M., and Waterfield, M. D., 1978, Purification of the fusion protein of Sendai virus: Analysis of NH_2-terminal sequence generated during precursor activation. *Proc. Natl. Acad. Sci. U.S.A.* **75**:2737–2740.
Giga, Y., Sutoh, K., and Ikai, A., 1985, A new multimeric hemagglutinin from the coelomic fluid of the sea urchin *Anthocidaris crassispina*. *Biochemistry* **24**:4461–4467.
Giga, Y., Ikai, A., and Takahashi, K., 1987, The complete amino acid sequence of Echinoidin, a lectin from the coelomic fluid of the sea urchin, *Anthocidaris crassispina*. *J. Biol. Chem.* **262**:6197–6203.

Gilfix, B. M., and Sanwal, B. D., 1982, Lectin-resistant myoblasts, in *Muscle Development: Molecular and Cellular Control* (M. L. Pearson and H. F. Epstein, eds.), pp. 329–336, Cold Spring Harbor Laboratory, Cold Spring Harbor, New York.

Glabe, C. G. and Lennarz, W. J., 1981, Isolation and partial characterization of a high molecular weight glycoconjugate, derived from the egg surface, which is implicated in sperm-egg adhesion, *J. Supramol. Struct. Cell. Biochem.* **15**:387–394.

Glabe, C. G., Grabel, L. B., Vacquier, V. D., and Rosen, S. D., 1982, Carbohydrate specificity of sea urchin sperm bindin: A cell surface lectin mediating sperm-egg adhesion. *J. Cell Biol.* **94**:123–128.

Goldstein, I. J., and Hayes, C. E., 1978, The lectins: Carbohydrate-binding proteins of plants and animals. *Adv. Carbohydr. Chem. Biochem.* **35**:127–340.

Goldstein, I. J., and Poretz, R. D., 1986, Isolation, physicochemical characterization, and carbohydrate-binding specificity of lectins, in *The Lectins: Properties, Functions and Applications in Biology and Medicine* (I. E. Liener, N. Sharon, and I. J. Goldstein, eds.), pp. 33–247, Academic Press, Orlando, FL.

Goldstein, I. J., Reichert, C. M., Misaki, A., and Gorin, P. A. J., 1973, An "extension" of the carbohydrate binding specificity of concanavalin A. *Biochim. Biophys. Acta* **317**:500–504.

Goodwin, G. C., Hammond, K., Lyle, I. G., and Jones, M. N., 1982. Lectin-mediated agglutinin of liposomes containing glycophorin. Effects of acyl chain length. *Biochim. Biophys. Acta* **689**:80–88.

Grafl, R., Lang, K., Vogl, H., and Schmid, F. X., 1987, The mechanism of folding of pancreatic ribounucleases is independent of the presence of covalently linked carbohydrate. *J. Biol. Chem.* **262**:10624–10629.

Grant, C. W. M., and Peters, M. W., 1984, Lectin–membrane interactions. Information from model systems. *Biochim. Biophys. Acta* **779**:403–422.

Green, E. D., Brodbeck, R. M., and Baenziger, J. U., 1987, Lectin affinity high-performance liquid chromatography. *J. Biol. Chem.* **262**:12030–12039.

Hakomori, S.-I., 1981, Glycosphingolipids in cellular interaction, differentiation and oncogenesis. *Annu. Rev. Biochem.* **50**:733–764.

Hakomori, S.-I., 1986, Glycosphingolipids. *Sci. Am.* **254**:32–41.

Hakomori, S.-I., and Kannagi, R., 1983, Glycosphingolipids as tumor-associated and differentiation markers, *J. Natl. Cancer Inst.* **71**:231–251.

Halberg, D. F., Wager, R. E., Farrell, D. C., Hildreth, J., Quesenberry, M. S., Loeb, J. A., Holland, E. C., and Drickamer, K., 1987, Major and minor forms of the rat liver asialoglycoprotein receptor are independent galactose-binding proteins. *J. Biol. Chem.* **262**:9828–9838.

Hammarström, S., Murphy, L. A., Goldstein, I. J., and Etzler, M. E., 1977, Carbohydrate binding specificity of four *N*-acetyl-D-galactosamine-"specific" lectins: *Helix pomatia* A hemagglutinin, soy bean agglutinin, lima bean lectin and *Dolichos biflorus* lectin. *Biochemistry* **16**:2750–2755.

Hampton, R. Y., Holz, R. W., and Goldstein, I. J., 1980, Phospholipid, glycolipid and ion dependencies of Concanavalin A- and *Ricinus communis* agglutinin I-induced agglutination of lipid vesicles. *J. Biol. Chem.* **255**:6766–6771.

Hardman, K. D., and Ainsworth, C. F., 1972, Structure of concanavalin A at 2.4-Å resolution. *Biochemistry* **11**:4910–4919.

Harrington, P. C., Moreno, R., and Wilkins, R. G., 1981, Metal ion interactions with apo-concanavalin A and some observations on metal ion requirements and sugar binding by *Bandeirea simplifolia* I lectin. *Isr. J. Chem.* **21**:48–51.

Hayman, M. J., Skehel, J. J., and Crumpton, M. J., 1973, Purification of virus glycoproteins by affinity chromatography using *Lens culinaris* phytohaemagglutinin, *FEBS Lett.* **29**:185–188.

Hedo, J. A., Harrison, L. C., and Roth, J., 1981, Binding of insulin receptors to lectins: Evidence for common carbohydrate determination on several membrane receptors. *Biochemistry* **20**:3385–3393.

Hersey, P., Schibeci, S. D., Townsend, P., Burns, C., and Cheresh, D. A., 1986, Potentiation of lymphocyte responses by monoclonal antibodies to the ganglioside GD3. *Cancer Res.* **46**:6083–6090.

Hesketh, T. R., Moore, J. P., Morris, J. D. H., Taylor, M. V., Rogers, J., Smith, G. A., and Metcalfe, J. C., 1985, A common sequence of calcium and pH signals in the mitogenic stimulation of eukaryotic cells. *Nature (London)* **313**:481–484.

Higgens, T. J. V., Chandler, P. M., Zurawski, G., Button, S. C., and Spencer, D., 1983, The biosynthesis and primary structure of pea seed lectin. *J. Biol. Chem.* **258**:9544–9549.

Hoekstra, D., 1982, Fluorescence method for measuring the kinetics of Ca^{2+}-induced phase separations in phosphatidylserine-containing lipid vesicles. *Biochemistry* **21**:1055–1061.

Hoekstra, D., and Düzgüneş, N., 1986, *Ricinus communis* agglutinin-mediated agglutination and fusion of glycolipid-containing phospholipid vesicles: Effect of carbohydrate head group size, calcium ions and spermine. *Biochemistry* **25**:1321–1330.

Hoekstra, D., and Klappe, K., 1986, Sendai virus–erythrocyte membrane interaction: Quantitative and kinetic analysis of viral binding, dissociation and fusion. *J. Virol.* **58**:87–95.

Hoekstra, D., and Wilschut, J., 1988, Membrane fusion of artificial and biological membranes. Role of local membrane dehydration, in *Water Transport in Biological Membranes* (G. Benga, ed.) CRC Press, Boca Raton, FL.

Hoekstra, D., Tomasini, R., and Scherphof, G., 1980, Interactions of phospholipid vesicles with rat hepatocytes in vitro. Influence of vesicle-incorporated glycolipids. *Biochim. Biophys. Acta* **603**:336–346.

Hoekstra, D., Düzgüneş, N., and Wilschut, J., 1985, Agglutination and fusion of globoside GL-4 containing phospholipid vesicles mediated by lectins and calcium ions. *Biochemistry* **24**:565–572.

Hoekstra, D., Klappe, K., Stegmann, T., and Nir, S., 1988, Parameters affecting the fusion of viruses with artificial and biological membranes, in *Molecular Mechanisms of Membrane Fusion* (S. Ohki, ed.), pp. 399–412, Plenum Press, New York.

Hoffman, L. M., Ma, Y., and Barker, R. F., 1982, Molecular cloning of *Phaseolus vulgaris* lectin mRNA and use of cDNA as a probe to estimate lectin transcript levels in various tissues. *Nucleic Acids Res.* **10**:7819–7828.

Hoflack, B., and Kornfeld, S., 1985, Purification and characterization of a cation-dependent mannose 6-phosphate receptor from murine P388D$_1$ macrophages and bovine liver. *J. Biol. Chem.* **260**:12008–12014.

Holmgren, J., Svennerholm, L., Elwing, H., Fredman, P., and Strannegärd, O., 1980, Sendai virus receptor: Proposed recognition structure based on binding to plastic-adsorbed gangliosides. *Proc. Natl. Acad. Sci. U.S.A.* **77**:1947–1950.

Hong, K., Düzgüneş, N., Meers, P. R., and Paphadjopoulos, D., 1987, Protein modulation of liposome fusion in *Cell Fusion* (A. E. Sowers, ed.), pp. 269–284, Plenum Press, New York.

Järnefelt, J., Rush, J., Li, Y.-T., and Laine, R. A., 1978, Erythroglycan, a high molecular weight glycopeptide with the repeating structure [galactosyl-(1-4)-2-deoxy-2-acetamido-glucosyl(1-3)] comprising more than one-third of the protein-bound carbohydrate of human erythrocyte stroma. *J. Biol. Chem.* **253**:8006–8009.

Jirgensons, B., 1980, Circular dichroism tests on the effect of alkali on conformation of lectins. *Biochim. Biophys. Acta* **625**:193–201.

Jones, G. W., and Isaacson, E., 1983, Proteinaceous bacterial adhesion and their receptors. *CRC Crit. Rev. Microbiol.* **10**:229–260.

Juliano, R. L., and Stamp, D., 1976, Lectin-mediated attachment of glycoprotein-bearing liposomes to cells. *Nature* **261**:235–238.

Junqua, S., Wils, P., Mishal, Z., and Le Pecq, J. B., 1987, Comparison of inhibitory effect of galactose analogs on the binding and cytotoxicity of an anti-globotriaosylceramide monoclonal antibody coupled or not coupled to pokeweed antiviral protein. *Eur. J. Immunol.* **17**:459–464.

Kabat, E. A., 1978, Dimensions and specificities of recognition sites on lectins and antibodies. *J. Supramol. Struct.* **8**:79–88.

Kahn, M. I., Sastry, M. V. K., and Surolia, A., 1986, Thermodynamic and kinetic analysis of carbohydrate binding to the basic lectin from winged bean *(Phosphocarpus tetragonolobus)*. *J. Biol. Chem.* **261**:3013–3019.

Kannagi, R., Nudelman, E., and Hakomori, S.-I., 1982, Possible role of ceramide in defining structure and function of membrane glycolipids. *Proc. Natl. Acad. Sci. U.S.A.* **79**:3470–3474.

Kelly, R. B., Deutsch, J. D., Carlson, S. S. and Wagner, J. A., 1979, Biochemistry of neurotransmitter release, *Ann. Rev. Neurosci.* **2**:399–446.

Ketis, N. V., and Grant, C. W. M., 1983, Control of high affinity lectin binding to an integral membrane glycoprotein in lipid bilayers. *Biochim. Biophys. Acta* **685**:347–354.

Ketis, N. V., and Grant, C. W. M., 1983, Time-dependent lectin binding to isolated receptors in model membranes. *Biochim. Biophys. Acta* **730**:359–368.

Kinsey, W. H., and Lennarz, W. J., 1981, Isolation of a glycopeptide fraction from the surface of the sea urchin egg that inhibits sperm–egg binding and fertilization. *J. Cell Biol.* **91**:325–331.

Knudsen, K. A., 1985, The calcium-dependent myoblast adhesion that precedes cell fusion is mediated by glycoproteins. *J. Cell Biol.* **101**:891–897.

Koch, O., and Uhlenbruck, G., 1982, From invertebrate to "in vertebrate" lectins: A review, in *Biotechs News*, E. Y. Laboratories Inc. Mateo, CA, USA, Autumn '82-1.

Kokourek, J., 1986, Historical background, in *The Lectins. Properties, Functions and Applications in Biology and Medicine* (I. E. Liener, N. Sharon, and I. J. Goldstein, eds.), pp. 1–32, Academic Press, Orlando, FL.

Kornfeld, R., and Kornfeld, S., 1985, Assembly of asparagine-linked oligosaccharides. *Annu. Rev. Biochem.* **54**:631–664.

Kornfeld, S., 1982, Oligosaccharide processing during glycoprotein biosynthesis in *The Glycoconjugates* (M. I. Horowirtz, ed.) Vol. IIIA, pp. 3–23, Academic Press, Orlando, FL.

Laferte, S., Fukuda, M. N., Fukuda, M., Dell, A., and Dennis, J. W., 1987, Glycosphingolipids of lectin-resistant mutants of the highly metastatic mouse tumor cell line MDAY-D2. *Cancer Res.* **47**:150–159.

Lee, P. M., and Grant, C. W. M., 1980, Headgroup oligosaccharide dynamics of a transmembrane glycoprotein. *Can. J. Biochem.* **58**:1197–1205.

Li, E., Gibson, R., and Kornfeld, S., 1980, Structure of an unusual complex-type oligosaccharide isolated from Chinese hamster ovary cells. *Arch. Biochem. Biophys.* **199**:393–399.

Liener, I. E., Sharon, N., and Goldstein, I. J (eds.), 1986, *The Lectins: Properties, Functions and Application in Biology and Medicine*, Academic Press, Orlando, FL.

Ling, N., Zeytin, F., Böhlen, P., Esch, F., Brazeau, P., Wehrenberg, W. B., Baird, A., and Guillemin, R., 1985, Growth hormone releasing factors. *Annu. Rev. Biochem.* **54**:403–423.

Lis, H., and Sharon N., 1986, Lectins as molecules and as tools. *Annu. Rev. Biochem.* **55**:35–67.

MacDonald, R. I., and MacDonald, R. C., 1975, Assembly of phospholipid vesicles bearing sialoglycoprotein from erythrocyte membrane. *J. Biol. Chem.* **250**:9206–9214.

Maggio, B., 1985, Geometric and thermodynamic restrictions for the self-assembly of glycosphingolipid–phospholipid systems. *Biochim. Biophys. Acta* **815**:245–258.

Maggio, B., Cumar, F. A., and Caputto, R., 1978, Induction of membrane fusion by polysialogangliosides. *FEBS Lett.* **90**:149–152.

Maggio, B., Cumar, F. A., and Caputto, R., 1981, Molecular behaviour of glycosphingolipids in interfaces. Possible participation in some properties of nerve membranes. *Biochim. Biophys. Acta* **650**:69–87.

Maggio, B., Ariga, T., Sturtevant, J. M., and Yu, R. K., 1985, Thermotropic behavior of binary mixtures of dipalmitoylphosphatidylcholine and glycosphingolipids in aqueous dispersions. *Biochim. Biophys. Acta* **818**:1–12.

Mansson, J., and Olofsson, S., 1983, Binding specificities of the lectins from *Helix pomatia*, soybean and peanut against different glycosphingolipids in liposome membranes. *FEBS Lett.* **156**:249–252.

Makita, A., and Taniguchi, N., 1985, Glycosphingolipids, in *Glycolipids* (H. Wiegandt, ed.), pp. 1–100, Elsevier Science Publishers, Amsterdam.

Marcus, D. M., Dustira, A., Diego, i., Osovitz, S., and Lewis, D. E., 1987, Studies of the mechanism by which gangliosides inhibit the proliferative response of murine splenocytes to concanavalin A. *Cell. Immunol.* **104**:71–78.

Markwell, M. K., and Fox, C. F., 1980, Protein–protein interactions within paramyxoviruses identified by native disulfide binding or reversible chemical cross-linking. *J. Virol.* **33**:152–166.

Meehan, E. J., Jr., McDuffie, J., Einspahr, H., Bugg, C. E., and Suddath, F. L., 1982, The crystal structure of pea lectin at 6-Å resolution. *J. Biol. Chem.* **257**:13278–13282.

Metcalfe, J. C., Hesketh, T. R., Smith, G. A., Morris, J. D., Corps, A. N., and Moore, J. P., 1985, Early response pattern analysis of the mitogenic parthway in lymphocytes and fibroblasts. *J. Cell Sci.* **3**:199–228.

Mir-Lechaire, F. J., and Barondes, S. H., 1978, Two distinct developmentally regulated lectins in chick embryo muscle. *Nature (London)* **272**:256–258.

Molin, K., Fredman, P., and Svennerholm, L., 1986, Binding specificities of the lectins PNA, WGA and UEAI to polyvinylchloride-adsorbed glycosphingolipids. *FEBS Lett.* **205**:51–55.

Momoi, T., Tokunaga, T., and Nagai, Y., 1982, Specific interaction of peanut agglutinin with the glycolipid asialo GM1. *FEBS Lett.* **141**:6–10.

Morell, P. (ed.), 1984, *Myelin*, Plenum Press, New York.

Myers, M., Wortman, C., and Freire, E., 1984, Modulation of neuraminidase activity by the physical state of phospholipid bilayers containing gangliosides GD_{1a} and GT_{1b}. *Biochemistry* **23**:1442–1448.

Nakajima, Y., Suzuki, H., Sakakibara, F., Kawauchi, H., Mizuno, D., and Yamazaki, M., 1986, Induction of a cytotoxin from murine macrophages by an animal lectin. *Jpn. J. Exp. Med.* **56**:19–25.

Neurohr, K. J., Mantsch, H. H., Young, N. M., and Bundle, D. R., 1982, Carbon-13 nuclear magnetic resonance studies on lectin–carbohydrate interactions: Binding of specifically carbon-13-labeled methyl (β-1)-lactoside to peanut agglutinin. *Biochemistry* **21**:498–503.

Neville, D. M., and Hudson, T. H., 1986, Transmembrane transport of diphtheria toxin related toxins and colicins. *Annu. Rev. Biochem.* **55**:195–224.

Newton, C., Pangborn, W., Nir, S., and Paphadjopoulos, D., 1978, Specificity of Ca^{2+} and Mg^{2+} binding to phosphatidylserine vesicles and resultant phase changes of bilayer membrane structure. *Biochim. Biophys. Acta* **506**:281–287.

Nowak, T. P., Haywood, P. C., and Barondes, S. H., 1976, Developmentally regulated lectin in embryonic chick muscle and a miogenic cell line. *Biochem. Biophys. Res. Commun.* **68**:650–657.

Nudelman, E., Kannagi, R., Hakomori, S., Parsons, M., Lipinski, J., Wiels, J., Fellous, M., and Tursz, T., 1983, A glycolipid antigen associated with Burkitt lymphoma defined by a monoclonal antibody, *Science* **220**:509–511.

Ollmann, M., Schwarzmann, G., Sandhoff, K., and Gallo, H.-J., 1987, Pyrene-labeled gangliosides: Micelle formation in aqueous solution, lateral diffusion and thermotropic behaviour in phosphatidylcholine bilayers. *Biochemistry* **26**:5943–5952.

Orr, G. A., Rando, R. R., and Bangerter, W. F., 1979, Synthetic glycolipids and the lectin-mediated aggregation of liposomes. *J. Biol. Chem.* **254**:4721–4725.

Parfett, C. L. J., Jamieson, J. C., and Wright, J. A., 1983, Changes in cell surface glycoproteins on non-differentiating L6 rat myoblasts selected for resistance to concanavalin A. *Exp. Cell Res.* **144**:405–415.

Parise, L. V., and Phillips, D. R., 1986, Fibronectin-binding properties of the purified platelet glycoprotein IIb-IIIA complex. *J. Biol. Chem.* **261**:14011–14017.

Papadimitriou, J. M., 1978, Macrophage fusion in vivo and in vitro: A review. *Cell Surf. Rev.* **5**:182–218.

Pasher, I., 1976, Molecular arrangements in sphingolipids. Conformation and hydrogen bonding of ceramide and their implication on membrane stability and permeability. *Biochim. Biophys. Acta* **455**:433–451.

Pauw, P. G., and David, J. D., 1979, Alterations in surface proteins during myogenesis of a rat myoblast cell line. *Dev. Biol.* **70**:27–38.

Pereira, M. E. A., Kabat, E. A., Lotan, R., and Sharon, N., 1976, Immunochemical studies on the specificity of the peanut *(Arachis hypogaca)* agglutinin. *Carbohydr. Res.* **51**:107–118.

Peters, M. W., Barber, K. R., and Grant, C. W. M., 1982, Headgroup behaviour of an uncharged complex glycolipid. *Biochim. Biophys. Acta* **693**:417–424.

Peters, M. W., and Grant, C. W. M., 1984, Freeze-etch study of an unmodified lectin interacting with its receptors in model membranes. *Biochim. Biophys. Acta* **775**:273–282.

Portis, A., Newton, C., Pangborn, W., and Paphadjopoulos, D., 1979, Studies on the mechanism of membrane fusion: Evidence for an intermembrane Ca^{2+}–phospholipid complex, snergism with Mg^{2+}, and inhibition by spectrin. *Biochemistry* **18**:780–790.

Pytela, R., Pierschbacher, M. D., Ginsberg, M. H., Plow, E. F., and Ruoslahti, E., 1986, Platelet membrane glycoprotein IIb/IIIa: Member of a family of Arg-Gly-Asp-specific adhesion receptors. *Science* **231**:1559–1562.

Rand, P. R., and Parsegian, V. A., 1984, Physical force considerations in model and biological membranes. *Can. J. Biochem. Cell. Biol.* **62**:752–759.

Rando, R. R., and Bangerter, F. W., 1979, Threshold effects on the lectin-mediated aggregation of synthetic glycolipid-containing liposomes. *J. Supramol. Struct.* **11**:259–309.

Rando, R. R., Slama, J., and Bangerter, F. W., 1980, Functional incorporation of synthetic glycolipids into cells. *Proc. Natl. Acad. Sci. U.S.A.* **77**:2510–2513.

Rao, V. S. R., and Biswas, M., 1982, The nature and size of the binding sites of blood group-specific, L-fucose-binding lectins: A conformational approach, in *Conformation in Biology* (R. Srinivasan and R. H. Sarma, eds.), pp. 183–190, Adenine Press, New York.

Rauvala, H., 1983, Cell surface carbohydrates and cell adhesion. *Trends Biochem. Sci.* **8**:323–325.

Redwood, W. R., and Polefka, T. G., 1976, Lectin–receptor interactions in liposomes. II. Interaction of wheat germ agglutinin with phosphatidylcholine liposomes containing incorporated monosialoganglioside. *Biochim. Biophys. Acta* **455**:631–643.

Redwood, W. R., Jansons, V. K., and Patel, B. C., 1975, Lectin–receptor interactions in liposomes. *Biochim. Biophys. Acta* **406**:347–361.

Rendi, R., Vatter, A. E., and Gordon, J. A., 1979, Divalent cation enhancement of the agglutinability by soy bean lectin of liposomes prepared from total lipid of erythrocytes and of erythrocyte membranes. *Biochim. Biophys. Acta* **550**:318–327.

Rintoul, D. A., Redd, M. B., and Wendelburg, B., 1986, N-parinaroyl glycosphingolipids: Synthesis and characterization of novel fluorescent probes of membrane structure. *Biochemistry* **25**:1574–1579.

Roberson, M. M., Wolffe, A. P., Tata, J. R., and Barondes, S. H., 1985, Galactoside-binding serum lectin of *Xenopus laevis*. *J. Biol. Chem.* **260**:11027–11032.

Rogers, G. N., and Paulson, J. C., 1983, Receptor determinants of human and animal influenza virus isolates: Differences in receptor specificity of the H3 hemagglutinin based on species of origin. *Virology* **127**:361–373.

Rosoff, P. W., Burakoff, S. J., and Greenstein, J. L., 1987, The role of the L3T4 molecule in mitogen and antigen-activated signal transduction. *Cell* **49**:845–853.

Ross, E. M., and Gilman, A. G., 1980, Biochemical properties of hormone-sensitive adenylate cyclase. *Annu. Rev. Biochem.* **49**:533–564.

Rossignol, D. P., Earles, B. J., Decker, G. L., and Lennarz, W. J., 194, Characterization of the sperm receptor on the surface of eggs of *Strongylocentrotus purpuratus*. *Dev. Biol.* **104**:308–321.

Rothman, J. E., and Lodish, H. F., 1977, Synchronised transmembrane insertion and glycosylation of a nascent membrane protein. *Nature* **269**:775–780.

Ruggeri, Z. M., De Marco, L., Gatti, L., Bader, R., and Montgomery, R. R., 1983, Platelets have more than one binding site for von Willebrand factor. *J. Clin. Invest.* **72**:1–12.

Ruocco, M. J., and Shipley, G. G., 1983, Hydration of *N*-palmitoyl-galactosylsphingosine compared to monosaccharide hydration. *Biochim. Biophys. Acta* **735**:305–308.

Schachter, H., Narasimhan, S., Gleeson, P., Vella, G. J., and Brockhausen, I., 1982, Oligosaccharide branching of glycoproteins: Biosynthetic mechanisms and possible biological functions. *Philos. Trans. R. Soc. London Ser. B* **300**:145–159.

Schnell, D. J., and Etzler, M. E., 1987, Primary structure of the *Dolichos biflorus* seed lectin. *J. Biol. Chem.* **262**:7220–7225.

Schwarz, R. T., and Datema, R., 1982, The lipid pathway of protein glycosylation and its inhibitors: The biological significance of protein-bound carbohydrates. *Adv. Carbohydr. Chem. Biochem.* **40**:287–379.

Sekharudu, Y. C., Biswas, M., and Rao, V. S. R., 1986, Complex carbohydrates: 2. The modes of binding of complex carbohydrates to Concanavalin A—a computer modelling approach. *Int. J. Biol. Macromol.* **8**:9–19.

Sharon, N., 1984a, Glycoproteins. *Trends Biochem. Sci.* **9**:198–202.

Sharon, N., 1984b, Surface carbohydrates and surface lectins are recognition determinants in phagocytosis. *Immunol. Today* **5**:143–147.

Sharom, F. J., Barratt, D. G., Thede, A. E., and Grant, C. W. M, 1976, Glycolipids in model membranes: Spin label and freeze-etch studies. *Biochim. Biophys. Acta* **455**:485–492.

Shimizu, T., and Hatano, M., 1983, ^{43}Ca and ^{67}Zn NMR spectra of Ca^{2+}, Zn^{2+}–Concanavalin A solutions. *Biochem. Biophys. Res. Commun.* **115**:22–28.

Slama, J. S., and Rando, R. R., 1980, Lectin-mediated aggregation of liposomes containing glycolipids with variable hydrophilic spacer arms. *Biochemistry* **19**:4595–4600.

Spohr, U., Hindsgaul, O., and Lemieux, R. U., 1985, Molecular recognition. II. The binding of the Lewis b and Y human blood group determinants by the lectin IV of *Griffonia simplicifolia*. *Can. J. Chem.* **63**:2644–2652.

Springer, W. R., Cooper, D. N. W., and Barondes, S. H., 1984, Discoidin I is implicated in cell–substratum attachment and ordered cell migration of *Dictostelium discoideum* and resembles fibronectin. *Cell* **39**:557–564.

Strosberg, A. D., Buffard, D., Lauwereys, M., and Foriers, A., 1986, Legume lectins: A large family of homologous proteins, in *The Lectins. Properties, Functions and Applications in Biology and Medicine* (I. E. Liener, N. Sharon, and I. J. Goldstein, eds.), pp. 249–264, Academic Press, Orlando, FL.

Sundler, R., 1982, Agglutination of glycolipid–phospholipid vesicles by Concanavalin. A. *FEBS Lett.* **141**:11–13.

Sundler, R., 1984, Studies on the effective size of phospholipid head groups in bilayer vesicles using lectin–glycolipid interaction as a steric probe. *Biochim. Biophys. Acta* **771**:59–67.

Sundler, R., and Wijkander, J., 1983, Protein-mediated intermembrane contact specifically enhances Ca^{2+}-induced fusion of phosphatidate-containing membranes. *Biochim. Biophys. Acta* **730**:391–394.

Surolia, A., Bachhawat, B. K., and Podder, S. K., 1975, Interaction between lectin from *Ricinus communis* and liposomes containing gangliosides. *Nature* **257**:802-804.

Suzuki, T., Inoue, K., Nojima, S., and Wiegandt, H., 1983, Interaction of concanavalin A with spin-labeled glycolipid incorporated into liposomes. *J. Biochem.* **94**:373-377.

Symington, F. W., Bernstein, I. D., and Hakomori, S.-I., 1984, Monoclonal antibody specific for lactosylceramide. *J. Biol. Chem.* **259**:6008-6012.

Symington, F. W., Murray, W. A., Bearman, S. I., and Hakomori, S.-I., 1987, Intracellular localization of lactosylceramide, the major human neurophil glycosphingolipid. *J. Biol. Chem.* **262**:11356-11363.

Taraschi, T. F., De Kruijff, B., Verkleij, A., and Van Echteld, C. J. A., 1982a, Effect of glycophorin on lipid polymorphism. A ^{31}P-NMR study. *Biochim. Biophys. Acta* **685**:153-161.

Taraschi, T. F., Van der Steen, A. T. M., De Kruijff, B., Tellier, C., and Verkleij, A. J., 1982b, Lectin-receptor interactions in liposomes: Evidence that binding of wheat germ agglutinin to glycoprotein-phosphatidylethanolamine vesicles induces non-bilayer structures. *Biochemistry* **21**:5756-5764.

Tillack, T. W., Allietta, M., Moran, R. E., and Young, W. W., 1983, Localization of globoside and Forssman glycolipids on erythrocyte membranes. *Biochim. Biophys. Acta* **733**:15-24.

Titani, K., Takio, K., Kuwada, M., Nitta, K., Sakakibara, F., Kawauchi, H., Takayanagi, G., and Hakomori, S.-I., 1987, Amino acid sequence of sialic acid binding lectin from frog *(Rana catesbeiana)* eggs. *Biochemistry* **26**:2189-2194.

Tsuji, S., Nakajima, S., Sasaki, T., and Nagai, Y., 1985, Bioactive gangliosides. IV. Ganglioside GQ_{1b}/Ca^{2+} dependent protein kinase activity exists in the plasma membrane fraction of neuroblastoma cell line GOTO. *J. Biochem.* **97**:969-972.

Uchida, T., Nagai, Y., Kawasaki, Y., and Wakayama, N., 1981, Fluorospectroscopic studies of various gangliosides and ganglioside-lecithin dispersions. Steady-state and time resolved fluorescence measurements with 1,6 diphenyl-1,3,5-hexatriene. *Biochemistry* **20**:161-169.

Umeda, M., Nojima, S., and Inoue, K., 1984, Activity of human erythrocyte gangliosides as a receptor to HVJ. *Virology* **133**:172-182.

Utsumi, H., Suzuki, T., Inoue, K., and Nojima, S., 1984, Haptenic activity of galactosyl ceramide and its topographical distribution on liposomal membranes. Effects of temperature and phospholipid composition. *J. Biochem.* **96**:97-105.

Utsumi, T., Aizono, Y., and Funatsu, G., 1987, Receptor-mediated interaction of ricin with the lipid bilayer of ganglioside GM1-liposomes. *FEBS Lett.* **216**:99-103.

Vacquier, V. D., and Moy, G. W., 1977, Isolation of bindin: The protein responsible for adhesion of sperm to sea urchin eggs. *Proc. Natl. Acad. Sci. U.S.A.* **74**:2456-2460.

Van Blitterswijk, W. J., 1983, Membrane fluidity in normal and malignant lymphoid cells, in *Membrane Fluidity in Biophysics*, Vol. 3, *Disease Processes* (R. C. Aloia and J. M. Boggs, eds.), pp. 85-159, Academic Press, Orlando, FL.

Van den Bosch, J., and McConnell, H. M., 1975, Fusion of dipalmitoylphosphatidylcholine vesicle membranes induced by Concanavalin A. *Proc. Natl. Acad. Sci. U.S.A.* **72**:4409-4413.

Verpoorte, J. A., 1975, Purification and characterization of glycoprotein from human erythrocyte membranes. *Int. J. Biochem.* **6**:855-862.

Villalta, F., and Kierszenbaum, F., 1983, Role of cell surface mannose residues in host cell invasion by *Trypanosoma cruzi*. *Biochim. Biophys. Acta* **736**:39-44.

Vodkin, L. O., Rhodes, P. R., and Goldberg, R. B., 1983, cA lectin gene insertion has the structural features of a transposable element. *Cell* **34**:1023-1031.

Von Figura, K., and Hasilik, A., 1986, Lysosomal enzymes and their receptors. *Annu. Rev. Biochem.* **55**:167-193.

Wakelam, M. J. O., and Pette, D., 1984, Myoblast fusion and inositol phospholipid breakdown:

Causal relationship or coincidence, in *Cell Fusion* (D. Evered and J. Whelan, eds.), pp. 100–118, Pitman, London.

Walsh, F. S., and Phillips, E., 1981, Specific changes in cellular glycoproteins and surface proteins during myogenesis in clonal muscle cells. *Dev. Biol.* **81**:229–237.

Wang, J. L., Cunningham, B. A., Waxdal, M. J., and Edelman, G. M., 1975, The covalent and three-dimensional structure of concanavalin A. I. Amino acid sequence of cyanogen bromide fragments F_1 and F_2. *J. Biol. Chem.* **250**:1490–1502.

Westrick, M. A., Lee, W. M. F., Goff, B., and Macher, B. A., 1983, Gangliosides of human acute leukemia cells. *Biochim. Biophys. Acta* **750**:141–148.

White, J., Kielian, M., and Helenius, A., 1983, Membrane fusion proteins of enveloped animal viruses. *Rev. Biophys.* **16**:151–195.

Wiegandt, H. (ed.), 1985, *New Comprehensive Biochemistry*, Vol. 10, *Glycolipids*, Elsevier, Amsterdam.

Williams, T. J., Plessas, N. R., Goldstein, I. J., and Lönngren, J., 1979, A new class of model glycolipids: Synthesis, characterization and interaction with lectins. *Arch. Biochem. Biophys.* **195**:145–151.

Wilschut, J., and Hoekstra, D., 1986, Membrane fusion: Lipid vesicles as a model system. *Chem. Phys. Lipids* **40**:145–166.

Wright, C. S., 1980, Crystallographic elucidation of the saccharide binding mode in wheat germ agglutinin and its biological significance. *J. Mol. Biol.* **141**:267–291.

Wright, C. S., 1981, Multi-domain structure of the dimeric lectin wheat germ agglutinin, in *Biomolecular Structure, Conformation, Functions and Evolution* (R. Srinivasan, E. Subramanian, and N. Yathindra, eds.), Vol. 1, pp. 9–17, Pergamon Press, Oxford.

Wright, C. S., Gavilanes, F., and Peterson, D. L., 1984, Primary structure of wheat germ agglutinin isolectin. 2. Peptide order deduced from X-ray structure. *Biochemistry* **23**:280–287.

Young, N. M., 1983, Magnesium as a natural substitute for manganese in concanavalin A and other lectins. *FEBS Lett.* **161**:247–250.

Zeeuws, R., and Strosberg, A. D., 1978, The use of methanol in high-performance liquid chromatography of phenylthiohydantoinamino acids. *FEBS Lett.* **85**:68–72.

Chapter 7

Energy-Transducing Complexes in Bacterial Respiratory Chains

Nobuhito Sone

1. INTRODUCTION

Oxidative phosphorylation in mitochondria and photophosphorylation in chloroplasts are highly organized systems that yield living energy for eukaryotes. By introducing the concept of an electrochemical H^+ gradient across the membrane ($\Delta\bar{\mu}H^+$)—that is, a proton motive force (pmf) composed of $\Delta\Psi$ and ΔpH as $\Delta\bar{\mu}H^+_{(mV)} = \Delta\Psi - Z \Delta pH$ ($Z \approx 60$ mV)—the chemiosmotic hypothesis (Mitchell, 1966) has played a central role in solving the problem of how organelles synthesize ATP using oxidative or photo energy. According to this hypothesis, the components of the electron transfer chain translocate protons across the membrane unidirectionally. The pmf thus established drives protons through an anisotropic reversible proton-translocating ATPase (ATP synthase) to synthesize ATP from ADP and Pi. There are other translocators (H^+ symporters such as Pi carrier and electrogenic translocators such as ATP/ADP translocator) in energy-transducing biomembranes that transport substrates using a pmf. These energy-transducing membranes include the cytoplasmic membranes of bacteria. Eubacteria at least (Archaebacteria may have a slightly different ATP synthase) contain the anisotropic $F_0 \cdot F_1$-type H^+-translocating ATPase (ATP synthase) that has very similar molecular architecture and characteristics to those of the

---

Nobuhito Sone   Department of Biochemistry, Jichi Medical School, Minamikawachi-machi, Tochigi-ken 329-04, Japan.

ATP synthases of mitochondria and chloroplasts (Amzel and Pedersen, 1983), as well as having several translocators that utilize pmf. In contrast, the bacterial electron transfer chains that produce a pmf show great variety, even if among those in aerobic respiratory systems (Jones, 1977). Earlier experimental data were mostly obtained by studies on whole cells or fragmented membrane vesicles, but recently, energy-transducing components of the electron transfer chains of bacteria have been purified and reconstituted into lipid bilayers for measurement of their function. Studies on the reconstitution of energy-transducing enzymes of oxidative phosphorylation into liposomal vesicles have been reviewed (Casey, 1984). A review on the reconstitution of various membrane proteins is also available (Etemadi, 1985), but, at the time of the review's publication, information on energy-transducing components of the electron transfer chain (respiratory complexes, I, III, and IV) of bacteria was fragmentary. This chapter reports methods for studies on energy-transducing components (Sections 2 and 3); the purification and reconstitution work of bacterial redox enzymes, especially respiratory complexes (Sections 4–7), and tentative conclusions deduced from work in this field (Section 8).

2. ELECTRON TRANSFER CHAIN AND ENERGY COUPLING

2.1. Bacterial Respiratory Chains

Great variety is seen among bacterial electron transfer systems, even among those for aerobic respiration utilizing molecular oxygen. Figure 1 shows a scheme of the aerobic electron transfer chain of *Paracoccus denitrificans*, which is quite similar to that of mitochondria (John and Whatley, 1977; Stouthamer, 1980). This scheme does not cover the bacterial respiratory chains but may be helpful in indicating the following points. (1) NADH (and NADPH) in the cytoplasma is usually a supplier of reducing power. (2) NADH : quinone oxidoreductase (complex I) is an energy-transducing enzyme that produces a pmf by electrogenic H^+ movement coupled with a downhill redox reaction. (3) There are several dehydrogenases (quinone reductases) that reduce quinone but do not themselves cause energy transduction. They include succinate dehydrogenase (also called complex II), NAD-independent malate dehydrogenase, and glycerol 3-phosphate dehydrogenase. Several bacteria have NADH dehydrogenase (NADH : quinone oxidoreductase) of this type instead of, or in addition to, the complex I type. (4) Quinones always participate in the bacterial electron transfer chain. Some bacteria use UQ (ubiquinone), other (mainly gram-positive ones) use MK (menaquinone), and a few others, such as *Escherichia coli*, use both (see Table I and review by Collins and Jones, 1981). (5) Quinol : cytochrome *c* oxidoreductase (complex III), known as the cytochrome

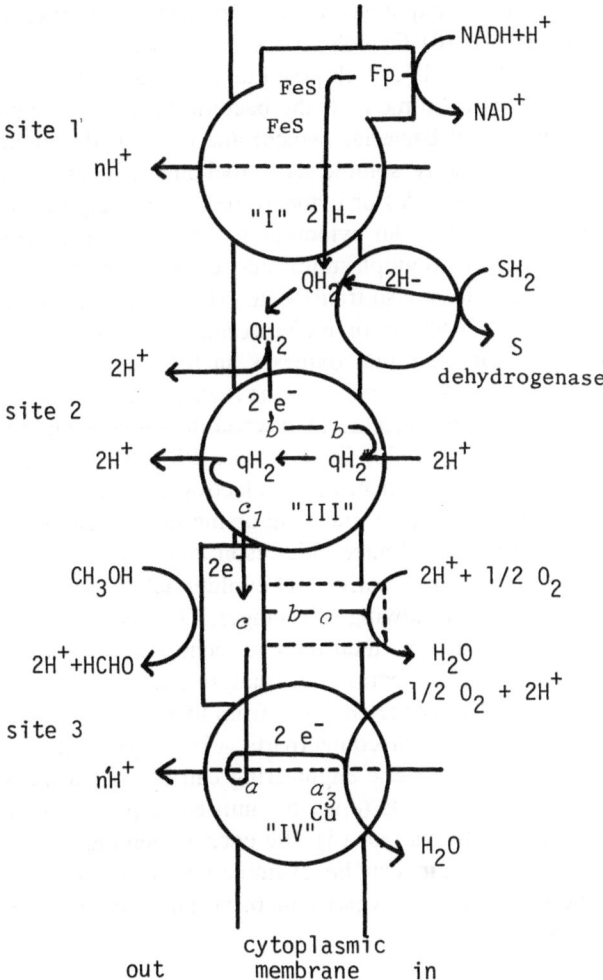

FIGURE 1. The respiratory chain of an aerobic bacterium, an example of *P. denitrificans*. Abbreviations: Fp, flavin center; Q, quinone; q, transmembrane hydrogen-carrying quinone derivatives in Q loop; FeS, iron–sulfur center; c, c_1, b, o, a, and a_3, cytochromes, I, III, and IV; the traditional number of the respiratory complexes. The stoichiometry of H^+/e^- at that site, n or n', is 2–4. In *P. denitrificans* n and n' are claimed to be 3 and 2, respectively (Boogerd *et al.*, 1984).

bc_1 complex, is also energy transducing. This complex is probably more common than any other complex in bacteria, and it is found in chloroplasts (named cytochrome b_6f complex) as well as mitochondria. But bacteria such as *E. coli* do not have this complex (Ingledew and Poole, 1984). The mechanism of proton translocation may be explained by a loop mechanism such as a Q-cycle model (Mitchell, 1976). (6) Cytochrome c is a mediator of electron transfer between complexes III and IV. High potential reductants such as methanol and methylamine reduce cytochrome c, if the bacteria have the enzyme in the periplasmic space. In several bacteria, cytochrome c is a hydrophobic membrane protein and is not as readily soluble as mitochondrial counterpart. (7) Cytochrome c oxidase (complex IV) or cytochrome aa_3 is energy transducing. The complex receives electrons from cytochrome c and transfers them to oxygen, probably on the opposite (cytoplasmic) side of the membrane. Complex IV from several bacteria at least also translocates H^+ across the membrane (proton pumping). (8) In some bacteria or under certain conditions, this cytochrome aa_3 is replaced by other terminal oxidases such as cytochrome o and cytochrome d. Many aerobes use cytochrome aa_3 when the O_2 concentration in the culture medium is high and cytochrome o when the O_2 concentration is reduced (Sapshead and Wimpener, 1972; Sone et al., 1983a). Thus, even when the oxidant is O_2 the terminal segment of the electron transfer chain varies. Some bacteria use fumarate, sulfate, nitrite, and/or nitrate as terminal oxidant(s)—a type of electron transfer called anaerobic respiration (Thauer et al., 1977).

Besides its value in comparative and evolutionary studies, research on bacterial respiration has the following advantages: (1) The enzymes (complexes) have simpler subunit structures than their mitochondrial counterparts but have similar active centers (chromophores) and catalyze analogous reactions. (2) Some bacterial enzymes, especially the enzymes from thermophilic bacteria, are more stable than their mitochondrial counterparts. (3) Enzymes in mutants can be studied and techniques of genetic engineering can be applied rather easily. (4) The conditions of bacterial cells (e.g., the nutrient supply, degree of aeration, and carbon source) can be varied. (5) The energy-yielding process in the bacterial electron transfer chain can be examined in whole cells. We can also analyze the energy-yielding processes in bacterial electron transfer chain by using whole cells.

2.2. Two Approaches Using Whole Cells or Membrane Vesicles

2.2.1. H^+/O Ratio Measurement with Cells Using Endogenous Substrates

As described in Section 1, the redox reaction in respiratory complexes, which translocates electrons and H^+ vectorially across the membrane, resulted in the formation of a pmf composed of $\Delta\Psi$ and ΔpH. When bacterial cells are in the resting state, in which the pmf is not utilized rapidly and permeant ions such as SCN^- or K^+ *plus* valinomycin are present to prevent formation of a

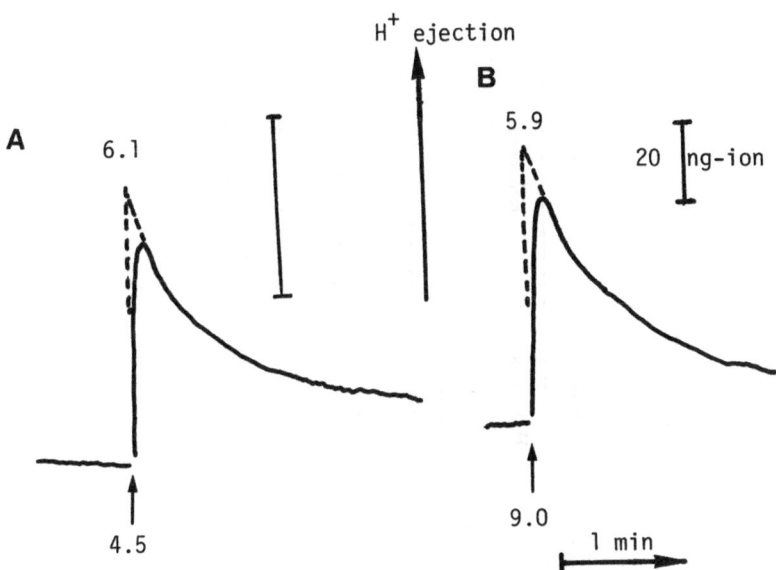

FIGURE 2. Measurement of H^+/O ratio using resting cells with endogenous substrates. Resting cells were incubated in 140 mM KCl containing 66 mM KSCN and 0.1 μM valinomycin at pH 6.3–6.6. After anaerobiosis and equilibrium of ΔH^+, an indicated ng-atom of O_2 was pulsed at the arrow. Numbers in the figure show H^+/O ratio obtained (see Sone et al., 1987a). A, the thermophilic bacterium PS3; B, Ps. AM1.

membrane potential, addition of a small volume of air-saturated KCl to an anoxic suspension of cells results in a burst of respiration and a transient pH change, as shown in Figure 2. The H^+/O ratio is calculated as the amount of H^+ ejected, which is the value extrapolated to 0 time, dividing by the amount of O_2 introduced. Table I summarizes results of H^+/O ratio measurement in relation to the electron transfer component of the bacteria. The table includes both the summary of Jones (1977) and recent data but is by no means a complete list. The highest H^+/O ratios were observed in *P. denitrificans;* the ratios for *Alcaligenes eutrophus* and thermophilic bacteria were as high as 8, but *Pseudomonas* AM1, which has a similar electron transfer chain, showed an H^+/O ratio of about 6, and *B. subtilis* and *B. megaterium* had ratios as low as 4 (Jones et al., 1975). The latter values were as low as that of *E. coli*, which does not have either the cytochrome bc_1 complex or cytochrome *c* oxidase. Although Jones (1977) pointed out that the reason for the low H^+/O ratios for *B. subtilis* and *B. megaterium* may be due to the absence of cytochrome *c* in these bacteria, cytochrome *c* is present in the respiratory chain of *B. subtilis* W23 and probably also in *B. megaterium* (de Vrij, 1986). If so, all these bacteria have similar respiratory components, such as Ndh (NADH dehydrogenase

Table I
Relationship between Respiratory Component and H$^+$/O Ratio

Organism	Respiratory components[a]				H$^+$/O ratio	References
Bacillus megaterium D440	Ndh	—	MK	b	—	aa_3 (o)	4.0	Downs and Jones, 1975
Bacillus subtilis D473	Ndh	—	MK	b	—	aa_3 (o)	4.0	Jones et al., 1975
Bacillus stearothermophilus	Ndh	—	MK	b	c	aa_3 (o)	7.3	Chicken et al., 1981
Methylophilus lysodeikticus	Ndh	—	MK	b	—	aa_3 (o)	5.0	Jones et al., 1975
Escherichia coli W	Ndh	Q	(MK)	b	—	o (d)	4.0	Brice et al., 1974
Paracoccus denitrificans	Ndh	Q	—	b	c	aa_3 (o)	8.0[b]	Scholes and Mitchell, 1970
Alcaligenes eutrophus H16	Ndh	Q	—	b	c	aa_3 (o)	7.8	Jones et al., 1975
Pseudomonas AM1	Ndh	—	—	b	c	aa_3 (o)	5.9	Keevil and Anthony, 1979
Thermus thermophilus	Ndh	—	MK	b	c	aa_3 (o)	7.9	McKay et al., 1982

[a] Respiratory chain components present at low concentrations or exhibiting low activity are shown in brackets. See also van Verseveld et al. (1981) and Hitchens and Kell (1984).
[b]

or NADH : quinone oxidoreductase) for complex I, cytochrome b and cytochrome c for complex III, and cytochrome aa_3 for complex IV, but some showed a low H^+/O ratio, while the others showed higher ratios. Thus, more precise studies are necessary to explain the data of the H^+/O ratio with endogenous substrate.

2.2.2. H^+/O Ratio Measurement at Site 3

The rationale of the lower H^+/O ratio such as 4 or 6 is that the proton pumping activity at site 3 (cytochrome oxidase level) may be lacking. If the cells can be depleted in endogenous substrate and/or their respiration blocked before cytochrome c by a certain inhibitor, such as antimycin A, the H^+/O ratio at site 3 is measurable more directly. Figure 3 shows oxygen pulse experiments with *Ps.* AM1 and *P. denitrificans*. Upon addition of an oxygen pulse, a proton ejection (apparent H^+/O ratio of 2) took place using *Ps.* AM1 with ascorbate and TMPD (N,N,N',N'-tetramethyl p-phenelenediamine) as a substrate, and this apparent proton ejection disappeared in the presence of a potent protonophore FCCP (carbonylcyanide p-trifluoromethoxy phenylhydrazone). In the case of *P. denitrificans*, an apparent proton ejection of 2.8 in H^+/O ratio and net alkalinization in the presence of FCCP were observed. Ascorbate ($pK_a = 4.2$) is known to release 1 H^+ upon oxidation to dehydroascorbate at a neutral pH. But several bacteria, such as *E. coli*, hydrate dehydroascorbate to

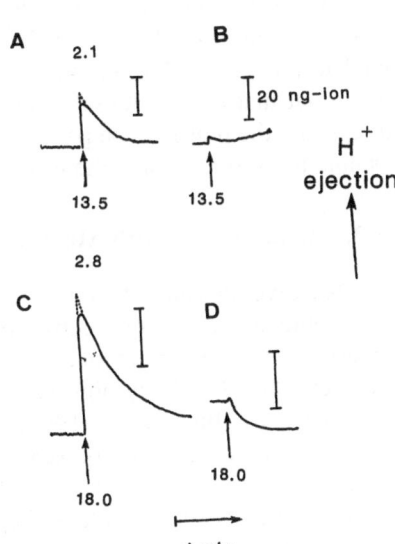

FIGURE 3. Measurement of H^+/O ratio using starved cells with ascorbate *plus* TMPD as substrate (see Sone *et al.*, 1987a). A, starved *Ps.* AM1 cells; B, as A but with FCCP; C, starved *P. denitrificans* cells in the presence of antimycin A; D, as C but with FCCP.

2,3-ketogulonate. Thus, it is likely that the following reaction occurs with *Ps.* AM1:

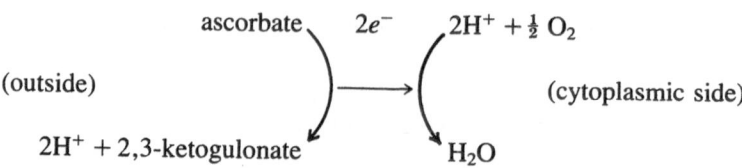

On the contrary, the reaction with *P. denitrificans* is as follows:

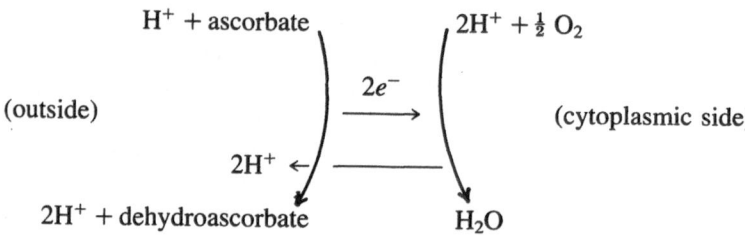

where $2H^+$ movement occurs as a vectorial process indicating that this cytochrome oxidase pumps protons in addition to vectorial electron transport. Cytochrome oxidase of *Ps.* AM1 apparently does not pump protons even in resting cells, indicating that the enzyme does not pump protons intrinsically.

These resting cell experiments are useful for testing the proton pump activity of the cytochrome oxidase level. However, the prerequisite to prepare endogenous substrate-depleted cells or to inhibit endogenous respiration nearly completely by a suitable inhibitor is often difficult to fulfill. It should also be pointed out that this H^+/O ratio measurement with resting cells has the disadvantage that the pmf may be partially discharged by some rapid processes, such as solute transport, even if there is no sign of this process.

2.2.3. Experiments with Membrane Vesicles

Oxidative phosphorylation activity of bacteria can be measured by preparing membrane vesicles that correspond to submitochondrial particles of mitochondria. However, membrane vesicles from bacteria did not usually give a high P/O ratio. Relatively high values of about 1 were observed with those of *P. denitrificans* (Imai *et al.*, 1967) and *Mycobacterium phlei* (Brodie, 1959), but even in these cases, no phosphorylation activity was measured at the third site (cytochrome *c* oxidase level). The reason for these low P/O ratios is unknown but may be attributable to the fact that bacterial cytoplasmic membranes do not produce well-sealed, small vesicles. On the other hand, membrane enzymes, even when they are crude mixtures such as the octylglucoside-soluble fraction of whole membranes, can be reconstituted into H^+-tight liposomal vesicles with P lipids, in which they transduce energy actively (Sone, 1986).

3. RECONSTITUTION INTO LIPOSOMES

3.1. A Tool to Study Energy-Transducing Enzymes

The third approach to the problem of bacterial oxidative phosphorylation is at the enzyme level. This approach needs two steps: purification of the membrane enzyme and preparation of proteoliposomes from the purified enzyme and polar lipids, in which the enzyme can function and the pmf formed can be maintained. Methods for incorporation of enzyme into a planar bilayer and measurement of the enzyme's activities have now been developed. The chemiosmotic theory of oxidative phosphorylation has been proved experimentally by this approach. Kagawa and Racker (1971) reported the recovery of the ATP-Pi exchange reaction upon formation of proteoliposomes on removal of cholate from a mixture of the crude ATP synthase fraction and soybean P lipids solubilized with cholate. We have reported the synthesis of ATP by proteoliposomes containing only ATP synthase and P lipids upon imposition of an artificial pmf (Sone et al., 1977b). For this, we used purified ATP synthase to exclude the possibility of interaction between respiratory components and ATP synthase. The enzyme was solubilized from membranes of a thermophilic bacterium with Triton X-100 and purified by DEAE–cellulose chromatography and gel filtration on Sepharose 6B in the presence of Triton X-100 (Sone et al., 1975). Then the purified enzyme, composed of eight kinds of subunits, was reconstituted into proteoliposomes by dialysis of a mixture of the enzyme and P lipids in the presence of cholate *plus* deoxycholate (Sone et al., 1976). The resulting proteoliposomes were then equilibrated with buffer of pH 5.5 and mixed with buffer of pH 8.4 containing 0.15 M KCl in the presence of valinomycin (K^+ ionophore) to impose an inside-negative K^+ diffusion potential. Transient ATP formation dependent on the size of the pmf (over 180 mV) indicates that ATP synthase forms ATP upon H^+ movement in the presence of a sufficient pmf (Sone et al., 1977b). Since the reverse process, ATP-dependent proton movement, and the resulting formation of a pmf of as much as 200 mV (negative inside) have also been shown (Sone et al., 1976), the chemiosmotic theory of Mitchell was confirmed in a pure system. In this experiment ATP synthase ($F_0 \cdot F_1$) was prepared as an eight-subunit pure form for the first time. This enzyme from the thermophilic bacterium PS3 is sufficiently stable for its purification at room temperature.

This reconstitution technique into liposomes opens a new way for the de-

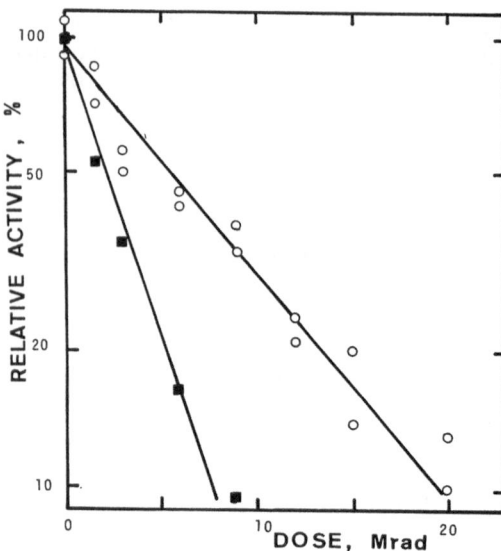

FIGURE 4. Decay curves for radiation inactivation of PS3 aa_3-type cytochrome oxidase. H^+ pump activity (■) and cytochrome c oxidase (○) were measured after reconstitution into liposomes (see Sone and Kosako, 1986).

termination of the functional molecular weight of an energy-transducing enzyme by radiation-inactivation experiments. Although the target theory is generally applicable to determine a functional molecular weight of the enzyme (see Kempner and Schlegel, 1979, for review), it is not possible to determine a target size for a vectorial reaction directly. In fact, very low I_{50} doses (0.1–0.6 Mrad) for 50% inhibition were obtained for the respiratory control ratio, $^{32}P_i$-ATP exchange reaction and Ca^{2+} accumulation activities, while those of ATPase and cytochrome c oxidase activities were 1.9 and 7.4 Mrad, giving reasonable target sizes of 290 and 67 kDa, respectively (Kagawa, 1967).

Cytochrome c oxidase from the thermophilic bacterium PS3 or from bovine heart was irradiated with a high-energy electron beam. Then the irradiated enzyme was reconstituted into proteoliposomes with P lipids, and proton pump and oxidase activities were measured. As shown in Figure 4, both activities by the PS3 enzyme (and also by the bovine enzyme not shown) decreased as exponential functions of the radiation doses. If we apply the target theory, the following functional molecular weights are obtained for oxidation and proton pumping: 63–73 kDa for the bovine enzyme and 80–100 and 190–230 kDa for the PS3 enzyme (Sone and Kosako, 1986). These results suggest that a dimer structure is necessary for the proton pumping activity of the PS3 enzyme, while cytochrome c oxidation can be catalyzed by a monomer form, since the PS3 enzyme is composed of each one of three subunits (56, 38, and 22 kDa; see also Table II). It is also likely that a dimer structure composed of the core subunits (major three subunits encoded for in the mitochondrial DNA) is also necessary for proton pumping in the mitochondrial enzyme.

3.2. Purification of Respiratory Complexes

Membrane enzymes can be purified by ion-exchange column chromatography in the presence of a nonionic detergent such as Triton X-100 or polyoxyethylene lauryl ether ($C_{12}E_8$). Since membrane enzymes sometimes lose activity, especially on dilution, as during chromatography, addition of protecting reagents such as dithiothreitol and glycerol may be helpful, and care is required in choosing a suitable pH and ionic strength of the medium. Although expensive, alkyl sugar detergents sometimes give good results without causing denaturation. In particular, lauryl maltoside is recommended for the purification of cytochrome c oxidase (Rosevear *et al.*, 1980) and cytochrome bc_1 complex (Ljungdahl *et al.*, 1987). If there is a good ligand, affinity chromatography is a rapid and effective purification step. Bacterial cytochrome c oxidase was purified from *Rhodobacter sphaeroides* by this method on a yeast cytochrome c-bound Sepharose column (Gennis *et al.*, 1982).

An alternative method for purifying cytochrome bc_1 complex from various sources without enzyme dilution has been reported (Hurt and Hauska, 1981; Hauska, 1986). The procedure consists of solubilization of membranes with cholate *plus* octylglucoside (or N-methyl-N-nonanoylglucamide), fractionation of the extract with ammonium sulfate, and centrifugation in a sucrose density gradient (14–30%) containing cholate and octylglucoside (or N-methyl-N-nonanoylglucamide).

It is noteworthy to point out that the respiratory complexes are so hydrophobic that they seem to span the membrane. Thus, purification of the membranes to remove peripheral proteins increases the content of respiratory complexes and facilitates their subsequent purification.

3.3. Reconstitution Methods

Most membrane enzymes, especially respiratory complexes, require polar lipids for exhibiting catalytic activity. Polar lipids in the membrane also act as barriers against hydrophilic substances and ions and maintain the electrochemical ion gradient across the membrane. For reconstitution of a function of any energy-transducing process, the enzyme involved must span the lipid bilayer properly and a closed space must have been formed between bilayer lipids. Several methods have been devised to achieve these conditions (see Racker, 1979), and the four methods most frequently used are described in Sections 3.3.1 to 3.3.4. In most cases, soybean P lipids (asolectin) can be used as provided commercially or after partial purification. However, since asolectin failed to activate cytochrome bc_1 complex of the thermophilic bacterium PS3, it is worth trying to use P lipids from the host bacterium also. Membrane enzymes often require P lipids to be active.

3.3.1. Detergent Dialysis Methods

Dialyzable detergents such as cholate, deoxycholate, and octylglucoside are used to solubilize (or suspend) a membrane enzyme and P lipids. Since bacterial respiratory complexes are more hydrophobic than the complexes of mitochondria (see below), cholate, which was used in the earlier studies (Kagawa and Racker, 1971), seems unsuitable for some bacterial enzymes, and deoxycholate or octylglucoside must be used in place of or in addition to cholate. When deoxycholate was present, addition of Mg^{2+} at about 2–5 mM to the dialysis buffer greatly increased formation of active proteoliposomes (Sone et al., 1977a).

3.3.2. Detergent Dilution Methods

The mixture of P lipids and membrane enzyme with a detergent is diluted with medium by stirring (Racker et al., 1979). The resulting proteoliposomes can be collected by ultracentrifugation. Usually, a detergent with a high critical micelle concentration (CMC) alone or in combination, such as cholate *plus* octylglucoside, has been used.

3.3.3. Incorporation Methods

Membrane enzymes can be incorporated into preformed liposomes by sonication. Satisfactory results have been obtained when a stable enzyme such as bacteriorhodopsin is used. In some cases a membrane enzyme can be incorporated into preformed liposomes in the presence of a detergent such as the lysophospholipids (Eytan and Racker, 1977).

3.3.4. Freeze–Thaw–Sonication Methods

An enzyme is added to a liposome suspension, usually prepared by sonication, and the mixture is frozen for several minutes. The resultant aggregated mixture is sonicated briefly (e.g., for 5–10 sec) until the suspension becomes nearly clear (Kasahara and Hinkle, 1977). A much shorter sonication period is required than in the sonic incorporation method (5 min or more), and incorporation of membrane protein should occur upon freezing. Usually two or three cycles of freeze–thaw–sonication give better results than a single cycle. The presence of small amounts of detergent, such as octylglucoside, accelerated the process slightly (Sone, 1986). By this method it is possible to detect H^+ translocation with proteoliposomes prepared from the heptylthioglucoside- (or octylglucoside-) soluble fraction without purifying cytochrome c oxidase (Sone, 1986). This freeze–thaw–sonication process is also applicable to proteoliposomes reconstituted by other methods, indicating that these vesicles can be kept frozen. However, it should be remembered that the orientation (sidedness) of the enzyme may be scrambled. The detergent–dialysis method and incorporation into preformed liposomes with a detergent give more or less oriented proteoliposomes, whereas scrambled ones are obtained by sonication or freeze–thaw–sonication methods. Results seem to depend on the nature of the membrane protein.

This method is also useful for loading some solutes into the lumina of liposomes. KCl was loaded by this method (Sone et al., 1981) to produce a membrane potential with valinomycin to measure H^+ permeability through the H^+ channel of TF_0 (membrane sector of the thermophilic ATP synthase).

4. CYTOCHROME OXIDASE

4.1. Bacterial Cytochrome Oxidase

Many aerobic bacteria contain cytochrome aa_3 as a terminal oxidase, as in the mitochondrial respiratory chain. This respiratory complex IV also seems to be a coupling site (site 3) in bacteria, where pmf ($\Delta\bar{\mu}_H^+$) is formed upon oxidation. Since 1979 (Sone et al., 1979; Yamanaka et al., 1979), several of these cytochrome oxidases have been highly purified from various bacteria. These bacterial enzymes are characterized by their simple subunit structure and the similarity of their prosthetic groups, which consist of two a-hemes (cytochrome aa_3) and two Cu atoms (Cu_A and Cu_B), like mitochondrial enzymes.

Table II summarizes the subunit structures and proton pump activities of aa_3-type cytochrome oxidases. The bacterial enzymes are composed of 2 or 3 subunits and thus are much simpler than mitochondrial enzymes: that of yeast has 9 subunits (Power et al., 1984) and that of bovine heart has 13 (Kadenbach et al., 1986; Takamiya et al., 1987). These bacterial enzymes oxidize cytochrome c with dioxygen and concomitantly form a pmf. The enzymes catalyze at least electron transfer across the membrane, since proteoliposomes properly reconstituted with any of these enzymes show "respiratory control," that is, an increased oxidation rate on addition of an uncoupler (proton ionophore). Some of these enzymes, such as those from two types of thermophilic bacteria, PS3 (Sone and Hinkle, 1982) and T. thermophilus (Sone et al., 1983b; Yoshida and Fee 1984), are known to pump H^+ in addition to transferring electrons

Table II
Subunit Structures and Proton Pump Activities of aa_3-Type Bacterial Cytochrome Oxidases

Organism	Subunit M_r (kDa)			H^+ pump activity	References
	I	II	III		
Thermophilic bacterium PS3	56[a]	38 (c)	22	+	Sone and Yanagita, 1982; Sone and Hinkle, 1982
Bacillus subtilis	57	37	21		de Vrij et al., 1983
Bacillus firmus	56	40	14 (c)		Kitada and Krulwich, 1984
Thermus thermophilus	55	33 (c)			Fee et al., 1980; Sone et al., 1983;
	71[a]	34 (c)		+	Hon-nami and Oshima, 1984
Paracoccus denitrificans	45	28			Ludwig and Schatz, 1980; Solioz et al., 1983
	62[b]	33[b]	31[b]		
Nitrobacter agilis	45	31		+	Yamanaka et al., 1981; Sone et al., 1983b
Thiobacillus novellus	32	23		−	Yamanaka and Fujii, 1980
Rhodobacter sphaeroides	45	37 (35)		−	Gennis et al., 1982
Pseudomonas AM1	50	32		−	Fukumori et al., 1985a; Sone et al., 1987a

[a]Obtained from Ferguson plot.
[b]Obtained from DNA sequencing (Raitio et al., 1987).

across the membrane. Of the enzymes from mesophilic bacteria, only that from *P. denitrificans* was reported to show pumping H^+ (Solioz *et al.*, 1982), after several failures (Ludwig, 1980).

Several aerobic and facultative anaerobic bacteria are known to use alternative oxidases containing cytochromes, such as cytochrome *o* and cytochrome *d*. These enzymes are also called cytochrome oxidases.

4.2. Reconstitution of Proton Pump Activity

The thermophilic *Bacillus* PS3, isolated from a hot spring in Japan, contains aa_3-type cytochrome oxidases, which are composed of three subunits. This enzyme was reconstituted into proteoliposomes with soybean phospholipids by the freeze–thaw–sonication method (Sone and Hinkle, 1982; Sone and Yanagita, 1984). As shown in Figure 5, H^+ ejection coupled with oxidation of cytochrome *c* was measurable in the presence of K^+ *plus* valinomycin. The H^+/e^- ratio obtained (maximal amount of ejected H^+/ferrocytochrome *c* added) was close to 1, but simultaneous measurement of H^+ ejection and cytochrome *c* oxidation with a relatively small amount of the enzyme gave an H^+/e^- ratio of 1.2–1.4 at the beginning of the reaction. In the presence of a proton ionophore, the pH change (not shown) was almost similar to the spectrophotometer trace (B), indicating cytochrome *c* oxidation. This was due to scalar H^+ uptake in the reaction, since $2H^+$ are utilized for the formation of H_2O. The fact that the velocity of oxidation did not differ in the presence and absence of the proton ionophore (not shown) indicated that control of respiration did not take place on addition of such a small amount of substrate in the presence of valinomycin. The proton pump activity was inhibited by DCCD without significantly changing oxidase activity (Sone and Hinkle, 1982), as was the mitochondrial enzyme (Casey *et al.*, 1980).

The enzymes from *T. thermophilus* (Sone *et al.*, 1983b; Yoshida and Fee, 1984) pump protons, while those from the *N. agilis* (Sone *et al.*, 1983b) and *Ps.* AM1 (Sone *et al.*, 1987a) do not. Gennis *et al.* (1982) reported that when the purified *R. sphaeroides* enzyme was reconstituted into proteoliposomes by a cholate dialysis method, it did not pump H^+ in spite of a high respiratory control ratio. Both proton and oxidase activities of the *T. thermophilus* enzyme were not inhibited by DCCD (Hon-nami and Oshima, 1984).

It has been shown that after the bovine heart enzyme (Sone and Nicholls, 1984) or the PS3 enzyme had been preincubated at a certain temperature (about 43°C for the bovine enzyme and 50–55°C for the thermophilic enzyme), they did not show H^+ pump activity when reconstituted into proteoliposomes, whereas their protonophore-stimulatable oxidation activity proceeded almost as before pretreatment (showing respiratory control). The fact that the H^+ pump activity was more labile than oxidase activity suggests that some cytochrome oxidases of mesophiles may lose H^+ pump activity during purification, as well as suggesting that the proton pump activity is not a direct result of electron transfer, such as postulated by Mitchell's loop theory (Mitchell, 1966, 1976). One method to test the possibility of inactivation of the proton pump during preparation of the enzyme is to solubilize the membrane fraction from the bacterium of interest with heptylthioglucoside (or octylglucoside) and then to reconstitute the resulting soluble fraction into vesicles (Sone, 1986). Figure 6 shows the traces with a pH meter upon pulse with a reduced cytochrome c pulse of liposomes reconstituted from the extract of *B. caldolyticus*, *Ps.* AM1, *N. agilis*, and air-limited PS3 cells. No H^+ pumping activity was observed in the case of *Ps.* AM1 or *N. agilis*, while the membrane extract of *B. caldolyticus* appeared to pump H^+. The membrane extract of vigorously aerated PS3 cells containing

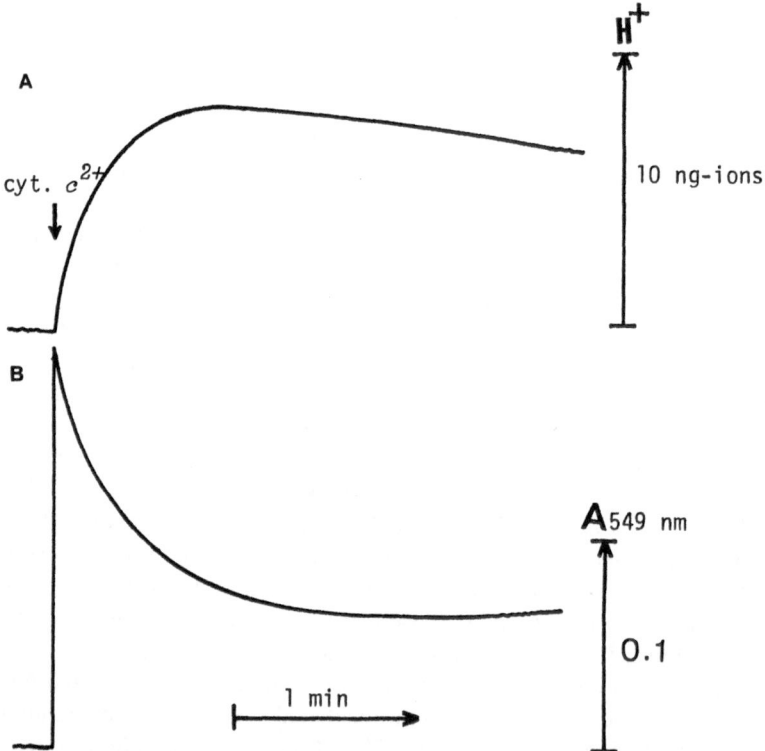

FIGURE 5. Time course of H^+ transport (A) and cytochrome c oxidation (B), when proteoliposomes reconstituted from purified PS3 cytochrome aa_3-type oxidase were pulsed with ferrocytochrome c (8.2 nmol from *Candida krusei*) in the presence of K^+ and valinomycin (see Sone and Yanagita, 1984).

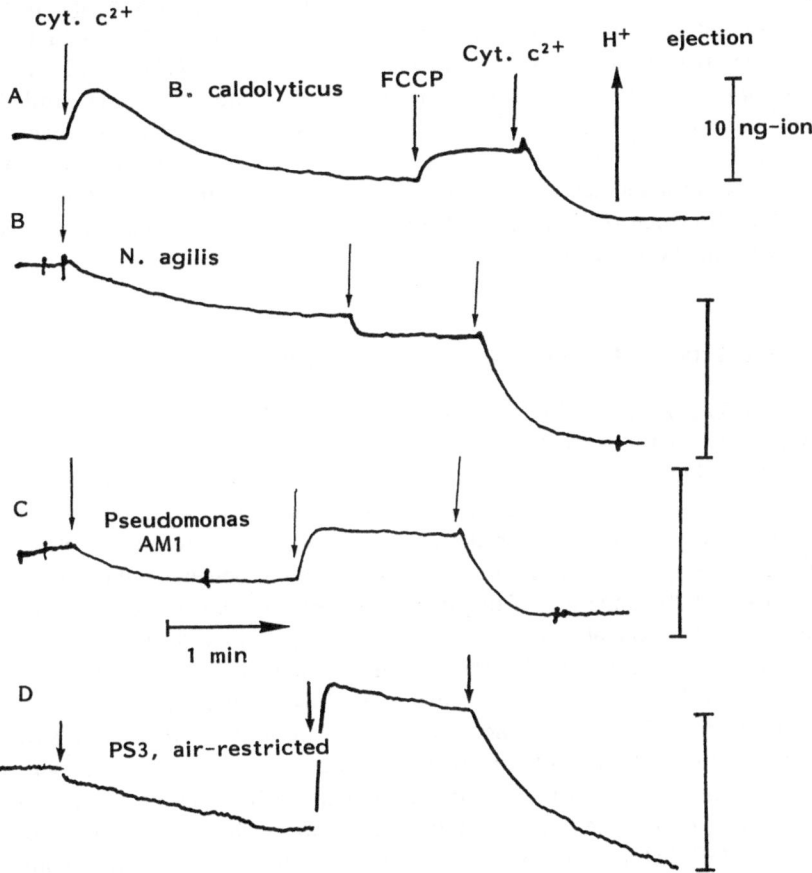

FIGURE 6. Measurement of proton pump activity of liposomes reconstituted from heptylthioglucoside-soluble fraction of membranes. Traces of the external pH on a ferrocytochrome c pulse in the presence of K^+ and valinomycin are shown (see Sone, 1986).

cytochrome aa_3-type oxidase showed H^+ pumping, while that of air-limited PS3 cells containing o-type oxidase did not (Sone, 1986). These vesicles showed a three- to four-fold increase in respiration on addition of FCCP. These results indicate that the enzymes of *N. agilis* and *Ps.* AM1 lose H^+ pumping activity even upon solubilization with heptyl thioglucoside, or that they do not pump H^+ intrinsically. The studies measuring H^+/O ratio (Section 2) showed that *Ps.* AM1 does not pump H^+ intrinsically. Thus, it seems likely that some bacterial aa_3-type cytochrome oxidases translocate electrons across the membrane without concomitant proton pumping, while others contain both functions. A similar difference is also found in the bacilli *B. subtilis* D473 and *B. magaterium* D440, which appears not to pump H^+ at site 3 (Table I), while *T. thermophilus* and PS3 pump H^+. Hence, the question as to why some bacteria have lost (or did not gain) the energy-yielding step at site 3 arises. For photosynthetic bacteria and several *Bacillus* species living in a nutrient-rich environment, the efficiency of utilization of nutrients may not be so important as the velocity to uptake nutrients.

4.3. Properties of Cytochrome aa_3-Type Oxidases

The PS3 enzyme is composed of three subunits. These three subunits seem to correspond to the three largest subunits of mitochondrial cytochrome oxidase : PS3 subunit I has a very similar amino acid composition to mitochondrial subunit I and exhibits abnormal behaviors on SDS-PAGE, such as high dependency of its mobility upon the gel concentration and formation of a soluble aggregate upon heating in the presence of sodium dodecyl sulfate (Ludwig and Schatz, 1980; Sone and Yanagita, 1982). PS3 subunit II, to which cytochrome *c* binds covalently, cross-reacts immunologically with yeast mitochondrial subunit II (Ludwig, 1980). PS3 subunit III binds to DCCD with concomitant loss of H^+ pumping activity (Sone and Hinkle, 1982), as observed with subunit III of the bovine enzyme. The enzyme from *B. subtilis* has a similar subunit structure but does not contain heme c (de Vrij *et al.*, 1983). In contrast, all purified enzymes from mesophilic gram-negative bacteria are composed of two subunits, except the enzyme of *R. sphaeroides,* which gives a third band on SDS-PAGE that appears to be a proteolytic product of subunit II (Gennis *et al.*, 1982). In these two-subunit enzymes, the larger subunit corresponds to mitochondrial subunit I and the smaller one to subunit II judging from their properties. Then the question arises of whether these enzymes consist intrinsically of two subunits. Saraste *et al.* (1986) reported the DNA sequence for subunit III (corresponding to the mitochondrial subunit III) in the cytochrome oxidase gene of *P. denitrificans*. Loss of subunit III during the purification procedure may be responsible for the low or negligible H^+ pumping activity. Raitio *et al.* (1987) also reported that the genes for the three subunits are in two operons with four unidentified open reading frames, as well as their DNA sequences. The two-subunit *T. thermophilus* enzyme that pumps protons (Sone *et al.*, 1983b; Yoshida and Fee, 1984) is dicyclohexyl carbodiimide-resistant (Hon-nami and Oshima, 1984), unlike the mitochondrial and PS3 enzymes. Fee *et al.* (1986) claim that subunit I of *Thermus*, which is larger than subunit I from other sources, contains all the chromophores (two heme a's, Cu_A and Cu_B), and 33-34 kDa component (subunit II) is cytochrome c_1-like protein.

The spectra of the oxidized and reduced forms of the PS3 enzyme are very similar to those of the 1 : 1 cytochrome c–cytochrome aa_3 complex of mitochondrial enzymes, except that a small and broad peak of the oxidized form at 840 nm, known to be due to one of the copper atoms, is replaced by a smaller peak at about 780 nm. Most of the bacterial enzymes show the band at 840 nm. The spectrum of the CO-reduced form was also similar to that of the mitochondrial enzyme (Sone and Yanagita, 1982; Sone et al., 1983a). Ligands such as NaN_3 and HCN inhibit the PS3 enzyme with concomitant spectral shifts with slightly lower affinities (Sone and Nicholls, 1984). Analysis by flash photolysis of the CO-reduced form of the PS3 enzyme at low temperature (triple trap experiment of Chance et al., 1975) showed a compound A-like spectrum at $-113°C$ (Sone et al., 1984a). The PS3 enzyme also exhibits resting-pulse conversion (Sone et al., 1984b). These studies indicate that bacterial enzymes correspond to mitochondrial core subunits, and the intermediate steps and the reaction mechanism are probably the same. An excellent short review on the properties of bacterial cytochrome oxidase has recently been published (Ludwig, 1987), in addition to previous ones (Ludwig, 1980; Poole, 1983).

4.4. Cytochrome o-Type and d-Type Oxidases

Instead of aa_3-type cytochrome oxidase, several bacteria produce cytochrome o, especially when the oxygen concentration of the culture medium is reduced. Cytochrome o is a group of B-type cytochromes which utilize O_2 as the terminal oxidant, and thus shows a characteristic absorption spectrum upon binding CO. Bacteria such as E. coli synthesize cytochrome o as the terminal oxidase, even under highly aerobic conditions, and cytochrome d under semi aerobic conditions.

Cytochrome o-type oxidase has been prepared as a cytochrome b_{562}–cytochrome o complex with two subunits (55 and 33 kDa) or four subunits (55, 35, 22, and 17 kDa). Kita et al. (1984a), using E. coli K12, extracted their membranes with Triton X-100 and purified the complex by chromatography on ion-exchange and gel filtration columns, while Matsushita et al. (1984) used E. coli B and extracted their membranes with octylglucoside and purified the oxidase by chromatography on ion-exchange and gel filtration columns. Matsushita et al. found four bands after electrophoresis in the presence of dodecyl sulfate when visualized by silver stain, while two bands appeared by usual coomassie blue stain.

The enzyme catalyzed ubiquinol oxidation with concomitant formation of a membrane potential of as much as 140–150 mV, when reconstituted into liposomes by the freeze–thaw–sonication method (Kita et al., 1982) or by the octylglucoside-dilution method *plus* freeze–thaw–sonication (Matsushita et al., 1984). The fact that $E'm$ of this cytochrome b_{562}–cytochrome o complex is 120 mV (pH 7.4) and shows pH dependency of -60 mV/pH suggests that the enzyme induces translocation of H^+, like aa_3-type oxidase, in addition to transfer of electrons across the membrane (Kita et al., 1984a). However, from measurements of the external and internal pH, Matsushita et al. (1984) concluded that no vectorial H^+ translocation takes place in proteoliposomes with this enzyme. The energetics with whole cells, especially the H^+/O ratio (Section 2), appears to be consistent with this result, since $2H^+$ translocation probably takes place at site 1 (NADH : quinone oxidoreductase level). This cytochrome o-type oxidase was reconstituted into the planar lipid bilayer and open-circuit membrane potentials were measured directly. The membrane potential generated during ubiquinol-1 oxidation became zero at an applied voltage of 150 mV (Hamamoto et al., 1985). With PS3 aa_3-type oxidase, this equilibrium voltage is above 200 mV, which also breaks the lipid bilayer (T. Hamamoto and N. Sone, unpublished observation).

Membrane-bound cytochrome o was solubilized and purified from *Methylophilus methylotrophus* (Carver and Jones, 1983), PS3 (Baines et al., 1984), and *Gluconobacter suboxydans* (Matsushita et al., 1987b). The *G. suboxydans* enzyme was reported to generate a membrane potential similar to the *E. coli* enzyme. The enzyme from *R. capsulata* was claimed to pump protons, since the enzyme in liposomes shows a proton extrusion of 0.24 H^+/e^- with ferrocytochrome c pulse in the presence of K^+ *plus* valinomycin (Hüdig and Drews, 1984). However, the reaction was carried out at pH 7.2 and this H^+ extrusion may be due to pK_a change upon oxidation of cytochrome c.

Cytochrome d-type oxidase was prepared from semiaerobically grown *E. coli* (Kita et al., 1984b; Lawrence et al, 1986; Miller and Gennis, 1983). This enzyme has two kinds of heme—protoheme and d heme (chlorin), as prothetic groups, and was composed of two kinds of subunit (50–51 and 26–28 kDa). Proteoliposomes reconstituted from this enzyme by a freeze–thaw–sonication procedure generate a membrane potential (-150 mV) coupled with ubiquinol-1 oxidation and dissipatable by addition of a proton ionophore (Kita et al., 1984b).

In *E. coli*, in which cytochrome bc_1 complex has not been found, cytochrome o-type or d-type oxidase catalyzes ubiquinol oxidation. Many flavoprotein dehydrogenases and several dehydrogenases requiring pyrroloquinoline quinone as a coenzyme, such as D-glucose dehydrogenase of *E. coli* (Matsushita et al., 1987a), reduce UQ. However, it is not known what is the natural substrate for cytochrome o-type oxidase in bacteria, such as *P. denitrificans* and PS3, which produce it with a higher concentration of B-type and C-type cytochromes including cytochrome bc_1 complex instead of the aa_3-type oxidase under an air-limited condition (Sone et al., 1983a). Thus, in these bacteria one plausible candidate for the substrate for o-type oxidase is a c-type cytochrome (as shown in Figure 1). A reduced cytochrome c pulse with proteoliposomes containing crude cytochrome o-type oxidase (Figure 6D) showed rapid oxidation of cytochrome c with concomitant formation of a membrane potential, without translocating H^+ (Sone, 1986). Meanwhile, it is worth mentioning that if there is cytochrome c-linked dehydrogenase such as methanol dehydrogenase in *M. methylotrophus*, a proton translocating loop operates as follows:

$$\text{(periplasmic side)} \quad \begin{array}{c} CH_3OH \\ 2H^+ + CO_2 \end{array} \Bigg\rangle \xrightarrow[c\ b\ o]{2e^-} \Bigg\langle \begin{array}{c} \tfrac{1}{2}O_2 + 2H^+ \\ H_2O \end{array} \quad \text{(cytoplasmic side)}$$

In summary, these cytochrome o-type and d-type oxidases are likely to catalyze transmembrane electron transfer, which results in generation of a pmf by virtue of e^- movement and alkalinization of cytoplasmic space due to H_2O formation from $2H^+ + 2e^- + \tfrac{1}{2}O_2$.

5. CYTOCHROME bc_1 COMPLEXES

5.1. Properties

The second segment of the electron transfer chain (complex III) catalyzes quinol-dependent cytochrome c reduction. The mitochondrial enzyme is composed of nine or eleven subunits, including cytochrome b and cytochrome c_1, and iron sulfur (FeS) protein, and is thus called cytochrome bc_1 complex (see Weiss, 1987, for review). As also reviewed by Hauska et al. (1983), very similar enzymes with much simpler subunit structures were purified from photosynthetic bacteria, cyanobacteria, and spinach chloroplasts. Although the latter two enzymes are usually called cytochrome b_6–cytochrome f complexes because of the slightly different properties of their b-type and c-type cytochromes, their homology is clear, especially because the amino acid sequences of the enzymes of chloroplast (Widger et al., 1984), *R. sphaeroides* (Gabellini and Sebald, 1986; Gabellini et al., 1985), and *P. denitrificans* (Kurowski and Ludwig, 1987) are homologous. Recently, bc_1 complexes have been purified from *P. denitrificans* (Yang and Trumpower, 1986) and from the thermophilic bacterium PS3 (Kuto and Sone, 1988). These bc_2 complexes were shown to form supercomplexes with aa_3-type cytochrome oxidases (Berry and Trumpower, 1985; Sone et al., 19887b).

Figure 7 shows the subunit structure of PS3 enzyme. Protein staining showed the presence of four polypeptides (29, 23, 21, and 14 kDa) and heme staining; the largest (29 kDa) and the third largest (21 kDa) subunits carry hemes. The spectra of these bands on application of large amounts of sample are shown in Figure 8. The largest subunit (29 kDa) has an α band at 553 nm with a shoulder at 547 nm, indicating that it is cytochrome c_1 having the spectral properties known for cytochrome f, while the third band shows an oxy hemoglobin type spectrum, suggesting that it is denatured cytochrome b (b_6). There is some evidence that the second largest subunit (23 kDa) has a Riesk-type FeS center. The *Paracoccus* enzyme has only three subunits, but its cytochrome c_1 has a large molecular mass (62 kDa on SDS-PAGE, 44,654 Da from the DNA sequence). Table III summarizes the numbers of subunits and molecular masses of the subunits carrying chromophores. With the exception of *Paracoccus* cytochrome c_1, the molecular masses of cytochrome c_1 and FeS protein are very similar. On the contrary, cytochrome b (b_6) can be classified into two groups; the b group (39–44 kDa) including the enzymes of mitochondria and of gram-negative bacteria, and the b_6 group (21–23 kDa) including those of chloroplasts, cyanobacteria, and probably gram-positive bacteria. A recent sequence study showed that the 23 kDa chloroplast cytochrome b_6 appears to have five membrane-spanning segments that are evidently homologous to the first five (of nine) segments of cytochrome b of mitochondria (Widger *et al.*, 1984). Four histidine residues for two protohemes are also conserved. The rest (C-terminal

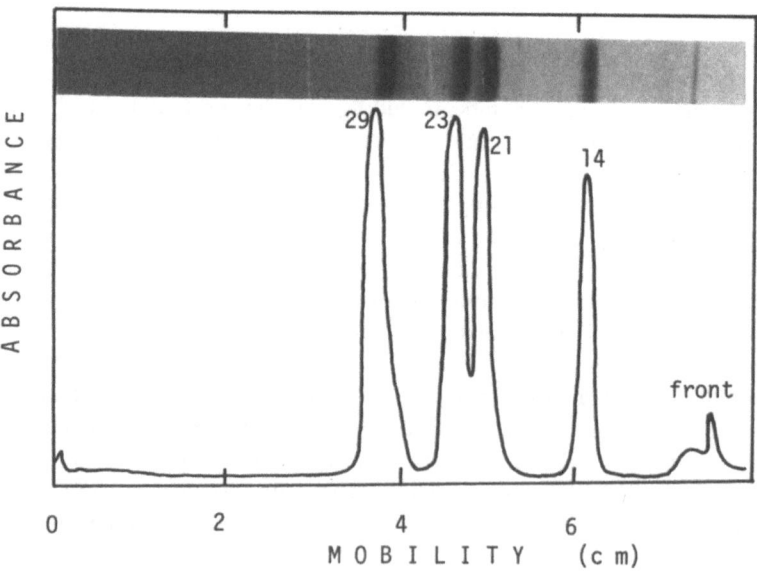

FIGURE 7. SDS–gel electrophoresis showing subunit pattern of purified PS3 cytochrome bc_1 complex. The number in the figure shows M_r of each subunit in kDa.

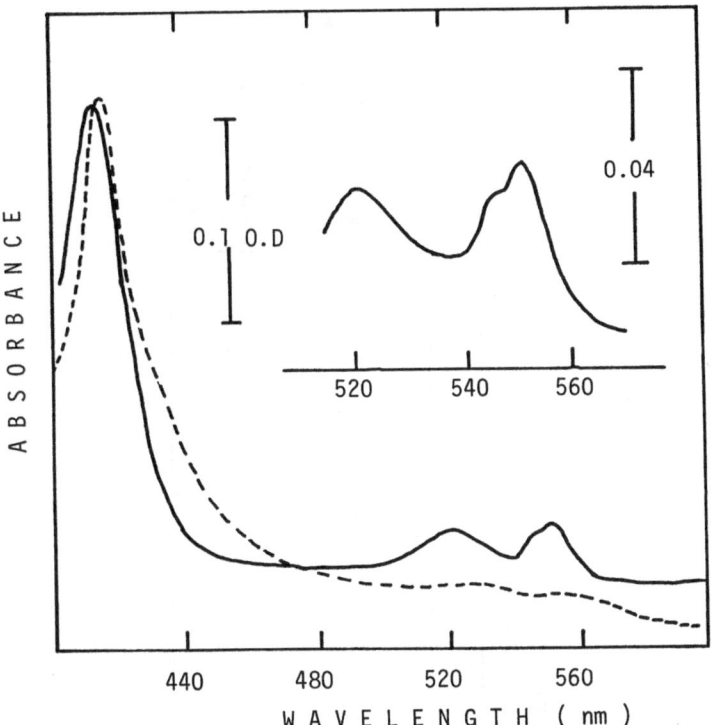

FIGURE 8. Absorption spectra of the bands separated on the SDS–gel electrophoresis. ——, The 29 kDa subunit showing cytochrome c_1; ---, the 21 kDa subunit showing denatured cytochrome b.

part) of the mitochondrial cytochrome b protein has homology to the smallest 16 kDa polypeptide. The chromophore content and the subunit stoichiometry show that the monomer enzyme contains one cytochrome b protein with two protohemes, one cytochrome c_1, and one Fe_2S_2-type iron sulfur protein and possibly another subunit(s). Iba *et al.* (1985), however, isolated a bc_1 complex with a smaller cytochrome b (17 kDa) from *R. sphaeroides*. The sequence of this cytochrome b has only two histidine residues for protoheme binding and is not homologous to the cytochrome b of higher molecular weight (Iba *et al.*, 1987).

The midpoint potential at pH 7.0 *($E'm$)* and species of quinone are summarized in Table IV. The mitochondrial cytochrome b and that of *R. sphaeroides* show two $E'm$ values of components named b_l (b-565) and b_h (b-562) of mitochondria. The chloroplast cytochrome b_6 has also been reported to show two $E'm$ values, which are much lower than those for mitochondria. The PS3

Table III
Subunit Structures of Cytochrome bc_1 Complex

Organism	Number of subunits	M_r (kDa)			References
		Cytochrome b	Cytochrome c_1	FeS	
PS3	4	21	29	23	Sone et al., 1987b
Paracoccus denitrificans	3	39	62	20	Yang and Trumpower, 1986
Rhodobacter sphaeroides GA	3	42	34	25	Hauska et al., 1983
Rhodobacter sphaeroides R26	4	43	30	24	Yu et al., 1984
Spinach chloroplast	4	23	34	20	Hurt and Hauska, 1981
Anabaena variabilis	4	23	31	22	Krinner et al., 1982
Bovine mitochondria	11	44	28	25	Schägger et al., 1986

Table IV
Midpoint Potentials of Redox Components and Species of Quinone in Cytochrome bc_1 (b_6f) Complexes[a]

Component	Bovine heart mitochondria	*Rhodobacter sphaeroides*	Spinach chloroplast	Thermophile PS3
Cytochrome b_l	−34	−60	−170	−190
Cytochrome b_h	93	50	−50	−130
Cytochrome c_1 or f	232	285	340	200
Rieske FeS	280	285	290	
Type of quinone	UQ	UQ	PQ	MK

[a] The E'_m is given in mV. The values for PS3 cytochrome(s) b_l and b_h may be identical (in this case, −160 mV; Kutoh an Sone, 1988). The other data were taken from Hauska *et al.* (1983).

enzyme also shows a low $E'm$ value of −160 mV, or possibly −190 and 120 mV. In the cases of PS3 cytochrome b and chloroplast b_6, the difference in the α peaks, if any, is not prominent, being as small as 2 nm. PS3 cytochrome c_1 shows a very similar spectrum to cytochrome f (as already mentioned) but its $E'm$ is lower than that of chloroplasts. These characteristics of the bc_1 (b_6f) complexes suggest that these complexes are probably the most conserved energy-transducing complexes in the electron transfer chain and can be classified into two subgroups; the mitochondrial type and the chloroplast type. The enzyme from the gram-positive thermophile seems to belong to the chloroplast type, but a few of its properties are intermediate between those two types.

5.2. Reconstitution

Isolated bc_1 complexes from various sources, such as beef heart mitochondria, spinach chloroplasts, *Anabaena variabilis,* and *R. sphaeroides*, were reconstituted into proteoliposomes by a sonication procedure in the presence of octylglucoside and cholate (Hurt *et al.*, 1982, 1983). The proteoliposomes showed low quinol-dependent cytochrome c reduction which is stimulated by the addition of the protonophore FCCP or a combination of ionophores, for example, valinomycin *plus* nigericin, that had a similar action. Measurement of H^+ movement showed the release of about $2H^+$ upon transfer of one electron to cytochrome c in the presence of K^+ *plus* valinomycin. Addition of either nigericin, FCCP, or Triton X-100 reduced the H^+/e^- ratio to about 1, indicating scalar release of H^+ from reduced quinone. These data are summarized in Table V. All four bc_1 (b_6f) complexes showed vectorial H^+ translocation of $1H^-/e^+$ and scalar H^+ release of $1H^+/e^-$ ($2H^+$ release outside, cytochrome c side). This is the same stoichiometry as found in mitochondria (see review of

Table V
Proteoliposomes Containing Cytochrome $bc_1(b_6f)$ Complexes Showing Respiratory Control and Proton Translocation[a]

Proteoliposomes containing the complex from	Q:cytochrome c activity	Stimulation by val/nig	Apparent H^+/e^- ratio		
			Control	+val/nig	+FCCP	+Triton X-100
Bovine mitochondria	18.7	3.2	1.98	1.26	n.t.	0.92
Rhodobacter sphaeroides	7.7	2.2	2.2	1.57	1.07	n.t.
Anabaena variabilis	9.3	3.2	1.85	1.10	n.t.	n.t.
Spinach chloroplast	14.3	3.5	1.98	1.15	1.16	1.19

[a] Q:cytochrome c/plastcyanin reductase activity was expressed in μmol/hr per cytochrome c_1 (f), and for the H^+ ratios, the initial rates of H^+ release were devided by initial rates of cytochrome c/plastcyanin reduction; n.t. stands for not tested (see Hurt et al., 1982).

Boyer et al., 1977) and thus the same reaction mechanism as the Q-cycle model (Mitchell, 1976) may be applicable. Several difficulties in the Q-cycle model are removed in the newer semiquinone cycle model (Wikström and Krab, 1986) and the Q-cycling dimeric complex model (de Vries, 1986). Quinone may, however, bind firmly on the complex. The proton translocation at the Q binding center is supposed to be at the heart of proton pumping in this complex. The fact that DCCD inhibits proton translocation of the mitochondrial complex III, without changing electron transfer activity much as in the case of complex IV, may support this idea (Beattie, 1986). The cytochrome bc_1 complex from gram-positive thermophile PS3 can be reconstituted into proteoliposomes with phospholipids of this bacterium by a freeze–thaw–sonication procedure (N. Sone, unpublished observation). These proteoliposomes showed generation of a membrane potential and scalar and vectorial H^+ movements, findings that indicate that bc_1 complexes, including b_6f complexes, are very similar not only in electron transfer but also in energy transduction.

6. NADH DEHYDROGENASES AND COMPLEX I

6.1. Membrane-Bound NADH Dehydrogenases

Mitochondrial NADH : quinone oxidoreductase (complex I) is known to be a very complicated membrane enzyme containing more than 20 subunits with FMN and several FeS centers (see Hatefi, 1985, for review). There are several NADH dehydrogenases, which reduce quinone (NADH : quinone oxidoreductases), free in the cytosol or loosely attached to cytoplasmic membranes, but the bacterial counterparts of complex I should be hydrophobic membrane proteins for energy transduction and for quinone reduction. NADH dehydrogenases have been purified from several bacteria such as *E. coli, Rhodobacter capsulata,* and several *Bacillus* species. The *ndh* gene of *E. coli* has been cloned and amplified, and the overexpressed NADH dehydrogenase has been purified. The enzyme is a single polypeptide with a molecular weight of 47,303 Da with tightly bound FAD but without Fe (Jaworowski *et al.*, 1981a). The primary amino acid sequence has a polarity of 43.4%. This value does not seem sufficient for an intrinsic membrane component such as a respiratory complex, but it is sufficient for attachment or anchoring of the enzyme to the membrane, since there are some hydrophobic portions (Jaworowski *et al.* 1981a, b). The membrane-bound *R. capsulata* enzyme solubilized with deoxycholate and purified by chromatography in the presence of cholate is an oligomer of about 97 kDa composed of 15 kDa monomer (Oshima and Drews, 1981).

NADH dehydrogenases have been purified from mesophilic *B. subtilis*

(Bergsma et al., 1982b), thermophilic B. stearothermophilus (Mains et al., 1980) and B. caldotenax (Kawada et al., 1981), and alkalophilic Bacillus YN-1 (Hisae et al., 1983). All these enzymes have FAD as a prosthetic group except the enzyme from B. stearothermophilus which has FMN instead (Mains et al., 1980). No FeS center has been reported. The subunit structures of these enzymes are simple: 63 kDa in B. subtilis, 43 kDa in B. stearothermophilus, 44 kDa in B. caldotenax, and 65 kDa in the alkalophile YN-1. These enzymes are all membrane bound and the (partially) purified enzymes require phospholipids but do not appear very hydrophobic. Moreover, proteolytic digestion of NADH dehydrogenase from the alkalophile YN-1 (Xuemin et al., 1985) resulted in loss of its ability to bind to liposomes. Since even after extensive proteolysis an energy-transducing complex such as PS3 cytochrome oxidase was still incorporated into liposomes and showed both oxidase and H^+ pump activities (Yanagita et al., 1983), this feature of NADH dehydrogenase suggests that these enzymes are not coupled with H^+ translocation like complex I. In fact there is no evidence available that any of these bacterial NADH dehydrogenases pump H^+ or generate a membrane potential. Thus, these NADH dehydrogenases probably reduce only quinone, as do the high potential membrane-bound dehydrogenases, such as succinate and glycerol-3-phosphate dehydrogenases, which bypass site 1.

6.2. Evidence for Proton Pumping NADH : Quinone Oxidoreductase

Some membrane-bound bacterial NADH dehydrogenases should be counterparts of mitochondrial complex I. An H^+ pumping NADH dehydrogenase is most likely to be present in bacteria whose respiratory chains are very similar to that of mitochondria, such as P. denitrificans, and in fact this bacterium shows a high H^+/O ratio (Section 2). The gram-negative thermophilic bacterium T. thermophilus also shows high H^+/O ratio (McKay et al., 1982; Quilter and Jones, 1982); its respiratory chain is reported to be similar to that of mitochondria (Fee et al., 1986). An H^+ pumping NADH dehydrogenase might also be present in some bacteria such as E. coli which show an H^+/O ratio of as low as 4, if sites 2 and 3 are not present or are incomplete. E. coli, which has no bc_1 complex or aa_3-type cytochrome oxidase, may have an NADH dehydrogenase containing FMN and FeS centers as a counterpart of complex I.

Experiments using inside-out membrane vesicles from E. coli clearly showed that energy-transducing NADH : ubiquinone oxidoreductase is present besides the non-energy-transducing type (described in Section 6.1) and is characterized by its requirement for deamino-NADH and its high sensitivity to respiratory inhibitors, such as 3-undecyl-2-hydroxy-1,4-naphthoquinone, piericidin A, and myxothiazol (Matsushita et al., 1987c).

Recently, an NADH dehydrogenase composed of two subunits (48 and 25 kDa) was purified from *P. denitrificans* (George and Ferguson, 1984). An antibody raised against this preparation cross-reacted with a 51 kDa subunit of mitochondrial complex I (George et al., 1986). Moreover, an antibody against a 49 kDa subunit of the mitochondrial complex I cross-reacted with a 46 kDa polypeptide in the whole membrane of *P. denitrificans*, suggesting that the two-subunit preparation described above may have lost this component during the preparation. Yagi (1986) also purified NADH : ubiquinone reductase from *P. denitrificans*. He used NaBr for solubilizing the enzyme and a column of NAD–agarose for chromatographic purification. The final preparation contained six different kinds of subunits (73, 70, 66, 55, 48, and 25 kDa) with four minor (contaminating?) polypeptides, which contained one FMN and about twelve iron atoms and acid-labile sulfur. The antisera against bovine complex I and its subunit (iron protein) cross-reacted with 70 and 48 kDa components and 48 and 25kDa components, respectively. These NADH : ubiquinol reductase preparations, however, were not rotenone sensitive, and reconstitution work into liposomes and measurement of proton pump activity have not been reported. More work is necessary to identify the bacterial complex I, although these preliminary data seem promising.

6.3. Na^+ Pumping NADH : Quinone Oxidoreductase

A marine bacterium, *Vibrio alginolyticus*, generates a membrane potential even when its membrane is made permeable to protons by addition of a proton ionophore under slightly alkaline conditions, pH 8.5 (Tokuda and Unemoto, 1982). Its respiratory chain NADH oxidation requires Na^+ for maximum activity, whereas its glycerol-3-phosphate oxidation does not. One of the membrane NADH dehydrogenases of this bacterium was solubilized with an alkylpolyoxyethylene ether detergent and was purified by ion-exchange and gel filtration chromatographies (Hayashi and Unemoto, 1987). The purified enzyme, composed of three subunits (52, 46, and 32 kDa), contains one FAD and one FMN per 150 kDa and shows Na^+-dependent NADH : quinone oxidoreductase activity. A slightly less pure preparation of the same enzyme could be reconstituted into liposomes with soybean phospholipids by the octylglucoside-dilution method (Tokuda, 1984). On addition of NADH, these proteoliposomes take up Na^+ and generated a membrane potential of up to 150 mV. These results clearly indicate that this three-subunit NADH dehydrogenase containing both FAD and FMN as chromophores is an electrogenic Na^+ pump. Tokuda and Unemoto (1982, 1984) also found that mutants lacking this enzyme still showed NADH–quinol reductase activity. The latter activity was not Na^+ dependent and was not inhibited by 2-heptyl-4-hydroxyquinoline-*N*-oxide (HOQNO), which inhibits Na^+-dependent NADH oxidation. Thus, *V. alginolyticus* retains two kinds of NADH : quinone oxidoreductases. Na^+ pumping NADH dehydrogenase is reported to be widely distributed in marine bacteria (Udagawa et al., 1986), and the gene may be located in plasmids (Tokuda et al., 1987). The relationship between this enzyme and H^+ pumping complex I is not known at present.

7. ENERGY-TRANSDUCING COMPONENTS OTHER THAN COMPLEXES I–IV

7.1. Anaerobic Electron Transfer Systems

Some bacteria can grow anaerobically with fumarate as the terminal oxidant. The electron transfer chain of *Wolinella succinogenes* has been analyzed in detail by Kröger et al. (1986). It is composed mainly of hydrogenase, formate dehydrogenase, MK, and fumarate reductase, as shown in Figure 9, and thus this is a typical example of a system for the extracytoplasmic oxidation of simple reductants, as described by Hooper and DiSpirito (1985).

Fumarate reductase from *W. succinogenes* is composed of one 76 kDa subunit bearing one FAD and one Fe_4S_4 center, one 31 kDa subunit bearing one Fe_2S_2 center, and two 25 kDa subunits bearing one protoheme, respectively (Unden and Kröger, 1981, 1982). Formate dehydrogenase is composed of three different subunits of 110, 25, and 20 kDa (Unden et al., 1983) in a ratio of 1 : 1 : 2. The 25 kDa subunit is a low potential cytochrome *b*, and the 110 kDa component, containing 9 nmole molybdenum and 160 nmole nonheme iron and sulfide per mg protein, reduced 2,3-dimethyl-1,4-naphthoquinone but not vitamin K_1 in liposomes (Unden and Kröger, 1983). The proteoliposomes catalyzing formate-dependent fumarate reduction were reconstituted from purified formate dehydrogenase and fumarate reductase with soybean P-lipids and vitamin K_1 (Unden et al., 1983).

In both enzymes, the cytochrome *b* subunits are hydrophobic, suggesting that these components span the membrane. Soluble fumarate reductase composed of subunits of 76 and 31 kDa reduced fumarate by viologen radicals but not by quinol (Unden and Kröger, 1981).

Formate-dependent fumarate reduction by inverted membrane vesicles prepared from *W. succinogenes* cells in a French press formed a pmf of as much as 180 mV (Mell et al., 1986). Measurement of pH in the presence of K^+ *plus* valinomycin (to reduce $\Delta\Psi$) and DCCD (to block ATP synthase) showed acidification of the external buffer, which gave an H^+/e^- ratio of about 1.0 (1.4 when extrapolated to zero concentration by changing the size of the fumarate pulse). Two explanations of these findings are possible; one for an H^+/e^- ratio of 1 and the other for that of 2. In the former case, MKH_2 reduced by formate dehydrogenase diffuses in the membrane and donates electrons to cytochrome *b* of fumarate reductase at the positive external side of the membrane (Figure

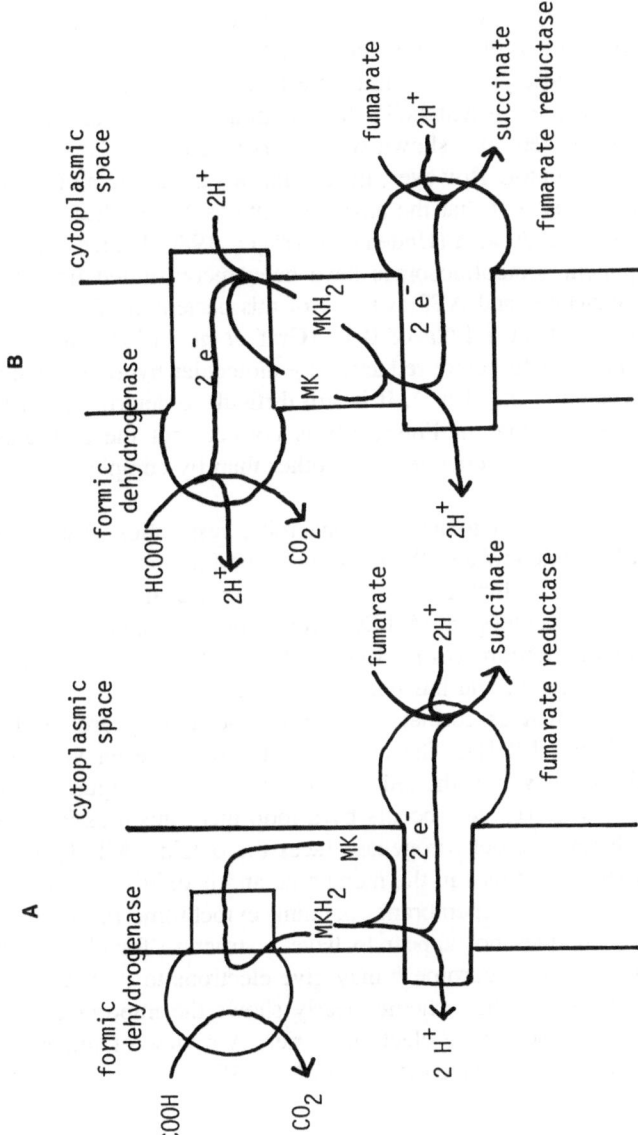

FIGURE 9. An anaerobic respiratory chain of *Wolinella succinogenes* showing proton translocation by loop mechanism (modified from Kröger et al., 1986). A, a scheme explaining $1H^+/e^-$ translocation; B, a scheme explaining $2H^+/e^-$.

9A). It is noteworthy that in this case, any MK reducing dehydrogenases can generate a pmf, irrespective of whether the reducing site of the enzyme faces the positive side or not. In the latter case, to explain the H^+/e^- of 2, MK should be reduced on the cytoplasmic side by electron transfer and should be oxidized by cytochrome b of the fumarate reductase on the positive side (Figure 9B). In both cases, MK plays a central role as a membrane-mobile hydrogen carrier, just as UQ does in Mitchell's loop hypothesis. Similar stoichiometry and electron transfer were also shown with H_2 as an electron donor and fumarate as an electron acceptor. However, little is known about the molecular structure of this enzyme, except that the enzyme contains Ni and has at least three subunits of 60, 58, and 30 kDa (Unden and Kröger, 1986; Unden et al., 1982).

Multicomponent proteoliposomes have been reconstituted from hydrogenase, fumarate reductase, and ATP synthase of this bacterium. These proteoliposomes synthesize ATP at a P/2e of 0.11 (Graf et al., 1985) and show MK (vitamin K_1)-dependent fumarate reduction by molecular hydrogen (Unden and Kröger, 1982; Unden et al., 1983). It is still difficult to determine whether the route in Figure 9A or that in Figure 9B is correct, but these studies show clearly that a third way to generate a pmf other than by complexes I and III is possible.

E. coli is also known to have an anaerobic respiration system utilizing fumarate and MK (Ingledew and Poole, 1984). The fumarate reductase is composed of four subunits, which are the 66 kDa subunit bearing FAD, a 27 kDa subunit bearing FeS centers, a 15 kDa hydrophobic subunit, and a 13 kDa hydrophobic subunit (Lemire et al., 1982). The DNA encoding for these subunits forms the *frd* operon and has been sequenced (Cole et al., 1985). Menaquinone reductases such as formate dehydrogenase and glycerol 3-phosphate dehydrogenase supply MKH_2. The catalytic sites of these enzymes face the cytoplasmic side. However, if the redox reactions shown in Figure 9A operate, proton translocation of $1H^+/e^-$ occurs by a loop mechanism as in *W. succinogenes*, in which formate dehydrogenase faces the outside; MKH_2 reduced by formate dehydrogenase moves in the membrane and is oxidized on the external side and gives electrons to membrane-spanning cytochrome b, since fumarate reductase of *E. coli* does not appear to have a structure for electron transfer across the membrane. Cytochrome b may give electrons to fumarate reductase on the cytoplasmic side. This scheme clearly shows the important role of cytochrome b as a transmembrane electron carrier. A review of fumarate reductase of *E. coli* has been published (Cole et al., 1985).

7.2. Transhydrogenase

Transhydrogenase catalyzes exchange of reducing equivalents between the two pyridine nucleotide couples, $NADH/NAD^+$ and $NADPH/NADP^+$. Because of their different degree of reducing states, oxidation of NADPH by NAD^+ generates a pmf, while the reverse reaction proceeds only when a pmf is imposed. This energy-transducing transhydrogenase has been found in several bacteria such as *E. coli* (Chetkauskaite and Grinius, 1979) and *P. denitrificans* (Asano *et al.*, 1967). Both sites for substrates are on the cytoplasmic side. The mitochondrial enzyme was reported to be composed of a single 97 kDa polypeptide but it has no known chromophore (Rydström, 1981). The proton-translocating *E. coli* enzyme was purified, which is composed of two subunits of 50 and 47 kDa (Clarke and Bragg, 1985). The nucleotide sequence was recently determined (Clarke *et al.*, 1986). Since the enzyme is energy transducing, this reaction site is sometimes called site 0. The stoichiometry is generally thought to be $1H^+/e$, but this is not certain.

In resting cells, NADPH is also used as an endogenous substrate as NADH, and the H^+/O ratio may be the sum of those of sites 0–3, rather than sites 1–3. Although Scholes and Mitchell (1970) formerly explained the ratio of 8 obtained for *P. denitrificans*, cytochrome *c* oxidase has now been shown to pump protons in addition to transferring electrons across the membrane at site 3. Since it is rather difficult to quantitate the contribution of NADPH oxidation to the O_2 pulse in resting cells, the contribution of site 0 will not be considered in the following section.

8. CONCLUDING REMARKS AND OUTSTANDING PROBLEMS

Only a few purified bacterial cytochrome aa_3-type oxidases (complex IV) have been shown to pump protons; the others, composed of two subunits, did not show proton pump activity but retained transmembrane electron-transferring activity. Many of these enzymes seemed to lose proton pump activity during purification, due to loss of subunit III. However, some of these enzymes, such as that of *Ps.* AM1, may not pump protons intrinsically.

The cytochrome bc_1 complexes (complex III) are the most conserved complexes in the bacterial electron transfer chain and can be classified into two subgroups, a mitochondrial type and a chloroplast type. All these bacterial complex III's seem to eject $4H^+$ to the outside and take up $2H^+$ on the cytoplasmic side. A few facultative anaerobes such as *E. coli* lack this complex but can oxidize UQ by a cytochrome *o*-type (or *d*-type) oxidase with concomitant $2H^+/O^-$ transport.

NADH dehydrogenases, even if they are membrane-bound NADH : Q oxidoreductases, are of two types: energy-transducing ones that are counterparts of complex I, and nonenergy conserving enzymes that merely reduce quinone and are not so hydrophobic.

These variations in proton translocation ability among the complexes forming

Table VI
H^+/O Ratios in Various Bacterial Respiration Systems with Endogenous Substrates in Comparison with Those of Beef Heart Mitochondria

Systems	Coupling site			H^+/O ratio[a]	
	1	2	3	Theoretical	Experimental
Mitochondria					
Wikström and Hinkle	2–3	4	2	9–10	—
Lehninger and Azzone	4	4	4	12	—
Bacteria					
Type 1 (e.g., *P. denitrificans*)	+	+	+	9–12	7.5–10.5
Type 2 (e.g., *Ps.* AM1)	+	+	−	6–8	5.9
Type 3 (e.g., *B. stearothermophilus*)	−	+	+	6–8	7.3
Type 4 (e.g., *B. subtilis*)	−	4	−	4	4.0
Type 5 (e.g., *E. coli*)	2	− 2	−	4–6	4.0

[a] Experimental H^+/O ratios are taken from that in Table I, and mitochondrial theoretical values mainly from the review by Wikström and Saraste (1984).

coupling sites and combinations of them may explain the H^+/O ratios shown in Table I, as summarized in Table VI. This summary may not explain all the data but seems to be the most comprehensive summary at present.

These respiratory complexes in bacteria (complexes III and IV) with or without (complex I) have much simpler subunit structures than their mitochondrial counterparts. However, the constituents and structures of the chromophores in complexes III and IV are almost the same as those of the mitochondrial complexes.

Besides these electron transfer complexes, intrinsic membrane proteins with electron carriers, such as cytochrome o-type and d-type oxidases, have been found in some bacteria. These components catalyze oxidation of substrates on the outside with release of H^+, electron transfer across the cytoplasmic membrane, and reduction of oxygen with uptake of H^+ on the cytoplasmic side. This relatively simple mode of generating a pmf may be widely distributed in cells that use simple extracytoplasmic reductants. Complexes III and IV also show transmembrane electron transfer activity, an array of cytochrome b's or metal centers composed of cytochrome aa_3 and two copper atoms driving electrons transmembranously.

The mechanism of proton translocation is not known at present. A loop mechanism such as the Q-cycle model is possible at site 2, where lipophilic low-molecular-weight quinones are involved. However, since this quinone may be firmly bound to a quinone-binding protein, the molecular mechanism of proton translocation in complex III may be similar to those of complexes I and

IV, which are redox-driven H^+ pumps. Special attention should be paid to the H^+-channel structure in the proteins, the dimer structure of the enzyme, and changes of prothetic groups in relation to proton movements. Studies on bacterial enzymes, which have simple subunit structures, may help in elucidating the mechanism(s) of proton translocation in these electron transfer complexes.

9. REFERENCES

Amzel, L. M., and Pedersen, P. L., 1983, Proton ATPases: Structure and mechanism. *Annu. Rev. Biochem.* **52**:801–824.

Asano, A., Imai, K., and Sato, R., 1967, Oxidative phosphorylation in *Micrococcus denitrificans*. II. The properties of pyridine nucleotide transhydrogenase. *Biochim. Biophys. Acta* **143**:477–486.

Baines, B. S., Hubbard, J. A. M., and Poole, R. K., 1984, Purification and partial characterization of two cytochrome oxidases (caa_3 and o) from the thermophilic bacterium PS3. *Biochim. Biophys. Acta* **766**:438–445.

Beattie, D. S., 1986, Is cytochrome bc_1 complex a proton pump?, probably yes. *J. Bioenerg. Biomembr.* **18**:1–20.

Bergsma, J., Maarten, B. M., Dongen, V., and Konings, N. W., 1982a, Purification and characterization of NADH dehydrogenase from *Bacillus subtilis*. *Eur. J. Biochem.* **128**:151–157.

Bergsma, J., Meihuizen, E. K., Oeveren, V. W., and Konings, N. W., 1982b, Restoration of NADH oxidation with menaquinones and menaquinone analogues in membrane vesicles from the menaquinone-deficient *Bacillus subtilis aroD*. *Eur. J. Biochem.* **125**:651–657.

Berry, E. A., and Trumpower, B. L., 1985, Isolation of ubiquinol oxidase from *Paracoccus denitrificans* and resolution into cytochrome bc_1 and cytochrome caa_3 complexes. *J. Biol. Chem.* **260**:2458–2467.

Boogerd, F. C., van Verseveld, H. W., Torenvliet, D., Braster, M., and Stouthamer, A. H., 1984, Reconstitution of the efficiency of energy transduction in *Paracoccus denitrificans* during growth under a variety of culture conditions. *Arch. Microbiol.* **139**:344–350.

Boyer, P. D., Chance, B., Ernster, L., Mitchell, P., Racher, E., and Slater, E. C., 1977, Oxidative phosphorylation and photophosphorylation. *Annu. Rev. Biochem.* **46**:955–1025.

Brice, J. M., Law, J. F., Meyer, D. J., and Jones, C. W., 1974, Energy conservation in *Escherichia coli* and *Klebsiella pneumoniae*. *Biochem. Soc. Trans.* **2**:523–526.

Brodie, A. F., 1959, Oxidative phosphorylation in fractionated bacterial systems. 1. Role of soluble factors. *J. Biol. Chem.* **234**:398–404.

Carver, M. A., and Jones, C. W., 1983, The terminal respiratory chain of the methylotrophic bacterium *Methylophilus methylotrophus*. *FEBS Lett.* **155**:187–191.

Casey, R. P., 1984, Membrane reconstitution of the energy-conserving enzymes of oxidative phosphorylation. *Biochim. Biophys. Acta* **768**:319–347.

Casey, R. P., Thelen, M., and Azzi, A., 1980, Dicyclohexylcarbodiimide binds specifically and covalently to cytochrome c oxidase while inhibiting its H^+-translocating activity. *J. Biol. Chem.* **255**:3994–4000.

Chance, B., Saronio, C., and Leigh, J. S., 1975, Functional intermediates in reaction of cytochrome oxidase with oxygen. *Proc. Natl. Acad. Sci. U.S.A.* **72**:1635–1640.

Chetkauskaite, A. V., and Grinius, L. L., 1979, Transhydrogenase as an additional site of energy accumulation in the *Escherichia coli* respiratory chain. *Biokhimiya* **44**:869–876 (in English).

Chicken, E., Spode, J. A., and Jones, C. W., 1981, Respiration-linked proton translocation in the moderate thermophile *Bacillus stearothermophilus*. *FEMS Microbiol. Lett.* **11**:181–185.

Clarke, D. M., and Bragg, P. D., 1985, Purification and properties of reconstitutively active nicotinamide nucleotide transhydrogenase of *Escherichia coli. Eur. J. Biochem.* **149**:517–523.

Clarke, D. M., Loo, T. W., Gllam, S., and Bragg, P. D., 1986, Nucleotide sequence of *pnt A* and *pnt B* genes encoding the pyridine nucleotide transhydrogenase of *Escherichia coli. Eur. J. Biochem.* **158**:641–653.

Cole, S. T., Condon, C., Lemire, B. D., and Weiner, J. H., 1985, Molecular biology, biochemistry and bioenergetics of fumarate reductase, a complex membrane-bound iron–sulfur flavoenzyme of *Escherichia coli. Biochim. Biophys. Acta* **811**:381–403.

Collins, M. D., and Jones, D., 1981, Distribution of isoprenoid quinone structure types in bacteria and their taxonomic implications. *Microbiol. Rev.* **45**:316–354.

De Vries, S., 1986, The pathway of electron transfer in the dimeric QH_2 : cytochrome c oxidoreductase. *J. Bioenerg. Biomembr.* **18**:195–224.

De Vrij, W., 1986, Energy-transducing processes in membrane vesicles from *Bacillus subtilis*, Thesis, University of Groningen.

De Vrij, W., Azzi, A., and Konings, W. N., 1983, Structural and functional properties of cytochrome oxidase from *Bacillus subtilis* W23. *Eur. J. Biochem.* **131**:97–103.

Downs, A. J., and Jones, C. W., 1975, Energy conservation in *Bacillus megaterium. Arch. Microbiol.* **105**:159–167.

Drozd, J. W., and Jones, C. W., 1974, Oxidative phosphorylation in *Hydrogenomonas eutropha* H16 grown with and without iron. *Biochem. Soc. Trans.* **2**:529–531.

Etemadi, A.-H., 1985, Function and orientational features of protein molecules in reconstituted lipid membranes. *Adv. Lipid Res.* **21**:281–428.

Eytan, G. D., and Racker, E., 1977, Selective incorporation of membrane proteins into proteoliposomes of different composition. *J. Biol. Chem.* **252**:3208–3213.

Fee, J. A., Choc, M. G., Findling, M. L., Lorence, R., and Yoshida, T., 1980, Properties of a copper-containing cytochrome c_1aa_3 complex: A terminal oxidase of the extreme thermophile *Thermus thermophilus* HB8. *Proc. Natl. Acad. Soc. U.S.A.* **77**:147–151.

Fee, J. A., Kuila, D., Michael, W., and Yoshida, T., 1986, Respiratory proteins from extremely thermophilic, aerobic bacteria. *Biochim. Biophys. Acta* **853**:153–185.

Fukumori, Y., Nakayama, K., and Yamanaka, T., 1985a, Cytochrome c oxidase of *Pseudomonas* AM1: Purification and molecular and enzymatic properties. *J. Biochem.* **98**:493–499.

Fukumori, Y., Nakayama, K., and Yamanaka, T., 1985b, One of the two copper atoms is not necessary for the cytochrome c oxidase activity of *Pseudomonas* AM1 cytochrome aa_3. *J. Biochem.* **98**:1719–1722.

Gabellini, N., and Sebald, W., 1986, Nucleotide sequence and transcription of the *fbc* operon from *Rhodopseudomonas sphaeroides. Eur. J. Biochem.* **154**:569–579.

Gabellini, N., Harnisch, U., McCarthy, J. E. G., Hauska, G., and Sebald, W., 1985, Cloning and expression of the *fbc* operon encoding the FeS proton, cytochrome b and cytochrome c_1 from the *Rhodopseudomonas sphaeroides* b/c_1 complex. *EMBO J.* **4**:549–553.

Gennis, R. B., Ludwig, B., Casey, R. C., and Azzi, A., 1982, Purification and characterization of the cytochrome c oxidase from *Rhodopseudomonas sphaeroides. Eur. J. Biochem.* **125**:189–195.

George, C. L., and Ferguson, S. J., 1984, Immunological identification of a two subunit NADH–ubiquinone oxidoreductase. *Eur. J. Biochem.* **143**:567–573.

George, C. L., Ferguson, S. J., Cleeter, M. W. J., and Ragan, C. I., 1986, Structural relationships between the NADH dehydrogenases of *Paracoccus denitrificans* and bovine heart mitochondria as revealed by immunological cross-reactivities. *FEBS Lett.* **198**:135–139.

Graf, M., Bokranz, M., Bocher, R., Friedl, P., and Kroger, A., 1985, Electron transport driven phosphorylation catalyzed by proteoliposomes containing hydrogenase, fumarate reductase and ATP synthase. *FEBS Lett.* **184**:100–103.

Hamamoto, T., Carrasco, N., Matsushita, K., Kaback, H. R., and Montal, M., 1985, Direct measurement of the electrongenic activity of o-type cytochrome oxidase from $E.$ $coli$ reconstituted into planar lipid bilayers. Proc. Natl. Acad. Sci. U.S.A. 82:2570–2573.

Hatefi, Y., 1985, The mitochondrial electron transport and oxidative phosphorylation system. Annu. Rev. Biochem. 54:1015–1070.

Hauska, G., Hurt, E., Gabellini, N., and Lockau, W., 1983, Comparative aspects of quinol–cytochrome c/platcyanin oxidoreductases. Biochim. Biophys. Acta. 726:97–133.

Hayashi, M., and Unemoto, T., 1987, Subunit components and their roles in the sodium-transport NADH : quinone reductase of a marine bacterium, Vibrio alginolyticus. Biochim. Biophys. Acta 890:47–54.

Hisae, N., Aizawa, K., Koyama, N., Sekiguchi, T., and Nosoh, Y., 1983, Purification and properties of NADH dehydrogenase from an alkalophilic bacillus. Biochim. Biophys. Acta. 743:232–238.

Hitchens, G. D., and Kell, D. B., 1984, On the effects of thiocyanate and venturicidin on respiration-driven proton translocation in Paracoccus denitrificans. Biochim. Biophys. Acta 766:222–232.

Hon-nami, K., and Oshima, T., 1984, Purification and characterization of cytochrome c oxidase from Thermus thermophilus HB8. Biochemistry 23:454–466.

Hooper, A. B., and DiSpirito, A. L., 1985, In bacteria which grow on simple reductants, generation of a proton gradient involves extracytoplasmic oxidation of substrate. Microbiol. Rev. 49:140–157.

Hudig, H., and Drews, G., 1984, Reconstitution of b-type cytochrome oxidase from Rhodopseudomonas capsulata in liposomes and turnover studies of proton translocation. Biochim. Biophys. Acta 765:171–177.

Hurt, G. H., and Hauska, G., 1981, Ubiquinol : cytochrome c reductase isolated in Triton X-100 by hydroxylapatite and gel chromatography. Eur. J. Biochem. 117:591–599.

Hurt, E. C., Hauska, G., and Shahnk, Y., 1982, Electrogenic proton translocation by the chloroplast cytochrome $b_6 f$ complex reconstituted into phospholipid vesicles. FEBS Lett. 149:211–216.

Hurt, E. C., Gabellini, N., Shahak, Y., Lockau, W., and Hauska, G., 1983, Extra proton translocation and membrane potential generation. Universal properties of cytochrome $bc_1/b_6 f$ complexes reconstituted into liposomes. Arch. Biochem. Biophys. 225:879–885.

Iba, K., Takamiya, K., and Arata, H., 1985, Isolation and characterization of cytochrome b-562 from cytochrome bc_1 complex of the photosynthetic bacterium Rhodopseudomonas sphaeroides R-26. FEBS Lett. 183:151–154.

Iba, K., Morohashi, K., Miyata, T., and Takamiya, K., 1987, Structural gene of cytochrome b-562 from cytochrome bc_1 complex of Rhodobacter sphaeroides. J. Biochem. 102:1511–1518.

Imai, K., Asano, A., and Sato, R., 1967, Oxidative phosphorylation in Micrococcus denitrificans. I. Preparation and properties of phosphorylating membrane fragments. Biochim. Biophys. Acta 143:462–476.

Ingledew, J., and Poole, R. K., 1984, Respiratory chains of Escherichia coli. Microbiol. Rev. 48:222–271.

Jaworowski, A., Campbell, H. D., Poulis, M. I., and Young, I. G., 1981a, Genetic identification and purification of the respiratory NADH dehydrogenase of Escherichia coli. Biochemistry 20:2041–2047.

Jaworowski, A., Mayo, G., Shaw, D. C., Campbell, H. D., and Young, I. G., 1981b, Characterization of the respiratory NADH dehydrogenase of Escherichia coli and reconstitution of NADH dehydrogenase in ndh mutant membrane vesicles. Biochemistry 20:3621–3628.

John, P., and Whatley, F. R., 1977, Bioenergetics of Paracoccus denitrificans. Biochim. Biophys. Acta 463:129–153.

Jones, C. W., 1977, Aerobic respiratory systems in bacteria. *Symp. Soc. Gen. Microbiol.* **27**:23–59.

Jones, C. W., Brice, J. M., Downs, A. J., and Drozd, J. W., 1975, Bacterial respiration-linked proton translocation and its relationship to respiratory chain composition. *Eur. J. Biochem.* **52**:265–271.

Kadenbach, B., Stroh, A., Ungibauer, M., Kuhn-Nentwig, L., Büge, U., Jarausch, J., 1986, Isozymes of cytochrome *C* oxidase: characterization and isolation from different tissues, *Methods Enzymol.* **126**:32–45.

Kagawa, Y., 1967, Target of components oxidative phosphorylation. Studies with a linear accelerator. *Biochim. Biophys. Acta* **131**:526–588.

Kagawa, Y., and Racker, E., 1971, Partial resolution of the enzymes catalyzing oxidative phosphorylation, XXV. Reconstitution of vesicles catalyzing ^{32}Pi-ATP exchange. *J. Biol. Chem.* **246**:5477–5487.

Kasahara, M., and Hinkle, P. C., 1977, Reconstitution and purification of the *D*-glucose transporter from human erythrocytes. *J. Biol. Chem.* **252**:7384–7390.

Kawada, N., Takada, K., and Nosoh, Y., 1981, Effect of lipids on a membrane-bound NADH dehydrogenase from *Bacillus caldotenax*. *J. Biochem.* **89**:1017–1027.

Keevil, C. W. and Anthony, C., 1979, Effect of growth conditions on the involvement of cytochrome *c* in electron transport, proton translocation and ATP synthesis in the facultative methylotroph *Pseudomonas* AM1. *Biochem. J.* **182**:71–79.

Kempner, E. S., and Schlegel, W., 1979, Size determination of enzymes by radiation inactivation. *Anal. Biochem.* **92**:2–10.

Kita, K., Kasahara, M., and Anraku, Y., 1982, Formation of a membrane potential by reconstituted liposomes made with cytochrome b_{562}-^{o}complex, a terminal oxidase of *E. coli*. *J. Biol. Chem.* **257**:7933–7935.

Kita, K., Konishi, K., and Anraku, Y., 1984a, Terminal oxidases of *E. coli* aerobic respiratory chain. I. *J. Biol. Chem.* **259**:3368–3374.

Kita, K., Konishi, K., and Anraku, Y., 1984b, Terminal oxidases of *E. coli* aerobic respiratory chain. II. *J. Biol. Chem.* **259**:3375–3381.

Kitada, M., and Krulwich, T. A., 1984, Purification and characterization of the cytochrome oxidase from alkalophilic *Bacillus firmus* RAB. *J. Bacteriol.* **158**:963–966.

Krinner, M., Hauska, G., Hurt, E., and Lockau, W., 1982, A cytochrome *f-b*$_6$ complex with plastoquinol–cytochrome *c* oxidoreductase activity from *Anabaena variabilis*. *Biochim. Biophys. Acta* **681**:110–117.

Kröger, A., Paulsen, J., and Schröder, I., 1986, Phosphorylative electron transport chains lacking a cytochrome bc_1 complex. *J. Bioenerg. Biomembr.* **18**:225–234.

Kurowski, B., and Ludwig, B., 1987, The genes of the *Paracoccus denitrificans* bc_1 complex. *J. Biol. Chem.* **262**:13805–13811.

Kutoh, E. and Sone, N., 1988, Quinol-cytochrome *c* oxido-reductase from the thermophilic bacterium PS3, *J. Biol. Chem.* **263**:9020–9026.

Lemire, B. D., Robinson, J. J., and Weiner, J. H., 1982, Identification of membrane anchor polypeptides of *Escherichia coli* fumarate reductase. *J. Bacteriol* **152**:1126–1131.

Ljungdahl, P. O., Pennoyer, J. D., Robertson, D. E., and Trumpower, B. L., 1987, Purification of highly active cytochrome bc_1 complexes from phylogenically diverse species by a single chromatographic procedure. *Biochim. Biophys. Acta* **891**:227–241.

Lorence, R. M., Koland, J. G., and Gennis, R. B., 1986, Coulometric and spectroscopic analysis of the purified cytochrome *d* complex: Evidence for the identification of "cytochrome a_1" as cytochrome b-595. *Biochemistry* **25**:2314–2321.

Ludwig, B., 1980, Heme aa_3-type cytochrome *c* oxidases from bacteria. *Biochim. Biophys. Acta* **594**:177–189.

Ludwig, B., 1987, Cytochrome *c* oxidase in prokaryotes. *FEMS Microbiol. Rev.* **46**:41–56.

Ludwig, B., and Schatz, G., 1980, A two-subunit cytochrome c oxidase (cytochrome aa_3) from *Paracoccus denitrificans*. *Proc. Natl. Acad. Sci. U.S.A.* **77**:196–200.

Mains, I., Power, D. M., and Thomas, E. W., 1980, Purification of an NADH: Dichlorophenol–indophenol oxidoreductase from *Bacillus stearothermophilus*. *Biochem. J.* **191**:457–465.

Matsushita, K., Patel, L., and Kaback, H. R., 1984, Cytochrome o type oxidase from *E. coli*; characterization of the enzyme and mechanism of electrochemical proton gradient generation. *Biochemistry* **23**:4703–4714.

Matsushita, K., Nonobe, M., Shinagawa, E., Adachi, O., and Ameyama, M., 1987a, Reconstitution of pyrroloquinoline quinol-dependent D-glucose oxidase respiratory chain of *Escherichia coli* with cytochrome o oxidase. *J. Bacteriol.* **169**:205–209.

Matsushita, K., Shinagawa, E., Adachi, O., and Ameyama, M., 1987b, Purification, characterization and reconstitution of cytochrome o-type oxidase from *Gluconobacter suboxydans*. *Biochim. Biophys. Acta* **894**:304–312.

Matsushita, K., Ohnishi, T., and Kaback, H. R., 1987c, NADH–ubiquinone oxidoreductases of the *Escherichia coli* aerobic respiratory chain. *Biochemistry* **26**:7732–7737.

McKay, A., Quilter, J., and Jones, C. W., 1982, Energy conservation in extreme thermophile *Thermus thermophilus* HB8. *Arch. Microbiol.* **131**:43–50.

Mell, H., Wellnitz, C., and Kröger, A., 1986, The electrochemical proton potential and the proton/electron ratio of the electron transport with fumarate in *Wollinella succinogenes*. *Biochim. Biophys. Acta.* **852**:212–221.

Miller, M. J., and Gennis, R. B., 1983, The purification and characterization of the cytochrome d terminal oxidase complex of the *Escherichia coli* aerobic respiratory chain. *J. Biol. Chem.* **258**:9159–9165.

Mitchell, P., 1966, Chemiosmotic coupling in oxidative and photosynthetic phosphorylation. *Biol. Rev.* **41**:445–502.

Mitchell, P., 1976, Possible molecular mechanism of the proton motive function of cytochrome systems. *J. Theor. Biol.* **62**:327–367.

Ogura, T., Sone, N., Tagawa, K., and Kitagawa, T., 1984, Resonance Raman study of the aa_3-type cytochrome oxidase of thermophilic bacterium PS3. *Biochemistry* **23**:2826–2831.

Oshima, T., and Drews, G., 1981, Isolation and partial characterization of the membrane-bound NADH dehydrogenase from the phototrophic bacterium *Rhodopseudomonas capsulata*. *Z. Naturforsch.* **36**:400–406.

Poole, R. K., 1983, Bacterial cytochrome oxidases. A structurally and functionally diverse group of electron-transfer proteins. *Biochim. Biophys. Acta* **726**:205–243.

Power, S. D., Lochrie, M. A., Sevarino, K. A., Patterson, T. E., and Poyton, R. O., 1984, The nuclear-coded subunits of yeast cytochrome c oxidase. *J. Biol. Chem.* **259**:6564–6570.

Quilter, A. M. G., and Jones, C. W., 1982, Energy conservation in the extreme thermophile *Thermus thermophilus* HB8. *Arch. Microbiol.* **131**:43–50.

Racker, E., 1979, Reconstitution of membrane processes. *Meth. Enzymol.* **55**:699–711.

Racker, E., Violand, S., O'Neal, S., Alfonzo, M., and Telford, J., 1979, Reconstitution, a way of biochemical research, some new approaches to membrane-bound enzymes. *A. Biochem. Biophys.* **198**:470–477.

Raitio, M., Jalli, T., and Saraste, M., 1987, Isolation and analysis of the genes for cytochrome c oxidase in *Paracoccus denitrificans*. *EMBO J.* **6**:2825–2833.

Rosevear, P., van Aken, T., Baxter, J., and Fergusson-Miller, S., 1980, Alkyl glucoside detergents: A simpler synthesis and their effects on kinetics and physical properties of cytochrome c oxidase. *Biochemistry* **19**:4108–4115.

Rydström, J., 1981, Energy-linked nicotinamide nucleotide transhydrogenase, in *Chemiosmotic Proton Circuits in Biological Membranes* (V. P. Skulachev and P. C. Hinkle, eds.), pp. 483–508, Addison-Wesley, Reading, MA.

Sapshead, L. M., and Wimpenny, J. W. T., 1972, The influence of oxygen and nitrate on the formation of the cytochrome pigments of the aerobic and anaerobic respiratory chain of *Micrococcus denitrificans*. *Biochim. Biophys. Acta* **267**:388-397.

Saraste, M., Raitio, M., Jalli, T., and Peramaa, A., 1986, A gene in *Paracoccus* for subunit III of cytochrome oxidase. *FEBS Lett.* **206**:154-156.

Schägger, H., Link, Th. A., Engel, W. D., and von Jagow, G., 1986, Isolation of the eleven protein subunits of the bc_1 complex from beef heart. *Meth. Enzymol.* **126**:224-237.

Scholes, P. B., and Mitchell, P., 1970, Respiration-driven proton translocation in *Micrococcus denitrificans*. *J. Bioenerg.* **1**:309-323.

Solioz, M., Carafoli, E., and Ludwig, B., 1982, The cytochrome *c* oxidase from *Paracoccus denitrificans* pumps protons in a reconstituted system. *J. Biol. Chem.* **257**:1579-1582.

Sone, N., 1986, Measurement of proton pump activity of the thermophilic bacterium PS3 and *Nitrobacter agilis* at the cytochrome oxidase level using total membranes and heptyl thioglucoside. *J. Biochem.* **100**:1465-1470.

Sone, N., and Hinkle, P. C., 1982, Proton transport by cytochrome *c* oxidase from the thermophilic bacterium PS3 reconstituted in liposomes. *J. Biol. Chem.* **257**:12600-12604.

Sone, N., and Kosako, T., 1986, Evidence for dimer structure of proton-pumping cytochrome *c* oxidase, an analysis by radiation inactivation. *EMBO J.* **5**:1515-1519.

Sone, N., and Nicholls, P., 1984, Effect of heat treatment on oxidase activity and proton-pumping capability of proteoliposome-incorporated beef heart cytochrome aa_3. *Biochemistry* **23**:6550-6554.

Sone, N., and Yanagita, Y., 1982, A cytochrome aa_3-type terminal oxidase of a thermophilic bacterium. Purification, properties and proton pumping. *Biochim. Biophys. Acta* **682**:216-226.

Sone, N., and Yanagita, Y., 1984, High vectorial proton stoichiometry by cytochrome *c* oxidase from the thermophilic bacterium PS3 reconstituted in liposomes. *J. Biol. Chem.* **259**:1405-1408.

Sone, N., Yoshida, M., Hirata, H., and Kagawa, Y., 1975, Purification and properties of a dicyclohexylcarbodiimide-sensitive adenosine triphosphatase from a thermophilic bacterium. *J. Biol. Chem* **250**:7917-7923.

Sone, N., Yoshida, M., Hirata, H., Okamoto, H., and Kagawa, Y., 1976, Electrochemical potential of protons in vesicles reconstituted from purified, proto-translocating adenosine triphosphatase. *J. Membr. Biol.* **30**:121-134.

Sone, N., Yoshida, M., Hirata, H., and Kagawa, Y., 1977a, Reconstitution of vesicles capable of energy transformation from phospholipids and adenosine triphosphatase of a thermophilic bacterium. *J. Biochem* **81**:519-528.

Sone, N., Yoshida, M., Hirata, H., and Kagawa, Y., 1977b, Adenosine triphosphate synthesis by electrochemical proton gradient in vesicles reconstituted from purified adenosine triphosphatase and phospholipids of thermophilic bacterium. *J. Biol. Chem.* **252**:2956-2960.

Sone, N., Ohyama, T., and Kagawa, Y., 1979, Thermostable single band cytochrome oxidase. *FEBS. Lett.* **106**:39-42.

Sone, N., Hamamoto, T., and Kagawa, Y., 1981, pH dependence of H^+ conduction through the membrane moiety of the H^+-ATPase ($F_0 \cdot F_1$) and effects of tyrosyl residue modification. *J. Biol. Chem.* **256**:2873-2877.

Sone, N., Kagawa, Y., and Orii, Y., 1983a, Carbon monoxide-binding cytochromes in the respiratory chain of the thermophilic bacterium PS3 grown with sufficient or limited aeration. *J. Biochem.* **93**:1329-1336.

Sone, N., Yanagita, Y., Hon-Nami, K., Fukumori, Y., and Yamanaka, T., 1983b, Proton-pump activity of *Nitrobacter agilis* and *Thermus thermophilus* cytochrome *c* oxidases. *FEBS. Lett.* **155**:150-154.

Sone, N., Naqui, A., Kumar, C., and Chance, B., 1984a, Reaction of caa_3-type terminal cyto-

chrome oxidase from the thermophilic bacterium PS3 with oxygen and carbon monoxide at low temperatures. *Biochem. J.* **221**:529–533.
Sone, N., Naqui, A., Kumar, C., and Chance, B., 1984b, Pulsed cytochrome c oxidase from the thermophilic bacterium PS3. *Biochem. J.* **223**:809–813.
Sone, N., Sekimachi, M., Fukumori, Y., and Yamanaka, T., 1987a, Evidence against proton pump activity by cytochrome c oxidase of *Pseudomonas* AM1. *J. Biochem.* **102**:481–486.
Sone, N., Sekimachi, M., and Kutoh, E., 1987b, Identification and properties of a quinol oxidase super-complex composed of a bc_1 complex and cytochrome oxidase in the thermophilic bacterium PS3. *J. Biol. Chem.* **262**:15386–15391.
Stouthamer, A. H., 1980, Bioenergetic studies on *Paracoccus denitrificans*. *Trends Biochem. Soc.* **5**:164–166.
Takamiya, S., Lindorfer, M. A., and Capaldi, R. A., 1987, Purification of all thirteen polypeptides of bovine heart cytochrome c oxidase from one aliquot of enzyme. *FEBS Lett.* **218**:277–282.
Thauer, R. K., Jungerman, K. J., and Decker, K., 1977, Energy conservation in chemotrophic anaerobic bacteria. *Bacteriol. Rev.* **41**:100–180.
Tokuda, H., and Unemoto, T., 1982, Isolation of *Vibrio alginolyticus* mutants defective in the respiration-coupled Na^+ pump. *J. Biol. Chem.* **257**:10007–10014.
Tokuda, H., 1984, Solubilization and reconstitution of the Na^+-motive NADH oxidase activity from the marine bacterium *Vibrio alginolyticus*. *FEBS Lett.* **176**:125–128.
Tokuda, H., and Unemoto, T., 1984, Na^+ is translocated at NADH : quinone oxidoreductase segment in the respiratory chain of *Vibrio alginolyticus*. *J. Biol. Chem.* **259**:7785–7790.
Tokuda, H., Udagawa, T., Asano, M., Yamamoto, T., and Unemoto, T., 1987, Conjugation-dependent recovery of the Na^+ pump in a mutant of *Vibrio alginolyticus* lacking three subunits of the Na^+ pump. *FEBS Lett.* **215**:335–338.
Udagawa, T., Unemomoto, T., and Tokuda, H., 1986, Generation of Na^+ electrochemical potential by the Na^+-motive NADH oxidase and Na^+/H^+ antiport system of a moderately halophilic *Vibrio costicola*. *J. Biol. Chem.* **261**:2616–2622.
Unden, G., and Kröger, A., 1981, The function of subunits of the fumarate reductase complex of *Vibrio succinogenes*. *Biochim. Biophys. Acta* **120**:577–584.
Unden, G., and Kröger, A., 1982, Reconstitution in liposomes of the electron transport chain catalyzing fumarate reduction by formate. *Biochim. Biophys. Acta* **682**:258–263.
Unden, G. and Kröger, A., 1983, Low-potential cytochrome b as an essential electron-transport component of menaquinone reduction in Vibrio succinogenes, *Biochim. Biophys. Acta* **725**:325–331.
Unden, G., and Kröger, A., 1986, Reconstitution of a functional electron transport chain from isolated enzymes in liposomes. *Meth. Enzymol.* **126**:387–399.
Unden, G., Bocher, R., Knecht, J., and Kröger, A., 1982, Hydrogenase from *Vibrio succinogenes*, a nickel protein. *FEBS Lett.* **145**:230–234.
Unden, G., Morschel, E., Bokranx, M., and Kröger, A., 1983, Structural properties of the proteoliposomes catalyzing electron transport from formate to fumarate. *Biochim. Biophys. Acta* **725**:41–48.
VanVerseveld, H. W., Krab, K., and Stouthamer, A. H., 1981, Proton pump coupled cytochrome c oxidase in *Paracoccus denitrificans*. *Biochim. Biophys. Acta* **635**:525–534.
Weiss, H., 1987, Structure of mitochondrial ubiquinol–cytochrome c reductase. *Curr. Top. Bioenerg.* **15**:67–90.
Widger, W. R., Cramer, W. A., Hederrmann, R. G., and Trebst, A., 1984, Sequence homology and structural similarity between cytochrome b of mitochondrial complex III and chloroplast b_6-f complex: Position of the cytochrome b hemes in the membrane. *Proc. Natl. Acad. Sci. U.S.A.* **81**:674–678.

Wikström, M., and Krab, K., 1986, The semiquinone cycle, a hypothesis of electron transfer and proton translocation in cytochrome bc_1-type complex. *J. Bioenerg. Biomembr.* **18**:181–193.

Wikström, M., and Saraste, M., 1984, The mitochondrial respiratory chain, in *Bioenergetics* (L. Ernster, ed.), pp. 49–94, Elsevier, New York.

Xuemin, X., Hisae, N., Koyama, N., and Nosoh, Y., 1985, Loss of liposome binding of NADH dehydrogenase from alkalophilic bacillus on subtilisin digestion. *FEBS Lett.* **181**:313–317.

Yagi, T., 1986, Purification and characterization of NADH dehydrogenase complex from *Paracoccus denitrificans*. *Arch. Biochem. Biophys.* **250**:302–311.

Yamanaka, T., and Fujii, K., 1980, Cytochrome a-type terminal oxidase derived from *Thiobacillus novellus*, *Biochim. Biophys. Acta* **591**:53–62.

Yamanaka, T., Fujii, K., and Kamita, Y., 1979, Subunits of cytochrome a-type terminal oxidases derived from *Thiobacillus novellus* and *Nitrobacter agilis*. *J. Biochem.* **86**:821–824.

Yamanaka, T., Kamita, Y., and Fukumori, Y., 1981, Molecular and enzymatic properties of cytochrome aa_3-type terminal oxidase derived from *Nitrobacter agilis*. *J. Biochem.* **89**:265–273.

Yanagita, Y., Sone, N., Kagawa, Y., 1983, Proton pumping and oxidase activity of thermophilic cytochrome oxidase remain after its extensive proteolysis. *Biochem. Biophys. Res. Commun.* **113**:575–580.

Yang, X., and Trumpower, B. L., 1986, Purification of a three-subunit ubiquinol cytochrome c oxidoreductase complex from *Paracoccus denitrificans*. *J. Biol. Chem.* **261**:12282–12289.

Yoshida, T., and Fee, J. A., 1984, Studies on cytochrome c oxidase activity of the cytochrome c_1aa_3 complex from *Thermus thermophilus*. *J. Biol. Chem.* **259**:1031–1036.

Yu, L., Mei, Q., and Yu, C. A., 1984, Characterization of purified bc_1 complex from *Rhodopseudomonas sphaeroides* R-26. *J. Biol. Chem.* **259**:5752–5760.

Chapter 8

Reconstitution of the High-Affinity Receptor for Immunoglobulin E

Jean-Pierre Kinet and Bruce Grasberger

1. INTRODUCTION

This chapter focuses on the reconstitution of the Fcε receptor with high affinity for immunoglobulin E (FcεRI). This receptor is found exclusively on mast cells, basophils, and related cells. Its aggregation initiates the release of mediators responsible for the allergic reaction. Understanding its function will not only shed light on the field of Fc receptors in general but may also help in designing new treatments of allergy.

In order to study the functional activity of a reconstituted receptor, the obvious first step is to identify the function(s) in which this receptor is directly involved. The elements necessary for the function can then be purified and reconstituted into an artificial bilayer, where ideally each variable can be controlled and the activity restored.

An intrinsic function has not yet been attributed to the receptor and therefore functional reconstitution is not possible. Nevertheless, reconstitution of the purified subunits has been accomplished in such a way that functional studies should be possible in the future.

Jean-Pierre Kinet and Bruce Grasberger   Section on Chemical Immunology, Arthritis and Rheumatism Branch, National Institute of Arthritis and Musculoskeletal and Skin Diseases, National Institutes of Health, Bethesda, Maryland 20892.

To fully understand the strategy of purification and of reconstitution of FcεRI, several important features about this receptor should be noted:

1. The binding of the ligand to its receptor ($K_a > 1 \times 10^9$) is completely passive, that is, monomeric IgE binds to mast cells and basophils without triggering these cells. Furthermore, there is now evidence that the tetrameric structure of the receptor and the stoichiometry between the different subunits is unaffected by the binding of IgE (Rivera et al., 1988). IgE can be viewed as part of a receptor for antigen and can be used to purify the receptor without concern that the IgE will modify the receptor's structure. The purification scheme should keep the receptor in its native form, avoiding conditions that could affect the receptor's ability to trigger an associated function. There is evidence that the protocol of purification described here fulfills the latter requirement.
2. The initial event in the degranulation of mast cells and basophils involves aggregation of the receptors on the plasma membrane of these cells. Therefore, in order to study the early reactions associated with activation, one must be able to prepare liposomes that individually contain multiple copies of receptor, or planar bilayers that contain a reasonable density of receptors.

2. PURIFICATION OF THE RECEPTOR

2.1. Sensitivity to Detergent

In the first attempts to purify the FcεRI, only one polypeptide was identified: a surface-iodinatable glycopeptide with an apparent M_r of 50,000 on sodium dodecyl sulfate polyacrylamide gels, eventually called the α subunit (Conrad and Froese, 1976; Kulczycki et al., 1976). Amino acid analysis and studies using the technique of radiation inactivation suggested that there was only one α subunit per bound IgE (Fewtrell et al., 1981; Kanellopoulos et al., 1980). However, sedimentation diffusion analysis of IgE binding activity revealed a M_r higher than 50,000 (Newman et al., 1977). The latter results suggested that the receptor might consist of more than a single polypeptide. Studies using chemical cross-linking reagents also indicated that additional components were associated with the α subunit and were being lost during purification (Holowka and Metzger, 1982; Holowka et al., 1980). Eventually, it was found that the association between the subunits of the receptor was sensitive to mild detergents. When an appropriate detergent to phospholipid ratio was used during the purification, the α subunit was found to be associated with one β and two disulfide linked γ subunits (Perez-Montfort et al., 1983). Subsequently, we developed a practical method of assaying the dissociation so that we could explore a variety of different factors that influence the detergent-induced dissociation (Kinet et al., 1985b). The results of that study can be summarized as follows:

1. Whatever chemical or physical conditions were used to induce the dissociation, β and γ_2 always dissociated from α in unison.
2. The detergents that dissociated the subunits of the receptor most efficiently were the same as those Umbreit and Strominger (1973) found to be the most efficient to disrupt cellular membranes. Such detergents have a hydrophilic/lipophilic balance (HLB number)* between 12 and 14.
3. The critical variable was the concentration of micellar detergent relative to the concentration of receptors:
   a. Among detergents with a similar HLB number, the dissociation was directly proportional to the concentration of micellar detergent; this correlation held even though detergents having a critical micelle concentration (CMC) that varied over a 100-fold range were tested. Similarly, whether the detergents were nonionic such as Triton X-100 and octylglucoside, or had a net charge such as sodium cholate, or were zwitterionic such as CHAPS, appeared irrelevant.
   b. Even below the CMC, a moderate amount of β and γ dissociated from α over a prolonged period of time. This appears contradictory only if one presumes that below the CMC no micelles are present. However, it is well known that the CMC cannot be considered as an absolute value above which micelles appear and below which no micelles are present (Tanford, 1973). Recently, we found that, under these mild conditions (below the CMC), β remains attached to the dimer of γ subunits after the complex $\beta\gamma_2$ dissociates from the α (Rivera et al., 1988). Under harsher conditions (above the CMC), β and γ_2 dissociate from each other.
   c. The dissociation is also a function of the concentration of receptor; the lower the concentration of receptors, the greater the dissociation. But this is true only above a defined threshold: in a fixed volume, the molar ratio of micellar detergent to receptor must be greater than 10^5 to induce dissociation. This may explain why in order to get complete dissociation with moderate concentrations of detergent at neutral pH and at 4°C, large volumes of detergents are required. This is, of course, easily performed by exposure of the receptor to detergent in an *open system* such as an affinity column and explains the past failure to purify the tetrameric re-

---

*The HLB number is an empirical parameter used by cosmetics chemists for the classification of emulsifying reagents (Griffin, 1949). Most detergents used in contemporary cell biology research have an HLB number between 12 and 14 because these efficiently emulsify the phospholipids of a cell membrane (Egan et al., 1976; Umbreit and Strominger, 1973).

4. Conditions that destabilize electrostatic interactions, such as high salt, high temperature, and extremes of pH, promoted dissociation.
5. Interfering with hydrophobic interactions by using chaotropic ions also induced dissociation.
6. Under "physiological conditions," dissociation proceeded slowly and was proportional to the time of exposure. In a *closed system*, it appeared to stabilize at an intermediate value as if an equilibrium was being established. Since the receptor was at a very low concentration compared to the detergent, the dissociation should have followed pseudo-first-order kinetics if the receptor was homogeneous and the dissociation involved only receptors and micelles of detergent. This was clearly not the case. Therefore, one has to postulate either that the receptors are heterogeneous or that other components (e.g., lipids) are involved or both.

2.2. Protective Effect of Lipids

A search for the best conditions for solubilization and reincorporation of $[^{125}I]IgE$–receptor complexes suggested that the ratio of the concentration of micellar detergent to the concentration of phospholipid was important. This ratio was dubbed ρ (Rivnay and Metzger, 1982; Rivnay et al., 1982). The minimum ρ at which maximal solubilization occurs is $\rho = 2$. When these conditions are used, the interaction between the subunits is relatively stable. Using an assay similar to that used to study dissociation, we explored the conditions required for protection of the tetrameric structure of the receptor.

The conclusions of that study were as follows:

1. The phospholipids do not protect the receptor simply by neutralizing the detergent micelles because purified phospholipids that can prevent solubilization of cell membranes failed to prevent dissociation of the receptor.
2. The phospholipids required for protection seemed to be specific: we failed to find simple lipid mixtures that were effective, and only crude mixtures of animal cell phospholipids were maximally protective.
3. When such complex mixtures of lipids were used, a $\rho < 5$ was required for maximal protection.
4. The protective effect varied with the type of detergent.

The distribution of phospholipids in the microenvironment of the IgE–receptor has been analyzed by Rivnay and Fischer (1986). The receptor-bound lipids were shown to contain phosphatidylcholine (PC) and sphingomyelin (SM). Synthetic PC and SM, either alone or together, failed to protect the association of the subunits (Kinet et al., 1985a). However, it is possible that the effectiveness of the crude mixture of phospholipids is related to the specific fatty acid side chains of the phospholipids. For example, the location and the extent of unsaturation of fatty acids could be critical. It is interesting to note that the subunits of the receptor contain ester-linked fatty acids (Kinet et al., 1985a). These covalently bound lipids might interact with tightly but noncovalently bound phospholipids necessary for the interaction of the subunits of the receptor. We have postulated a model for solubilized receptors in which receptor-bound lipids are in equilibrium with mixed micelles of lipid and detergent. The receptor irreversibly dissociates when a critical amount of tightly bound and specific lipids is stripped from the receptor by detergent micelles. If specific lipids corresponding to these stripped lipids are added to the mixture, the equilibrium is shifted and the association of the subunits is protected. However, analysis of the variability of the data suggests that the mechanism of dissociation and of protection of the receptor is more complicated than our simple model would predict.

It should be stressed that all the data described above were obtained from preparations of purified receptors. It was assumed that the receptors in these preparations were not dissociated and had been fully protected because the purification was performed under "protective conditions." This was checked by examining intrinsically labeled receptors purified under the same conditions. The ratios of the counts in the incorporated amino acids between $\alpha : \beta : \gamma$ were generally consistent with the 1 : 1 : 2 stoichiometry suggested by cross-linking studies. Under dissociating conditions, the ratios of $\alpha : \beta$ and of $\alpha : \gamma$ increased dramatically. Most recently, we have cloned the α and β subunit of the receptor (Kinet et al., 1987, 1988). This has allowed us to determine the precise ratio of a particular amino acid in α and β. We have compared these ratios with the corresponding ratios obtained in labeling studies with ^{35}S methionine ($n=8$), ^3H tryptophan ($n=3$), ^{35}S cysteine ($n=10$), and ^3H histidine ($n=2$). These recent analyses suggest that 25–30% dissociation of β occurs even though the purification is performed under relatively protective conditions. A similar conclusion can be drawn from experiments performed with a monoclonal anti-β antibody (Rivera et al., 1988).

2.3. Protocol for Purification

Growing Rat Basophilic Leukemia Cells. RBL-2H3 cells are grown in tissue culture flasks as adherent cells (Barsumian et al., 1981) or in spinner flasks as nonadherent cells (Isersky et al., 1975). The growth in spinner is the method of choice when large amounts of receptor are required. From a 2-liter culture about 2×10^9 cells can be harvested. This represents a binding capacity of 200 µg of IgE (1 nmol) for cells expressing 3×10^5 receptors/cell.

Binding of IgE. To remove cell debris, the cells are washed three times in medium, resuspended at a density of 1×10^7 cells/ml in a small spinner flask, and incubated at 37°C for 1 hr in presence of 4 µg/ml of monoclonal mouse antidinitrophenyl-IgE (Liu et al., 1980) whose final specific activity of iodine-125 after labeling with chloramine-T is around 2000 cpm/µg.

Solubilization. After washing away unbound IgE, the pellet of cells is resuspended in 10 mM recrystallized CHAPS* in borate buffered saline pH 8 (BBS) at a cell density of 5×10^7 cells/ml. This concentration of cells will yield a concentration of 2 mM phospholipids when solubilization is complete and a ρ of approximately 2. As mentioned above, these conditions largely maintain the integrity of the receptors. About 95% of the receptor-bound [^{125}I]IgE is recovered from the supernatant after 1 hr centrifugation at 30,000g.

Affinity Chromatography and Elution. The cell extract containing [^{125}I] antidinitrophenyl-IgE–receptor complexes is added to trinitrophenyl lysine beads (Holowka and Metzger, 1982) and incubated overnight at 4°C with gentle shaking. The effluent is collected and counted for ^{125}I. The beads bind at least 80% of the counts and are then poured into a column and washed with 50 volumes of 10 mM CHAPS and 2mM tumor phospholipids. This wash should be followed by a "stripping" of the lipids loosely bound to the receptor (Kinet et al., 1985a). This is accomplished by a very rapid wash with 10 mM CHAPS in BBS, using increased pressure if necessary to keep the time of exposure of the receptor to 10 mM CHAPS around 3 min. The volume of 10 mM CHAPS applied to the column is also important: the ratio of micelle/receptor should be around 3×10^4. At a lower ratio the stripping is incomplete, and at a higher ratio the receptor starts to dissociate substantially. Approximately 90 ml of 10 mM CHAPS is necessary to "strip" adequately about 1 nmol of IgE–receptor complexes. The "stripping" is followed by a wash with 2000 bed volumes of 2 mM CHAPS. The overall washing procedure can be accomplished in 1 hr. The IgE-receptor complexes are eluted from the column three times with two bed volumes of 10 mM (dinitrophenyl)-ε-aminocaproate in 22 mM CHAPS for 15 min, 12 hr, and 2 hr, respectively. If extremely clean preparations of receptors are needed, double affinity chromatography can be used. Receptors bound to arsonylated antidinitrophenyl-IgE are purified on an antibenzenearsonate column (Kanellopoulos et al., 1979) following elution from the trinitrophenylsine column. Since IgE–receptor complexes are eluted in 2 mM CHAPS (a concen-

---

*Recent preparations of CHAPS bought from different sources were found to be contaminated with deoxyCHAPS. The latter has a much lower CMC than CHAPS, therefore inducing substantial dissociation of the receptor. The recrystallization of CHAPS is performed as follows: 5 g of commercial grade CHAPS are added gradually to 200 ml of boiling methanol in a 500 ml Erlenmeyer flask with continuous stirring. After the material is dissolved, it is allowed to cool at room temperature and stored at $-20°C$ overnight. CHAPS, but not deoxyCHAPS, will precipitate. The precipitate is filtered under vacuum, and then rinsed with 4 volumes of $-20°C$ precooled methanol. The material is then dried by lyophilization for 48 hr.

tration of CHAPS below its CMC), they can easily be concentrated on a Centricon-10 device (Amicon, Danvers, MA) without fear that the detergent concentration will be increased. The concentrated preparation of receptors can then be diluted into a detergent–lipid buffer to achieve a protein to lipid ratio appropriate for reconstitution.

3. MECHANISMS OF ACTION OF THE RECEPTOR: POSSIBLE FUNCTIONS TO RECONSTITUTE

3.1. Role of Aggregation

As mentioned in the introductory note, binding of monomeric IgE to the cell causes no known stimulation. When receptor-bound IgE interacts with a multivalent antigen, degranulation of the cells occurs. When unliganded receptors are aggregated by multivalent antireceptor antibodies, the cells are stimulated as well (Ishizaka and Ishizaka, 1978; Isersky et al., 1978). This observation demonstrates that a critical event in triggering the cell is the aggregation of receptors.

3.2. Functions Associated with the Receptor-Mediated Triggering of Mast Cells

Two extensive reviews have been published recently which analyze in detail the controversial data about the receptor-mediated activation (Beavan and Cunha-Melo, 1988; Metzger et al., 1986). So far, no intrinsic function has been attributed to the receptor itself and the cloning of the α and β subunits has not been helpful in that regard.

However, many phenomena have been shown to be linked to receptor-mediated activation. These include immobilization of receptors, internalization of receptors, activation of serine protease, stimulation of lipid methylation, stimulation of adenylate cyclase, hydrolysis of phosphatidylinositides, and possible interaction with a GTP-binding protein.

Another function linked to receptor-mediated activation deserves special discussion: the rise in intracytoplasmic Ca^{2+}. One group of investigators have isolated a molecule called cromolyn binding protein which they propose interacts with aggregated receptors, leads to the formation of a Ca^{2+} channel (Mazurek et al., 1982), and thereby causes a rise in intracytoplasmic Ca^{2+}. Some of the problems related to these studies are discussed in the context of the reconstitution studies.

4. RECONSTITUTION STUDIES

When working with membrane proteins, the activity to be reconstituted generally dictates which of two major methodologies is employed. In order to study channel and transport activities using electrical measurements, the proteins of interest are commonly inserted into a planar bilayer separating two electrically isolated compartments (see also Chapters 3 and 9). Biochemical activities are more readily studied in an aqueous vesicle suspension where the proteins are inserted into the lipid bilayers comprising the vesicle boundaries. Both techniques have been employed to study FcεRI.

It should be noted, however, that some of the properties of the FcεRI system detailed above present problems for reconstitution not present in many of the procedures described in the literature. Specifically, the receptor's sensitivity to dissociation by detergent puts constraints on the composition of the solutions that can be safely used, and the activation by aggregation dictates several requirements for the final configuration of the system's components. The former limits the time of exposure of the receptor to the high concentrations of detergent that are desirable for good initial dispersion of the protein–lipid–detergent mixtures, if the integrity of the receptor is to be maintained. The latter assumes the aggregation signal leads to orientation-specific interactions of the receptors in the bilayer. Lipid bilayers are important for providing a nativelike microenvironment for the transmembrane regions of membrane proteins and are necessary for restoring the activity of many solubilized proteins for just this reason. Additionally, in systems where membrane proteins interact with each other, insertion in a bilayer provides a great enhancement of protein–protein association (Grasberger *et al.*, 1986). Reconstitution of FcεRI activity is likely to require insertion in a bilayer for both of these reasons. If the type of protein–protein contacts that are made in the intact cell are to be reproduced in the artificial bilayer, the interacting receptors must be oriented in the same direction in a continuous bilayer. For reincorporation into sealed vesicles, this means multiple copies of receptors per vesicle with the IgE binding domain facing out so that bound IgE is exposed to aggregating stimuli. Oppositely oriented receptors should not interfere but would not be stimulated.

4.1. Planar Bilayers

The FcεRI receptor has been reincorporated into planar bilayers in several experiments to test the properties of protein preparations for channel activity. Masurek, Pecht, and co-workers have described the isolation of a cromolyn binding protein (CBP) that putatively functions as a partially selective calcium channel in membranes. Cromolyn is an antiasthmatic that chelates calcium and inhibits histamine release from stimulated mast cells (Mazurek *et al.*, 1980).

The purification procedures use a protein–cromolyn conjugate formed by an organic synthesis (Mazurek et al., 1982) that is reported to give low yields (Corcia et al., 1986) in the early literature (Mazurek et al., 1982, 1983a,b, 1984, 1986). Later reports use a modified procedure to synthesize the conjugate, the details of which have yet to be published (Corcia et al., 1986; Pecht et al., 1986). The purified preparations were initially found to be a single 60 kDa (unreduced polyacrylamide gels) protein with a pI of 4.5–5 and 20–25% acidic amino acid (Mazurek et al., 1983a), with less than 1% contamination (Mazurek et al., 1984). More recently, the purification was reported to give proteins of 68, 47, and 24 kDa in variable yields (Corcia et al., 1986). These preparations have been used to restore the ability to degranulate to an RBL-2H3 variant cell line (Mazurek et al., 1983a). The line was derived from cells selected for the absence of cromolyn conjugate binding and was found to have lost the ability to release internal stores of histamine in response to IgE aggregation. The cells bound IgE normally and could be stimulated to release by the calcium ionophore A23187 (Mazurek et al., 1983b). Sendai virus envelop glycoproteins NH and F solubilized with Triton X-100 were added to CBP and dialyzed to form vesicles that, when fused with the variant line, restored the ability to degranulate.

Direct measurements of channel activity were done initially using RBL membrane or purified CBP (Mazurek et al., 1984). Subsequently, bilayers containing only purified FcεRI and CBP were found to exhibit the same conductances. Two methods for forming planar bilayers were used. The first used a Teflon septum with a 100 μm diameter hole to separate two chambers (Schindler, 1980). The proteins were first reincorporated into vesicles using the "fast dilution method" (Schindler et al., 1984) as follows: 20 mg of lipid mixture (soybean lecithin/egg PC/cholesterol 9 : 10 : 1) in a chloroform methanol are dried under nitrogen in a round bottom flask. Twenty milliters of buffer containing 1 mg of soybean lecithin and the protein were added and the lipid film resuspended with the aid of glass beads. The resultant suspension is dialyzed and filtered through 0.4 μm polycarbonate filters. Vesicles with and without protein are added to the chambers on opposite sides of the septum to a level below the hole. After approximately 1 min, monolayers form spontaneously at the air–water interface. Buffer is then added to both chambers to raise the level above the hole, where a bilayer is deposited. The bilayer forms a high resistance seal but contains channel activity when triggered by IgE aggregating stimulants (multivalent antigen or anti-IgE antibodies) or by anti-CBP antibodies.

The second method is better adapted for single-channel recording as the bilayer is formed on a patch-clamp pipet (Corcia et al., 1986; Coronado and Latorre, 1983). Again, the protein is first incorporated into vesicles. Two milligrams of mixed lipid (soybean lecithin/egg PC/cholesterol 8 : 11 : 1) in chloroform methanol were dried under nitrogen and then vacuum. Next, 770 µl of buffer containing 50 µg of RBL tumor lipid, 12 pmoles of FcεRI, with or without CBP were dialyzed for 4 hr, then added to the lipids and sonicated 30 sec. Seventy-five microliters of this mixture were added to a microtiter plate well into which a patch-clamp pipet is introduced. A monolayer forms at the air–water surface. The pipet is withdrawn, and a monolayer forms across the tip. On reintroduction of the pipet through the interface, a bilayer is formed. If only FcεRI is present, aggregation of the IgE by anti-IgE or antigen does not show any channel activity. However, if both FcεRI and CBP are present, aggregation leads to the opening of channels with a conductance of 1–2 pS in the presence of 1.8 mM calcium, and 4–5 pS in the presence of 100 mM calcium.

A major weakness of this work is the isolation procedure for the CBP. The published method for synthesizing the cromolyn–protein conjugate used in the purification does not work in our laboratory, and the details of the modified protocol are still unavailable. The papers do present a large amount of data that indicate that there is stimulated channel activity in the RBL cell membranes. The experiments using purified FcεRI indicate that during the purification at least some of the native function is retained and that reincorporation into bilayers can restore that activity. The affinity purified CBP must be further characterized to distinguish between those proteins that contain the channel activity, those that function as accessory proteins, and those that are irrelevant. Once these assignments are made, one can start to elucidate the nature of the FcεRI–CBP interaction.

4.2. Vesicles

Studies of the reconstitution of FcεRI into lipid vesicles are still limited by the absence of a function that can be assigned to the receptor itself. Indeed, none of the functions that have been associated with receptor aggregation in whole cells, with the possible exception of calcium flux, have been assigned to specific proteins and hence are not available for reincorporation studies. However, progress is being made in fulfilling the criteria for an activatable reconstituted FcεRI system. As mentioned previously, these criteria include intact receptors, multiple copies per vesicle, and correct orientation in the membrane.

Reconstitution studies were initiated before the tetrameric subunit structure of the FcεRI had been fully elucidated. They were in fact contemporaneous with the discovery of the detergent sensitivity of the receptor. Since the existence of the β subunit was known before the conditions had been devised to maintain its association with the α chain during purification, initial reconstitution studies were done with unpurified detergent extracts. The basic scheme for the experiments was to solubilize cell membrane bound [^{125}I]IgE with octylglucoside, adjust the lipid and detergent composition and concentration to the desired values, and dialyze extensively (Rivnay and Metzger, 1982). Reincorporation was assessed by ultracentrifugation of the sample on a sucrose gradient; lipid vesicles float near the top and the unreincorporated proteins run into the gradient. Using a low detergent to lipid ratio of 0.6, some of the receptors migrated at a peak intermediate to the unincorporated protein and lipid peaks, possibly representing a population of small vesicles with high protein content. The pattern remains unchanged if one uses concentrations of egg yolk lecithin between 33 and 85 mM (constant detergent to lipid ratio of 0.6), soybean lipids, and fast or slow dialysis. Raising the detergent to lipid ratio to 1.0, 2.2, 3.8, and 15 at a constant phospholipid concentration of 16 mM soybean lecithin resolves the receptor pattern into two peaks, one comigrating with the lipid peak, and one farther down in the gradient. The lighter lipid-associated peak contained 40–50% of the total protein counts. The rate of detergent removal appears to be critical at these higher detergent concentrations, as rapid removal using a Sepharose G-25 column gave a pattern with little association of IgE and lipid. These findings are consistent with the data of Almog et al. (1986), which indicate that vesiculation proceeds as a function of the micellar detergent/phospholipid ratio. It is possible that the higher detergent concentrations expose the vesicles in the intermediate peak to the detergent/phospholipid ratio critical for vesicle growth for a longer period of time, allowing them to fuse into larger vesicles.

The search for a gentler detergent extraction that might yield a more active (incorporatable) receptor led to the observation that solubilization of membrane proteins is a function of the micellar detergent to lipid ratio, the ρ value described previously. Using a concentration of detergent that is suboptimal for solubilizing the receptors ($\rho = 1.25$) did not increase the fraction of counts associated with the lipid peak in the sucrose gradients. However, changing the source of added lipids to extracted tumor cells led to almost 80% of the counts migrating with the lipid peak.

Parallel studies showed that maintaining a low ρ value is critical for retaining the subunit structure of the FcϵRI. Subsequent studies have verified the importance of the lipid and detergent used—the combination of CHAPS and tumor lipids causing the least subunit dissociation. Rivnay et al. (1984) demonstrated that the loss of the β and γ subunits as a result of exposure to detergent at a high ρ value abrogated the ability of the α chains to associate with the lipid vesicles. In those experiments, purified, iodinated FcϵRI was mixed with tumor lipids to a final phospholipid concentration of 8–10 mM, and CHAPS was added to achieve a ρ value of 2.5. This mixture was incubated 1–2 hr on ice and then dialyzed extensively. The receptors were purified on affinity columns under different conditions. Receptors that were washed with detergent alone (no added lipid) lost their β and γ subunits and failed to reincorporate. Receptors that were washed with a mixture of detergent and lipid that maintained a ρ value of 2.5, or exposed to chemical cross-linking reagents and washed with detergent alone, retained their β and γ subunits and migrated with the lipid peak on sucrose gradient analysis. The conclusion of these studies is that the α subunit alone does not spontaneously reincorporate on detergent dialysis, but that receptors with a full complement of subunits do.

Maintaining a ρ value of approximately 2 during the receptor purification involves adding exogenous lipids to the washing and elution buffers. This procedure limits the degree to which one can lower the phospholipid to protein ratio. Assuming a constant area per phospholipid in a bilayer (60 $Å^2$), the phospholipid to protein ratio and the vesicle size determine the number of receptors per vesicle. Using receptors purified with buffers of $\rho = 2$, and tumor lipids for vesicle formation, the phospholipid to protein ratio is too high and the average vesicle diameter too low to achieve the goal of multiple copies per vesicle (Metzger et al., 1984). However, the observation that the receptor is relatively stable in the presence of submicellar detergent (Kinet et al., 1985b) now allows one to adjust freely the phospholipid to protein ratio as well as the lipid composition. Experiments are currently underway to characterize the products of a reincorporation procedure that yields a population of larger liposomes.

Preliminary results show that egg yolk lecithin at millimolar concentrations, using detergent dialysis, yields vesicles with diameters greater than 2500 Å. The vesicles are routinely sized using a calibrated Sephacryl S1000 column (Reynolds et al., 1983). A typical elution profile is shown in Figure 1. Receptors loaded with ^{125}I-labeled IgE were added to PC and CHAPS (final concentrations 2.7 and 18 mM, respectively, giving a ρ value of 5) and dialyzed against 330 volumes of buffer (0.25 M KCl, 10 mM TRIS pH 8.5), changed three times in 2 days. The lipid mixture contained amounts of [^{3}H]dipalmityl phosphatidylcholine. The receptors in peak A are associated with the vesicles in peak C that elute with the void volume of the column. Assuming a diameter of 2500 Å, one can calculate from the phospholipid to protein ratio that the vesicles in peak C each contain an average of 13 receptors. The exclusion limit for S1000 is approximately 2500 Å, so that this value for the diameter should be an underestimate, because the void volume will contain vesicles this size and larger. This is confirmed by trapped volume measurements which routinely estimate average vesicle size to be greater than 2500 Å in similar preparations. However, vesicles stained with uranyl acetate and viewed with an electron microscope indicate average diameters of approximately 1500 Å, suggesting that some of the lipid that migrates in the void volume is due to aggregation of the vesicles. Even using this smaller value, there would still be multiple copies in these large vesicles. Note that this does not represent the limit of the method, as the added protein could be increased by at least an order of magnitude. The electron micrographs also indicate that the vesicles are largely unilamellar.

Reconstitution of Immunoglobulin E Receptor

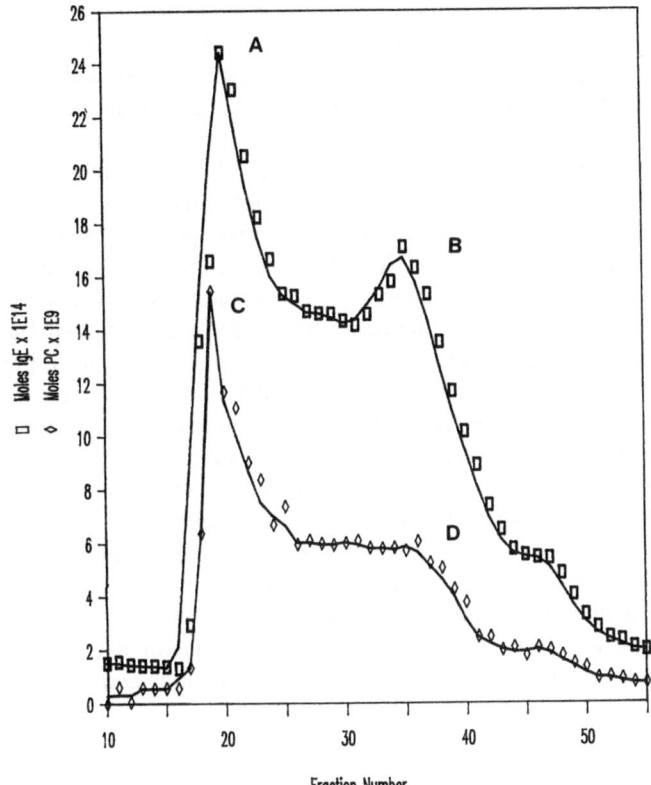

FIGURE 1. Elution pattern from a Sephacryl S1000 sizing column. Vesicles were made by dialyzing 3 ml of 2.7 mM [^{3}H]PC (9.22 × 10^{11} cpm/mol), 18 mM CHAPS, and 10 μg/ml [^{125}I]IgE bound to purified FcεRI (1.69 × 10^{17} cpm/mol), against 1 liter of 0.25 M KCl, 10 mM TRIS, pH 8.5. The buffer was changed three times in 2 days. Two hundred microliters of this material were loaded on a 0.7 × 45 cm packed S1000 column. Fractions of 0.5 ml each were collected and counted for ^{3}H (◇) and ^{125}I (□). Peaks A and C elute with or near the void volume, and peak B near the position for free IgE.

The receptors in peak B elute at a position that is consistent with them being free in solution, but they may also be associated with the vesicles of small diameter in peak D. The latter possibility is supported by the migration of the majority of receptors with the phospholipid in sucrose gradients. The amount of phospholipid appearing at position D (peak height) varies from being a shoulder of the main peak, as shown, to being a minor second peak in some preparations. There are reports of preferential association of membrane proteins with small vesicles which could explain the bimodal distribution of the protein (Greenhut *et al.*, 1986). This suggests that the receptors initially insert into small vesicles which then fuse to form large-diameter vesicles.

The exposure of the IgE on the vesicle surface is being explored by binding B1E3, a rat monoclonal anti-mouse IgE (gift from D. Conrad) and by surface iodination. Preliminary results indicate that the number of B1E3 antibodies bound per IgE on vesicles with reincorporated FcεRI is about half that on IgE-loaded RBL cells, suggesting that the receptors are randomly oriented in the bilayer. This is consistent with the finding that surface iodination of the vesicles labels the β and γ subunits. Surface iodination of intact cells labels only the α subunit, but lysing the cells allows all three subunits to be iodinated. The hydrophobic labeling reagent [^{125}I]TID (Amersham) seems to preferentially label the β subunit in soluble receptors and therefore may not be useful in distinguishing between reincorporated and soluble receptors. Experiments to be done in the near future include the generation of monovalent B1E3 Fab fragments to see if the B1E3/IgE stoichiometry can give information about the cross-linking of receptors reincorporated into vesicles.

Experiments that look for a biochemical perturbation associated with FcεRI aggregation are not difficult to design, but negative results will be uninformative. Once a function associated with the receptor or with interacting proteins is found, the techniques described here will be useful in studying the function in a pure form, isolated from complex interactions and occurring in a cell membrane.

5. CONCLUSIONS

The FcεRI consists of a multisubunit complex in cell membranes. Conditions have been found which allow the purification of receptors with their tetrameric structure ($\alpha\beta\gamma_2$). These purified receptors spontaneously associate with phospholipid in solution. Planar bilayers containing FcεRI can be formed, presumably through a monolayer intermediate, from the phospholipid–protein mixture. Other proteins may be added to this system that give evidence of calcium channel activity when the FcεRI are aggregated. FcεRI can also be reconstituted into phospholipid vesicles by detergent dialysis. The number of FcεRI per vesicle can be made sufficiently large to allow receptor aggregation, and thus the system can be used to study possible functions that may be associated with this aggregation.

It should be noted that there may exist critical functions that will not be easily reconstituted in either the vesicle or planar bilayer systems. Menon *et al.* (1984, 1986a,b) have shown that aggregated receptors are rapidly immobilized on RBL cell surfaces, and the immobilization correlates well with degranulation. This is presumably the result of receptor–cytoskeletal interactions. The reconstitution of such interactions is a technology that has not yet been developed and is likely to involve many difficulties. If these interactions are important in the signal transduction pathway that leads from the aggregation to degranulation, they will provide a major stumbling block to the reconstitution of such a pathway. Current efforts focus on the earliest biochemical events associated with FcεRI aggregation. Once these functions are identified, the requisite molecules can be purified and added to the reconstitution systems, in order to study in detail the steps in the signal transduction pathway.

6. REFERENCES

Almog, S., Kushnir, T., Nir, S., and Lichtenberg, D., 1986, Kinetic and structural aspects of reconstitution of phosphatidylcholine vesicles by dilution of phosphatidylcholine–sodium cholate mixed micelles. *Biochemistry* **25**:2597–2605.

Barsumian, E. L., Isersky, C., Petrino, M. G., and Siraganian, R. P., 1981, IgE-induced histamine release from rat basophilic leukemia cell lines: Isolation of releasing and nonreleasing clones. *Eur. J. Immunol.* **11**:317–323.

Beavan, M. A., and Cunha-Melo, J. R., 1988, Membrane phosphoinositide-activated signals in mast cells and basophils. *Prog. Allergy* **42**:123–184.

Conrad, D. H., and Froese, A., 1976, Characterization of the target cell receptor for IgE. II. Polyacrylamide gel analysis of the surface IgE receptor from normal rat mast cells and from rat basophilic leukemia cells. *J. Immunol.* **116**:319–326.

Corcia, A., Schweitzer-Stenner, R., Pecht, I., and Rivnay, B., 1986, Characterization of the ion channel activity in planar bilayers containing IgE-Fc epsilon receptor and the cromolyn-binding protein. *EMBO J.* **5**:849–854.

Coronado, R., and Latorre, R., 1983, Phospholipid bilayers made from monolayers on patch-clamp pipettes. *Biophys. J.* **43**:231–236.

Egan, R. W., Jones, M. A., and Lehninger, A. L., 1976, Hydrophile–lipophile balance and critical micelle concentration as key factors influencing surfactant disruption of mitochondrial membranes. *J. Biol. Chem.* **251**:4442–4447.

Fewtrell, C., Kempner, E., Poy, G., and Metzger, H., 1981, Unexpected findings from target analysis of immunoglobulin E and its receptor. *Biochemistry* **20**:6589–6594.

Grasberger, B., Minton, A. P., DeLisi, C., and Metzger, H., 1986, Interaction between proteins localized in membranes. *Proc. Natl. Acad. Sci. U.S.A.* **83**:6258–6262.

Greenhut, S. F., Bourgeois, V. R., and Roseman, M. A., 1986, Distribution of cytochrome b5 between small and large unilamellar phospholipid vesicles. *J. Biol. Chem.* **261**:3670–3675.

Griffin, W. C., 1949, Classification of surface-active agents by "HLB," *J. Soc. Cosmet. Chem.* **1**:311–319.

Holowka, D., Hartmann, H., Kanellopoulos, J., and Metzger, H., 1980, Association of the receptor for immunoglobulin E with an endogenous polypeptide on rat basophilic leukemia cells. *J. Recept. Res.* **1**:41–68.

Holowka, D., and Metzger, H., 1982, Further characterization of the beta-component of the receptor for immunoglobulin E. *Mol. Immunol.* **19**:219–227.

Isersky, C., Metzger, H., and Buell, D. N., 1975, Cell cycle-associated changes in receptors for IgE during growth and differentiation of a rat basophilic leukemia cell line. *J. Exp. Med.* **141**:1147–1162.

Isersky, C., Taurog, J. D., Poy, G., and Metzger, H., 1978, Triggering of cultured neoplastic mast cells by antibodies to the receptor for IgE. *J. Immunol.* **121**:549–558.

Ishizaka, T., and Ishizaka, K., 1978, Triggering of histamine release from rat mast cells by divalent antibodies against IgE-receptors. *J. Immunol.* **120**:800–805.

Kanellopoulos, J., Rossi, G., and Metzger, H., 1979, Preparative isolation of the cell receptor for immunoglobulin E. *J. Biol. Chem.* **254**:7691–7697.

Kanellopoulos, J. M., Liu, T. Y., Poy, G., and Metzger, H., 1980, Composition and subunit structure of the cell receptor for immunoglobulin E. *J. Biol. Chem.* **255**:9060–9066.

Kinet, J. P., Quarto, R., Perez-Montfort, R., and Metzger, H., 1985a, Noncovalently and covalently bound lipid on the receptor for immunoglobulin E. *Biochemistry* **24**:7342–7348.

Kinet, J. P., Alcaraz, G., Leonard, A., Wank, S., and Metzger, H., 1985b, Dissociation of the receptor for immunoglobulin E in mild detergents. *Biochemistry* **24**:4117–4124.

Kinet, J. P., Metzger, H., Hakimi, J., and Kochan, J., 1987, A cDNA presumptively coding for the alpha subunit of the receptor with high affinity for immunoglobulin E. *Biochemistry* **26**:4605–4610.

Kinet, J. P., Blank, U., Ra, C., White, K., Metzger, H., and Kochan, J., 1988, Isolation and characterization of cDNA coding for the β subunit of the rat high affinity receptor for immunoglobulin E, *Proc. Natl. Acad. Sci. USA.* **85**:6483–6487.

Kulczycki, A. Jr., McNearney, T. A., and Parker, C. W., 1976, The rat basophilic leukemia cell receptor for IgE I. Characterization as a glycoprotein. *J. Immunol.* **117**:661–665.

Liu, F. T., Bohn, J. W., Ferry, E. L., Yamamoto, H., Molinaro, C. A., Sherman, L. A., Klinman, N. R., and Katz, D. H., 1980, Monoclonal dinitrophenyl-specific murine IgE antibody: Preparation, isolation, and characterization. *J. Immunol.* **124**:2728–2737.

Mazurek, N., Berger, G., and Pecht, I., 1980, A binding site on mast cells and basophils for the anti-allergic drug cromolyn. *Nature (London)* **286**:722–723.

Mazurek, N., Bashkin, P., and Pecht, I., 1982, Isolation of a basophilic membrane protein binding the anti-allergic drug cromolyn. *EMBO J.* **1**:585–590.

Mazurek, N., Bashkin, P., Loyter, A., and Pecht, I., 1983a, Restoration of Ca^{2+} influx and degranulation capacity of variant RBL-2H3 cells upon implantation of isolated cromolyn binding protein. *Proc. Natl. Acad. Sci. U.S.A.* **80**:6014–6018.

Mazurek, N., Bashkin, P., Petrank, A., and Pecht, I., 1983b, Basophil variants with impaired cromoglycate binding do not respond to an immunological degranulation stimulus. *Nature* **303**:528–530.

Mazurek, N., Schindler, H., Schurholz, T., and Pecht, I., 1984, The cromolyn binding protein constitutes the Ca^{2+} channel of basophils opening upon immunoglobulin stimulus. *Proc. Natl. Acad. Sci. U.S.A.* **81**:6841–6845.

Mazurek, N., Dulic, V., Pecht, I., Schindler, H. G., and Rivnay, B., 1986, The role of the Fc epsilon receptor in calcium channel opening in rat basophilic leukemia cells. *Immunol. Lett.* **12**:31–35.

Menon, A. K., Holowka, D., and Biard, B., 1984, Small oligomers of immunoglobulin E (IgE) cause large-scale clustering of IgE receptors on the surface of rat basophilic leukemia cells. *J. Cell Biol.* **98**:577–583.

Menon, A. K., Holowka, D., Webb, W. W., and Baird, B., 1986a, Cross-linking of receptor-bound IgE to aggregates larger than dimers leads to rapid immobilization. *J. Cell Biol.* **102**:541–550.

Menon, A. K., Holowka, D., Webb, W. W., and Baird, B., 1986b, Clustering, mobility, and triggering activity of small oligomers of immunoglobulin E on rat basophilic leukemia cells. *J. Cell Biol.* **102**:534–540.

Metzger, H., Rivnay, B., Henkart, M., Kanner, B., Kinet, J. P., and Perez-Montfort, R., 1984,

Analysis of the structure and function of the receptor for immunoglobulin E. *Mol. Immunol.* **21**:1167–1173.

Metzger, H., Alcaraz, G., Hohman, R., Kinet, J. P., Pribluda, V., and Quarto, R., 1986, The receptor with high affinity for immunoglobulin E. *Annu. Rev. Immunol.* **4**:419–470.

Newman, S. A., Rossi, G., and Metzger, H., 1977, Molecular weight and valence of the cell-surface receptor for immunoglobulin E. *Proc. Natl. Acad. Sci. U.S.A.* **74**:869–872.

Pecht, I., Dulic, V., Rivnay, B., and Corcia, A., 1986, Transmembrane signaling in basophils: Ion conductance measurements on planar bilayers reconstituted with purified Fcε receptor and the cromolyn binding protein, in *Mast Cell Differentiation and Heterogeneity* (A. D. Befus, ed.), pp. 301–312, Raven Press, New York.

Perez-Montfort, R., Kinet, J. P., and Metzger, H., 1983, A previously unrecognized subunit of the receptor for immunoglobulin E. *Biochemistry* **22**:5722–5728.

Reynolds, J. A., Nozaki, Y., and Tanford, C., 1983, Gel-exclusion chromatography on S1000 Sephacryl: Application to phospholipid vesicles. *Anal. Biochem.* **130**:471–474.

Rivera, J., Kinet, J. P., Kim, J., Pucillo, C., and Metzger, H., 1988, Studies with a monoclonal antibody to the β-subunit of the receptor with high affinity for immunoglobulin E. *Mol. Immunol.* **25**:647–661.

Rivnay, B., and Fischer, G., 1986, Phospholipid distribution in the microenvironment of the immunoglobulin E-receptor from rat basophilic leukemia cell membrane. *Biochemistry* **25**:5686–5693.

Rivnay, B., and Metzger, H., 1982, Reconstitution of the receptor for immunoglobulin E into liposomes. Conditions for incorporation of the receptor into vesicles. *J. Biol. Chem.* **257**:12800–12808.

Rivnay, B., Wank, S. A., Poy, G., and Metzger, H., 1982, Phospholipids stabilize the interaction between the alpha and beta subunits of the solubilized receptor for immunoglobulin, E. *Biochemistry* **21**:6922–6927.

Rivnay, B., Rossi, G., Henkart, M., and Metzger, H., 1984, Reconstitution of the receptor for immunoglobulin E into liposomes. Reincorporation of purified receptors. *J. Biol. Chem.* **259**:1212–1217.

Schindler, H., 1980, Formation of planar bilayers from artificial or native membrane vesicles. *FEBS Lett.* **122**:77–79.

Schindler, H., Spillecke, F., and Neumann, E., 1984, Different channel properties of *Torpedo* acetycholine receptor monomers and dimers reconstituted in planar membranes. *Proc. Natl. Acad. Sci. U.S.A.* **81**:6222–6226.

Tanford, C., 1973, *The Hydrophobic Effect*, Wiley, New York.

Umbreit, J. N., and Strominger, J. L., 1973, Relation of detergent HLB number to solubilization and stabilization of D-alanine carboxypeptidase from *Bacillus subtilis* membranes. *Proc. Natl. Acad. Sci. U.S.A.* **70**:2997–3001.

Chapter 9

Reconstitution of Acetylcholine Receptors into Planar Lipid Bilayers

Wolfgang Hanke and Heinz Breer

1. INTRODUCTION

The electrical activity of membranes is produced and regulated by the coordinated gating of specific ion channels. Accordingly, the exploration of the molecular basis of membrane excitation and synaptic transmission has required considerable investigation of ion channels, the molecular units that mainly control ion permeabilities of cell membranes. Studies on the mechanisms of action for channel proteins depend on the convergence of knowledge about the structure of the membrane proteins, together with a detailed characterization of their function at the molecular level. The nicotinic acetylcholine receptor (AChR), one of the most detailed investigated channel-forming proteins, provides some unique opportunities, justifying the pursuit of this goal.

The nicotinic AChR of postsynaptic membranes it the link between the release of acetylcholine (ACh) at nerve terminals and the postsynaptic membrane conductance change that occurs at cholinergic junctions in nerve and muscle. It converts the binding of the ligand (ACh) into a transient opening of a cation-selective channel. The binding of ACh is thus transduced into a large increase in the cation permeability of postsynaptic membranes and the resulting membrane depolarization can then trigger an action potential.

Wolfgang Hanke and Heinz Breer   Department of Biology and Biophysics, Osnabrück University, Osnabrück D-4500, Federal Republic of Germany.

Nicotinic AChRs are found in muscle as well as muscle-derived electric tissue, in addition to nervous tissue (ganglia and brain) of vertebrates. Rather high concentrations of AChR with nicotinic pharmacology also exist in invertebrate nervous tissue; notably, the ganglia of arthropods appear very rich in nicotinic AChR. However, AChRs from all those different sources obviously differ in their pharmacology as well as their physiological and biochemical properties.

A major goal in neurochemistry has been to characterize the molecular elements involved in ACh- and AChR-mediated signal transmission. In particular, the identification of the AChR molecules from vertebrates, notably from the electric tissue of fish, has been the subject of intensive and successful investigations throughout the last decade. As a result, the nicotinic AChR from electrocytes and muscle cells of vertebrates is considered as one of the best-characterized integral membrane proteins (Conti-Tronconi and Raftery, 1982; Hucho, 1986; Maelicke, 1984; Popot and Changeaux, 1984).

While it has been completely established that the AChR of electrocytes and muscle cells is composed of four different subunits (polypeptides), such that its subunit structure is a_2bcd, much less is known about nicotinic AChRs in nervous tissue. However, recent immunological and molecular approaches have revealed that the nicotinic AChR from vertebrate brain is quite different from the muscle receptor; the neuronal receptor is apparently composed of only two different subunits (polypeptides) (Boulter et al., 1987; Whiting and Lindstrom, 1987). An analysis of receptors from invertebrate ganglia suggests that the neuronal nicotinic AChR from locust may even be formed by a single class of polypeptides (Breer et al., 1985; Hanke and Breer, 1986).

The availability of receptor preparations from different sources with different pharmacological and molecular properties now allows comparative studies on the functions of the ACh-gated cation channels, for example, by analyzing channel function after reconstituting the isolated proteins into artificial membrane systems, such as liposomes and planar lipid bilayers. These approaches may contribute to the exploration of the structure–function relationship of this particular channel type. A comparative function analysis of identified channels from different sources will contribute to the correlation of structure and function of these highly specialized molecules.

Reconstitution of functional channel proteins has never been accomplished in planar lipid bilayers containing decane (black lipid membranes) (Mueller et al., 1962), but in virtually solvent-free membranes, folded bilayers (Montal and Mueller, 1972) and bilayers made on the tip of a glass micropipette, the fluctuations of ion channels activated by cholinergic agonists have been recorded (Hanke, 1985; Montal et al., 1986). Additionally, large liposomes, containing the receptor protein in their membranes, have been investigated by use of the patch-clamp technique (Tank et al., 1983). These bilayer techniques and the preparations used with them are now widely accepted as valid model systems in which the functional properties and integrity of isolated and purified channel proteins (AChRs) can be assayed.

The fact that integral membrane proteins, which form specific ion channels in plasma membranes, can safely be transferred to artificial membranes, such as planar lipid bilayers, not only opens an excellent possibility for the assay of the functional integrity of a purified channel protein, but also permits study of physiologically relevant channels that are not otherwise accessible for electrophysiological techniques. Furthermore, after reconstitution, both sides of the membrane and thus of the channel are easily accessible, which is another advantage of patch-clamp and classical electrophysiology.

The techniques of reconstitution are particularly powerful in cooperation with approaches from classical electrophysiology, biochemistry, and molecular genetics, as well as with ion flux and binding studies. For this reason the results from reconstitution experiments should be seen in a larger context and a strategy should be established, to use each technique for what it can best provide. The principles of such a strategy are given in Figure 1. Gating data necessary

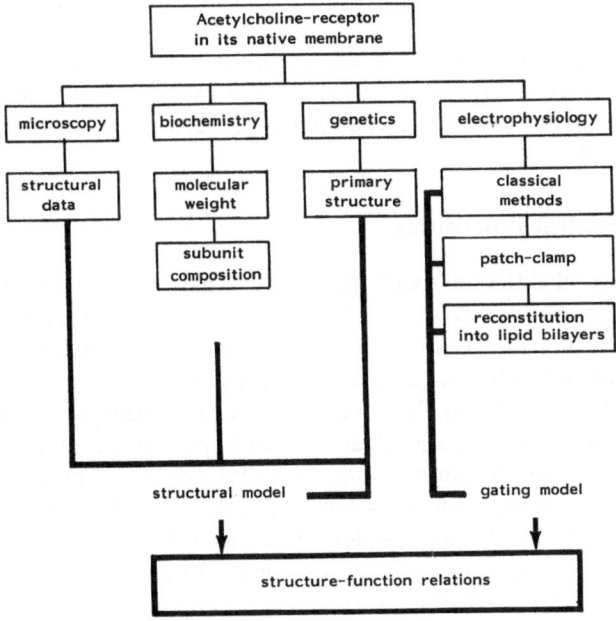

FIGURE 1. Scheme of a strategy on how to investigate acetylcholine-receptor channels using different techniques. The aim of such investigations is to explore the structure of the protein, its function in the membrane, and structure–function relationships by modifying the protein.

for structure–function models and understanding of drug action can only be obtained by electrophysiological (better single-channel) methods in this system; furthermore, the dependence of AChR channels on physiological parameters of the membranes must be studied (almost exclusively) in reconstitution experiments, which allow electrophysiological analysis, with purified proteins incorporated in artificial membranes of known composition and physical state. In addition, bilayer reconstitution techniques are often used to access membrane proteins and ion channels that cannot be reached by other electrophysiological techniques.

2. TECHNICAL ASPECTS OF ACETYLCHOLINE-RECEPTOR RECONSTITUTION

As a prelude to our surveying recent work on reconstituted acetylcholine-receptor channels, it is valuable to introduce some technical aspects of reconstitution experiments. These are comments on technical steps involved in reconstituting proteins into planar lipid bilayers and some particular problems even the advanced experimenter will have to deal with in planar lipid bilayer experiments.

Adequate preparation is an essential prerequisite for any reconstitution experiment. Biochemists involved in this type of experiment will have to do the work of constructing adequate preparations—a crucial point of all the work (see also Chapter 3).

2.1. Biochemical Procedures

A variety of different AChR preparations have been used for bilayer reconstitution experiments. Kasai and Changeaux (1971) demonstrated that AChR-rich vesicle membranes from electric tissue membranes undergo permeability changes induced by cholinergic agonists. These flux studies on vesicles from native membranes suffered, however, from a large background and the low time resolution. Schindler and Quast (1980) improved the time reconstitution by making receptor-containing planar lipid bilayers from monolayers spread from AChR-containing vesicles of native membranes—thus making this preparation accessible to electrophysiology and the higher time resolution of this technique. The incorporation of receptor-enriched membrane fragment vesicles in planar lipid bilayers is still a very valuable procedure used to explore basic properties of ACh-gated channels that may otherwise not be accessible to electrophysiological procedures.

Epstein and Racker (1978) introduced a procedure for incorporating detergent-extracted AChR into soybean lipid vesicles. By using this preparation, an assay system was available for the first time to monitor the activity of channels following the purification of AChRs. Nicotinic AChRs have subsequently been purified to homogeneity by affinity chromatography in nondenaturing detergent solutions, using α toxins or cholinergic analogues as covalently bound column ligands. These highly purified preparations were then used in bilayer reconstitution experiments (Boheim *et al.*, 1981; Nelson *et al.*, 1980). In many cases it was found that stabilizing amounts of additional phospholipids have to be present during the complete purification procedure (Montal *et al.*, 1986). Recently, α-toxin binding protein has also been isolated from the nervous tissue of insects using α-bungarotoxin as affinity ligand (Breer *et al.*, 1985; Hanke and Breer, 1986). This protein appeared to be a functional agonist-activated cation channel with nicotinic pharmacology (Hanke and Breer, 1986, 1987). Biochemical analysis revealed that this protein is composed of four or five

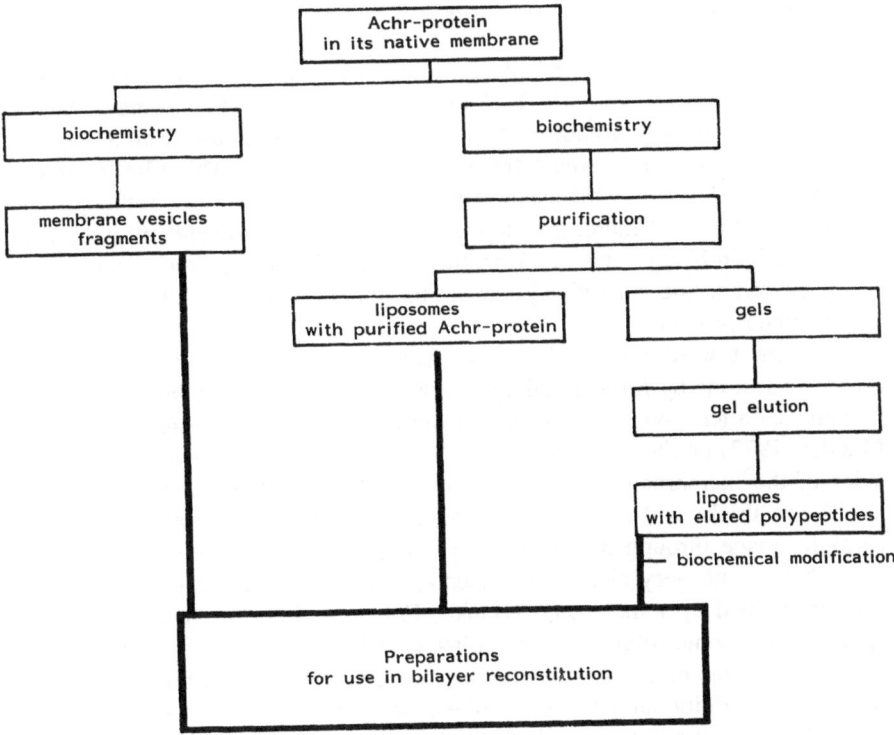

FIGURE 2. Schematic presentation of a way to prepare acetylcholine-receptor protein in different degrees of purity for use in bilayer reconstitution experiments. Any source of AChRs may deliver starting material for this procedure.

identical or very similar subunits (polypeptides) and thus is very different from the hetero-oligomeric AChR from muscle cells and electrocytes. The possible homo-oligomeric structure of this neuronal AChR from insect brain encouraged a further step of purification: eluting receptor polypeptides from SDS–polyacrylamide gels after electrophoretic separation, renaturing them in an appropriate buffer, and incorporating them into liposomes for use with planar lipid bilayer reconstitution experiments. Very recently, experimental conditions have been established, by use of which functional (cholinergic) agonist-activated channels could be renatured from such SDS–polyacrylamide gel eluted polypeptides of insect AChR and were found in planar lipid bilayers after fusing with liposomes containing the proteins.

To summarize, of the different preparations that may be used in bilayer reconstitution experiments, only the above are to be discussed here; a scheme for preparing AChRs for such reconstitution experiments is given in Figure 2.

2.2. Formation of Planar Lipid Bilayers

The first technique to form planar lipid bilayers was developed by Mueller and co-workers in the early 1960s (Mueller et al., 1962). They formed bilayers on partitions between two aqueous solutions by painting a film from a decane–lipid solution on a small hole (100- to 500-μm diameter). This film became a "black lipid bilayer" by diffusion of the decane out of the film. Although this technique is easy to handle, the large amount of organic solvent still present in the planar lipid bilayer after its formation is a disadvantage. When using bilayers of this type, one must always ensure that the decane does not influence the reconstituted proteins.

A step toward more solvent-free bilayers, and thus toward real membranes, was done by Montal and co-workers, who assembled bilayers from two monolayers after having spread these from hexane–lipid solutions (Montal and Mueller, 1972) on the air–water interfaces of the two aqueous solutions on both sides of the membrane-carrying partition. Such a partition must be very thin (5–10 μm thick) and a hole with a diameter from 100 to 300 μm must be carefully made through it. Advanced techniques, especially those for making small holes with very clean edges (septa) through preformed partitions, have been developed by some groups (Hanke, 1985). Schindler went a step further, spreading liposome solutions as monolayers and assembling bilayers from these truly solvent-free monolayers (Schindler and Quast, 1980). In both techniques, however, an additional amount of solvent may come into the bilayer from the pretreatment of the hole, on which the bilayer is formed. This pretreatment is necessary to stabilize the bilayer and is usually done with hexadecane or squalene. Whether this organic solvent affects the reconstituted proteins or not is another issue.

Reconstitution of Acetylcholine Receptors

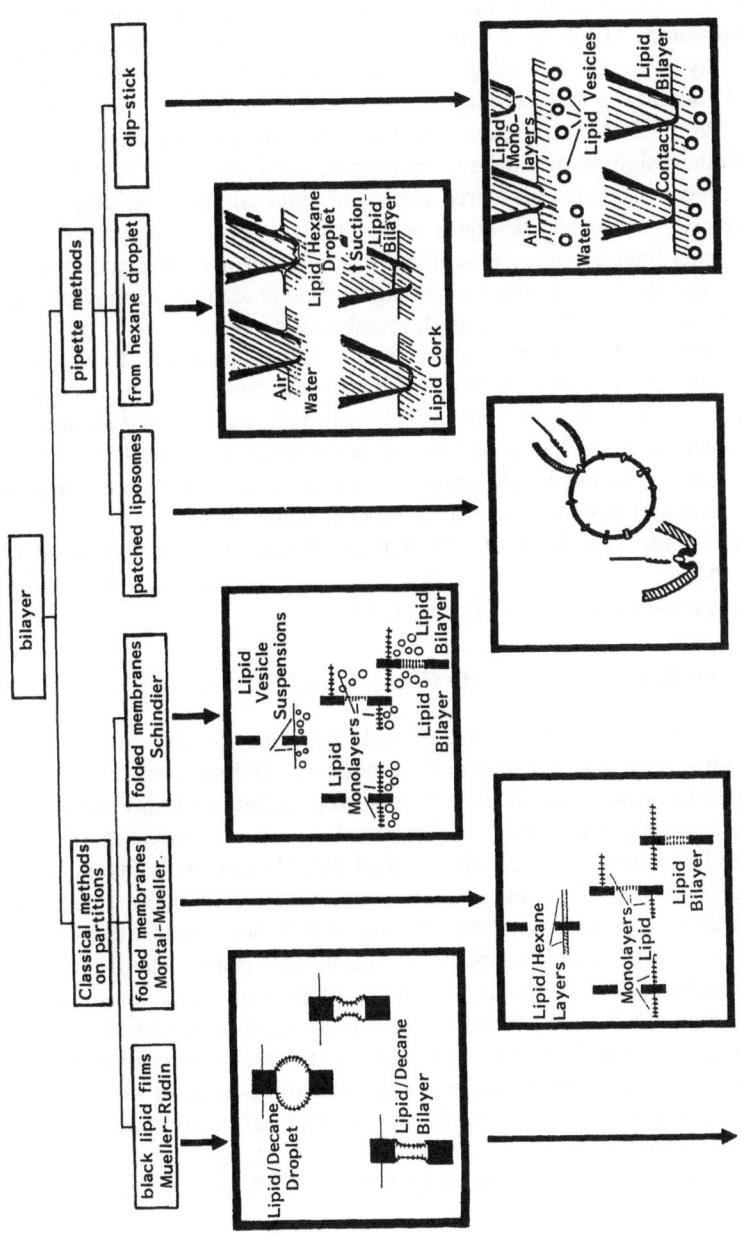

FIGURE 3. In this figure all important techniques to make planar lipid bilayers are summarized. A simplified sketch of each technique is given.

Other approaches to the formation of planar lipid bilayers were influenced by the patch-clamp technique. Tank and co-workers successfully patched large preformed liposomes (Tank et al., 1983), and several other groups succeeded in forming bilayers on the tips of glass pipettes (Coronado and Latorre, 1983; Hanke, 1985; Schürholz and Schindler, 1983; Suarez-Isla et al., 1983; Wilmsen et al., 1983). Both techniques combine some advantages of the patch-clamp technique—high resolution and membrane stability—with the main advantages of bilayer experiments—free choice of lipid composition and use of purified proteins. Additionally, bilayers on tips of glass pipettes may be formed without any additional organic solvent, which helps in avoiding artificial effects. Besides the above-mentioned advantages, pipette techniques may also be used to investigate channels at very high membrane potentials.

An interesting aspect of these glass pipette techniques is that they can be performed on any patch-clamp, setup, making reconstitution techniques accessible to a larger number of electrophysiologists; in fact, patch-clamp techniques have been established in a large number of laboratories over the last decade. Difficulty in gaining access to the aqueous solution inside the pipette must be considered as a major disadvantage of all pipette techniques.

In Figure 3 the principles of different procedures for the formation of bilayers are shown. The interested reader may find more detailed information in some recent reviews (e.g., see Hanke, 1985).

2.3. Incorporation of Proteins into Planar Lipid Bilayers

Any of the bilayer techniques mentioned above may be the starting point for incorporating proteins into planar lipid bilayers. Different strategies have been developed to do so. The purified protein preparations (but also membrane fragment preparations) are usually incorporated into the membranes of liposomes. These liposomes may be either spread directly into monolayers to assemble bilayers as first done by Schindler and Quast (1980), or fused into preformed bilayers (see Cohen, 1986; Hanke, 1986); here again different approaches can be used to establish appropriate fusion rates (Hanke, 1986). Whereas the first technique suffers from the possibility of proteins denaturation at the air-water interface, the second technique has as its main drawback the control of the fusion process itself (Hanke, 1986). Nevertheless, both techniques have been used successfully to reconstitute the acetylcholine-receptor channel into planar lipid bilayers. However, no successful attempt has been reported to date for the reconstitution of functional acetylcholine-receptor channels in decane-containing planar lipid bilayers, indicating that the AChR may be very sensitive to organic solvent.

Alternatively, the acetylcholine receptor has been investigated in patched liposomes (Tank et al., 1983) and in bilayers on tips of glass pipettes (Montal

Reconstitution of Acetylcholine Receptors

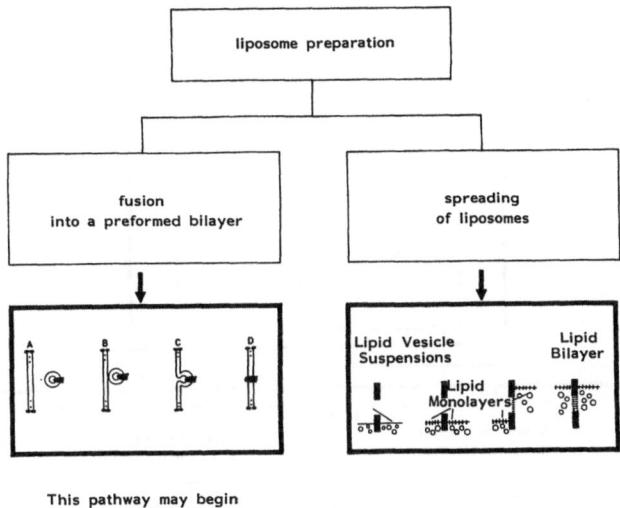

FIGURE 4. Principles of the most important techniques for incorporation of membrane-integral proteins in planar lipid bilayers are given here. They may be used together with classical as well as with dip-stick bilayers.

et al., 1986). By using these approaches, better experimental resolution was obtained, due to the small area of the bilayer used. The data obtained by these techniques are in agreement with those from more classical approaches.

An overview of techniques employed to incorporate proteins into planar lipid bilayers is given in Figure 4.

2.4. How to Collect and Process Data

Figure 5 shows a scheme of a bilayer setup (electrical and mechanical equipment) that is useful for nearly all types of bilayer experiments. Extensions may be necessary for some specific types of investigation, especially when different aqueous or physical environments are needed. Multi- and single-channel experimentals are usually done with such a setup and the data (current recordings of multichannel systems or single-channel current fluctuations) are collected on a tape recorder or other storage medium. Principles of mechanical and electrical components of the setup have been described in detail in the literature (e.g., see Hanke, 1985). Data evaluation, including advanced statistical procedures, depends on specific scientific problems under investigation. A

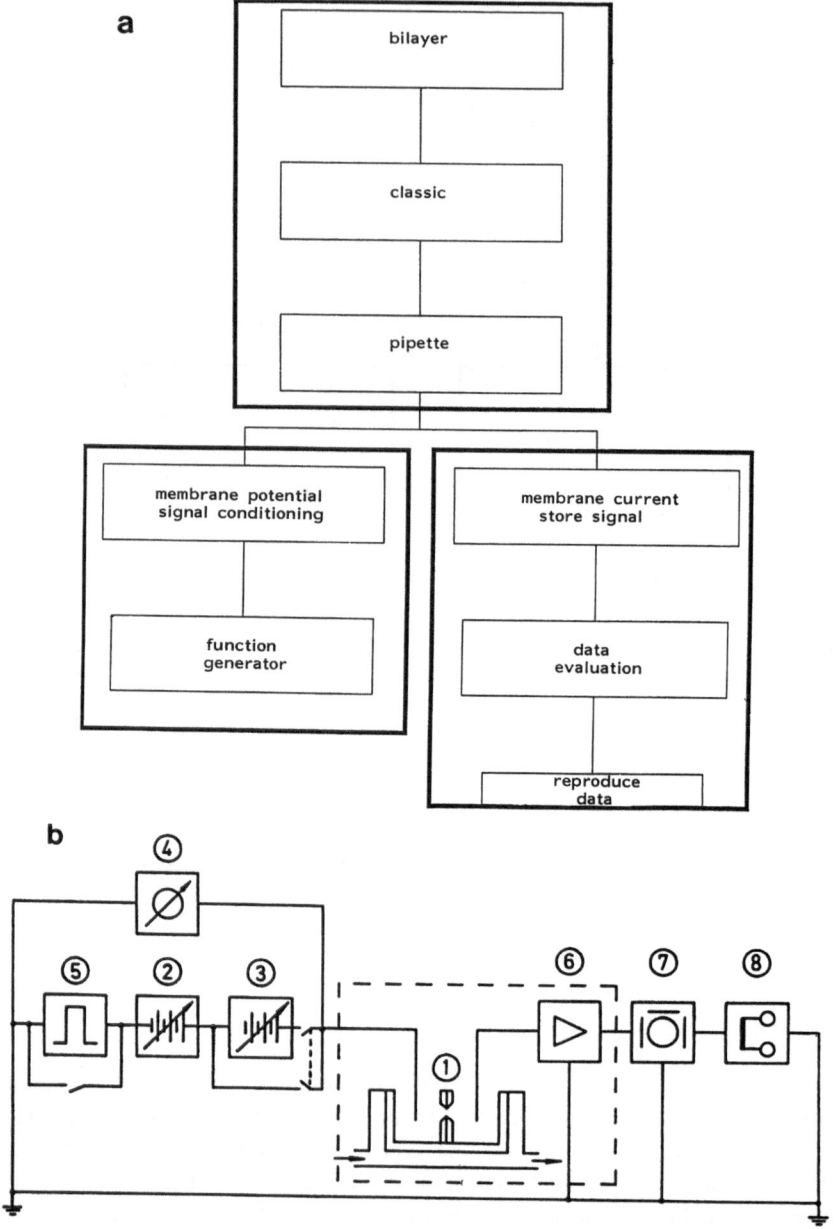

FIGURE 5. Principles of a bilayer setup (b) (electrical and mechanical) are presented together with an overview of how to construct such a setup (a). Any data evaluation may be carried out following the storing device: (1) bilayer (chamber or dip-stick) together with temperature regulation and measurement, electrodes, and other mechanical equipment; (2) voltage source; (3) offset regulation; (4) voltmeter; (5) function generator; (6) current–voltage converter; (7) oscilloscope; (8) data-storage device.

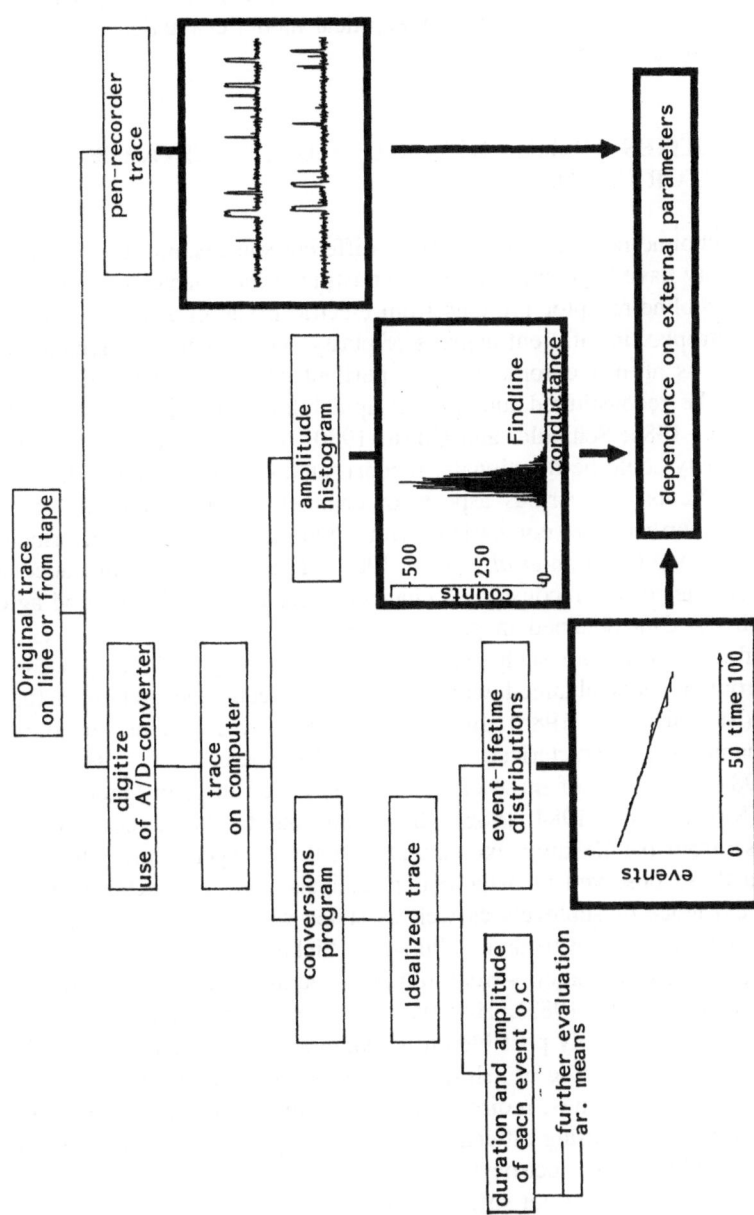

FIGURE 6. Scheme of evaluating single-channel data on a computer. This evaluation may be extended for special needs. Commercial software is available to do the data evaluation.

more general procedure for evaluating single-channel data is given in Figure 6. This procedure can currently be done with commercially available patch-clamp software and, based on such data, a theoretical model evaluation may follow, as explained in some detail below.

3. PROPERTIES OF RECONSTITUTED ACETYLCHOLINE-RECEPTOR CHANNELS

Acetylcholine-receptor proteins from different sources and in different degrees of purity have been successfully reconstituted into planar lipid bilayers.

Acetylcholine-receptor proteins from electric tissue of *Torpedo* and electric eel electroplax in different degrees of purity and membrane fragment vesicles as well as affinity-chromatography purified protein were the first AChR channels to be reconstituted into planar lipid bilayers (Boheim *et al.*, 1981; Nelson *et al.*, 1980; Schindler and Quast, 1980). It was demonstrated that the channel activity could be functionally reconstituted into planar lipid bilayers by these experiments and various aspects of channel function were analyzed in some detail. Furthermore, comparison with patch-clamp data was performed in an additional step (Boheim *et al.*, 1981, 1982). The observed conductance and gating properties of the reconstituted channels were found to be in good agreement with the data obtained in patch-clamp experimentals. Traces of single-channel fluctuations from such experimentals are in Figure 7; they were recorded from bilayers obtained with the dip-stick technique and from folded bilayers (Montal *et al.*, 1986). In more advanced investigations, the purified AChR channel was characterized in added detail (Labarca *et al.*, 1984; Montal *et al.*, 1984). The data of these experiments have led to gating models of the channel (Montal *et al.*, 1984, 1986) which were found to be in good agreement with those from patch-clamp experiments. In these experiments, phenomena similar to those observed in patch-clamp analysis—like conductance parameters, the existence of sublevels as well as states with lower conductance, the ion selectivity and gating behavior, bursting, flickering, and desensitization at higher agonist concentration—were found (Colquhoun and Sakmann, 1985; Sakmann *et al.*, 1980, 1983). However, a comparison of reconstitution and patch-clamp data was only possible on preparations from different sources, using data from electric tissue of fish (reconstitution) and data from muscle membranes (patch-clamp). Thus, a direct comparison of data for one type of AChR from identical material using different techniques is still open. Data about patch-clamping membranes of electrocytes have only recently been published (Hesse *et al.*, 1985), but they are not sufficient for a more detailed discussion.

Up to 1986, the impossibility of reconstituting AChRs from sources other than electric tissue resulted in the problems previously mentioned. Thus, for

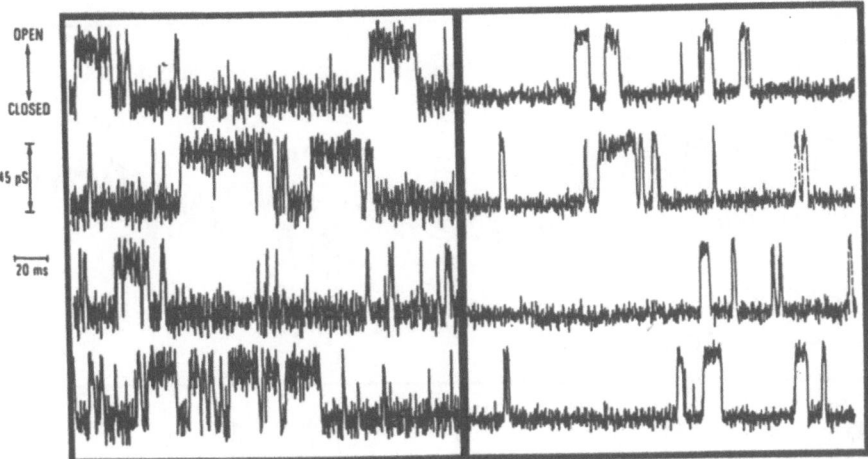

FIGURE 7. Single-channel recordings of the purified *Torpedo* acetylcholine-receptor protein, reconstituted into planar lipid bilayers: left side folded bilayer, right side dip-stick technique (otherwise identical conditions; channels were activated by suberyldicholine). The better resolution of the dip-stick technique is obvious; however, the channel parameters are not changed significantly by the technique used. This figure was reproduced from Montal *et al.* (1986).

quite a long time it seemed to be clear that only AChRs from the exceptional source, the electric tissue, could be purified and reconstituted. This was changed and another interesting field of reconstitution experiments was opened when, after purification, Breer *et al.* (1985) first reconstituted an AChR from insects into planar lipid bilayers (Hanke and Breer, 1986). After a more detailed functional characterization of this protein (Hanke and Breer, 1986, 1987), some interesting features of this neuronal AChR channel emerged. Comparison to AChR channels from *Torpedo* electroplax revealed differences in pharmacology as well as similarities in gating and conducting parameters (Hanke and Breer, 1987). A trace of channel fluctuations from neuronal insect AChR reconstituted into planar lipid bilayers and activated by carbamylcholine is shown in Figure 8b. In Figure 8a, a multichannel recording is given, demonstrating an increase of conductance as channels are activated by an agonist (Carb) and a decrease in conductance after channel inactivation by an antagonist (curarine), to verify the pharmacological integrity of the purified and reconstituted protein. In Table I various parameters of different reconstituted AChR channels are summarized; in parallel, data from other sources (patch-clamp data from muscle cells) are also listed. The first patch-clamp data from isolated nerve cells of insects, the source of purified protein, have been published (Beadle and Lees, 1986; Satelle, 1986) or are in the process of being obtained (E. Tareilus and W. Hanke,

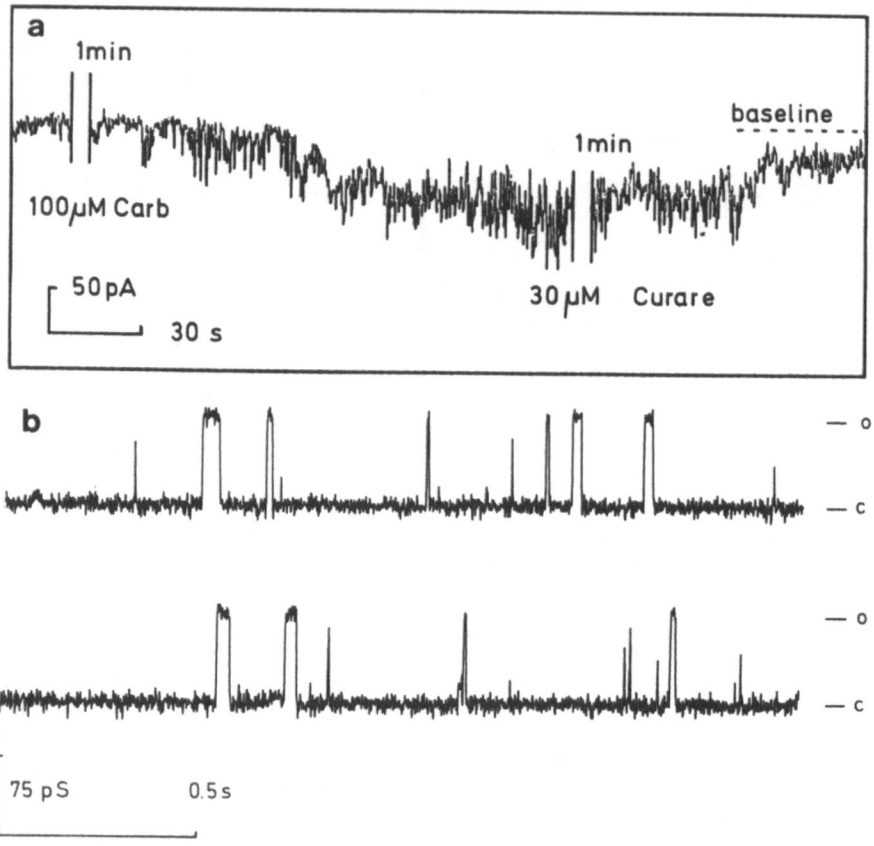

FIGURE 8. (a) Single-channel recording of the purified locust neuronal acetylcholine receptor activated by carbamylcholine in a planar lipid bilayer. Protein was incorporated in the bilayer by the fusion technique. (b) Multichannel recording of neuronal AChRs in planar lipid bilayers, showing activation by Carb and inactivation by curarine, to demonstrate the pharmacological integrity of the purified protein.

in preparation). Additionally, some data from patch-clamp experiments with neuronal AChR from chick ciliary ganglions have been published very recently (Margiotta *et al.*, 1987). Extending these investigations will permit comparison of data obtained by different experimental approaches in analyzing an AChR from one and the same source.

The AChR channel is defined by its pharmacological properties, notably from its activation by cholinergic agonists. Thus, each investigation of this channel will have to deal with its characterization by the use of a variety of different drugs. In bilayer reconstitution experiments the *Torpedo* AChR chan-

Table I[a,b]
Parameters of Reconstituted AChR Channels

Channel properties	Locusta		Torpedo		Muscle	
	Reconstitution	Patch	Reconstitution	Patch	Reconstitution	Patch
Maximum conductance (pS)	100	—	95	—	—	100
Na$^+$ concentration for half maximum conductance (mM)	50	—	400	—	—	?
Approximate cooperativity of agonists	1	—	2	—	—	2
Mean open time at low agonist concentration (msec)	3	?	5	?	—	1
Block by curarine	++	—	++	—	—	++
Block by hexamethonium	+	—	++	—	—	++
Potency of some agonists	S-A-C	—	S-A-C	—	—	S-A-C
Subunit composition	*eventually or probably homo-oligomeric		a$_2$bcd		a$_2$bcd	

[a] In this table some single-channel data of different acetylcholine-receptor channels are summarized together with some pharmacological aspects. Patch-clamp and reconstitution data are taken into account. All data are single-channel data.

[b] — = no data available; ? = no sufficient data available, but the data published are in agreement with the data summarized here; S = Suberyldicholine; A = acetylcholine; C = carbamylcholine; * = data are preliminary; the protein may be a homo-oligomer or may be composed from two very similar subunits. In case of homo-oligomeric protein, from none denaturing gels a number of five subunits may be estimated.

nel has been found to be activated by different agonists, such as acetylcholine (ACh), suberyldicholine (Sub), and carbamylcholine (Carb), and blocked by some antagonists, such as hexamethonium and d-tubocurarine. Additionally, the action of local anesthetics such as QX 222 (Boheim et al., 1981) has been demonstrated. Comparable pharmacological experiments have recently been performed with the reconstituted neuronal AChR from insects (Hanke and Breer, 1986, 1987), resulting in interesting differences in its response to agonist concentration and in particular to hexamethonium. The results on hexamethonium action are nevertheless in good agreement with pharmacological action in binding studies. In Table I some pharmacological aspects of the different reconstituted AChRs are given.

Since AChR-linked ion channels are activated by specific agonists, their response to various concentrations of the stimulating ligands is of particular interest—not only to explore the ligand affinity of the receptor but also to obtain some information on the kinetics of channel activation. However, at least for reconstitution experiments, there is still a considerable lack of appropriate data on the concentration dependence of channel activity. This is partly due to problems concerning desensitization of AChRs at high agonist concentration and also due to problems in evaluating the data obtained in such experiments. The verification of the number of active channels in the bilayer, for example, is very crucial for data evaluation, but this number is very difficult to estimate exactly (Hanke and Breer, 1988).

In order to investigate the dependence of channel properties on agonist concentration, however, the number of active channels in the bilayer must be determined and this can only be achieved by statistical approaches. Nevertheless, only when all such data are available can a model for receptor activation be developed.

A major advantage of reconstituting channel proteins into planar lipid bilayers is the incorporation and functional analysis of purified proteins in a defined lipid environment (Hanke, 1985). Only the use of a planar lipid bilayer made from one defined lipid containing one purified protein facilitates the construction of a two-component system that may eventually be described theoretically. For a native membrane with all the many different lipid and protein components, it will be very difficult if not impossible to understand all the interactions between these components and to describe the complete system or even to select and understand in detail one particular functional molecule.

In the literature, a great deal of data dealing with purified AChR proteins in planar lipid bilayers have been presented. Most of the data were obtained using receptor proteins from the electric organ of fish; some data were obtained using receptor preparations from insect nervous tissue. Some additional data were recently collected using preparations from the nervous and muscle tissue of earthworms (H. Breer and W. Hanke, unpublished observation). An exten-

Reconstitution of Acetylcholine Receptors

sion of these investigations to AChRs from other sources and a detailed comparison of the results obtained may be a desirable goal for future experiments.

Very recently, data have been published on the successful incorporation of proteins, which were eluted from SDS–polyacrylamide gels, into planar lipid bilayers (Young *et al.*, 1987). This approach has never previously been successfully applied to the subunits of AChR from electric organs of fish. This failure may have been due to the rather complicated substructure of the AChR. We have recently tried to reconstitute the receptor polypeptides from insect ganglia into planar lipid bilayers, after separation of SDS–polyacrylamide gel followed by electroelution (H. Breer and W. Hanke, unpublished observation). In Figure 9 a trace of Carb-activated channel fluctuations is shown, recorded from a preparation containing polypeptides eluted from SDS–polyacrylamide gels reconstituted into a planar lipid bilayer. Data evaluation of such traces revealed that the channel activity recorded very closely resembles or is even identical to the channel formed by the native planar lipid bilayer. Thus, the polypeptides eluted from SDS–polyacrylamide gels apparently do form functional channels activated by cholinergic agonists. Following this finding, it can be concluded that polypeptides isolated from insect nervous tissue, which mi-

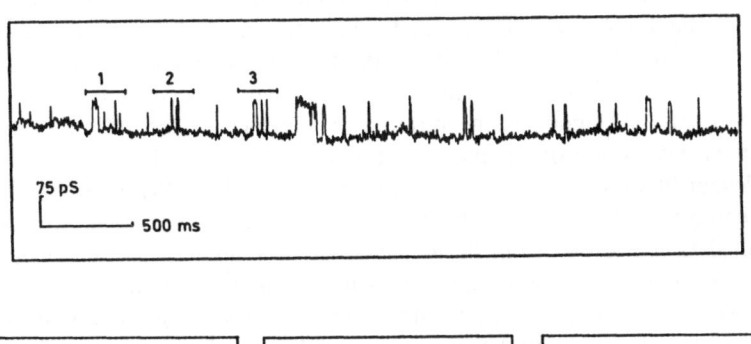

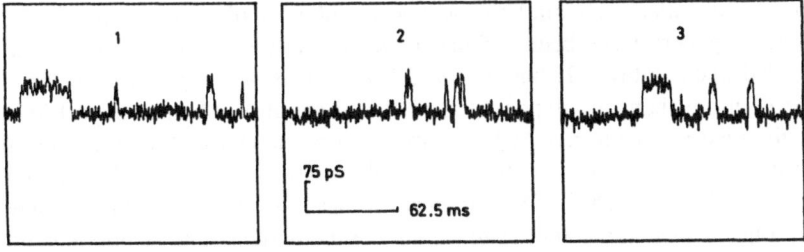

FIGURE 9. Single-channel recording of SDS–polyacrylamide gel eluted neuronal acetylcholine-receptor polypeptide after reconstitution into a planar lipid bilayer. The channel fluctuations activated by carbamylcholine are very similar or even identical with those from the native protein in conductance and lifetime.

grate as a single band on SDS–polyacrylamide are sufficient to form functional AChR channels in planar lipid bilayers.

A consequent application of this technique will now allow for the investigation of chemically modified proteins; furthermore, modifications of this approach may successfully be employed to investigate the functional relevance and the possibility of substitutions of the various subunits forming the peripheral AChR of vertebrates. This type of question can presently only be approached by the use and application of techniques of molecular genetics, that is, expression of AChR subunits in oocyte membranes after injection of subunit-specific mRNA (Mishina et al., 1984, 1985; Sakmann et al., 1985). Detailed knowledge on the relevance of different subunits for receptor action, however, is essential in evaluating the principal requirements for a functional ACh-activated ion channel.

4. GATING MODELS AND THEORETICAL ASPECTS

The results from electrophysiological, biochemical, and ion-flux studies may lead to models explaining the function of AChR channels. These models can help to explain some basic properties of the AChR, such as the dependence of the channel properties on agonist concentration and membrane potential (Colquhoun and Hawkes, 1977, 1981, 1982) or the selectivity and conductance parameters of the channel. In this chapter we focus on questions dealing with gating models and agonist-concentration dependence of the AChRs. Detailed theoretical description of channel gating is avoided, since this can be found in the relevant literature (e.g., see Colquhoun and Hawkes, 1977, 1981, 1982).

Gating models of AChR channels may be derived especially from single-channel data as well as from the dependence of single-channel parameters on agonist concentration. The peripheral AChR channel has been demonstrated to be activated by two agonist molecules, and it has been suggested that the two agonist molecules may bind to the two α subunits of this protein. Recently, however, papers have been published demonstrating the presence of "single liganded open states" (Labarca et al., 1985). This was derived from the presence of more than one exponential in the distribution of open-state lifetimes. For the nicotinic neuronal AChR from insect brain, it was found in concentration-dependent experiments, that the resulting Hill plot gave a slope of *only 1* (Hanke and Beer, 1987, 1988), indicating that this AChR may be activated by only one agonist molecule. This finding is consistent with a possible homo-oligomeric structure for the protein and indicates the possibility of reconstituting a functional receptor from a single SDS–polyacrylamide-eluted polypeptide band (see Figure 10). In addition to the above findings, in patch-clamp exper-

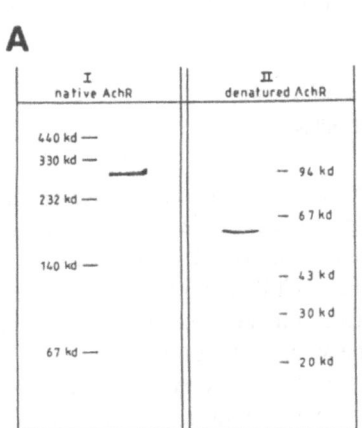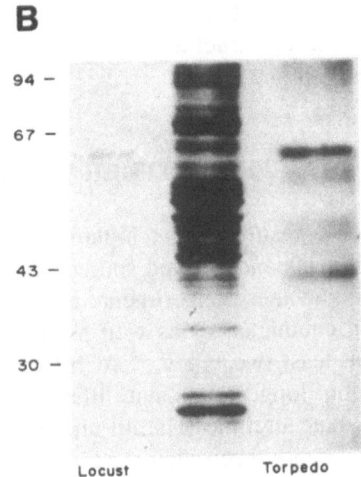

FIGURE 10. Gels of different acetylcholine receptor preparations for use in bilayer reconstitution experiments. (A) Native (I) and SDS–polyacrylamide gels (II) of neuronal AChR from locust tissue. (B) An SDS–polyacrylamide gel of neuronal AChR from locust compared with the peripheral AChR from *Torpedo*.

iments even spontaneous activity (i.e., ion channel fluctuations) has been reported without an agonist being present, for the peripheral AChR channel (Jackson, 1984, 1986). Similar observations in bilayer reconstitution experiments are not so convincing, due to the problems with artifacts in this kind of experiment.

Nevertheless, all these data suggest a minimum channel-gating model, given as

$$R + 2L \rightleftharpoons RL_1 + L \rightleftharpoons RL_2$$
$$\updownarrow \qquad \updownarrow \qquad \updownarrow$$
$$R^* + 2L \rightleftharpoons R^*L_1 + L \rightleftharpoons R^*L_2$$

This model does not include any desensitization and other additional states; it is a minimum model for receptor activation by up to two agonist molecules, as postulated from the experimental data. Dependence of open-state probability P_o and state lifetimes t_i on agonist concentration may be recalculated from this model, and statistical tests can be used to verify each part of it.

A portion of the available data, especially those for the neuronal AChR from insect, may be fitted by an even simpler model with the assumption that the main part of the channel activity can be explained by an open state with only one bound agonist molecule. Collection and evaluation of the data needed to construct a more "real" model will be one of the problems of future investigations.

5. DATA FROM OTHER EXPERIMENTS

The results of patch-clamp experiments on peripheral AChRs in muscle membranes and derived sources again support the previously mentioned gating model. In these experiments additional classes of AChR open states with different conductances as well as sublevels for AChR channels have been found and at least two types of AChR can be discriminated—one with low conductance but longer open-state lifetime and one with high conductance but shorter open-state lifetime (Hamill and Sakmann, 1981). These two classes of AChR have recently been identified as receptor complexes with different subunit composition (Mishina et al., 1986).

After injection of subunit-specific mRNA, different classes of AChRs have been detected in the membrane of *Xenopus* oocytes (Mishina et al., 1986; Sakmann et al., 1985) and it has been shown that some functional AChR-channel fluctuations could already be found if all four subunit-specific mRNAs had not been injected (Mishina et al., 1986). Furthermore, for the neuronal AChR it has recently been demonstrated that mRNA for a_3 or a_4 and b are sufficient to induce the functional expression of nicotinic AChRs in *Xenopus* oocytes (Boulter et al., 1987).

6. PRINCIPAL REQUIREMENTS FOR A FUNCTIONAL ACETYLCHOLINE-RECEPTOR CHANNEL

As discussed above, AChRs have been described with a biochemical subunit composition a_2bcd (muscle AChR and AChR of electric organs) and neuronal AChRs, which are supposed to be composed of two different subunits in vertebrates, or even of identical subunits in insect neuronal tissue. Additional evidence for the formation of homo-oligomeric receptor complexes comes from reconstitution experiments using neuronal receptor polypeptides from insects eluted from SDS–polyacrylamide gels. Furthermore, open states of the AChR with one and two ligands bound have been described, and even spontaneous activity without agonist being present was found. Thus, in spite of the enormous efforts to explore the nicotinic AChR, it is still an open question what elements are essential to form a functional AChR channel. A large amount of work will be necessary, especially in structure–function research, to verify a minimum model. New data toward this goal can come from patch-clamp analysis of AChR channels in oocyte membranes after injection of (modified) subunit-specific mRNA, as well as from reconstitution experiments using SDS–polyacrylamide gel-separated and -eluted (modified) polypeptides.

7. SUMMARY AND CONCLUSION

Obviously, bilayer reconstitution experiments have largely contributed to the understanding of the AChR-channel function. Nevertheless, at present there are many unanswered questions concerning the minimum structural requirements for AChR-channel function, agonist cooperativity, and different types of AChR. Another complex of parameters important for receptor function which must be explored in much more detail, is the dependence of AChR-channel function on membrane composition and its physical state. This important but rather neglected field is predestined to be explored by reconstitution techniques.

All the results on AChRs reconstituted in planar lipid bilayers cannot adequately be discussed without the data obtained by other techniques, thus coming back to the statements already mentioned in the introductory section about strategies for investigating ion channels in general. Only such a strategy can lead to a molecular understanding of channel function.

ACKNOWLEDGMENTS. This work was supported by the Deutsche Forschungsgemeinschaft, SFB 171, C5, and C11. We thank Dr. W. Junge for continuous support.

8. REFERENCES

Beadle, D. J., and Lees, G., 1986, Insect neuronal cultures—a new tool in insect neuropharmacology, in *Neuropharmacology and Pesticide Action* (M. G. Ford, G. G. Lunt, R. C. Reay, and P. N. R. Usherwood, eds.), pp. 423–444, Ellis Horwood, Chichester, England.

Boheim, G., Hanke, W., Barrantes, F. J., Eibl, H., Sakmann, B., Fels, G., and Maelicke, A., 1981, Agonist-activated ionic channels in acetylcholine receptor reconstituted into planar lipid bilayers. *Proc. Natl. Acad. Sci. U.S.A.* **78**:3586–3590.

Boheim, G., Hanke, W., Methfessel, C., Eibl, H., Kaupp, U. B., Maelicke, A., and Schultz, J. E., 1982, Membrane reconstitution below lipid phase transition temperature, in *Transport in Biomembranes: Model Systems and Reconstitution* (R. Antolini, ed.), pp. 87–97, Raven Press, New York.

Boulter, J., Evans, K., Goldman, D., Martin, G., Treco, D., Heineman, S., and Patrick, J., 1987, Isolation of cDNA clone coding for a possible neuronal nicotinic acetylcholine receptor α-subunit. *Nature (London)* **319**:368–374.

Breer, H., Kleene, R., and Hinz, G., 1985, Molecular forms and subunit structure of the acetylcholine receptor in the nervous system of insects. *J. Neurosci.* **5**:3386–3392.

Cohen, F. S., 1986, Fusion of liposomes to planar bilayers, in *Ion Channel Reconstitution* (C. Miller, ed.), pp. 131–139, Plenum Press, New York.
Colquhoun, D., and Hawkes, A. G., 1977, Relaxation and fluctuation of membrane currents that flow through drug operated channels. *Proc. R. Soc. London (Biol.)* **119**:231–247.
Colquhoun, D., and Hawkes, A. G., 1981, On the stochastic properties of single ion channels. *Proc. R. Soc. London (Biol.)* **211**:205–235.
Colquhoun, D., and Hawkes, A. G., 1982, on the stochastic properties of single ion channel openings and of clusters of bursts. *Philos. Trans. R. Soc. London (Biol.)* **300**:1–59.
Colquhoun, D., and Sakmann, B., 1985, Fast events in single channel currents activated by acetylcholine and its analogues at the frog-muscle end-plate. *J. Physiol.* **369**:501–557.
Conti-Tronconi, B. M., and Raftery, M. A., 1982, The nicotinic cholinergic receptor: Correlation of molecular structure with functional properties. *Annu. Rev. Biochem.* **51**:491–530.
Coronado, R., and Latorre, R., 1983, Phospholipid bilayers made from monolayers on patch-clamp pipettes. *Biophys. J.* **43**:231–236.
Epstein, M., and Racker, E., 1978, Reconstitution of carbamylcholine-dependent sodium ion flux and desensitization of the acetylcholine receptor from *Torpedo californica*. *J. Biol. Chem.* **253**:6660–6662.
Hamill, O. P., and Sakmann, B., 1981, Multiple conductance states of single acetylcholine receptor channels in embryonic muscle cells. *Nature (London)* **294**:962–964.
Hanke, W., 1985, Reconstitution of ion channels. *CRC Crit. Rev. Biochem.* **19**:1–44.
Hanke, W., 1986, Incorporation of ion channels by fusion, in *Ion Channel Reconstitution* (C. Miller, ed.), pp. 141–157, Plenum Press, New York.
Hanke, W., and Breer, H., 1986, Channel properties of an insect neuronal acetylcholine receptor protein reconstituted in planar lipid bilayers. *Nature (London)* **321**:171–174.
Hanke, W., and Breer, H., 1987, Characterization of the channel properties of a neuronal acetylcholine receptor reconstituted into planar lipid bilayers. *J. Gen. Physiol.* **90**:855–879.
Hanke, W., and Breer, H., 1988, Gating properties of a neuronal acetylcholine receptor. *J. Gen. Physiol.*, submitted.
Hess, G. P., Kolb, H.-A., Läuger, P., Schoffeniels, P., Schwarze, W., Ugaonkar, J. B., and Pasquale, E. B., 1985, Acetylcholine receptor from *Electropherus electricus*: A comparison of single channel current recordings and chemical measurements, in *Molecular Basis of Nerve Activity* (J.-P. Changeaux, F. Hucho, M. Maelicke, and E. Neumann, eds.), pp. 317–331, de Gruyter, Berlin.
Hucho, F., 1986, The nicotinic acetylcholine receptor and its ion channel. *Eur. J. Biochem.* **158**:(2):211–326.
Jackson, M. B., 1984, Spontaneous openings of the acetylcholine receptor channel. *Proc. Natl. Acad. Sci. U.S.A.* **81**:3901–3904.
Jackson, M. B., 1986, Kinetics of unliganded acetylcholine receptor channel gating. *Biophys. J.* **49**:663–672.
Kasai, M., and Changeaux, J.-P., 1971, In vitro excitation of purified membrane fragments by cholinergic agonists. *J. Membr. Biol.* **6**:1–80.
Labarca, P., Lindstrom, J., and Montal, M., 1984, Acetylcholine receptor in planar lipid bilayers: Characterization of the channel properties of the purified nicotinic acetylcholine receptor from *Torpedo californica* reconstituted in planar lipid bilayers. *J. Gen. Physiol.* **83**:473–496.
Labarca, P., Montal, M. S., Lindstrom, J. M., and Montal, M., 1985, The occurrence of long openings in the purified cholinergic receptor channel increases with acetylcholine concentration. *J. Neurosci.* **5**:3409–3413.
Maelicke A., 1984, Biochemical aspects of cholinergic excitation. *Angew. Chem.* **23**:195–221.
Margiotta, J. F., Darwin, K. B., and Berg, E. D., 1987, The functional properties and regulation

of functional acetylcholine receptor on chick ciliary ganglion neurons. *J. Neurosci.* **7**(11):3612–3622.

Mishina, M., Kurosaki, T., Tobimatsu, T., Morimoto, Y., Noda, M., Yamamoto, T., Terao, M., Lindstrom, J., Takahashi, T., Kuno, M., and Numa, S., 1984, Expression of functional acetylcholine receptor from cloned cDNAs. *Nature (London)* **307**:604–608.

Mishina, M., Tobimatsu, T., Imoto, K., Tanaka, K., Fujitsu, Y., Fukuda, K., Kurasaki, M., Takahashi, H., Morimoto, Y., Hirose, T., Inayama, S., Takahashi, T., Kuno, M., and Numa, S., 1985, Location of functional regions of acetylcholine receptor α-subunit by site-directed mutagenesis. *Nature (London)* **313**:364–369.

Mishina, M., Takai, T., Imoto, K., Noda, M., Takahashi, T., Numa, S., Methfessel, C., and Sakmann, B., 1986, Molecular distinction between fetal and adult forms of muscle acetylcholine receptor. *Nature (London)* **321**:406–411.

Montal, M., and Mueller, P., 1972, Formation of bimolecular membranes from lipid monolayers and a study of their electric properties. *Proc. Natl. Acad. Sci. U.S.A.* **69**:3561–3566.

Montal, M., Labarca, P., Fredkin, D. R., Suarez-Isla, B. A., and Lindstrom, J., 1984, Channel properties of the purified acetylcholine receptor from *Torpedo californica* reconstituted in planar lipid bilayer membranes. *Biophys. J.* **45**:165–174.

Montal, M., Anholt, R., and Labarca, P., 1986, The reconstituted acetylcholine receptor, in *Ion Channel Reconstitution* (C. Miller, ed.), pp. 157–204, Plenum Press, New York.

Mueller, P., Rudin, D., Tien, H. T., and Wescott, W. C., 1962, Reconstitution of excitable cell membrane structure in vitro. *Circulation* **26**:1167–1171.

Nelson, N., Lindstrom, J., and Montal, M., 1980, Reconstitution of purified acetylcholine receptors with functional ion channels in planar lipid bilayers. *Proc. Natl. Acad. Sci. U.S.A.* **77**:3057–3061.

Popot, J. L., and Changeaux, J.-P., 1984, The nicotinic receptor of acetylcholine: Structure of an oligomeric integral membrane protein. *Physiolog. Rev.* **64**:1162–1239.

Sakmann, B., Patlak, J., and Neher, E., 1980, Single acetylcholine activated channels show burst kinetics in presence of desensitizing concentrations of agonist. *Nature (London)* **286**:71–73.

Sakmann, B., Bormann, J., and Hamill, O. P., 1983, Ion transport by single receptor channels, in *Cold Spring Harbor Symposia on Quantitative Biology*, Vol. XLVIII, pp. 247–257, Cold Spring Harbor Laboratory, Cold Spring Harbor, NY.

Sakmann, B., Methfessel, C., Mishina, M., Takahashi, T., Takai, T., Kurasaki, M., Fukuda, K., and Numa, S., 1985, Role of acetylcholine receptor subunits in gating of the channel. *Nature (London)* **318**:538–543.

Satelle, D. B., 1986, Insect acetylcholine receptors—biochemical and physiological approaches, in *Neuropharmacology and Pesticide Action* (M. G. Ford, G. G. Lunt, R. C. Reay, and P. N. R. Usherwood, eds.), pp. 445–497, Ellis Horwood, Chichester, England.

Schindler, H., and Quast, U., 1980, Functional acetylcholine receptor from *Torpedo californica* in planar bilayers. *Proc. Natl. Acad. Sci. U.S.A.* **77**:3052–3056.

Schürholz, T., and Schindler, H., 1983, Formation of lipid–protein bilayers by micropipette guided contact of two monolayers. *FEBS Lett.* **152**:187–190.

Suarez-Isla, B. A., Wan, K., Lindstrom, J., and Montal, M., 1983, Single channel recordings from purified acetylcholine receptors reconstituted in bilayers formed at the tip of patch-pipettes. *Biochemistry* **22**:2319–2323.

Tank, D. W., Huganir, R. L., Greengard, P., and Webb, W. W., 1983, Patch-recorded single-channel currents of the purified and reconstituted *Torpedo* acetylcholine receptor. *Proc. Natl. Acad. Sci. U.S.A.* **80**:5129–5133.

Whiting, P., and Lindstrom, J., 1987, Purification and characterization of a nicotinic acetylcholine receptor from rat brain. *Proc. Natl. Acad. Sci. U.S.A.* **84**:595–599.

Wilmsen, U., Methfessel, C., Hanke, W., and Boheim, G., 1983, Channel current fluctuation studies with solvent free lipid bilayers using Neher–Sakmann pipettes, in *Physical Chemistry of Transmembrane Ion Motion* (G. Spach, ed.), pp. 479–485, Elsevier, Amsterdam.

Young, J. D.-E., Cohn, Z. A., and Norton, B. G., 1987, Functional assembly of gap-junction conductance in lipid bilayers: Demonstration that the major 27 kD protein forms the junctional channel. *Cell* **48:**733–743.

Chapter 10

Liposomes as Carriers of Drugs
Observations on Vesicle Fate after Injection and Its Control

Gregory Gregoriadis

1. INTRODUCTION

Detailed knowledge of the behavior of liposomes (phospholipid vesicles) after injection is a prerequisite for their successful application as drug carriers in the therapy, prevention, or detection of disease (Gregoriadis, 1980). This discussion deals with two aspects: (1) the extent to which vesicles retain entrapped agents in the presence of biological fluids, namely, blood and (2) the rate by which vesicles are cleared from the circulation and their distribution in tissues. Evaluation of the first is likely to lead to methods of preventing or reducing agent leakage from the carrier en route to its target site. Study of the second aspect will contribute to the elucidation of the mechanism(s) through which vesicles leave the circulation to end up in tissues and also allow manipulations controlling vesicle clearance or uptake by tissues in need of pharmacological intervention. Both aspects have been discussed in detail elsewhere (Gregoriadis, 1983; Gregoriadis *et al.*, 1983; Senior, 1987).

Two general factors influence the behavior of liposomes *in vivo:* the first relates to structural characteristics of the carrier system and the other to the biological milieu in contact with it. This, however, is a rather superficial division of factors since any behavior that originates from a particular structural property of liposomes does so because of the environment in which they exist.

Gregory Gregoriadis Medical Research Council Group, Academic Department of Medicine, Royal Free Hospital School of Medicine, London NW3 2QG, United Kingdom.

For instance, liposomes with a negative charge are cleared from the blood more rapidly than neutral ones because, presumably, blood contains components that adhere to the vesicles' negative surface. In turn, vesicles become more attracted to the phagocytic cells. Because liposomes will ultimately be applied in a variety of normal and diseased states, understanding of the influence of the milieu on the behavior of vesicles is thus essential. Progress made in this laboratory during the last 8 years in understanding liposomal behavior (as already defined) *in vivo* and ways by which we have succeeded in controlling it will be outlined.

2. RETENTION OF DRUGS BY LIPOSOMES IN CONTACT WITH BLOOD

Work (Gregoriadis and Ryman, 1972a,b) on the fate of liposomes and entrapped agents began in 1970. It dealt with events occurring following intravenous injection of rats with multilamellar liposomes (MLVs) entrapping albumin (Gregoriadis and Ryman, 1972a). Liposomes were composed of egg phosphatidylcholine (PC), cholesterol, and dicetylphosphate (molar ratio of 7 : 2 : 1) and labeled with appropriate markers both in the lipid (bilayer) and the aqueous phase. Study of the clearance of the two labels from the circulation supported quantitative retention of the aqueous (protein) marker by liposomes, which at that time were thought to be stable. We now know of course that had a small molecular weight solute been used with the cholesterol-poor liposomes instead of the protein, conclusions of vesicle stability would have been different, namely, that we would have seen leakage of solute, hence bilayer instability.

Indications of solute loss from cholesterol-poor liposomes in the presence of blood first appeared in experiments with liposomal penicillin (Gregoriadis, 1973). Following intravenous injection of rats with MLVs containing radiolabeled protein (albumin) and penicillin, the rate of albumin clearance was much slower than that of the drug, suggesting loss of penicillin from the circulating vesicles. Such loss, however, was attributed (Gregoriadis, 1973) to a passive outward diffusion of penicillin rather than destabilization of the liposomal membrane. Krupp *et al.* (1977), Tall and Small (1977), and Scherphof *et al.* (1978) provided evidence that liposomes interact with plasma high-density lipoproteins (HDLs), which remove phospholipid and thus destabilize the vesicle's bilayer. Although the mechanism of vesicle destabilization by HDLs was unknown, we nevertheless reasoned (Gregoriadis and Davis, 1979) that such action of HDL could be counteracted by rendering the membranes more packed through the incorporation of excess cholesterol. Experiments (Gregoriadis and Davis, 1979) to test the hypothesis were therefore carried out with MLVs incorporating various amounts of cholesterol and containing β-fructofuranosidase or carboxyfluorescein (CF). The enzyme marker, shown in earlier work to be retained even by cholesterol-poor liposomes in blood (Gregoriadis and Ryman, 1972b), served as an indicator of solute leakage into the vesicles exposed to sucrose (a substrate for the entrapped enzyme) following their incubation in blood. CF, on the other hand, had already been used (Weinstein et al., 1978) in its quenched form as a marker for liposomes to distinguish between intact vesicles and vesicles releasing their dye content within the cells. Quenched CF was adopted in our experiments as a marker of liposomal stability (in terms of solute leakage out of vesicles) in the presence of biological fluids. Stability of liposomes on exposure to given environment was estimated as %(liposomal)CF latency and derived from $100(Dye_t - Dye_f)/Dye_t$, where t and f denote, respectively, total (measured by the presence of Triton X-100) and free dye. For instance, in the absence of CF leakage ($f=0$) latency acquires a value of 100 (denoting unchanged liposomal stability) while total CF leakage ($t=f$) would reduce CF latency value to nil. In our experience, Triton X-100 as used (Gregoriadis and Davis, 1979; Senior and Gregoriadis, 1984) releases entrapped CF completely even though brief boiling of samples may be necessary with some cholesterol-rich MLVs (e.g., those made of high melting phospholipids) (Senior and Gregoriadis, 1984). A study of the ability of Triton X-100 to release liposomal CF has been published recently (Sila et al., 1986).

Using MLVs containing the enzyme or CF, we found that after exposure to rat blood or blood serum (but not to buffer alone) at 37°C for 2.5 hr, diffusion of sucrose into or loss of CF from the vesicles was significantly diminished in direct relation to the liposomal cholesterol content. We (Gregoriadis and Davis, 1979) also observed (confirmed later by Lelkes and Tandeter, 1982) that bilayer permeability to the solutes in the presence of whole blood (a more realistic milieu in terms of *in vivo* applications) was less pronounced than in serum. This was tentatively explained (Gregoriadis and Davis, 1979) either in terms of a HDL–erythrocyte interaction taking precedence over that between liposomes and HDLs or as a result of liposomes with little or no cholesterol improving their stability through acceptance of cholesterol from blood cells. A similar effect of cholesterol in improving stability of liposomes was observed *in vivo* after intravenous injection of rats (Gregoriadis and Davis, 1979).

Attention was subsequently turned to small unilamellar vesicles (SUVs). Since SUVs, compared to MLVs, exhibit longer circulation times after intravenous injection (Juliano and Stamp, 1975), these vesicles were considered (and eventually shown to be) (Gregoriadis and Senior, 1984; Wolff and Gregoriadis, 1984) a more appropriate carrier for *in vivo* targeting. For instance, a long half-life would delay interception of ligand-bearing liposomes by the reticuloendothelial system (RES) and thus allow such vesicles to interact effectively with cells expressing the appropriate receptor, within the vascular compartment (Wolff and Gregoriadis, 1984) or in areas accessible to SUVs

(Gregoriadis and Senior, 1984). A series of experiments established (Kirby and Gregoriadis, 1980; Kirby *et al.*, 1980a,b) that cholesterol enrichment of PC SUVs, regardless of surface charge, rendered from stable (high CF latency values) in the presence of whole mouse blood and mouse plasma in that order. As before, the effect of cholesterol in inducing liposomal stability was confirmed *in vivo* on intravenous injection into mice. Furthermore, the stability of SUVs composed of equimolar PC and cholesterol was retained fully within the peritoneal cavity (after intraperitoneal injection), with the vesicles recovered quantitatively in the circulating blood (Kirby *et al.*, 1980a). Cholesterol-rich SUVs were also found to enter the blood circulation intact after footpad injection (Tumer *et al.*, 1983). Data from a number of laboratories have added support to the notion of a stabilizing role for liposomal cholesterol (e.g., Allen and Cleland, 1980; Damen *et al.*, 1981; Guo *et al.*, 1980).

Results (Gregoriadis and Davis, 1979; Kirby *et al.*, 1980a) with liposomes of varying cholesterol content were thus consistent with the hypothesis that excess sterol would interfere with phospholipid removal by HDLs. Indeed, this was demonstrated in subsequent work (Kirby *et al.*, 1980b) using neutral SUVs with or without cholesterol, labeled with radioactive PC, and containing CF. On incubation in the presence of human serum at 37°C, there was, as expected, quantitative release of CF in the case of cholesterol-free SUVs. At the same time, however, much (about half) of the radioactive PC was recovered (after molecular sieve chromatography) in fractions in which HDLs were also recovered (see Senior *et al.*, 1983). On the other hand, with cholesterol-rich SUVs, most of both CF and PC were coeluted with liposomes. These findings were confirmed in an experiment (Kirby *et al.*, 1980b) in which SUVs as above were incubated with human serum, subsequently fractionated into various lipoprotein species: of the PC radioactivity from cholesterol-free and cholesterol-rich SUV preparations, about 44% and 21%, respectively, were recovered with HDLs. When serum was heated at 55°C (to inactivate lecithin–cholesterol acyltransferase prior to mixing with SUVs), lecithin movement to HDLs from cholesterol-free SUVs was retained, thereby excluding a significant role for this enzyme in liposome destabilization. Results similar to those seen *in vitro* (Kirby *et al.*, 1980b) were obtained on chromatography of serum from mice intravenously injected with radiolabeled SUVs and killed at time intervals. Considerable movement of PC to HDLs for cholesterol-free SUVs (but not for cholesterol-rich SUVs) was observed with increasing time between injection and serum sampling (Kirby and Gregoriadis, 1980). Similar findings were reported independently by Tall (1980) from *in vivo* experiments in rats.

Further substantiation of a role for HDLs in the destabilization of liposomes came from experiments in which cholesterol-free SUVs (containing quenched CF) injected intravenously into lipoprotein-deficient mice retained much more of their CF latency than when injected in intact mice (Senior *et al.*,

1983). In the same work (Senior *et al.*, 1983), plasma from lipoprotein-deficient mice was mixed with increasing amounts of HDLs, low and intermediate density (LDLs + IDLs) or very-low-density lipoproteins (VLDLs) (to simulate the physiological range of lipoprotein concentrations in mouse blood) before the addition of cholesterol-free PC SUVs, and then incubated at 37°C. Among the lipoprotein species studied, only HDLs were able to reduce liposomal stability. Of interest was also the observation that addition of HDLs to lipoprotein-deficient plasma (previously heated at 56°C for 30 min) did not restore the plasma's full activity in destabilizing SUVs. This confirms reports by Damen *et al.* (1981) and Tall *et al.* (1983) that HDL action in removing phospholipids is facilitated by a heat labile factor exhibiting phosphatidylcholine transfer activity, apparently a protein of molecular weight 41,000 (Tall *et al.*, 1983). Finally, exposure of PC SUVs to plasma from a patient with congenital lecithin–cholesterol acyltransferase deficiency, characterized by low HDL levels, resulted in reduced loss of vesicle stability compared to the loss seen on exposure to normal human plasma (Senior *et al.*, 1983).

The effect of cholesterol content of PC liposomes on their stability was also shown to apply to SUV composed of a variety of other phospholipids (Gregoriadis and Senior, 1980). Thus, CF-containing SUVs made of dilauroyl phosphatidylcholine (DLPC), dimyristoyl phosphatidylcholine (DMPC), dioleoyl phosphatidylcholine (DOPC), or sphingomyelin (SM), all became leaky to the dye on incubation for 60 min with mouse blood at 37°C. With some of the preparations (e.g., DLPC, DMPC, and DOPC liposomes) loss of CF was total, whereas a considerable proportion of it was retained by SM SUVs. The presence of equimolar (to the phospholipid) cholesterol, however, in the SUV preparations resulted in CF retention, the extent of which depended on the nature of their phospholipid component. For instance, CF retention was nearly full for DMPC and SM liposomes even after incubation (in blood) for 6 and 12 hr, respectively. Retention was much less pronounced for DLPC vesicles. Similar findings were noted for liposomes of larger sizes (Gregoriadis, 1985; Senior *et al.*, 1985).

While reduction in solute loss can be quantitative (up to 100% retention) for small polar solutes such as CF, sucrose, and bleomycin (Gregoriadis, 1985) and undoubtedly many others, it appears to be only partial for less polar solutes (e.g., melphalan and vincristine), even for liposomes composed of equimolar SM and cholesterol (Kirby and Gregoriadis, 1983). Alternative approaches to improving liposomal stability in blood have been reported and include the incorporation of tocopherol (in place of cholesterol) (Halks-Miller *et al.*, 1985), of mono-, di-, or trisialo gangliosides which can act synergistically with cholesterol (Allen et al., 1985), or of carbamyl analogues of egg PC (Gupta *et al.*, 1981) and the polymerization of vesicles (Krause *et al.*, 1987).

Interaction of plasma HDLs with cholesterol-free liposomes leads, according to Scherphof *et al.* (1978), to the dissolution of the vesicles with transfer of their phospholipids to the lipoprotein. On the other hand, experiments (Kirby and Gregoriadis, 1981) with PC SUVs free of cholesterol and containing solutes of increasing molecular weight (i.e., sucrose, inulin, and polyvinyl pyrrolidone) have supported pore formation on the vesicle bilayer (at least for SUVs that have retained a sufficient mass of phospholipid) rather than total vesicle disintegration. Thus, decreasing loss of solutes on incubation in plasma at 37°C was observed with increasing molecular weight of solute. Furthermore, "empty" PC SUVs mixed with plasma, into which CF had been previously added, contained some of the dye after chromatographic separation of the vesicles (Kirby and Gregoriadis, 1981). We proposed (Gregoriadis *et al.*, 1983) that the structural entity of such porous SUVs is retained by means of hydrophobic regions of plasma proteins interacting with the fatty acid chains of the remainder phospholipid on the surface of the pore channels. Yoshioka *et al.* (1984), on the other hand, reported that exposure of large unilamellar vesicles to plasma is followed by an initial increase of their permeability (effected by phospholipid loss to HDLs), which according to the authors is subsequently abolished as a result of vesicle "shrinkage" to smaller entities with a reduced aqueous volume. It thus appears that "permanent" pore formation (proposed for SUVs) in such large vesicles does not occur. Scherphof and Morselt (1984) have recently carried out a systematic study on plasma-induced destabilization of SUVs and concluded that susceptibility of vesicles to the action of HDLs increases as the vesicle size decreases. Results by Yoshioka *et al.* (1984) and Scherphof and Morselt (1984) may offer an explanation to the observations by Kirby and Gregoriadis (1981): it is feasible that because SUVs used in our studies consisted of a relatively heterogeneous mixture of vesicles (up to 60 nm in diameter; Kirby and Gregoriadis, 1980), there could have been a selective destruction of the smaller vesicles with larger ones surviving either as porous entities or by acquiring a reduced size.

3. CIRCULATION OF LIPOSOMES IN THE BLOOD

It was shown by Gregoriadis and Ryman (1972a,b) that on intravenous injection into rats, multilamellar liposomes are removed from the circulation rapidly (within minutes) and recovered in the RES, mostly the liver and spleen (Segal *et al.*, 1974). In terms of applying liposomes as a drug delivery system, rapid vesicle clearance is desirable for agents destined for action in these tissues (agents are less likely to leak excessively from the carrier during the latter's short presence in the intravascular space), although there are instances where liposomal agents are meant to act in alternative sites in the body. This could be achieved by delaying vesicle interception by the liver and spleen, thus allowing such liposomes to come into effective contact with target cells intravascularly or with extravascular sites to which access by liposomes is slow. In other cases, prolonged circulation of liposomes in the blood would be a prerequisite for effective agent action intravascularly (e.g., their use as hemosomes; see Gregoriadis, 1988).

Early findings of a nonlinear (curved) liposomal clearance pattern from the circulation (Gregoriadis and Ryman, 1972a) implied that vesicle populations of varying sizes, in the heterogeneous multilamellar preparation used, exhibited varying rates of clearance, with the smaller vesicles being removed at a slower rate. Juliano and Stamp (1975) were the first, however, to attribute a longer half-life to the SUVs. An additional early observation (Gregoriadis and Neerunjun, 1974), confirmed later by Juliano and Stamp (1975), was that the rate of removal of liposomes from the circulation after intravenous injection was also dependent on vesicle surface charge, with neutral and positive liposomes being cleared less rapidly than negative ones. It became apparent that by combining appropriate vesicle size and surface charge one could control vesicle half-lives in the circulation. Prior to a discussion of half-lives, however, it is perhaps worth considering that reported values for half-lives of vesicles are of little value and can also be misleading (especially in comparative experiments) unless the amount of liposomal lipid and total vesicle surface or effective mean vesicle diameter in the injected dose are also taken into consideration. For instance, it has been known for some time (Abra and Hunt, 1981; Gregoriadis and Neerunjun, 1974; Gregoriadis and Ryman, 1972a) that increasing amounts of liposomal lipid in the injected dose will lead to progressively slower rates of vesicle clearance until the dose is so large that the capacity of vesicle removal by tissues is saturated. Saturation of such capacity is achieved with different amounts of the same liposomal lipid for preparations of differing vesicle size and appears to be related to the number of vesicles and total vesicle surface in the respective preparations (Abra and Hunt, 1981).

Choice of methods to measure vesicle concentration in the blood at any given time is a major difficulty in monitoring liposomal clearance from the circulation. One can either measure a liposome-associated water-soluble marker, a lipid component of the carrier itself, or both. With any methodology, however, monitored markers must remain fully associated with the carrier during the latter's entire period of presence in the circulation. On the other hand, verification of marker association with liposomes in blood samples removed after injection is likely to involve chromatography or some other separation procedure to ensure that none (or at least very little) of the marker exists in its free form. This approach, in addition to being time consuming, can lead to marker dissociation from the carrier during the verification procedure. In addition, leaked water-soluble markers (e.g., CF, inulin) can be cleared from the circulation very rapidly and lipid marker molecules are likely to exchange with lipid molecules on cell surfaces or lipoproteins. Both such events would reduce the radioactive pool of the vesicles to give falsely accelerated rates of vesicle clearance. Finally, even if one were to use a marker that remains fully associated with the bilayers (e.g., a covalently linked marker), there is always the possibility that some vesicles will disintegrate or fuse to form fragments or vesicles with altered sizes, still retaining the marker. These new entities, owing to size changes, will assume different rates of clearance. It does seem therefore that, in this respect, perfection with as complex (and vulnerable) a system as liposomes cannot be achieved.

The use of quenched CF to monitor vesicle (average) clearance was proposed (Gregoriadis and Davis, 1979; Gregoriadis and Senior, 1980; Kirby *et al.*, 1980a,b; Senior and Gregoriadis, 1982a,b) on the grounds that assay of quenched dye in serum samples is equivalent to monitoring original intact vesicles (any transformation of vesicles to smaller or larger entities is likely to involve CF loss). Obviously, however, monitoring of *all* vesicles present in blood at any time is dependent on the absence of or minimal CF leakage. Therefore, the approach is only applicable to very stable liposomes, that is, vesicles that do not lose a significant proportion of entrapped CF while they are circulating in the blood. As it is not practical to measure CF leakage *in vivo* (the dye is excreted very rapidly through the kidneys), leakage is first tested *in vitro* by incubating liposomes containing quenched CF in whole blood at 37°C. Time of incubation will depend on the "expected" half-life of vesicles following intravenous injection. For instance, for MLV preparations expected to have a half-life measured in minutes (e.g., 30 min), a 2 hr incubation period (to compensate for the latter, slower portion of the nonlinear clearance pattern) is likely to exceed the time after which practically no vesicles are present in the circulation. Similarly, for SUVs with an expected half-life of hours (e.g., 2 hr) incubation for, say, 10 hr (5 half-lives) would probably be sufficient.

Having established already the half-life (about 2 hr in mice for the amount of liposomal phospholipid injected) of SUVs made of equimolar PC and cholesterol (Kirby *et al.*, 1980a), a systematic study of the rates of clearance of SUVs composed of various phospholipids was undertaken. Markers used were quenched CF (its latency retains a value of nearly 100% on incubation with blood at 37°C; Kirby *et al.*, 1980a) and ^{14}C-labeled cholesteryl oleate, practically all of which remains associated with the vesicles after intravenous injection (Kirby and Gregoriadis, 1980). Indeed, the ratio of the two labels in the blood of mice up to 150 min after injection retained a value that was identical to that in the preparation before injection (Kirby *et al.*, 1980a). Preparations (SUVs) composed of equimolar phospholipid and cholesterol and shown to retain their CF content quantitatively in the presence of blood at 37°C for 1–12 hr (Gregoriadis and Senior, 1980) were injected intravenously into mice and clearance rates monitored. It was of interest to observe that all clearance patterns were essentially linear and that half-lives varied (1–16 hr) depending on the choice of the liposomal phospholipid component. SUVs composed of equimolar SM and cholesterol exhibited the longest half-life (16 hr). A long half-life for cholesterol-rich SM SUVs was also reported independently by Hwang *et al.* (1980) and, more recently, confirmed by Spanjer *et al.* (1986).

It was previously observed that SUV stability *in vitro* for cholesterol-rich vesicles, although always pronounced during relatively short (e.g., up to 2 hr) periods of incubation in the presence of blood (37°C), was reduced considerably for some SUV compositions (e.g., PC SUVs) after longer periods of incubation, while for others (e.g., SM SUVs) it remained high, even after much longer periods (e.g., 24 hr). It appeared that this phenomenon correlated with the half-lives of the corresponding vesicles in the circulation; that is, the greater the vesicle stability (over prolonged periods of incubation in blood) the longer the half-life. We therefore attempted (Senior and Gregoriadis, 1982a) to establish whether such a relationship between stability and clearance did indeed exist. SUV compositions were so arranged in a single experiment that vesicle stability in a variety of preparations would differ. Thus, SUVs (always containing cholesterol equimolar to the total phospholipid used) were made of PC, SM, or a number of combinations of the two phospholipids (varying molar ratios). All SUV preparations studied were found stable (minimal loss of CF latency) during the first 2 hr of incubation in the presence of blood at 37°C. As the time of incubation increased, however, vesicle stabilities behaved differently: SUVs remained fully stable after 24 hr for compositions of SM only (as expected) or SM combined with a relatively small proportion (23% of total phospholipid) of PC. With the concentration of PC increasing, stability over 24 hr incubation was gradually reduced to reach half its initial (CF latency) value for vesicles made of PC only. All preparations (also labeled with radioactive phospholipid) were subsequently injected intravenously into mice and half-lives monitored on the basis of both quenched CF and radioactivity. As anticipated, there was a direct relationship between vesicle half-life (very similar values were obtained using the two labels) and CF latency *in vitro* during the 24 hr incubation period (Senior and Gregoriadis, 1982a). Further confirmation of this relationship was obtained (Senior and Gregoriadis, 1982a) with SUVs made of equimolar distearoyl phosphatidylcholine (DSPC) and cholesterol or DSPC only. The former retained CF fully (100% latency value) even after 48 hr of incubation while the latter's latency value declined to 40% for the same period of time. Half-lives of 20 and 1.3 hr, respectively, were established in mice for such SUVs injected in amounts similar to those used for experiments (see above) with other phospholipid compositions. Furthermore, the possibility that liposomes (regardless of phospholipid composition) are removed from the circulation at comparable rates and that differences in half-lives observed are artifacts reflecting varying degrees of CF leakage *in situ* was investigated and con-

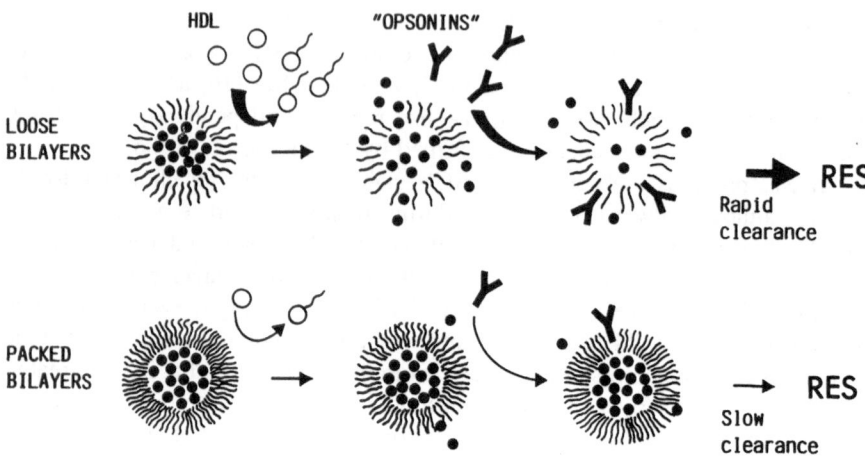

FIGURE 1. A proposed mechanism relating liposome stability in blood and clearance. *Upper:* vesicles with "loose" bilayers (little or no cholesterol content) are attacked by plasma HDLs, which remove phospholipid molecules to produce gaps. Entrapped solutes diffuse out of vesicles to an extent dependent on solute and gap size. Opsonins (or other plasma proteins normally responsible for the removal of foreign particles) can insert themselves into such gaps. *Lower:* Because bilayers are packed and/or rigid (excess cholesterol content and/or phospholipid component with high Tc), HDL action, solute leakage, and opsonin insertion are reduced or abolished. On comparing the two types of vesicle, the greater availability of opsonins on the surface of liposomes with loose bilayers promotes more rapid vesicle uptake by the RES.

sidered unlikely in work discussed elsewhere (Senior and Gregoriadis, 1982a). The effect of phospholipid composition and/or cholesterol content of liposomes on their clearance from the circulation has now been confirmed by others (e.g., Agarwal *et al.*, 1986; Patel *et al.*, 1983; Roerdink *et al.*, 1986; Spanjer *et al.*, 1986).

To unify observations of the influence of cholesterol and phospholipid species on blood-induced vesicle permeability to entrapped solute (CF) and vesicle clearance rates from the circulation, a hypothesis was put forward (Gregoriadis *et al.*, 1983). It was proposed (Figure 1) that, for vesicles surviving HDL-induced disintegration, bilayer destabilization (leading to entrapped solute loss) promotes the adsorption onto or insertion into the bilayer of plasma factors (opsonins?), which in turn facilitate vesicle uptake by the RES. Such uptake gradually decreases as the lipid composition of the vesicles (e.g., excess cholesterol) reduces the action of HDL, bilayer destabilization, and the adsorption or insertion of opsonin. On the other hand, Patel *et al.* (1983) have attributed the effect of excess cholesterol on delaying liposomal clearance to an inhibitory role of the sterol on the uptake of vesicles by the liver. Their proposition, however, was not substantiated by direct evidence and is, in addition, incompatible with the observation (discussed above) by us and others that cholesterol-rich SUVs of varying phospholipid composition (shown to retain their entrapped solute marker fully on incubation in blood) exhibit widely different rates of clearance and therefore (as shown below) different rates of vesicle uptake by tissues.

The direct relationship between vesicle stability and half-life in the circulation applies, at least in an obvious and practically useful way, to neutral vesicles of small sizes (e.g., up to 80 nm diameter in our experiments). For large neutral vesicles (e.g., 0.2 μm diameter), there is a progressive increase in clearance rates even though vesicle stability in the presence of blood is retained fully on prolonged incubation. For example, after intravenous injection of mice with similar amounts of CF-containing neutral cholesterol-rich DSPC SUVs (64 ± 5.9 nm diameter) and large unilamellar DSPC vesicles (LUVs) passed twice through a nucleopore filter of 0.4 μm diameter, half-lives were 7.5 and 0.2 hr, respectively. Yet the same vesicles exhibited CF latency values greater than 98.5% on incubation in blood for 24 hr at 37°C (Senior et al., 1985). The same observation was made for negatively (distearoyl phosphatidic acid) charged cholesterol-rich DSPC SUVs when compared to neutral DSPC SUVs. Although both preparations were stable in blood (CF latency values greater than 90% at the end of a 45 hr incubation at 37°C), half-lives after intravenous injection in mice were 1 and 10 hr, respectively (Senior et al., 1985). It is thus reasonable to conclude that vesicle size and surface (negative) charge override the state of membrane stability in determining vesicle clearance, even for liposomes that are exceedingly stable.

4. DISTRIBUTION OF LIPOSOMES IN TISSUES

Precise evaluation of the extent to which tissues participate in the uptake of liposomes requires that the marker chosen is retained quantitatively both by the carrier until it reaches the tissue and by the latter until most of the injected liposomes have been eliminated from the blood. Over 10 years ago we (Gregoriadis et al., 1977) were able to demonstrate that ^{111}In maintained its localization in the liver of mice for several days at the maximum levels achieved following intravenous injection of liposomal ^{111}In-labeled bleomycin. An additional advantage of ^{111}In-labeled bleomycin is that, when in its free form (as, for instance, after leakage from liposomes), it is excreted rapidly in the urine and fails to accumulate in tissues.

Delivery of drugs to target cells via liposomes is generally dependent on (1) minimal milieu-induced drug leakage from the vesicles and (2) accessibility to and effective (in terms of drug action) interaction with the cells. The former requirement has already been discussed and shown to be possible with drug solutes which, like CF, sucrose, and insulin, remain entrapped within the vesicles in the absence of significant HDL-mediated loss of bilayer phospholipid. Numerous groups (see Gregoriadis, 1988), on the other hand, have shown that the second requirement can be satisfied by the phagocytic cells of the RES (mostly for large liposomes) and the hepatic parenchymal cells, which, as reported by Spanjer et al. (1986), take up neutral SUVs quantitatively. Variations in vesicle structural characteristics, amount of liposomal lipid in the injected dose, saturation of the RES with excess liposomes (Dave and Patel, 1986; Gregoriadis and Neerunjun, 1974), and route of administration are known to influence the degree of participation in uptake by tissue cells (particularly those of the liver and spleen) (e.g., Abra and Hunt, 1981; Ryman and Tyrrell, 1980; Spanjer et al., 1986). Recently, we were able to present evidence of considerable participation of the bone marrow macrophages as well, provided that hepatic and splenic uptake is curtailed (Senior et al., 1985). This was achieved in intravenously injected mice by the use of SUVs with a long half-life, namely, cholesterol-rich DSPC: total hepatic uptake of the [^{111}In]bleomycin marker, at a time when no significant amount of radioactivity could be detected in blood, was reduced from 56% of the dose (for the relatively short-lived cholesterol-rich PC SUVs) to 25% observed for the long-lived cholesterol-rich DSPC SUVs. Much (about 35%) of the remainder dose was recovered in the carcass. In similarly treated rats, scans of carcasses (from which all internal organs had previously been removed) revealed that practically all of the liposomal ^{111}In marker was localized in the bones, presumably the marrow macrophages. These results, in conjunction with those by Spanjer et al. (1986), suggest that liver cells (probably parenchymal cells), as opposed to those in the bone marrow, take up SUVs more avidly. This could be attributed to a greater accessibility of the former cells to the vesicles. On the other hand, an imposed delay in hepatic uptake of the vesicles (e.g., by increasing their half-life) enables large numbers of them to eventually reach and interact with the more phagocytic bone marrow macrophages.

Prolonged vesicle circulation has facilitated targeting to cells or other entities within the vasculature. Thus, long-lived liposomes coated with ligands that recognize respective receptors on targets have ample time to interact quantitatively with such receptors and the feasibility of *in vivo* targeting has been demonstrated already (e.g., Gregoriadis et al., 1981; Leserman, et al., 1983; Spanjer et al., 1985; Wolff and Gregoriadis, 1984). Cell types in the intravascular space, which are potential targets for ligand-bearing liposomes, include erythrocytes, subpopulations of lymphocytes, and vascular endothelial cells. Although antibodies as a carrier system may be preferable for drug targeting *in vivo*, there are instances where liposomes would be the vesicle of choice and supporting arguments have been discussed elsewhere (Gregoriadis, 1983). A good example is the use of mannosylated liposomes containing either macrophage activating factors (Barratt *et al.*, 1986) or antigens, with the vesicles in the latter case acting as targeted immunological adjuvants (Garcon *et al.*, 1988; Gregoriadis *et al.*, 1988). Prolonging the circulation time of large unilamellar liposomes is a greater challenge, however, as these will accommodate much larger quantities of pharmacologically active agents. In this respect, understanding of mechanisms by which blood cells avoid rapid clearance by tissues could provide clues for the design of long-lived large vesicles.

5. CONCLUSIONS

Rapid progress in both liposome technology (Gregoriadis, 1984) and elucidation of ways by which liposomes interact with the biological milieu has enabled us to tailor the system so that it (1) retains drugs quantitatively en route to the intended target and (2) circulates in the blood for controlled periods of time. This has facilitated uptake of liposomes by specific cells (including components of the RES) or, when in targeted form, their association with intravascular entities that normally do not take up the carrier effectively. Such advances have, in many cases, provided opportunities for the application of liposomes experimentally and clinically (Gregoriadis, 1988).

6. REFERENCES

Abra, R. M., and Hunt, A., 1981, Liposome disposition *in vivo:* Dose and vesicle size effects. *Biochim. Biophys. Acta* **666**:493–503.

Agarwal, K., Bali, A., and Gupta, C. M., 1986, Effect of phospholipid structure and survival times of liposomes in circulation. *Biochim. Biophys. Acta* **883**:468–475.

Allen, T. M., and Cleland, L. G., 1980, Serum-induced leakage of liposome contents. *Biochim. Biophys. Acta* **597**:418–426.

Allen, T. M., Ryan, J. L., and Papahadjopoulos, D., 1985, Gangliosides reduce leakage of aqueous-space markers from liposomes in the presence of human plasma. *Biochim. Biophys. Acta* **818**:205–210.

Barratt, G., Tenu, J. P., Yapo, A., and Petit, J.-F., 1986, Preparation and characterization of liposome containing mannosylated phospholipids capable of targeting drugs to macrophages. *Biochim. Biophys. Acta* **862**:153–164.

Damen, J., Regts, J., and Scherphof, G., 1981, Transfer and exchange of phospholipid between small unilamellar liposomes and rat plasma high density lipoproteins. *Biochim. Biophys. Acta* **665**:538–545.

Dave, J., and Patel, H. M., 1986, Differentiation in hepatic and splenic phagocytic activity during reticuloendothelial blockade with cholesterol-free and cholesterol-rich liposomes. *Biochim. Biophys. Acta* **888**:184–190.

Garcon, N., Gregoriadis, G., Taylor, M., and Summerfield, J., 1988, Mannose-mediated targeted immunoadjuvant action of liposomes. *Immunology* **64:**743–745.

Gregoriadis, G., 1973, Drug entrapment in liposomes. *FEBS Lett.* **36:**292–296.

Gregoriadis, G., 1980, Tailoring liposome structure. *Nature (London)* **283:**814–815.

Gregoriadis, G., 1983, Targeting of drugs with molecules, cells and liposomes. *Trends Pharmacol. Sci.* **4:**304–307.

Gregoriadis, G. (ed.), 1984, *Liposome Technology,* Vols. 1–3, CRC Press, Boca Raton, FL.

Gregoriadis, G., 1985, Liposomes as carriers of drugs and vaccines. *Trends Biotechnol.* **3:**235–241.

Gregoriadis, G. (ed.), 1988, *Liposomes as Drug Carriers: Recent Trends and Progress,* Wiley, Chichester.

Gregoriadis, G., and Davis, C., 1979, Stability of liposomes *in vivo* and *in vitro* is promoted by their cholesterol content and the presence of blood cells. *Biochim. Biophys. Res. Commun.* **89:**1287–1293.

Gregoriadis, G., and Neerunjun, D. E., 1974, Control of the rate of hepatic uptake and catabolism of liposome-entrapped proteins injected into rats. Possible therapeutic applications. *Eur. J. Biochem.* **47:**179–185.

Gregoriadis, G., and Ryman. B. E., 1972a, Fate of protein-containing liposomes injected into rats. An approach to the treatment of storage diseases. *Eur. J. Biochem.* **24:**485–491.

Gregoriadis, G., and Ryman, B. E., 1972b, Lysosomal localization of β-fructofuranosidase-containing liposomes injected into rats. *Biochem. J.* **129:**123–133.

Gregoriadis, G., and Senior, J., 1980, The phospholipid component of small unilamellar liposomes controls the rate of clearance of entrapped solutes from the circulation. *FEBS Lett.* **119:**43–46.

Gregoriadis, G., and Senior, J., 1984, Targeting of small unilamellar liposomes to the galactose receptor *in vivo. Biochem. Soc. Trans.* **12:**337–339.

Gregoriadis, G., Neerunjun, D., and Hunt, R., 1977, Fate of a liposome-associated agent injected into normal and tumour bearing rodents. Attempts to improve localization in tumour tissues. *Life Sci.* **21:**357–370.

Gregoriadis, G., Meehan, A., and Mah, M. M., 1981, Interaction of antibodies bearing small unilamellar liposomes with target free antigens *in vitro* and *in vitro:* Some influencing factors. *Biochem. J.* **200:**203–210.

Gregoriadis, G., Kirby, C., and Senior, J., 1983, Optimization of liposome behaviour *in vivo. Biol. Cell* **47:**11–18.

Gregoriadis, G., Garcon, N., Senior J., and Davis, D., 1988, The immunoadjuvant action of liposomes: Nature of immune response and influence of liposomal characteristics, in *Liposomes as Drug Carriers: Recent Trends and Progress* (G. Gregoriadis, ed.), pp. 279–307, Wiley, Chichester.

Guo, L. S. S., Hamilton, R. L., Goerke, J., Weinstein, J. N., and Havel, R. J., 1980, Interaction of unilamellar liposomes with serum lipoproteins and apoliproteins. *J. Lipid Res.* **21:**993–1003.

Gupta, C. M., Bali, A., and Dhawan, S., 1981, Modification of phospholipid structure results in greater stability of liposomes in serum. *Biochim. Biophys. Acta* **648:**192–198.

Halks-Miller, M., Guo, L. S. S., and Hamilton, R. L., 1985, Tocopherol–phospholipid liposomes: Maximum content and stability to serum proteins. *Lipids* **20:**195–200.

Hwang, K. J., Luke, K. F. S., and Beaumier, P. L., 1980, Hepatic uptake and degradation of unilamellar sphingomyelin/cholesterol liposomes; A kinetic study. *Proc. Natl. Acad. Sci. U.S.A.* **77:**4030–4034.

Juliano, R. L., and Stamp, D., 1975, Effects of particle size and charge on the clearance of liposomes and liposome-encapsulated drugs. *Biochem. Biophys. Res. Commun.* **63:**651–658.

Kirby, C., and Gregoriadis, G., 1980, The effect of the cholesterol content of small unilamellar liposomes on the fate of their lipid components *in vivo*. *Life Sci.* **27**:2223–2230.

Kirby, C., and Gregoriadis, G., 1981, Plasma-induced release of solutes from small unilamellar liposomes is associated with pore formation in the bilayers. *Biochem. J.* **199**:251–254.

Kirby, C., and Gregoriadis, G., 1983, The effect of lipid composition of small unilamellar liposomes containing melphalan and vincristine on drug clearance after injection into mice. *Biochem. Pharmacol.* **32**:609–615.

Kirby, C., Clarke, J., and Gregoriadis, G., 1980a, Effect of the cholesterol content of small unilamellar liposomes on their stability *in vivo* and *in vitro*. *Biochem. J.* **186**:591–598.

Kirby, C., Clarke, J., and Gregoriadis, G., 1980b, Cholesterol content of small unilamellar liposomes controls phospholipid loss to high density lipoproteins in the presence of serum. *FEBS Lett.* **111**:324–328.

Krause, H. J., Juliano, R. L., and Regen, S., 1987, In-vivo behaviour of polymerized lipid vesicles. *J. Pharm. Sci.* **76**:1–5.

Krupp, L., Chobanian, A. V., and Brecher, J. P., 1977, The in-vivo transformation of phospholipid vesicles to a particle resembling HDL in the rat. *Biochem. Biophys. Res. Commun.* **722**:1251–1258.

Lelkes, P. I., and Tandeter, H. B., 1982, Studies on the methodology of the carboxyfluorescein assay and on the mechanism of liposome stabilization by red blood cells *in vitro*. *Biochim. Biophys. Acta* **716**:410–419.

Leserman, L. D., Machy, P., Deraux, C., and Barbet, J., 1983, Antibody-bearing liposomes: Targeting in vivo. *Biol. Cell* **47**:111–116.

Patel, H. M., Tuzel, N. S., and Ryman, B. E., 1983, Inhibitory effect of cholesterol on the uptake of liposomes by liver and spleen. *Biochim. Biophys. Acta* **761**:142–151.

Roerdink, F., Regts, J., Daemen, T., Bakker-Woudenberg, I., and Scherphof, G., 1986, Liposomes as drug carriers to liver macrophages: Fundamental and therapeutic aspects, in *Targeting of Drugs with Synthetic Systems* (G. Gregoriadis, J. Senior, and G. Poste, eds.), pp. 193–206 Plenum Press, New York.

Ryman, B. E., and Tyrrell, D. A., 1980, Liposomes—Bags of potential, *Essays Biochem.* **16**:49–98.

Scherphof, G., and Morselt, H., 1984, On the size-dependent disintegration of small unilamellar phosphatidylcholine vesicles in rat plasma. *Biochem J.* **221**:423–429.

Scherphof, G., Roerdink, G., Waite, M., and Parks, J., 1978, Disintegration of phosphatidylcholine liposomes in plasma as a result of interaction with high-density lipoproteins. *Biochim. Biophys. Acta* **542**:296–307.

Segal, A. W., Willis, E. J., Richmond, J. E., Slavin, G., Black, C. D. V., and Gregoriadis, G., 1974, Morphological observations on the cellular and subcellular destination of intravenously administered liposomes. *Br. J. Exp. Pathol.* **55**:320–327.

Senior, J., 1987, Fate and behaviour of liposomes *in vivo*: A review of controlling factors. *CRC Crit. Rev. Therapeutic Drug Carrier Systems* **3**:123–193.

Senior, J., and Gregoriadis, G., 1982a, Is half-life of circulating small unilamellar liposomes determined by changes in their permeability? *FEBS Lett.* **145**:109–114.

Senior, J., and Gregoriadis, G., 1982b, Stability of small unilamellar liposomes in serum and clearance from the circulation: The effect of the phospholipid and cholesterol components. *Life Sci.* **30**:2123–2136.

Senior, J., and Gregoriadis, G., 1984, Methodology in assessing liposomal stability in the presence of blood, clearance from the circulation of injected animals and uptake by tissues, in *Liposome Technology* (G. Gregoriadis, ed.), pp. 264–282, CRC Press, Boca Raton, FL.

Senior, J., Gregoriadis, G., and Mitropoulos, K., 1983, Stability and clearance of small unilamellar liposomes: Studies with normal and lipoprotein-deficient mice. *Biochim. Biophys. Acta* **760**:111–118.

Senior, J., Crawley, J. C. W., and Gregoriadis, G., 1985, Tissue distribution of liposomes exhibiting long half-lives in the circulation after intravenous injection. *Biochim. Biophys. Acta* **839**:1–8.

Sila, M., Au, S., and Weiner, N., 1986, Effects of Triton X-100 concentration and incubation temperature on carboxyfluorescein release from multilamellar liposomes. *Biochim. Biophys. Acta* **859**:165–170.

Spanjer, H. H., Van Berkel, T. J. C., Scherphof, G. L., and Kempen, H. J. M., 1985, The effect of a water-soluble tris-galactoside terminated cholesterol derivative on the in-vivo fate of small unilamellar vesicle in rats. *Biochim. Biophys. Acta* **816**:396–402.

Spanjer, H. H., Van Galen, M., Roerdink, F. H., Regts, J., and Scherphof, G., 1986, Intrahepatic distribution of small unilamellar liposomes as a function of liposomal lipid composition. *Biochim. Biophys. Acta* **863**:224–230.

Tall, A. R., 1980, Studies on the transfer of phosphatidylcholine from unilamellar vesicles into plasma high density lipoproteins in the rat. *J. Lipid Res.* **21**:354–363.

Tall, A. R., and Small, D. M., 1977, Solubilisation of phospholipid membranes by human plasma high density lipoproteins. *Nature (London)* **265**:163–164.

Tall, A. R., Forester, L. R., and Bongiovanni, G. L., 1983, Facilitation of phosphatidylcholine transfer into high density lipoproteins by an apoliprotein in the denisty of 1.20–1.26 gm/l fraction of plasma. *J. Lipid Res.* **24**:277–289.

Tumer, A., Kirby, C., Senior, J., and Gregoriadis, G., 1983, Fate of cholesterol-rich unilamellar liposomes containing ^{111}In-labelled bleomycin after subcutaneous injection into rats. *Biochim. Biophys. Acta* **760**:119–125.

Weinstein, J. N., Blumenthal, R., Sharrow, S. O., and Henkart, P. A., 1978, Antibody-mediated targeting of liposomes: Binding to lymphocytes does not ensure incorporation of vesicle contents into the cells. *Biochim. Biophys. Acta* **509**:272–288.

Wolff, B., and Gregoriadis, G., 1984, The use of monoclonal anti-Thy$_1$ IgG$_1$ for the targeting of liposomes to AKR-A cells *in vitro* and *in vivo*. *Biochim. Biophys. Acta* **802**:259–273.

Yoshioka, S., Ishibashi, M., Shibazaki, T., and Uchiyama, M., 1984, Plasma induced instability of reverse-phase evaporation vesicles. *J. Parenter. Sci. Technol.* **38**:222–227.

Chapter 11

Reconstitution and Physiological Protein Translocation Processes

Abol-Hassan Etémadi

1. INTRODUCTION

Two interpretations can be attached to the word "reconstitution." In a restricted sense, it generally refers to incorporation of proteins or proteinaceous complexes into bilayer membranes, often with the aim of restoring the natural function of the incorporated material. In a more general sense it refers to the *in vitro* rebuilding or regenerating of natural biological processes with or without involvement of membranes (Etémadi, 1985). With regard to protein translocation processes, reconstitution experiments in both the restricted and the general senses have been attempted.

Interest in protein translocation processes stemmed from studies on protein secretion, which culminated in the hypothesis of "vectorial transport" by Palade (1975). It was suggested that this process occurs cotranslationally and involves a "signaling device" at the N termini of precursors of the proteins (Milstein *et al.*, 1972). This device is now referred to as the signal sequence (Blobel and Dobberstein, 1975). It had previously been proposed that secretory proteins crossed the hydrophobic core of the membrane through a channel (Redman and Sabatini, 1966). The "signal hypothesis" proposed formation of a proteinaceous *transient* channel or tunnel through which secretory proteins

---

Abol-Hassan Etémadi   Laboratory of Biochemistry and Molecular Biology of Bacterial Lipids, Faculty of Sciences of the University of Paris VI, Paris, France.

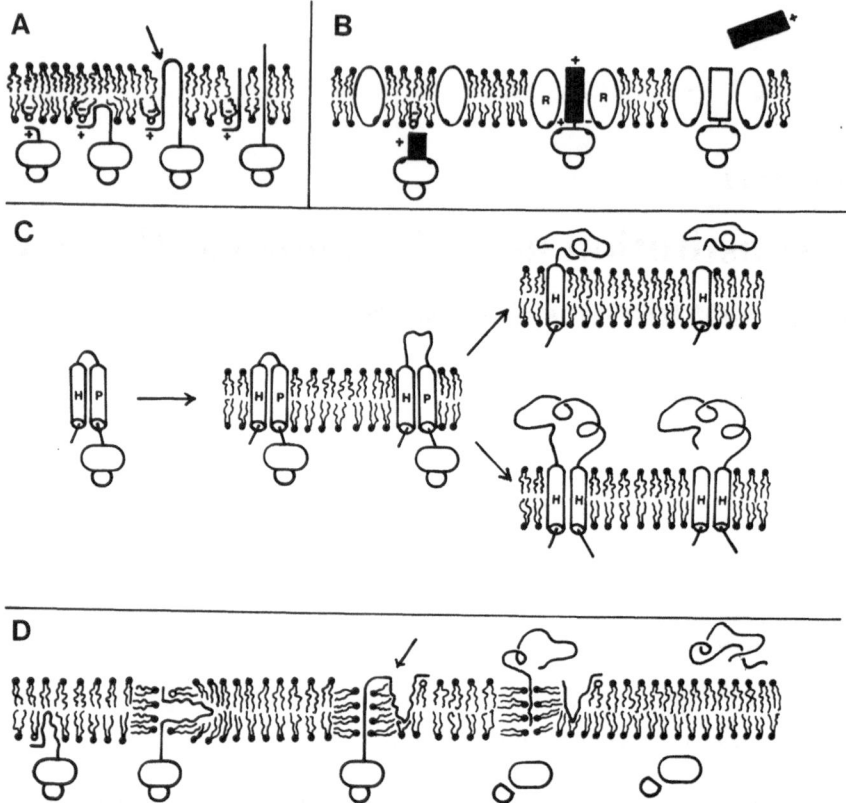

FIGURE 1. Some models for membrane–protein interaction during translocation of secretory and/or transmembrane integral proteins are reproduced. In all schemes, unless otherwise stated, the lower side and the upper side represent the cytoplasmic (cis) and the extracytoplasmic (trans) compartment, respectively. In the loop model A (DiRienzo et al. 1978; Inouye and Halegoua 1980), positively charged amino acid(s) at the N terminus of the signal sequence react with negatively charged head groups of membrane phospholipids. The hydrophobic part of the signal sequence would interact with and loop back in the hydrophobic core of the membrane, bending being due to the presence of a glycine or proline residue in the hydrophobic segment of the signal. Chain elongation would continue until the cleavage site reaches the side remote from the cytoplasm where processing occurs. In model B (Garnier et al., 1980) the signal sequence as a double amphipathic structure inserts in the membrane. Its hydrophobic core has an α-helical or extended conformation and its N terminus is extracytoplasmic. Only after this insertion ribophorin (R) assembly around the signal peptide occurs and interaction of ribophorins, with the ribosomes, leads to the formation of a channel through which the protein is translocated. In the helical hairpin model, C (Engelman and Steitz, 1981), the signal sequence and the N-terminal segment of the mature part of a nascent secretory or transmembrane integral protein form a hydrophobic (H) and a polar (P) helix, respectively, which together exhibit a helical hairpin conformation. The "helical hairpin" is inserted into the membrane in such a way that the N terminus of the signal sequence is cytoplasmic, as in the loop model. Continuation of translation extrudes residues of the polar helix in the trans side, while

would pass in the temporally restricted duration of their translation (Blobel and Dobberstein, 1975). It was soon realized that early stages of membrane insertion of a transmembrane integral protein, the polypeptide chain of the glycoprotein of the vesicular stomatitis virus (G protein), were similar to those of translocation of secretory proteins and that a "stop transfer sequence" abrogated the translocation through the microsomal membrane, allowing membrane anchoring (Lingappa et al., 1978; Rothman and Lenard, 1977). In a more recent version of the signal hypothesis, the "topogenic sequence hypothesis," emphasis is placed on specific sequences of precursor proteins: signal sequences, stop transfer sequences, insertion sequences, and sorting sequences; the interaction of these sequences with specific membrane receptors would trigger translocation and intracellular transport processes (Blobel, 1980). In contrast with the signal hypothesis, some authors envisaged a post-translational translocation process for membrane proteins (Bretscher, 1973; Singer, 1977) and the "membrane trigger hypothesis" proposed the post-translational process for secretion of secretory proteins and membrane insertion of membrane proteins (Wickner, 1979).

Involvement or noninvolvement of membrane proteins and/or the mode of their involvement in protein translocation processes constituted one essential point of divergence between various models and hpyotheses suggested concomitantly with or since the above proposals. Different authors noted the hydrophobicity and the potential richness in secondary structure of prepeptides of secretory and membrane proteins (Austen, 1979; Burstein and Schechter, 1978; Engleman and Steitz, 1981; Garnier et al., 1980; Nesmeyanova, 1982; Palmiter et al., 1978; Rosenblatt et al., 1980; Thibodeau et al., 1978; Von Heijne and Blomberg, 1979) and suggested their interaction with membrane lipids. However, the process of translocation was integrated into various models, some of which are shown in Figure 1. DiRienzo et al. (1978) and Inouye and Halegoua (1980) proposed the "loop model" (Figure 1A) in which the translocation occurs cotranslationally but does not involve membrane proteins (see legend to

---

the polar helix is constantly completed by new residues reaching the membrane. If a secretory protein is being synthesized (upper part of model C), the mature chain is released after cleavage of the signal sequence which remains membrane inserted, as in the loop model. If, however, a transmembrane protein is being synthesized (lower part of model C), a second hydrophobic helix would be formed, which anchors the protein in the membrane. Continuation of the synthesis leads to the cytoplasmic localization of the C terminus of the molecule. Model D (Nesmeyanova, 1982) admits the initial electrostatic interaction of positive charges in the signal sequence with negatively charged head groups of membrane phospholipids as in the loop model. Then this model diverges from all others (see, however, aspects in Garnier et al. 1980) in admitting a phase change in the membrane and translocation of phospholipid molecules concomitantly wtih translocation of a secretory protein. The signal sequence after being cleaved (arrow) remains in the membrane. Its N terminus is here extracytoplasmic as well as its C terminus.

Figure 1). The N-terminal segment of the signal sequence is cytoplasmic and, after cleavage, the signal peptide remains bound to the membrane, while in the signal hypothesis of Blobel and Dobberstein (1975), designated the "linear model" (DiRienzo et al., 1978), it is released in the extracytoplasmic compartment. Austen (1979) admitted an electrostatic interaction with membrane lipids similar to the one suggested by the loop model. However, he suggested the participation of the signal sequence itself and other proteins as channel formers in the process of translocation, while Garnier et al. (1980) (Figure 1B) proposed the interaction of the signal sequence as a double amphipathic structure with membrane lipids. Ribophorins, two transmembrane glycoproteins the presence of which in rough microsomes was previously reported (Kreibich et al., 1978a,b), would then be recruited to form a channel for translocation. No other protein would be involved. In this hypothesis the signal sequence is released in the extracytoplasmic compartment. Von Heijne and Blomberg (1979) and Von Heijne (1980) proposed, in a model designated "direct transfer model" and resembling the loop model, the membrane insertion of the signal sequence of nascent chains as two antiparallel helices, connected by a short loop. It was considered, however, that ribosomes would bind to the membrane presumably through ribophorins, but that ribophorins are not involved in formation of a tunnel; their interaction with ribosomes would be strong enough to force the entry during the translation of hydrophilic residues of the nascent chain into the hydrophobic phase of the membrane. Nevertheless, passage of a sufficiently charged and hydrophilic segment through the hydrophobic core would be so energetically unfavorable that the ribosome would detach from the membrane and the C terminus of the molecule would be translated and remain in the cytoplasm. A transmembrane protein would thus be generated. In another model, suggested by Engleman and Steitz (1981) and designated the "helical hairpin model" (Figure 1C), the N terminus of the nascent chain inserts into the membrane as two antiparallel helices. In this hypothesis no membrane proteins acting as receptors or translocators are involved, even a ribosome–membrane interaction is not envisaged. What plays the role of a stop transfer sequence is the hydrophobic segment, as in the "topogenic sequence hypothesis." However, here no membrane protein is involved in interaction with this sequence.

In another model, proposed by Nesmeyanova (1982) (Figure 1D), the secretion of proteins (through the bacterial cytoplasmic membrane) is proposed to occur in a process coupled with translocation of phospholipid molecules. Passage of the hydrophilic exported proteins occurs through a hydrophilic channel formed by phospholipid head groups, as a result of bilayer to hexagonal type II change of the membrane structure. Here again, no proteinaceous receptor–translocator is required. No energy additional to that used for synthetic processes is dissipated.

Thus, apart from the "membrane trigger hypothesis," all the models for translocation of secretory and transmembrane integral proteins envisage a cotranslational process. This is quite legitimate, particularly in the case of eukaryotic cells, and remains essentially valid, although as we shall see, recent interesting evidence shows the feasibility of a post-translational translocation of some of these proteins under particular experimental conditions. In fact, the presence of free and membrane-bound ribosomes in homogenates of eukaryotic cells has long been reported. Bound ribosomes are attached to the rough endoplasmic reticulum (see Leskes et al., 1971a,b; Palade and Siekevitz, 1956a,b), either loosely or tightly (Adelman et al., 1973), that is, they are detachable by high salt alone or requiring in addition puromycin treatment. Tightly bound ribosomes are linked to the endoplasmic reticulum through an electrostatic linkage, presumably to ribophorins (Kreibich et al., 1978a,b, 1980; see Section 2.1), in addition to being linked through nascent chains. Several studies indicate involvement of ribosomes bound to the endoplasmic reticulum in the genesis of secretory and transmembrane integral proteins. Binding of ribosomes involved in the genesis of these proteins occurs early in protein translation (see Blobel and Dobberstein, 1975; Boime et al., 1977; Burr and Burr, 1981; Eskridge and Schields, 1982; Meek et al., 1982; Scheele et al., 1980).

In the case of bacterial exported proteins, involvement of ribosomes bound to the inner membrane (Cancedda and Schlesinger, 1974; Inouye and Beckwith, 1977; Lory et al., 1983; Randall and Hardy, 1977; Randall et al., 1978; Rasmussen and Bassford, 1985; Smith et al., 1981) was also observed. However, bacterial-bound ribosomes involved in translation of exported proteins could be released by action of puromycin alone, with no requirement for high salt (Davis and Tai, 1980; Smith et al., 1978a,b, 1979). Thus, despite the proximity of ribosomes to the bacterial membrane (Davis and Tai, 1980) no electrostatic binding could be demonstrated in this case. Some genetic work once suggested involvement of bacterial ribosomes in cotranslational export of outer membrane and periplasmic proteins (Emr and Bassford, 1982; Emr et al., 1981). However, other work could not confirm this proposal (see Ito et al., 1984; Shiba et al., 1984).

With regard to proteins translocated to mitochondria (see Chua and Schmidt, 1979; Neupert and Schatz, 1981), chloroplasts (Chua and Schmidt, 1979; Grossman et al., 1980; Highfield and Ellis, 1978), and microbodies (Lazarow and De Duve, 1973; Lazarow et al., 1982), it is generally admitted that they are predominantly synthesized on free ribosomes. In some cases the contribution of bound ribosomes, in addition to free ribosomes, is reported in the genesis of these proteins. Organellar proteins of mitochondria, chloroplasts, and microbodies, synthesized under the direction of nuclear genome and translocated into the respective organelle by passage through at least one membrane, can generally be post-translationally translocated even in cases where bound ribosomes were reported as contributors in their genesis. I proposed calling these proteins "migratory proteins" (Etémadi, 1980a, 1985). Some migratory proteins in chloroplasts must cross through three membranes to reach their ultimate destination (Figure 2).

Besides the involvement or noninvolvement of proteinaceous material in the protein translocation processes and their co- or post-translational nature, the

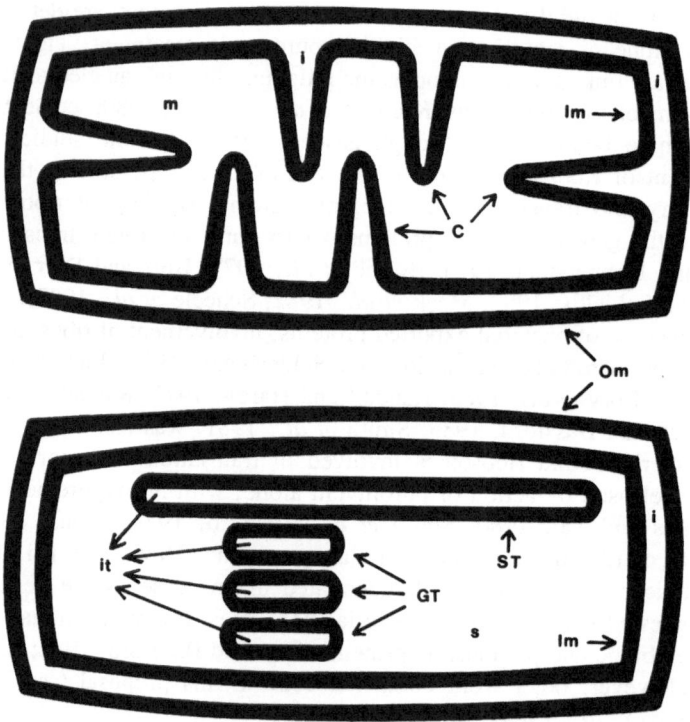

FIGURE 2. This figure shows different compartments of a mitochondrion (upper drawing) and a chloroplast (lower drawing). In both cases, organelles are enclosed in an outer and an inner membrane (Om and Im, respectively). The intermembrane space of the mitochondrion (i) and the matrix compartment (m) are shown. In the chloroplast, an intermembrane space (i) and a stroma compartment (s) are recognized. The inner membrane of mitochondria presents expansions plunging into the matrix; these are cristae (C). In the case of chloroplasts, two types of membrane structure exist in the stroma. These are stroma thylakoids (ST) and grana thylakoids (GT), the last being formed of stacked thylakoid membranes. Thylakoids enclose a single intrathylakoid compartment (it), due to the presence of interconnections. For a protein to reach this compartment it should traverse three membrane bilayers.

energy requirement remained an undecided point. The first clue as to the energy dependency of protein translocation came from studies in the laboratory of Neupert and his colleagues and Schatz and his collaborators on mitochondrial proteins of *Neurospora cràssa* (Hallermeyer and Neupert, 1976; Hallermeyer *et al.*, 1977; Harmey *et al.*, 1976, 1977) and *Saccharomyces cerevisiae* (Nelson *et al.*, 1979; Schatz, 1979), respectively. In the case of mitochondrial proteins, demonstration of an energy requirement proved to be easier than with proteins translocated through the endoplasmic reticulum, since in the last case processes of protein translation and translocation were most generally inseparable, translocation taking place cotranslationally. Furthermore, exactly what the energy requirement meant was not clear. As soon as the energy requirement was known, as a result of work in Neupert's and Schatz's laboratories, I assimilated translocation of migratory proteins to "some sort of active transport" (Etémadi, 1980a) and felt that the energy requirement should be considered in a general framework, crucial elements of which would be involvement of proteinaceous receptor–translocator systems in organellar membrane and requirement for energy for the process of translocation itself. Soon after, Neupert and Schatz (1981) made a similar proposal. At that time a generalization to the case of secretory and transmembrane proteins translocated through microsomal membrane was not possible because of the cotranslational nature of the process. However, as pointed out (Etémadi, 1980a), while the significance of the energy requirement even for post-translational protein translocation was overlooked (e.g., see Blobel, 1980), emphasis was placed on the usefulness of gaining this knowledge if one is to have an understanding of the general problems of protein translocation (Etémadi, 1980a). Recent progress, as a result of work in different laboratories (Hansen *et al.*, 1986; Rothblatt and Meyer, 1986b; Waters and Blobel, 1986) indicates the feasibility of a post-translational energy-dependent protein translocation through the microsomal membrane. The report by Maher and Singer (1986) of the possibility of post-translational translocation through this membrane of a protein reputed to be co-translationally translocated would have constituted the best support for this process. However, the evidence was recently questioned by Ibrahimi (1987), see below. Nevertheless, recent advances allow an extension to the case of secretory and transmembrane proteins of the proposal mentioned above concerning migratory proteins (Etemadi, 1980a, 1985) in stating that translocation of these proteins also involves "some sort of active transport" (Etemadi, 1986a, 1987, 1988).

The whole problem of physiological protein translocation can now be envisaged at two levels: (1) that of the general mechanism, which, with the exception of some specific cases, seems to involve "active transport," and (2) that of mechanistic details dealing with the steps and molecular events concerning the interaction of precursors and receptor–translocators, the exact path of proteins during translocation, the co- or post-translational occurrence and/or feasibility of translocation, and the way energy consumption is coupled to the above events.

Reconstitution in both the general and restricted sense has been attempted, and the most useful information to date has been derived from the first type of experiment; the path is now paved for fruitful liposomal reconstitution assays. The limited liposomal experiments made in the past were generally vitiated, since they relied mostly on the thermodynamics of protein–liposome interaction, ignoring involvement of receptor–translocators or the energy requirement. In what follows I first go through what we presently know of different aspects of physiological protein translocation processes. This allows a comparison to be made with assays using liposomal systems. Ultimately, a recent "unitary hypothesis" suggested by Singer *et al.* (1987a,b) is discussed and prospects for principles guiding further liposomal reconstitution experiments are reformulated.

2. MEMBRANE PROTEINS AND PROTEIN TRANSLOCATION PROCESSES

2.1. Microsomal Membrane Proteins

Various observations based on the inhibitory effect of salt washing (Meyer and Dobberstein, 1980a,b; Warren and Dobberstein, 1978), mild proteolysis (Walter *et al.*, 1979), or N-ethylmaleimide treatment (Katz *et al.*, 1977) of microsomes indicated involvement of proteinaceous material in the process of protein export through the endoplasmic reticulum. At the onset, such experiments were aimed essentially at probing the eventual involvement of proteins as channel or tunnel formers for translocation. They led to the discovery of the signal recognition particle (SRP) and docking protein. SRP, an 11S nucleoprotein complex, formed of six protein subunits of 72,000, 68,000, 54,000, 25,000, 14,000, and 9000 Da and an RNA molecule, identical to a cytoplasmic 7SL RNA, could be detached by salt wash of ribosomes (Walter and Blobel, 1980; Walter *et al.*, 1981). A similar complex (of M_r 250,000 Da) was found to be present as a soluble component in reticulocyte lysates, while the wheat germ extracts were devoid of this material (Meyer *et al.*, 1982).

During *in vitro* experiments on the precursor of a secretory protein (preprolactin) synthesized in the presence of wheat germ protein translation system, polysomes bind to the SRP and protein translation is stopped after a chain of 79 residues is synthesized (Meyer *et al.*, 1982; Walter and Blobel, 1981a; Walter *et al.*, 1981). Salt-washed microsomes that had lost their protein translocation capability, when added to such a translation-arrested system, released the translation arrest and recovered their translocation and processing aptitude (Walter and Blobel, 1981b). In other cases, protein translocation through (Bassuener et al., 1984; Erickson et al., 1983; Mueller et al., 1982) or insertion into (Anderson et al., 1982) the microsomal membrane was reported to be dependent on a signal recognition particle. The signal sequence was shown (Wiedmann et al., 1986a) to bind to the SRP, specifically to its 54,000 Da subunit as revealed by cross-linking experiments (Krieg et al., 1986; Kurzchalia et al., 1986; Wiedmann et al., 1987). More recent data (Siegel and Walter, 1988a) indicate that modification by N-ethylmaleimide of the 54,000-dalton subunit prevents binding of nascent chains to SRP, while that of 68,000- and 72,000-dalton subunits does not prevent the elongation arrest but binding to the docking protein and the subsequent translocation.

The exact significance of protein translation arrest by SRP is unclear, however. It was observed not to occur with some proteins (Anderson et al., 1982) that were devoid of a cleaved signal sequence. Nevertheless, it was observed in other cases of proteins with an uncleaved signal (Bonatti et al., 1984; Brown et al., 1984; Lipp and Dobberstein, 1986a; Rottier et al., 1985; Sakaguchi et al., 1984). Finally, a variability of arrest of protein synthesis, even with proteins endowed with a cleared signal, was observed (Siegel and Walter, 1985). Meyer (1985) reported that the arrest in translation is only observed with a wheat germ protein synthesizing system, not with reticulocyte lysates or a HeLa-cell-free system. Finally, association studies after dissociation of SRP revealed that omission of subunits of 9000 and 14,000 led to an incomplete particle, which, however, mediated translocation at a reduced level but without translation arrest. Thus, translation arrest is not required for translocation (Siegel and Walter, 1986).

Recently, Prehn et al. (1987) reported that the cotranslational translocation of precursor proteins of different origin [human placental lactogen, plant zein (a storage protein of protein bodies) and the bacterial plasmid β-lactamase] could occur through wheat germ microsomal membrane. A plant SRP-like complex was required. However, this particle not only did not arrest protein synthesis but increased the translation. This plant SRP did not reconstitute the translocation function of salt-washed canine microsomal membranes.

In canine SRP, the two smallest subunits of 9000 and 14,000 Da are bound to the 5' and 3' ends (Alu sequences) of the 7SL RNA, while other subunits are linked to the central (S) sequence (Guldenfinger et al., 1983). The most tightly bound subunit to the RNA appeared to be the 19,000 Da protein (Scoulica et al., 1987). Dissociation and reassembly studies indicated a functional role for the RNA part of the SRP complex in addition to its protein subunits (Walter and Blobel, 1983). In the case of yeast microsomes, despite the fact that neither the protease treatment nor the salt wash abrogates their translocation aptitude for yeast secretory proteins (Rothblatt and Meyer, 1986a), the presence of a postribosomal proteinaceous factor of 9.6S devoid of RNA and involved in post-translational protein translocation of yeast pre-pro-α-factor, was reported. The possibility that this complex may constitute a "denaturase" involved in unfolding protein precursors before they can be taken up post-translationally was envisaged (Waters et al., 1986). The factor, however, was more recently assimilated to an evolutionary precursor of SRP (Fecycz and Blobel, 1987). Thus, the function attributed to the RNA in SRP remains uncertain, since the yeast complex is devoid of RNA.

Another microsomal protein, involved in protein translocation is designated as "docking protein;" it was described as an integral membrane protein of about 72,000 Da (Meyer et al., 1982a) involved in the release of the arrest in protein translation resulting from interaction of SRP with the signal sequence of the nascent chain (Gilmore et al., 1982; Meyer, 1982; Meyer et al., 1982). This protein shows a 20-fold enrichment in the rough endoplasmic reticulum as compared to the smooth endoplasmic reticulum (Hortsch et al., 1985). Proteolysis of rough microsomes cleaves a fragment of approximately 60,000 Da from this protein (Meyer et al., 1982a,b). The length of the fragment of the protein cleaved by proteolysis of microsomes depends on the proteolytic enzyme used. A recent study indicated that elastase cleaves a fragment of 59,000 Da, leaving a segment of 14,000 Da anchored in the membrane, while trypsin releases a shorter 46,000-Da fragment. The part involved in releasing of the arrest of protein translation is located in the segment of 13,000 Da, the difference between the fragments cleaved by the two enzymes (Hortsch et al., 1985). Lauffer et al. (1985), reported an M_r of 69,684 Da for docking protein, which is shown as a transmembrane protein anchored in the microsomal membrane through its N terminus. More recently, Tajima et al., (1986) obtained evidence for the docking protein to be composed, in addition to the 70,000-Da polypeptide just described, of another polypeptide chain of approximately 30,000. Hortsch and Meyer (1988) report now that this 30,000-Da (α) subunit of the docking protein might be involved in binding of the 72,000-Da (β) subunit to the microsomal membrane. This binding would occur post-translationally without involvement of SRP and docking protein. The 72,000 dalton subunit would thus not be a transmembrane protein.

It is worthwhile at this point to consider two crucial aspects of the possible involvement of SRP and docking protein in the protein translocation process: (1) What is the bearing of their discovery on the co- or post-translational nature of microsomal protein translocation? (2) What are the molecular events accompanying and following the release of translation arrest and allowing translocation? With regard to the first point, the phenomenon of arrest of translation and its release in the presence of salt-washed microsomes appeared compatible with a strict coupling of translation with translocation. Then, when it was found that protein translation arrest does not occur in all cases, it was explicitly or implicitly admitted that protein translocation was still tightly associated with its translation. More recent experiments bring fundamental changes in this respect. Ainger and Meyer (1986) report that in the wheat germ protein translation system, SRP can still arrest translation if two-thirds of the nascent IgG light chain is completed, but that this chain can be translocated by addition of rough ribosomes. Experiments during which nascent bovine prolactin precursor was translocated led to a similar conclusion. A very recent report by Siegel and Walter (1988b) confirms these data and indicates that the discrepancy between these results and those reported above as to the length of nascent chains of pre-prolactin with which SRP can interact in a functional manner is correlated with different concentrations of SRP used. Thus, the possibility, under *in vitro* experimental conditions, of the lack of a strict coupling of translation with translocation is confirmed. As to the second point, neither the nascent chain nor the ribosome binds to the docking protein (Gilmore and Blobel, 1983). SRP is released and docking protein is present in microsomes in a far less stoichiometric ratio than bound ribosomes. Thus, it cannot apparently be directly involved in the actual translocation. An observation, the exact significance of which remains to be appreciated, is that in *Xenopus* oocytes expressing ovalbumin, neither the signal recognition particle nor the docking protein seemed to play a significant role in the synthesis and export of the protein (Richter *et al.*, 1985). In any event, the question is raised as to the nature of protein (s) of the microsomal membrane directly involved in the binding and translocation processes. Observations on *S. cerevisiae* mutants defective in protein translocation led to the suggestion that microsomal proteins other than SRP and docking protein are also involved in protein translocation (Ferro-Novick *et al.*, 1984a,b). However, further studies on this case did not allow the discovery of an additional transmembrane integral protein in protein translocation (Bernstein *et al.*, 1985). Nevertheless, more recently isolation of a thermosensitive yeast mutant (sec 61) was reported that accumulates multiple secretory and vacuolar precursor proteins. These are bound to the endoplasmic reticulum but are not processed and have a large cytoplasmic exposure indicating that their translocation is not completed (Deshaies and Schekman, 1987). It would be interesting to learn more about the way the mutated gene is involved in the translocation process.

Ribophorins, as mentioned above, were proposed to be implicated in binding of ribosomes (Kreibich *et al.*, 1978a,b); for a recent report on the structure and orientation of ribophorins see Crimaudo *et al.*, 1987). Such a role for these proteins was subsequently questioned (Bielinska *et al.*, 1979). However, further studies (Amar-Costesec *et al.*, 1984; Kreibich *et al.*, 1982; Marcantonio *et al.*, 1984) supported their possible correlation with ribosome binding, though besides their direct electrostatic binding to ribosomes, alternatives such as their implication in binding of nascent chains to the membrane or the transfer of these chains to the luminal side were not excluded (Marcantonio et al., 1984; see also Harnik-Ort et al., 1987). Recent controlled proteolytic experiments (Hortsch et al., 1986) carried out on microsomes led to recognition of conditions under which some 85% of ribophorin I and 100% of ribophorin II had resisted degradation; nonetheless, these vesicles had lost their aptitude for binding ribosomes, even after the function of docking protein (which was also cleaved) was reconstituted. Thus, proteins other than ribophorins should be involved in binding ribosomes to the membrane, although their contribution to the process cannot be excluded.

Different reports suggest the possible existence of binding sites for the signal sequence on the microsomal membrane. In fact, chemically synthesized prepeptides of a secretory protein inhibited competitively the *in vitro* translocation through microsomes of various secretory proteins (Majzoub et al., 1980; Rosenblatt et al., 1979). Interestingly, precursors of secretory proteins could bind to salt-washed microsomal membrane (Prehn et al., 1980, 1981). Studies by Austen and his colleagues (Austen and Ridd, 1983; Austen et al., 1984) show the binding to salt-washed microsomes of a synthetic peptide representative of signal sequences of precursors of secretory and transmembrane integral proteins. This binding inhibited translocation of secretory proteins. The synthetic peptide did not bind to the signal recognition particle. This point seems at variance, however, with a report by Duong et al. (1987), which does not exclude interaction of synthetic prepeptides with SRP. In any event, a recent report by Robinson et al. (1987) implicates a microsomal membrane protein of approximately 45,000 Da, distinct from all proteinaceous components mentioned above and from the signal peptidase, in binding of signal sequences to the microsomal membrane. Other authors, however, had suggested the possibility that the microsomal signal peptidase could constitute or be part of the translocation system (Evans et al., 1986). Recently also Wiedmann et al. (1987) reported the presence of a microsomal integral membrane glycoprotein of 35,000 daltons, which was designated as signal sequence receptor, to which the signal sequence could bind after release from the docking protein. Let us mention that a report by Gilmore and Blobel (1985), indicating that during their translocation secretory proteins cross an aqueous environment, was interpreted as favoring the involvement of a proteinaceous channel. However, such an interpretation is not unexceptionable since an aqueous channel can also be formed by head groups of lipids, as is inherent in the model proposed by Nesmeyanova (1982) (see Figure 1D). In summary, despite the fact that there are several hints to the involvement of proteinaceous components in the process of microsomal protein translocation, further work is required to fully establish their nature and details of their function.

2.2. Bacterial Membrane Proteins

Here again the inhibition of translocation of exported proteins by pronase treatment of inside-out inner *E. coli* membrane vesicles revealed involvement of membrane proteins in the process (Rhoads *et al.*, 1984; Smith, 1980). In other studies, compositional differences in protein content could be observed in ribosome-free membrane and membrane–ribosome complexes obtained from *Bacillus subtilis* (Caulfield *et al.*, 1983; Horiuchi *et al.*, 1983a,b; Marty-Mazars *et al.*, 1983). Some proteins unique to the membrane–ribosome complex were thought to be involved in protein translocation. Further analysis by Caulfield *et al.*, (1984) revealed that the membrane–ribosome complex could in fact be separated into two fractions, the first comprising monosomes and an S complex reminiscent of the signal recognition particle. This complex contained proteins of 64,000, 60,000, 41,000 and 36,000 Da. However, S complex was not present in the soluble fraction, was devoid of RNA, and could not lead to translation arrest. The second fraction contained polysomes attached to the membrane. This fraction comprised only the 64,000-Da protein of the complex (Caulfield *et al.*, 1984). S complex was thought to be involved in initiation of export. Its subunits, except the 64,000-Da entity, would detach upon contact of the protein-translating ribosome with the membrane; the complex would thus break up (Caulfield *et al.*, 1985). That bacterial 6S RNA is not involved in protein translocation is confirmed by other authors (Lee *et al.*, 1985). However, Mueller and Blobel (1984) reported the presence in *E. coli* homogenates of an "export factor" of 12S and suggested the possibility that it may contain an RNA species smaller than the 6S RNA. This factor and the one from the postribosomal yeast fraction, the function of which it could mimic interchangeably, were considered as evolutionary precursors of SRP (Fecycz and Blobel, 1987). Adler and Arvidson (1987) found the presence of a multiprotein complex on membrane-bound ribosomes of *Staphylococcus aureus*, correlated with exoprotein synthesis, which contains four tightly bound polypeptides of M_r 71,000, 60,000, 46,000, and 41,000 Da. The 60,000-Da entity was antigenically correlated with the 64,000-Da entity of *B. subtilis* mentioned above. However, the *S. aureus* complex is present both on membrane-bound ribosomes, where it is located between the membrane and ribosomes, and in solution. This complex is thus more similar to the microsomal SRP. However, it exhibits only four subunits and does not contain RNA.

Besides the above analytical data, other observations indicate the involvement of proteins in translocation of exported proteins through the bacterial cytoplasmic membrane. Studies on *E. coli* mutants defective in protein translocation (Wanner *et al.*, 1979) and several reports on *E. coli* expressing fusion proteins led to the suggestion of the presence of proteinaceous export devices.

In such assays the gene fusion concerns the N terminus of the precursor of an exportable protein, including the signal sequence, linked N terminally to a cytoplasmic protein such as β-galactosidase. Examination of localization of hybrid proteins expressed in *E. coli* and the inhibition of translocation of cellular exportable proteins as a result of jamming of the membrane are interpreted in terms of the existence of export machinery in the membrane (see Bassford *et al.*, 1979; Emr and Silhavy, 1982; Emr *et al.*, 1980; Ito *et al.*, 1981; Kiino and Silhavy, 1984). Studies on conditional *E. coli* mutants defective in protein export led to the proposal of involvement of a gene (sec A gene) product of 92,000 Da in protein translocation (Oliver and Beckwith, 1981, 1982a,b). The protein was assimilated to a component of a SRP-like complex or to docking protein. Interaction with a defective translocation system led to the arrest in the synthesis of transported protein (Kumamoto *et al.*, 1984). Also, the product of a sec B gene was involved in protein translocation in *E. coli* (Kumamoto and Beckwith, 1983). The product of this gene (Kumamoto and Beckwith, 1983) and that of another sec Y (or prl A) gene were suggested to interact with the product of sec A gene in the export machinery (Bickman *et al.*, 1984). However, it was very recently reported (Collier *et al.*, 1988) that sec B gene encodes a soluble cytoplasmic protein involved in unfolding of a periplasmic protein, maltose-binding protein, preventing its premature folding before translocation. The sec Y (prl A) gene product is identified to an integral protein of the inner *E. coli* membrane (Akiyama and Ito, 1985, 1987; Ito, 1984). Involvement of still another gene (prl D gene) product as well as that of sec Y gene in the process of translocation could be supported by observation of *E. coli* suppressor mutations in these genes, allowing translocation of exported proteins showing defective signal sequences. Products of these two genes were suggested to be assembled in a complex (Bankaitis and Bassford, 1985a). Extragenic suppressor mutations also suggested involvement of sec A, prl A, sec C, and other loci in protein translocation in *E. coli* which would possess a general mechanism of coupling of synthesis of exported proteins with secretion (Oliver, 1985; Oliver and Liss, 1985). Both prl A and sec A gene products were implicated in the indirect binding of ribosomes involved in the synthesis of exported proteins to the membrane (Rasmussen and Bassford, 1985). More recently, however, it was found difficult to identify components of export machinery by suppression of sec A mutations (Lee and Beckwith, 1986), and Strauch *et al.* (1986) could not confirm coupling between synthesis and translocation of exported proteins. For example, inhibition of the synthesis of maltose-binding protein in sec A (am) mutants was correlated with cAMP depletion resulting in a decreased transcription of mal E gene. Addition of cAMP led to the synthesis of the protein while translocation remained blocked. In addition, sec A (am) mutation did not affect synthesis of OmpA protein and lipoprotein, two other exported proteins. Furthermore, Shiba *et al.* (1986) find that the protein export defect as a result of sec Y 24 (ts) mutation can be suppressed extragenically by numerous (at least six) genes, they consequently draw attention to the unlikeliness that all these genes would specify components of the translocation system. In addition, indications suggest that sec Y gene product might be involved in post-translational translocation of an outer membrane protein, OmpA protein, not in coupling synthesis and translocation (Bacallao et al., 1986).

To summarize, while both analytical and genetic work indicate involvement of proteinaceous material in the process of protein translocation in bacteria, some interpretations of the genetic work, such as coupling between synthesis and translocation or significance attributed to suppressor mutations as revealing involvement of complexes in the translocation process, are questioned. Recently, further E. coli genes designated as sec D (Gardel et al., 1987) and sec E (Riggs et al., 1988) were reported to be correlated with protein export in E. coli. The localization of products of these genes remains to be elucidated.

2.3. Organellar Membrane Proteins

Involvement of proteinaceous material in organellar outer membranes both of mitochondria (Schatz, 1979) and chloroplasts (Chua and Schmidt, 1978, 1979; Highfield and Ellis, 1978) in translocation of migratory proteins has been envisaged. Direct evidence for involvement of proteins in the translocation process came from proteolytic experiments. Mild proteolysis of mitochondria (Argan et al., 1982, 1983; Gasser et al., 1982a) and chloroplasts (Chua and Schmidt, 1978; Cline et al., 1985) prevented post-translational translocation of migratory proteins. Binding studies using unenergized mitochondria or right-side-out vesicles of the outer mitochondrial membrane confirmed that proteinaceous material was required as receptor (Riezman, 1982; Riezman et al., 1983a). Studies on chloroplast envelope vesicles (Pfisterer et al., 1982) led to similar results. In these studies proteolysis prevented binding of precursors to the appropriate membranes.

That different receptors for different categories of migratory proteins of a given organelle might exist was suggested by the observation of Oda et al. (1984) that surface proteolysis of rat liver mitochondria abolished binding of preadrenodoxin, while it did not affect binding of presulfite oxidase. Likewise Zwizinski et al. (1984) reported the differential action of various proteolytic enzymes on receptors for different mitochondrial imported proteins. However, receptors were susceptible to one or the other protease, hence the necessity of using multiple proteolytic enzymes—and the eventual control by other techniques—before concluding a lack of receptor.

Competition studies on inhibition of binding of precursors to organelles favor the presence of receptors. Two types of study were carried out in which the inhibitor was constituted by receptors of the outer membranes or by ligands. Thus, Ono and Tuboi (1985) observed that import of the precursor for mitochondrial ornithine aminotransferase was inhibited by the presence of isolated outer membrane vesicles. After detergent solubilization of these vesicles and partial purification and reconstitution of the receptor in lecithin, the liposomes formed also inhibited import into mitochondria. Likewise, Bhat and Avadhani (1985) report that a membrane fraction obtained from mitoplasts inhibited import of carbamylphosphate synthetase by digitonin-treated mitochondria. With regard to inhibition by ligands, it was found that translocation of several mitochondrial proteins into rat liver mitochondria was inhibited by apocytochrome c (Arpin et al., 1980; Matsuura et al., 1981). The nature of proteins, the translocation of which was inhibited, remained undetermined. Recently, however, apocytochrome c was reported to compete with the uptake of pre-ornithine carbamyl transferase (Hernandez-Yago and Grisolia, 1987). A substantiation in this respect seems useful, since the report is contradictory with another one (see Mori et al., 1985). In *Neurospora crassa* the import of ADP-ATP carrier, or dicyclohexylcarbodiimide-binding protein (Henning and Neupert, 1981), as well as the insertion of porin into the outer membrane (Zwizinski et al., 1983), is not inhibited by apocytochrome c. Thus, different receptor types might be present (Henning and Neupert, 1981). Also, observations by Yaffe and Schatz (1984) with different yeast mutants confirm a differential inhibition of the import of mitochondrial proteins, hence the possible presence of different pathways.

Another set of experiments favoring the possibility of the existence of discriminatory routes for different (categories of) migratory proteins concerned uptake of precursors of these proteins specific to an organ by organelles of another organ. Two proteins of mitochondria of steroidogenic organs, that is, cytochromes P-450 *scc* (Matocha and Waterman, 1984, 1986; Ogishima et al., 1985) and P-450 11β (Matocha and Waterman, 1986) are not imported by heart mitochondria, which, however, can import adrenodoxin, another mitochondrial protein involved in steroidogenesis (Matocha and Waterman, 1986). Likewise, carbamylphosphate synthetase is taken up by mouse liver mitochondria while mitochondria of other tissues (kidney, brain, heart) are unable to import this matrix enzyme (Bhat and Avadhani, 1985). It is worthwhile mentioning at this point that Côté and Boulet (1985) place emphasis only on differences in efficiencies of receptor–translocators of mitochondria of different organs in translocating common or organ–specific proteins.

Besides organ specificity, species specificity was also noted. In numerous cases with mitochondria (Anderson, 1981; Cheng et al., 1987; Horwich et al., 1985a; Schleyer et al., 1982; Schmidt et al., 1983; Takiguchi et al., 1983; Teintze et al., 1982) and chloroplasts (Chua and Schmidt, 1978; Pfisterer et al., 1982), migratory proteins could exchangeably bind to or be taken up and processed, depending on the type of experiment carried out, by respective organelles of different species. Also, the firefly (*Photinus phyralis*) peroxisomal enzyme, luciferase, expressed in the transfected mammalian CV-1 (a monkey kidney) cell line was translocated to peroxisomes (Keller et al., 1987). Likewise expression in *S. cerevisiae* of a peroxisomal enzyme, alcohol oxidase, which this yeast is normally lacking, led to its peroxisomal localization (Distel et al., 1987). However, differences are reported; thus, the precursor of mammalian ornithine carbamyl transferase is not taken up by yeast mitochondria (Takiguchi et al. 1983), and apocytochrome *c* of *N. crassa* bound to mitochondria from the same origin could not be displaced by apocytochrome *c* from *Paracoccus denitrificans* (B. Henning et al., 1983). Also, the precursor of the small subunit of *Chlamydomonas reinhardtii* ribulose 1,5-bisphosphate carboxylase was not taken up by higher plant chloroplasts (Bedbrook et al., 1980; Coruzzi et al., 1983; Schmidt et al., 1979). In such cases the lack or deficiency of receptors or sequence differences in binding sites of migratory proteins might explain the lack of interspecies uptake.

The import in mitochondria presumably occurs through zones of contact between the outer and the inner mitochondrial membrane. The outer membrane at such zones was recently reported (Dorbani et al., 1987) to have compositional differences from other regions of the outer membrane. Ono and Ito (1984a,b) found that the receptor for an intermembrane compartment protein resisted the preparation of mitoplasts and was enriched in a part of the outer membrane firmly bound to the inner membrane. Bhat and Avadhani (1985) reported that digitonin-treated mitochondria from mouse liver containing 10–15% of outer membrane fragments are able to import 85–90% of the precursor of carbamylphosphate synthetase taken up by mitochondria. Schleyer and Neupert (1985) and Hartl et al., (1987) have made the interesting observation that in deenergized mitochondria the *in vitro* synthesized ADP-ATP carrier interacts with the membrane but is not translocated. However, it is kept in an aqueous environment, in a conformation unreactive with proteases (see Figure 5). Upon reenergization of mitochondria, the carrier protein is translocated to the inner membrane, even after trypsinization of mitochondria. Recently, Schwaiger et al., (1987) confirmed that mitoplasts were able to import mitochondrial proteins and that a proteinaceous material was involved in the uptake process. A factor removable by digitonin plus salt from mitoplasts was implicated in the import process. These authors also found that the preformed translocation intermediates remained with mitoplasts after subsequent fractionation. Furthermore, fractionation of mitochondria produced a fraction of intermediate density, as compared to the inner and outer membranes, which was enriched in translocation intermediates. Involvement of contact zones was also assessed by immunocytochemical studies. Recently Soellner et al. (1988) could entrap translocation intermediates of mitochondrial ADP-ATP protein and F_1-ATPase subunit β by decreasing the ATP level (or by lowering the temperature) in the *in vitro* uptake reactions. Such intermediates, which were in the stages of binding to the receptor or inserted in contact sites between the outer and the inner membrane, respectively, could interact with specific antibodies.

Information on the exact nature of the proteinaceous material involved in translocation of migratory proteins is rather scarce. Some observations suggest involvement of intrinsic membrane proteins in mitochondria (Argan *et al.*, 1982, 1983). Recently, Gillepsie (1987) found that a synthetic extrasequence of preornithine carbamyl transferase (residues 1–27) binds specifically in a saturable and reversible manner to a protein of 30,000 Da from the outer membrane of mitochondria as revealed by cross-linking experiments. Mild proteolysis abolished both the binding of the synthetic peptide to and the import of the precursor into mitochondria. The protein that constituted 4–10% of the outer membrane protein could not be removed by 1M KCl. Ohba and Schatz (1987) have also recently confirmed the involvement of proteinaceous material in mitochondrial translocation. Treatment of isolated yeast mitochondria with high concentrations of trypsin-inhibited protein import. Also, antibodies to outer membrane proteins or to a specific outer membrane protein of 45,000 Da inhibited the import. However, when mitoplasts were prepared from trypsin-treated mitochondria, import was restored and antibodies to the 45,000-Da protein did not affect the energy-dependent protein import. Thus, involvement of proteinaceous component(s) of the outer membrane and "one or more steps" of translocation would be bypassed when mitoplasts are used. A recent report by Pfanner *et al.* (1987b) indicated that mitochondrial precursor proteins are imported through a hydrophilic membrane environment. As mentioned above with regard to a similar report concerning translocation of proteins through the microsomal membrane, the observation as such, though highly interesting, does not necessarily prove involvement of membrane proteinaceous material.

With regard to chloroplasts, in order to gain insight into the proteinaceous material interacting with precursor proteins, Kloppstech and Bitsch (1986) labeled in the dark the *in vitro*-translated precursors of cytoplasmically synthesized pea chloroplast proteins with a heterobifunctional, cleavable, and photoactivatable cross-linker, that is, ($[^{35}S]$cyteamine) *N*-succinimidyl-3-((2-nitro-4-azidophenyl)-2-aminoethyldithio) propionate (designated as SNAP). Such labeled precursors when presented to isolated chloroplast envelope preparations, labeled envelope proteins of 14,000, 24,000, 31,000, 35,000, and 42,000 Da. As mentioned above, trypsin treatment of isolated envelopes prevents binding of precursor proteins. Such a treatment of isolated envelopes was shown in the present studies to cleave the 31,000- and 35,000-dalton entities. Recently Pain *et al.* (1988) confirmed the involvement of a protein of 30,000 daltons by using an antiidiotypic antibody approach which had previously found another application in the field of migratory proteins (see Sakakibara *et al.*, 1987). This approach allowed the localization of the 30,000-dalton protein at the contact sites between the two chloroplast membranes. Cornwell and Keegstra (1987), by the use of a heterobifunctional photoactivable cross-linker, were also able to bind the small subunit of ribulose 1,5-bisphosphate to a protein in the outer membrane of chloroplasts. The conjugate was of 86,000 Da. This would apparently mean that the membrane receptor protein is of approximately 66,000 Da, provided that a one-to-one stoichiometry of conjugation occurs. However, the efforts of these authors to stain such a protein in the electrophoretograms of the outer membrane remained unsuccessful.

Assays made by Lubben et al. (1987) are remarkable in that they incorporated two different stop transfer sequences of microsomally translocated proteins into precursors for ribulose 1,5-bisphosphate carboxylase; the chimeric precursors were translocated into isolated chloroplasts. The authors conclude that a direct passage through the lipid bilayer is not probable. In support, it can be mentioned that examples exist in which precursors of some hydrophobic proteins (seric apolipoproteins and melitten) cross the membrane and are processed, while they are able to aggregate with or insert into lipid bilayers. It should also be noted that signal sequences were even observed to cross the microsomal membranes when inserted in some chimeric constructs (see Section 4.1).

Several reports (Argan et al., 1983; Miura et al., 1983; Otha and Schatz, 1984; Takiguchi et al., 1983) also indicate the presence of cytosolic factor(s) that are proteinaceous in nature (Argan et al., 1983; Takiguchi et al., 1983) and involved in the translocation process. A low-molecular-weight material of less than 5000 Da (Argan et al., 1983), which is dialyzable and destructible by trypsin (Miura et al., 1983), as well as a 40,000-Da protein (Otha and Schatz, 1984) and ions [Mg^{2+} (Lewin and Norman, 1983), K^+ (Miura et al., 1983)] were involved in mitochondrial transport. Implication of a nucleoprotein complex of M_r 400,000 in this process was also mentioned (Firguira et al., 1984) and the presence of "import factor(s)" in reticulocyte lysates was reported. This factor(s) would bind the precursor forming a 5S complex. However, the factor(s) alone does not bind to mitochondria (Argan and Shore, 1985) and, importantly, RNA is not required for mitochondrial import (Argan and Shore, 1985; Burns and Lewin, 1986). Recently Joste et al. (1987) confirmed the presence of a stimulating "import factor" for three mitochondrial proteins in reticulocye lysates and Ono and Tuboi (1988) proposed that precursors of mitochondrial proteins would bind to and form a complex with the "import factor" which is recognized by the membrane receptor. Behra and Christen (1986) have presented evidence that both in vivo and in vitro the precursor of aspartate aminotransferase forms high-molecular-weight complexes presumably involved in translocation. It would thus seem clear that a simple interaction of precursors and lipids is far from representing the reality of migratory protein translocation.

3. ENERGY DEPENDENCE OF PROTEIN TRANSLOCATION PROCESSES

3.1. Secretory and Transmembrane Integral Proteins

Protein translocation through the microsomal membrane was thought not to require energy in addition to the amount used for chain translation. Since the process occurred cotranslationally, it appeared unpractical to experiment for the requirement for additional energy. Often, when under some experimental conditions proteolytic processing of exported proteins in prokaryotic systems did not occur, the lack of cleavage of the extrasequence was taken as revealing the lack of translocation, and conclusions as to energy requirement for the last process were drawn. It should, however, be borne in mind that for many secretory and transmembrane integral proteins a cleaved signal sequence does not exist. Furthermore, under various experimental circumstances, processing and translocation could be experimentally dissociated, so that processing did not occur or was delayed while translocation had occurred (DiRienzo and Inouye, 1979; Freudl et al., 1985; Giam et al., 1984; Halegoua and Inouye, 1979; Halegoua et al., 1976, 1977; Ichihara et al., 1982; Inouye et al., 1983a,b; Koshland et al., 1982; Lin et al., 1978, 1980; Lory et al., 1983; Russel and Model, 1981; Seehara and Khorana, 1984; Sekizawa et al., 1977; Wu et al., 1977).

Studies carried out on *E. coli*-exported proteins indicated that, even under normal conditions, while some of them were cotranslationally processed, others were processed co- and post-translationally and still others post-translationally (Josefsson and Randall, 1981a,b; Tweten and Iandolo, 1983). The observation that some *E. coli* exported proteins could be cleaved by an externally added proteinase K, only after translation of 80% (maltose-binding protein) or the full length (ribose-binding protein) of the chain, was interpreted to mean that a strict coupling between the translation and translocation does not exist (Randall, 1983). While such an interpretation is attractive, the alternative, that the translocated chains were not accessible to proteinase K, in the context of the membrane, cannot be excluded by such observations alone. In other words, problems of crypticity should be taken into account (see Etémadi, 1980a, 1985).

The possibility of translocation of exported proteins without the cleavage of signal sequence rendered equivocal the interpretation in terms of the lack of translocation of secretory and transmembrane proteins of bacteria that remained unprocessed as a result of treatment of cells with various drugs. In several cases uncouplers were observed to prevent processing (Daniels et al. 1981; Date et al., 1980; Enequist et al., 1981; Palva et al., 1981; Russel and Model, 1981; Tweten and Iandolo, 1983; Valusek et al., 1984). However, these drugs did not prevent translocation in all cases (Ghrayeb and Inouye, 1984; Russel and Model, 1981; Valusek et al., 1984; Zimmermann et al., 1982) and, in addition, the reported effect of uncouplers on the action of signal peptidase (Russel and Model, 1981; Zimmermann et al., 1982) rendered the knowledge of the exact cellular location of precursors in the presence of these drugs mandatory. In fact, in early works, the cytoplasmic localization was assessed only in some cases (Date et al., 1980; Ghrayeb and Inouye, 1984; Wolfe and Wickner, 1984; Zimmermann and Wickner, 1983) and a (slow) translocation after removal of uncouplers was shown (Ghrayeb and Inouye, 1984). The last experiment, translocation of the cytoplasmically located precursor after removal of the drug is in fact required for the demonstration of a post-translational energy-dependent translocation, since an effect of drugs on the membrane conformation preventing translocation cannot be excluded, as also pointed out by others (Muren and Randall, 1985). Despite these notes of caution, other *in vitro* experiments (Chen and Tai, 1985, 1987a; Mueller and Blobel, 1984) have shown the feasibility of the post-translational energy-dependent translocation of exported proteins in prokaryotic systems. With regard to eukaryotic systems, it was observed by different laboratories (Hansen et al., 1986; Rothblatt and Meyer, 1986b; Waters and Blobel, 1986) that yeast pre-pro-α-factor synthesized *in vitro* can be taken up post-translationally in an ATP-dependent manner into yeast microsomes. However, such an uptake was limited to the mentioned case and, for example, it did not work for invertase, another yeast secretory enzyme. A report by Maher and Singer (1986), stating that the precursor of a protein, the translocation of which was previously found to be co-translational, could occur post-translationally after dithiothreitol treatment would have constituted a clear-cut invaluable demonstration of the feasibility of post-translational microsomal uptake of proteins. Ibrahimi (1987), however, reports that the precursor in this case was still linked to ribosomes.

As to the nature of energy involved in translocation in prokaryotic cells, the involvement of membrane potential was first suggested by Date et al., (1980). Bakker and Randall (1984) did not admit this proposal and suggested instead the involvement of both the pH gradient (ΔpH) and membrane potential ($\Delta\psi$) components of the proton motive force, while whether the effect is at the level of translocation or only processing was not decided. Mueller and Blobel (1984) found that the presence of a functional F_0-F_1 ATPase, hence a proton motive force, was absolutely necessary for protein translocation in *E. coli*. Enequist et al., (1981) had previously correlated processing of exportable proteins in *E. coli* preferentially with the maintenance of the proton motive force, but a direct involvement of ATP had not been excluded. There are observations on bacteria, *Streptomyces fecalis* and strains of *E. coli* revealing that proton motive force is not required for growth (Harold and Van Brunt, 1977; Kinoshita et al., 1984), hence for synthesis and insertion of membrane components, in these cases. Chen and Tai (1985), having carried out studies on inside-out inner membrane vesicles, insist on the essentiality of ATP as an energy source, while a proton motive force would maintain the membrane topology only for a more efficient translocation. In recent work these authors (Chen and Tai, 1987a) observed that membrane perturbants such as ethanol, procaine, and phenethyl alcohol at low concentrations inhibit ATP-dependent protein translocation through *E. coli* inner membrane vesicles. Also, the optimal translocation of pre-OmpA in such a system was confirmed to require both the membrane potential and ATP (Geller *et al.*, 1986) and Yamane *et al.*, (1987) have reported that translocation on an OmpF-lipoprotein chimeric protein through inverted *E. coli* inner membrane vesicles involved both components of proton motive force and ATP. ATP had to be generated in the experiment medium; addition of ATP itself did not work. Other authors (Chen and Tai, 1987b) confirm the requirement for an ATP-generating system. They also present evidence indicating requirement for ATP under conditions of cotranslational protein translocation through *E. coli*-inverted inner membrane vesicles.

With regard to protein translocation in eukaryotic cells, *in vivo* experiments on yeast indicated that DNP inhibited a thorough translocation of the secretory protein invertase (Ferro-Novick *et al.*, 1984a,b). Since uncouplers were reported not to inhibit protein translocation at the level of microsomal membrane (Hansen *et al.*, 1986; Rothblatt and Meyer, 1986b; Waters and Blobel, 1986, see, however, Rees-Jones and Alawqati, 1984), it is possible to imagine an indirect *in vivo* effect of DNP at the level of mitochondrial ATP synthesis; a subsequent decrease of ATP level in the cytoplasm would be responsible for the defect in translocation. In line with such a possible explanation is the observation that *in vitro* post-translational protein translocation in microsomes under particular experimental conditions required the presence of GTP, creatine phosphate, and creatine phosphokinase (Perara *et al.*, 1986). It is also reported that cleavage of a phosphodiester is required for the post-translational insertion of the N-terminal segment of the glucose transporter comprising 8 out of 12 membrane-spanning domains of the molecule (Mueckler and Lodish, 1986).

3.2. Migratory Proteins

Since the early observations mentioned above, numerous authors have confirmed the energy requirement for protein translocation in mitochondria (e.g., see Anderson, 1981; Chien and Freeman, 1983; Fenton *et al.*, 1984; Gasser *et al.*, 1982a; Ikeda *et al.*, 1987; Kolansky *et al.*, 1982; Lewin and Norman, 1983; Maccecchini *et al.*, 1979a,b; Mori *et al.*, 1981, 1985; Morita *et al.*, 1982; Ono and Ito, 1984b; Ono *et al.*, 1985; Schleyer *et al.*, 1982; Sidhu and Beattie, 1983; Teintze *et al.*, 1982; Wandinger-Ness and Weiss, 1987; White and Sandalios, 1987; Zimmermann and Neupert, 1981; Zwizinski and Neupert, 1983; Zwizinski *et al.*, 1984) and chloroplasts (e.g., see Cline, 1986; Cline *et al.*, 1985; Coruzzi *et al.*, 1983; Fluegg and Hinz, 1986; Grossman *et al.*, 1980; Pfisterer *et al.*, 1982).

As to the step at which the energy is necessary in the process of protein translocation, it had been previously envisaged that either the processing or the translocation step could be involved. The suggestion that the post-translational protein translocation can be assimilated to "some sort of active transport" (Etémadi, 1980a) favored the essential implication of energy at the level of translocation, processing being a consequence of translocation. That the step of translocation, not that of cleavage, is energy requiring was confirmed by the observation of Zwizinski and Neupert (1983) that in the presence of o-phenanthroline, an inhibitor of mitochondrial processing enzyme, imported proteins were translocated without processing. Similar observations were made with bovine adrenal cortex mitochondria with regard to the cleavage of precytochrome P-450 *scc* and precytochrome P-450 11β (Ogishima *et al.*, 1985; Ou *et al.*, 1986). Also, import of cytochrome P-450 *scc* by bovine adrenal cortex mitochondria could occur at relatively low temperature at which processing by the cleavage enzyme was inhibited (Ou *et al.*, 1986). That processing is not an absolute requirement for translocation of migratory proteins could be confirmed during observations made on hybrid proteins bearing the extension peptide of the matrix enzyme ornithine carbamyl transferase plus 5 or 28 N terminal amino acids of the matrix enzyme and the full length of *E. coli* asparagine synthetase, a soluble protein. *In vitro* translocation experiments indicated that only the second hybrid was correctly processed. Presumably, the mature part contributes, for example, to the conformation required at the cleavage site (Nguyen *et al.*, 1986). Further work by Nguyen *et al.* (1987) has shown that when a deletion in residues 22–30 of the presequence of the enzyme was introduced, the mutant precursor could still be taken up into the matrix of isolated mitochondria while cleavage of the shortened signal did not occur despite the presence of the cleavage site (between Gly 32 and Ser 33). Vassarotti *et al.* (1987a) observed that small deletions distal to the cleavage site of the precursor of β-subunit of F_1-ATPase prevented maturation but that immature chains were, however, assembled in a functionally active enzyme, indicating that import and maturation have distinct requirements. The transit sequence of the precursor of the algal (*Chlamydomonas reinhardtii*) small subunit of ribulose 1,5-bisphosphate carboxylase is required for translocation through the envelope of vascular plant (pea, spinach) chloroplasts. However, cleavage to the mature size does not occur; only an intermediate is formed. It was suggested that processing of the precursor occurs in two steps, one of which is conserved during evolution (Mishkind *et al.*, 1985).

It is thus apparent that the energy-dependent translocation of mitochondrial and chloroplast migratory proteins and their processing have essentially distinct requirements, those of cleavage being principally governed by the activity of the cleavage enzyme in which the conformation of the cleavage site plays an important role (see also Miura *et al.*, 1986). A remarkable observation (Teintze *et al.*, 1982; Zwizinski *et al.*, 1984) is that in unenergized mitochondria, the binding of some mitochondrial proteins is significantly reduced or does not occur. Thus, an appropriate membrane conformation, presumably exposing the binding sites for some migratory proteins, is influenced by membrane energization.

As to the form under which the energy is used, it was first admitted that ATP as such is required (Nelson et al., 1979). However, further studies led to the conclusion that the membrane electrical potential was involved (Pfanner and Neupert, 1985; Schleyer et al., 1982) or that the proton motive force would intervene, although it was not decided if ΔpH or $\Delta\psi$ was used (Gasser et al., 1982a). More recently, import of subunit β of the mitochondrial F_1-ATPase was found to depend on both the membrane potential and ATP (Pfanner and Neupert, 1986). Other authors have found that ATP, in addition to an energized inner membrane, was required for import of proteins into mitochondria. GTP could replace ATP. In this action ATP exerted its effect from outside the mitochondria (Eilers et al., 1987). Nucleotide triphosphates were suggested to act on precursors of imported proteins (Chen and Douglas, 1987a,b; Pfanner et al., 1987a; Verner and Schatz, 1987) in unfolding them (Pfanner et al., 1987a, 1988) in a process involving the cleavage of a phosphodiester bond (Chen and Douglas, 1987a,b) conversely, however, Eilers et al. (1988) more recently proposed a requirement for membrane potential for unfolding a precursor at the surface of the membrane, while ATP would be required for import and not the membrane potential.

Studies on import of proteins to chloroplasts led to the proposal that the presence of ATP in the stroma compartment is required (Grossman et al., 1980; Pain and Blobel, 1987). However, Fluegg and Hinz (1986) report on the requirement for ATP from outside the envelope and suggest the involvement of a phosphorylation–dephosphorylation cycle. The reason(s) for the discrepancy is not clear. It would appear that the supply of ATP in the stroma compartment would be required for protein insertion into the thylakoid membrane and for uptake of intrathylakoid luminal proteins, since there are indications that proteins of thylakoid membrane (such as light-harvesting chlorophyll a/b proteins) and presumably those of the intrathylakoid compartment are first imported and released in the stroma, then taken up by thylakoid membranes in a process requiring ATP and soluble factor(s) from stroma (Cline, 1986). (For more information, see Section 4.2.2.)

The only mitochondrial migratory protein for which the translocation does not require an energized membrane is apocytochrome c (Henning and Neupert, 1981; Nelson et al., 1979; Zimmermann and Neupert, 1981). However, what is released in the intermembrane space is cytochrome c itself (Henning and Neupert, 1981). Thus, as previously discussed (Etémadi, 1985), a simple diffusion of apocytochrome c does not explain translocation, the case being reminiscent of group translocation processes. Considering, in addition, the fact that "proteins" of the outer membrane were found to bind the apocytochrome c (Zwizinski et al., 1984), I have also suggested that the completion of translocation itself can involve the enzyme "apocytochrome c heme lyase." Recently, different authors appreciated the exact localization of the enzyme. Enosawa and Ohashi (1986) locate it on the outer surface of the inner mitochondrial membrane, presumably in contact zones between the outer and the inner membrane, while Nicholson et al., (1987) find it in the intermembrane space. However, the last authors confirm that the enzyme, which is found not to exist in a stable complex with the binding protein, "plays a central role in transfer of apocytochrome c." As also suggested previously (Etémadi, 1985), it would appear useful to keep in mind this interesting example; maybe other specific examples of group translocation processes could be encountered in protein translocation processes, despite the fact that processing reactions are generally not coupled with or do not furnish the driving force for the act of translocation and can be dissociated from it.

Contradictory indications exist regarding the energy requirement for protein translocation to microbodies. Uptake of the precursor of glyoxysomal malate dehydrogenase from watermelon (*Citrullus vulgaris*) by glyoxysomes of castor bean (*R. communis*) was reported not to be energy requiring (Gietl and Hock, 1986). However, Goodman (1985) suggested the requirement for an energized state of peroxisomes, possibly through the presence of a proton pump, for translocation of peroxisomal enzyme, alcohol oxidase, in *Candida boidinii* and, recently, Douma et al. (1987) have reported the presence in purified peroxisomes from *Hansenula polymorpha* of a H^+-ATPase resembling the mitochondrial enzyme.

4. TRANSLOCATION AND MEMBRANE-ANCHORING SIGNALS

4.1. Signal and Membrane-Anchoring Sequences for Proteins Translocated through the Endoplasmic Reticulum and the Bacterial Cytoplasmic Membrane

4.1.1. Structural Features of Signal Sequences

Concentrating on active transport, the general mechanism governing cellular protein translocation processes, imposes the involvement of membrane proteinaceous components interacting with precursors on the one hand and energy consumption on the other. A detailed mechanistic knowledge of the process of translocation, in the framework of the active transport, should then be sought. The most fruitful part in the signal hypothesis (Blobel and Dobberstein, 1975; Milstein et al., 1972) has been the proposal of the involvement of signals at the N termini of exported proteins in triggering their translocation, despite the fact that early proposals on the community of (Blobel and Sabatini, 1971) or a high homology between (Devillers-Thiery *et al.*, 1975) signal sequences of different exported proteins was soon observed to be unverified.

Several secretory and transmembrane integral proteins are devoid of cleaved signal sequences as first shown for ovalbumin (Palmiter *et al.*, 1978), then observed in numerous other cases especially of membrane proteins (Bonatti and Blobel, 1979; Chiocchia and Dirckamer, 1984; Chyn *et al.*, 1979; Fields *et al.*, 1981; Ghersa *et al.*, 1986; Gorin *et al.*, 1984; Holland *et al.*, 1984; Kopito and Lodish, 1985; Mc Lennan *et al.*, 1985; Mueckler *et al.*, 1985; Noda *et al.*, 1984; Paul and Goodenough, 1983; Schechter *et al.*, 1979; Spiess and Lodish, 1985; Spiess *et al.*, 1985; Strubin *et al.*, 1984; Zerial *et al.*, 1987), including several microsomal transmembrane integral proteins (Armstrong *et al.*, 1984; Bar-Nun *et al.*, 1980; Black *et al.*, 1979; Brown and Simoni, 1984; Gonzalez and Kasper, 1980; Heinemann and Ozols, 1984; Kasper and Porter, 1985; Lauffer *et al.*, 1985; Okada *et al.*, 1982) and bacterial inner (Brusilow *et al.*, 1981; Büchel *et al.*, 1980; Ehring *et al.*, 1980; Higgins *et al.*, 1982; Innis *et al.*, 1984; Poulis *et al.*, 1981; Santos *et al.*, 1982; Wolfe *et al.*, 1983; Yazgu *et al.*, 1984) and some outer (De Geus *et al.*, 1984) membrane proteins. In such cases uncleaved signals are either located N terminally or, in some cases, internal to the chain.

Cleaved sequences can be of 15–36 amino acid residues (see Abrahamsen *et al.*, 1985; Watson, 1984). Inherent in the loop model (DiRienzo *et al.*, 1978; Inouye and Halegoua, 1980) is the recognition of three regions in signal sequences, an N-terminal segment comprising positively charged residues, followed by a hydrophobic segment divided by a proline or a glycine residue, then the region comprising the cleavage site at the C terminus. Table I shows signal sequences of some exported proteins of eukaryotic and prokaryotic cells. The N terminus (n region) is of variable length and composition comprising one or two positively charged residues in eukaryotic and prokaryotic sequences, respectively (Von Heijne, 1985). Often, signal sequences of gram-positive bacteria are "unusually" long—that of the staphylococcal protein A being 36 residues; they also have on average more than 3 positively charged amino acid residues at their N termini (Abrahamsen *et al.*, 1985; Watson, 1984). The minimal length of the hydrophobic segment (h-region) is of some 8 residues and the c region (C terminus) comprising the cleavage site, is formed of 5 or 6 residues in eukaryotic and prokaryotic signal sequences, respectively (Von Heijne, 1985). The C-terminal segment, comprising the cleavage site, was proposed to follow the "-3, -1" rule. The importance of the conformation of the polypeptide chain near the cleavage site of the prepeptide for the action of the signal peptidase, in particular the frequency of the breakage of repeating conformations, and the fact that the cleavage occurs, in general, at the C terminus of an uncharged or small amino acid residue were previously noted (Austen, 1979). The "-3, -1" rule defines the amino acid residues most frequently encountered

Table I
Primary Structure of Some Signal Sequences of Secretory and Transmembrane Proteins[a]

Protein	Primary structure	References
Ovine β casein	MKVLILACLVALALA - R	Garnier et al. (1980)
Vicilin (*Pisum sativum*)	MLIAIAFLASVCVSS - R	Lycett et al. (1983)
Rat prolactin	MNSQVSARKAGTLLLLMMSNLLFCQNVQT - L	McKean and Maurer (1978)
Rat growth hormone	MAADSQTPWLLTFSLLCLLWPQEAGA - F	Bancroft et al. (1980); see also Seeburg et al. (1978)
Human growth hormone	MATGSRTSLLLAFGLLCLPWLQEGSA - F	Martial et al. (1979)
Human proinsulin	MALWMRLLPLLALLALWGPDPAAA - F	Bell et al. (1979)
Bovine proparathyroid hormone	MMSAKDMVKVMIVMLAICFLARSDG - K	Rosenblatt et al. (1979)
Vesicular stomatitis virus glycoprotein (G protein)	MKCLLYLAFLFIHVNC - K	Lingappa et al. (1978)
E. coli maltose-binding protein	MKIKTGARILALSALTTMMFSASALA - K	Bedouelle et al. (1980)
E. coli alcaline phosphatase	MKQSTIALALLPLLFTPVTKA - R	Inouye et al. (1982)
E. coli plasmid pBR 322, *S. typhimurium* TM$_1$ β-lactamase	MSIQHFRVALIPFFAAFCLPVFA - H	Sutcliffe (1978); Koshland et al. (1982)
E. coli lipoprotein	MKATKLVLGAVILGSTLLAG - C	Inouye et al. (1980)
E. coli phage λ-receptor	MMITLRKLPLAVAVAAGVMSAQAMA - V	Emr and Silhavy (1983)
E. coli bacteriophage fd coat protein	MKKSLVLKASVAVATLVPMLSFA - A	Sugimoto et al. (1977)
Bacillus licheniformis β-lactamase	MKLWFSTLKLKKAAAVLLFSCVALAGCANNQTNA - S	Lai et al. (1981)

[a] Residues after hyphen (-) in the sequence are those of the N terminus of mature proteins. The correspondence between one-letter symbols for amino acids and the three-letter symbols is as follows: A = Ala, C = Cys, D = Asp, E = Glu, F = Phe, G = Gly, H = His, I = Ile, K = Lys, L = Leu, M = Met, N = Asn, P = Pro, Q = Gln, R = Arg, S = Ser, T = Thr, V = Val, W = Trp, Y = Tyr.

around the cleavage site (Perlman and Halvorson, 1983; Von Heijne, 1984). However, this rule suffers several exceptions (see references in Ibrahimi and Gentz, 1987).

A question of importance with regard to the detailed molecular events governing protein translocation concerns the part of the signal sequence entering in interaction. Some observations were interpreted as favoring the involvement of charged residues at the N terminus of the signal sequence in the export of *E. coli* lipoprotein (Inouye *et al.*, 1982; Inukai and Inouye, 1983). Other studies indicate that no stringent conditions as to the conformation of the proximal segment of signal sequences should exist (Brown *et al.*, 1984). However, there are data indicating that the presence of positively charged residues in the n region is dispensable for correct translocation (Kaiser *et al.*, 1987; Lipp and Dobberstein, 1986b). For a fusion precursor bearing the signal of lipoprotein and 9 amino acids of the N terminus of its mature chain plus β-lactamase sequence deleted of its own signal, it could be observed by Garcia *et al.*, (1987) that translocation through the mammalian microsomal membrane could occur. Mutations that suppressed positively charged residues did not affect translocation and processing, which, however, occurred at a potential cleavage site for signal peptidase I but not for signal peptidase II, which is absent in microsomes. These studies also showed that interaction with SRP was not affected as a result of substitutions in the positively charged residues. Of course, in this and other cases, involvement of initiating N-terminal methionine in eukaryotic systems cannot be excluded. With this reservation in mind, however, the lack of involvement of positively charged amino acid residues is also supported by the presence of rare translocated proteins such as vicilins (Lycett *et al.*, 1983) the signal sequences of which lack positively charged residues at their N termini or even the presence of signal sequences such as that of rat growth hormone (Bancroft *et al.*, 1980; Seeburg *et al.*, 1978) with a negatively charged residue in the N-terminal region (see Table I). Thus the requirement for positively charged residues within the signal sequence is not substantiated. This does not mean an insensitivity of the translocation process to manipulations in the charge of the signal. In a recent study, Freudl *et al.* (1988) have shown that depending on the nature of mutation introduced in the region coding the signal sequence of the OmpA-protein gene, this *E. coli* protein can either be translocated entirely post-translationally or entirely co-translationally: when a positive charge was added near the N-terminal and the hydrophobic region was shortened, the translation became post-translational; when the hydrophobic region was lengthened, the translocation was co-translational. During studies on interchangeability between signal sequences of translocated proteins of bacterial and eucaryotic cells (Tallmadge *et al.*, 1980a,b, 1981), as well as during other experiments (Perara and Lingappa, 1986; Rothstein *et al.*, 1984; Simon *et al.*, 1987; Wiedmann *et al.*, 1986b), it appeared that signal sequences extended at their N termini and thus made internal are still operative. For example, when a plasmid was constructed in such a way that it could code for preprolactin and an additional 250 amino acids (part of two globin chains in tandem) localized N terminally to the precursor chain, it was observed that not only prolactin but also its signal peptide together with the long N-terminally added sequence was translocated through the microsomal membrane *in vitro* and through the endoplasmic reticulum of *Xenopus laevis* oocytes during *in vivo* experiments, although in the last case the translocated extended signal sequence was mostly degraded (Simon *et al.*, 1987). A remarkable case is the one reported by Wiedmann *et al.* (1986b) in which the extension segment added N terminally was highly charged; it did not remain cytoplasmic, however, and was translocated to the side opposite to the cytoplasm, revealing again that either an electrostatic interaction with the membrane did not occur or, if it did occur, the interaction did not result in the cytoplasmic location of the N terminus as would have been expected from different models (see Figure 1).

Several studies concerned the central hydrophobic segment of signal sequences. Substitution (e.g., that of hydroxyleucine for leucine) (Gordon *et al.*, 1981; Hortin and Boime, 1980a,b) in the signal peptide of eukaryotic secretory proteins and mutational modifications on the signal sequence of *E. coli*-exported proteins (Bedouelle *et al.*, 1980; Emr and Silhavy, 1982, 1983; Emr *et al.*, 1980; Michaelis *et al.*, 1983a,b) confirmed the requirement for the hydrophobicity of this region which had to be able to exhibit the α-helical conformation (Emr and Silhavy 1983; Fikes and Silhavy, 1987). Kendall *et al.* (1986), using site-directed mutagenesis, generated various presequences in the precursor of *E. coli* alkaline phosphatase (see Table I); such modifications pointed again to the importance of a structural feature of the hydrophobic segment rather than the primary sequence. Interestingly, Ryan *et al.* (1986) found that the presence of a charged residue introduced in the hydrophobic region by mutational substitution does not necessarily alter the function of the signal sequence provided that by substitution of other residues the hydrophobicity level of the segment was retained. Other authors (Blachly-Dyson and Stevens, 1987) also report that introduction of an aspartate residue in the hydrophobic core of the signal sequence of yeast vacuolar enzyme, carboxypeptidase Y, does not affect its translocation. Furthermore, a recent remarkable report by Kaiser *et al.* (1987) indicates that, in the case of yeast invertase, many *random sequences* with no more than 2 or 3 contiguous hydrophobic amino acid residues can still function efficiently in translocation of the protein in yeast. Thus, a contiguous stretch of 7 or 8 hydrophobic residues is not an absolute requirement. Likewise, Smith *et al.* (1987) by "shotgun cloning" of restriction fragments of *B. subtilis* chromosomal DNA obtained a high frequency of "export signal coding regions." Such signals placed 5' to truncated genes, deleted from the signal sequence coding region of exported proteins, could direct translocation of respective proteins both in *B. subtilis* and *E. coli* harboring plasmid vectors. More striking is the observation made by Blachly-Dyson and Stevens (1987) that deletion of signal sequence of yeast vacuolar carboxypeptidase Y does not totally abolish its post-translational translocation and glycosylation. Among other possibilities the eventuality of the presence of an internal, but less efficient, signal was envisaged, although a typical signal could not be detected in the sequence. Finally, an important point should be emphasized: not only can translocated proteins have signals that naturally or under particular experimental circumstances remain uncleaved, but also the signal and the cleaved sequences are not always synonymous, since the cleaved sequences can have dispensable amino acid residues or fragments, or the significant part of the signal might not correspond to the cleaved segment. Thus, the N-terminal cleaved sequence of the 13 amino acids (see Seehara and Khorana, 1984) of bacteriorhodopsin presumably does not constitute (see Etémadi, 1986a) all of the translocation signal. Storm and Lory (1987) have recently reported that the cleaved sequence of pilin in *Pseudomonas aeruginosa* is formed only of 6 N-terminal amino acid residues. Translocation involves an additional sequence from the N terminus of the mature protein. Also, a regulatory role for the signal sequence, influencing (diminishing) the rate of synthesis of exported proteins through its interaction with the translation system, was recently reported by Ibrahimi and Gentz (1987).

4.1.2. Structural Features of Membrane-Anchoring Sequences of Transmembrane Integral Proteins

When truncated precursors of transmembrane proteins were expressed, if the deletion at the C terminus comprised the anchoring domain, proteins were secreted, confirming that early events of secretion and membrane insertion are similar (Bremer *et al.*, 1982; Boeke and Model, 1982; Gething and Sambrook, 1982; U. Henning *et al.*, 1983; Rose and Bergman, 1982; Sveda *et al.*, 1982). Conversely, when modified secretory proteins to which a membrane-anchoring domain of a transmembrane protein was added were expressed, the hybrid protein remained membrane inserted (Boeke and Model, 1982; Guan and Rose, 1984; Rizzolo *et al.*, 1985).

Transmembrane integral proteins often contain, at the cytoplasmic side, positively charged residues close to the membrane-anchored hydrophobic domain. Such residues were found to be dispensable with regard to membrane anchorage of these proteins (Davis and Model, 1985; Davis *et al.*, 1985). Other studies indicated that the hydrophobic domain can be shortened to 12 amino acid residues (instead of 20) in the G protein of the vesicular stomatitis virus or that a charged residue can be inserted in the hydrophobic domain with conservation of the membrane topology, if a stretch of 11 uninterrupted hydrophobic residues was maintained (Adams and Rose, 1985a,b). Culter and Garoff

(1986a,b) have found that, as a result of replacement of the cytoplasmic positively charged residues by neutral or negatively charged residues or substitution of a hydrophobic residue by an acidic residue in the hydrophobic core of the transmembrane segment, although the location of the protein is not affected, its stability in the membrane is diminished. Recently, in contrast with the above observations, experiments interpreted as revealing the requirement for the cytoplasmic carboxyl termini of some *E. coli* membrane proteins for their insertion in the membrane were reported (Dalbey and Wickner, 1986; Kuhn *et al.*, 1986). Unfortunately, the lack of insertion was not demonstrated but only conjectured because of inaccessibility to protease action from outside; that is, the notion of crypticity was overlooked. Remarkable is the observation by Watanabe *et al.* (1986) that the precursor of an *E. coli* exported protein could be incorporated in *E. coli* inner membrane, while it passed through the microsomal membrane and become luminal. However, since in the second case the protein is glycosylated (once or twice) in its polypeptide chain, it would appear interesting to learn what the location of the protein would be if glycosylation were inhibited by tunicamycin, since it cannot be excluded that glycosylation in this particular case might play a role on the protein location.

4.1.3. Can a Signal Sequence Act as a Membrane-Anchoring Sequence and Vice Versa?

In some cases, the possibility for a signal sequence to act as an anchoring sequence could be observed (Coleman *et al.*, 1985) or suspected (Seeburg *et al.*, 1978), while other authors insist on a functional difference between these two types of sequence (Yost *et al.*, 1983). Finidori *et al.* (1987) have recently observed that a signal sequence, that of the influenza virus hemagglutinin, when made internal in a chain, was neither cleaved nor led to membrane anchoring but was translocated. In contrast, the N terminus of cytochrome P-450, which normally anchors this cytochrome to the microsomal membrane with a cytoplasmic location of the C terminus of the protein, could anchor a chimeric protein in the microsomal membrane. In both these cases a genuine signal sequence at the N terminus of chimeric constructs initiated the translocation process. Tabe *et al.* (1984), however, report that the uncleaved signal sequence of ovalbumin can act as an anchoring sequence. In a different instance of membrane proteins with a cytoplasmically located N terminus and an extracytoplasmic C terminus, the signal sequence and the anchoring sequence may be the same segment of the protein, as suggested by Claesson *et al.* (1983) and Holland *et al.* (1984) (see also Bos *et al.*, 1984; Spiess and Lodish, 1986). Zerial *et al.* (1986) have shown that the hydrophobic membrane-spanning domain of such a protein, human transferrin receptor, was necessary and sufficient (in the case of hybrid constructs) for translocation and insertion, that is, for acting as both signal and membrane-anchoring sequence. Spiess and Handschin (1987) have likewise shown that for asialoglycoprotein receptor H_1, a hydrophobic stretch of 10–12 residues was essential for membrane insertion. A report by Lipp and Dobberstein (1986b) concerning another example of these proteins is remarkable. The authors worked on hybrids made of the N terminus of the human invariant γ chain of class II histocompatibility antigens and a soluble cytoplasmic protein, chloramphenicol acetyltransferase (CAT); the invariant γ (Iγ) has a short N-terminal segment exposed cytoplasmically, while the bulk of the protein is extracytoplasmic. When the hybrid (Iγ-CAT) was such that most of the exoplasmic domain of Iγ was replaced by CAT chain, it became inserted in the membrane (Figure 3A). The CAT chain became exoplasmic and glycosylated. When the cytoplasmic domain of 30 amino acids comprising positively charged residues was deleted, so that the hybrid comprised only the hydrophobic membrane-spanning segment (also approximately of 30 residues) of the invariant chain plus 9 residues of the N terminus of the mature chain and the CAT sequence, the enzyme was processed and secreted.

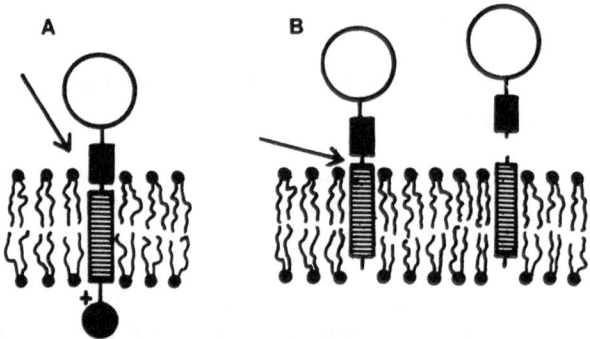

FIGURE 3. This figure (see Lipp and Dobberstein, 1986b) shows how the hydrophilic cytoplasmic domain of a putative uncleaved signal sequence might prevent access of its cleavage site to the cleavage enzyme by causing its burying in the membrane. In (A) a hybrid protein in which the N-terminal segment of the Iγ chain of the class II histocompatibility antigens—which normally exposes its N-terminal cytoplasmically and its C-terminal extracytoplasmically—is linked to a cytoplasmic soluble protein, chloramphenicol acetyltransferase. The plain disc and most of the dark rectangle show the soluble enzyme. The other part of the dark rectangle, the striated rectangle (the hydrophobic membrane embedded domain) and the dark disc (the short cytoplasmic domain) represent the N terminus of Iγ chain. The putative cleavage site is shown by an arrow and charges in the cytoplasmic domain close to the hydrophobic segment by +. Such a hybrid is inserted into the membrane with the orientation shown by Iγ itself, that is, N terminus in the cytoplasmic side. The charged residues would prevent the displacement of the hybrid in the direction perpendicular to the membrane surface. As a consqeuence, cleavage, presumably by signal peptidase, cannot occur. In (B) the cytoplasmic N terminus, including charged residues of Iγ, is omitted. Consequently, the putative cleavage site can reach the extracytoplasmic side in such a way that cleavage can now occur.

The interesting possibility was suggested that normally the cleavage site of the signal sequence, despite its presence, cannot reach the active site of the signal peptidase. Removal of the cytoplasmic segment comprising charged residues would allow the mobility of the membrane-inserted segment in the direction perpendicular to the plan of the membrane so that the active site of the cleavage enzyme can be reached. In another interesting report Zerial et al. (1987) indicate that replacement of the anchoring domain of human transferrin receptor by membrane-anchoring domains of either Semliki forest virus P 62 protein or hemagglutinin as well as a completely artificial uncharged peptide led to translocation and membrane anchoring, thus confirming that membrane-spanning domains of transmembrane proteins can also act as translocation signals. On the other hand, positively charged residues in the cytoplasmic domain are not required for signal or anchoring function.

The possibility for signal sequences to act in some cases as anchoring sequences and for anchoring sequences to act as signals for translocation justify serious doubts expressed elsewhere (Etémadi, 1986a) concerning the validity of an experimental procedure used (see for example Friedlander and Blobel, 1985; Dalbey and Wickner, 1988) to show the existence of internal signal sequences and stop transfer sequences in a protein crossing the membrane thickness more than once. Indeed in this procedure the behavior of segments outside their natural location is assayed, which can be misleading. The point was recently recognized by Audigier et al., 1987). The problem raised (Etémadi, 1986a) is of great theoretical importance. Since to date it remains uncertain whether recurrent use of signal recognition machinery is required for insertion of proteins crossing the membrane thickness several-fold. It is conceivable that once the precursor has been correctly guided and attached to the membrane and the translocation has begun, completion of insertion would follow without the necessity of further recognition events involving the recurrent use of SRP, at least when very short cytoplasmic loops exist.

4.1.4. The Nature of the Involvement of the Mature Part of Exported Proteins in the Translocation Process

The question can be raised as to the requirement for parts of the molecule other than the signal sequence in the translocation. Present data indicate that for most *natural* secretory and transmembrane integral proteins a functional signal is sufficient. However, when a cytoplasmic protein in a hybrid precursor is attached to a signal sequence, depending on the nature of the cytoplasmic protein, it can or cannot be translocated. In several experiments (Bassford et al., 1979; Emr and Silhavy, 1982; Emr et al., 1980; Kadonaga et al., 1984; Rothstein et al., 1985), translocation of hybrid proteins did not occur correctly. The report that the C terminus of a secretory protein was required for its translocation (Koshland and Botstein, 1980) was not confirmed (Ito *et al.*, 1981; Koshland and Botstein, 1982; Pollit and Zalkin, 1983). However, other reports were interpreted as favoring the requirement of segments of the mature part of the exported molecules in the process of translocation (Bankaitis and Bassford, 1985b; Bankaitis *et al.*, 1984; Rasmussen and Silhavy, 1987; Ryan *et al.*, 1986). Nevertheless, these observations might find alternative interpretations (see Etémadi, 1986a); in particular, a possible change of protein conformation can also explain the lack of translocation of proteins in which segments of the mature part were deleted. Conversely, a substitution in the mature part may retard folding and allow a partially defective signal sequence to act more efficiently. A recent report by Cover *et al.* (1987) furnishes an illustrative example in this respect. Also, studies on hybrid proteins constructed between the wild type or mutant presequence of lipoprotein, an *E. coli* outer membrane protein, and β-lactamase are remarkable (see Ghrayeb *et al.*, 1983, 1985; Lunn and Inouye, 1987). The efficiencies of translocation and processing were observed to be influenced by the mature parts. Such effects are presumably induced through conformational effects. Recently, McIntyre *et al.* (1987), working on translocation of the *E. coli* outer membrane protein OmpA and hybrids constructed of the N terminus of the precursor of this protein at different sites of which were linked a 253 amino acid-long fragment of the *E. coli* phage T_4 long tail fibers, arrived at the general conclusion that the export incompatibility of the soluble proteins might be responsible for cases of the inability of the signal sequences to facilitate export; that is, segments of the mature part of exported proteins are not required in triggering the translocation process. A remarkable case, reported by Fitts *et al.* (1987), discusses the secretion of defective mutants of β-lactamase in *S. typhimurium*. In this case mutation consists of substitution of tyrosine for cysteine in the mature part of the protein. Here, however, translocation and processing occurred but the protein remained bound to the outer aspect of the inner membrane.

4.2. Signal Sequences for Migratory Proteins

4.2.1. Location and Structural Features of Signal Sequences of Migratory Proteins

Involvement of extrasequences of mitochondrial migratory proteins in their uptake is supported by the fact that when mitochondrial and cytoplasmic isoenzymes exist, only mitochondrial forms have a precursor endowed with an extrasequence, even in cases in which both forms are encoded by a single gene, as is the case for the two forms of histidine tRNA synthetase (Natsoulis *et al.*, 1986) in *S. cerevisiae*. When isoenzymes are encoded by distinct genes, as is the case for yeast alcohol dehydrogenases (Young and Pilgrim, 1985) or citrate synthetases (Rosenkrantz *et al.*, 1986), only those coding for organellar isoenzymes provide an extrasequence. Note that such observations concerning natural products, which point to the high significance of *N*-terminal signals for translocation processes, can be extended to the example of a protein translocated through the endoplasmic reticulum. Carlson and Botstein (1982) have found that in *S. cerevisiae* the single gene for invertase encodes two mRNAs, one providing a full signal sequence for the secreted enzyme and the other for the intracellular enzyme.

Numerous observations indicate that extrasequences of migratory proteins, when present, are N terminal and their length much more variable than that of signal sequences of secretory and transmembrane integral proteins (see Table II). Recently, the involvement of the *N*-terminal extra-peptide in mitochondrial protein translocation was confirmed by Sakakibara *et al.* (1987) by observing that an antibody raised to the 29 residue-long signal of rat liver aspartate aminotransferase inhibited the import of the precursor into rat liver mitochondria. As we shall see, however, in rare cases the presence of an internal signal and, in the case of microbody proteins, the occurence of C-terminal signals are reported. Extra-sequences of migratory proteins contrarily to those of secretory and transmembrane integral proteins are generally devoid of a hydrophobic stretch, while being rich in basic amino acid residues. In fact, with some exceptions, such as the case of rat liver aspratate aminotransferase (Kamisaki *et al.*, 1982; see however Sakakihara *et al.*, 1987)—though not that of chicken liver enzyme (Jaussi *et al.*, 1985)—and the case of a subunit of 17,000 Da of the yeast complex III of electron transport system (Van Loon *et al.*, 1984), extrasequences of mitochondrial imported proteins were generally found to be basic. Several mitochondrial migratory proteins are devoid of a cleaved signal sequence (De haan *et al.*, 1983; Hampsey *et al.*, 1983; Mori *et al.*, 1985; Nakagawa *et al.*, 1987; Teintze *et al.*, 1982; Watanabe and Kubo, 1982; Zimmermann and Neupert, 1980a). The signal in such cases might be N-terminal but uncleaved or internal. In one such case, that of a 14,000 Da subunit of yeast mitochondrial complex III of electron transport, the N terminus of the protein was found to be rich in basic residues (see Table II) (De Haan *et al.*, 1983; see also Nakagawa *et al.*, 1987). In the case of 3-ketoacylCoA thiolase, a matrix enzyme, the presence of an internal sequence was envisaged (Mori *et al.*, 1985). However, more recently, the location of the signal at the N terminus was not excluded (Arakawa *et al.*, 1987). Adrian *et al.* (1986), using gene fusion techniques, found that for the case of a mitochondrial inner membrane protein, ADP-ATP carrier, which is also devoid of a cleaved extra-sequence, the information targeting the precursor to mitochondria was present in the first 115 residues of the molecule (out of 309 amino acid residues), in addition the amino terminal sequences prevented the membrane anchoring sequence (residues 78 to 98) from stopping transfer through the outer membrane. However, in a recent work, Pfanner *et al.* (1987c) have reported that a truncated ADP/ATP

Table II
Primary Structure of the Extrasequences or the N-Terminal Sequences of Some Migratory Proteins[a]

Protein	Primary structure	References
Human mitochondrial ornithine carbamyl transferase (m)	MLFNLRILLNNAAFRNGHNFMVRNFRCGQPLQ-NKVQLKG (with + marks over L, R, R, R, N, K)	(Horwich et al., 1984)
Rat mitochondrial ornithine carbamyl transferase (m)	MLSNLRILLNKAALRKAHTSMVRNFRYGKPVQ-S (with + marks over R, K, R, K, R, R, K)	(Lingelbach et al., 1986)
Chicken mitochondrial aspartate transaminase (m)	MALLQSRLLLSAPRRAAATARA- (with + marks over R, R, R)	(Jaussi et al., 1985)
N terminus of the precursor of 17,000 Da subunit of yeast complex III of electron transport (Im)	MDMLELVGEYWEQLKITVVPVVAAEDDDNEQHEEKAAEGEEKEE ENGDEDE- (with + marks over K, K, E marks with - signs over D, D, D, E, E, E, D, D, E)	(Van Loon et al., 1984)
Yeast mitochondrial cytochrome c peroxidase (im)	MTTAVRLLPSLGRTAHKRSLYLFSAAAAAAAATFAYSQSHKRS SSSPGGGSNHGWNNWGKAAALAS-TTP (with + marks over R, R, K, R, K, R, K)	(Kaput et al., 1982)
Yeast mitochondrial manganese superoxide dismutase (m)	MFAKTAAANLTKKGGLSLLSTTARRT-K (with + marks over K, K, K, R, R, K)	(Van Steeg et al., 1986)
Yeast mitochondrial subunit IV of cytochrome c oxidase (Im)	MLSLRQSIRFFKPATRTLCSSRYLL-Q (with + marks over R, R, K, R, R)	(Maarse et al., 1984)

Protein	Sequence	Reference
Yeast mitochondrial subunit β of F_1-ATPase (Im)	MVLPRLYTATSRAAFKAAK-Q (with + marks over L, K, K, K)	(Vassarotti et al., 1987a)
Yeast mitochondrial alcohol dehydrogenase III (m)	MLRTSTLFTRRVQPSLFSRNILRLQST-AAIP (with + marks over R, RR, R, R)	(Young and Pilgrim, 1985)
Yeast mitochondrial cytochrome c_1 (Im)	MFSNLSKRWAQRTLSKSFYSTATGAASKSGKLTQKLVTAGVAAAGIT ASTLLYADSLTAEA-MTAAEHGLH (with + and − marks)	(Van Loon et al., 1986)
		(Schmidt et al., 1979)
Chlamydomonas reinhardtii chloroplast small subunit of ribulose 1,5-bisphosphate carboxylase (s)	MAVIAKSSVSAAVARPARSSVRPMAALKPAVKAAPVAAPAEAND-M (with + and − marks)	
Pea chloroplast small subunit of ribulose 1,5-bisphosphate carboxylase (s)	MASMISSSAVTTVSRASRGQSAAVAPFGGLKSMTGFPVKKVNTDITS ITSNGGRVKC- (with ++ and − marks)	(Coruzzi et al., 1983; Karlin-Neumann and Tobin, 1986)
Wheat chloroplast chlorophyll a/b binding protein (T)	MAATTMSLSSSSFAGKAVKNLPSSALIGDARVNM-RKTAAKAK (with + and − marks)	(Lamppa et al., 1985)

[a] For the correspondence of the one-letter and the three-letter symbols for amino acids, see Table I footnote. For each protein the compartment where it is localized is indicated parenthetically, as in Figure 2. T = thylakoid membrane.

carrier lacking 103 N-terminal residues was able to be imported in mitochondria in a receptor- and energy-requiring manner. Nevertheless, for reasons mentioned repeatedly in the text, in such situations it is advisable to check the inefficiency of the *N*-terminal parts of the molecule.

The richness in basic residues was first suggested to be of importance for interaction of the precursor of a chloroplast migratory protein, ribulose 1,5-bisphosphate carboxylase, with chloroplasts (Schmidt *et al.*, 1979; see also Cashmore, 1983; Lamppa *et al.*, 1985) and then for precursors of numerous mitochondrial proteins with mitochondria (Horwich *et al.*, 1984; Hurt *et al.*, 1984; Kaput *et al.*, 1982; Kumamoto *et al.*, 1987; Maarse *et al.*, 1984; Miura *et al.*, 1982; Mori *et al.*, 1982; Viebrock *et al.*, 1982; see also Nagata *et al.*, 1983; Yoshida *et al.*, 1985). Several observations indicate that positively charged rhodamines, rhodamine 123 (Anderson, 1981; Mori *et al.*, 1982; Morita *et al.*, 1982; Oda *et al.*, 1984) and rhodamine 6G (Horwich *et al.*, 1985c), prevent import of mitochondrial proteins. Such observations were interpreted in terms of an action at the level of binding of positively charged precursors to negative charges on the mitochondrial membrane (Morita *et al.*, 1982). Rhodamine 123 (Johnson *et al.*, 1981) and its ethyl analogue (Severina and Skulachev, 1984) are vital dyes that also penetrate into the matrix of mitochondria and the cytoplasmic compartment of bacteria. Some authors (see Lubin *et al.*, 1987) do not exclude interaction of rhodamine 123 with mitochondrial membrane lipids.

Studies by Horwich *et al.* (1985a) in which a mitochondrial protein, ornithine carbamyl transferase, was allowed to be synthesized in the presence of canavanine instead of arginine in HeLa cells transfected with a plasmid containing the cDNA for mRNA of the enzyme indicated that the four arginine residues present in the extrasequence of 32 residues (see Table II) are required for import of the precursor. Also, *in vitro* experiments indicated that the precursor in which three distal arginine residues were replaced by glycine could not be imported (Horwich *et al.*, 1985b). Besides the basicity, emphasis is placed on the role of the secondary structure of the extrasequence in the process of binding and import. Horwich *et al.* (1986, 1987) found that deletions at the N and C termini of the extrasequence did not affect translocation of ornithine carbamyl transferase, while the middle portion was absolutely required. Furthermore, the presence of arginine residue at position 23 was of importance; its substitution by glycine led to a complete loss of translocation. Arg 23 was suggested to contribute to a local secondary structure, presumably α-helical, which would be essential for the function in the midpoint of the extension sequence. Further work (Sztul *et al.*, 1987), however, indicates that substitution of either aspartate or glycine for Arg 23 did not prevent translocation and processing which normally occurs in two steps (first an intermediate is formed in which the cleavage occurs between residues 24 and 25, then the further eight residues of the extra-sequences are removed). Such substitutions revealed another "cryptic" site of cleavage between residues 16 and 17. Remarkably enough residue 15 of the extra-sequence is also an arginine. Lingelbach et al. (1986), having observed that residues 15–19 of the extrapeptide of the same protein, including positively charged residues, could be deleted without the function of the signal being affected, suggested that "specific structural elements containing basic residues" are involved rather than the general basicity of the signal. Chu et al. (1987b), working on translocation of rat mitochondrial malate dehydrogenase precursor mutated in the region of the extra-sequence, confirmed a role for the positive charge (of arginine residues) in enhancing the efficiency of translocation. Interestingly, however, the authors (Chu et al., 1987a) found also that the uncharged residue leucine at position 13 of transit sequences plays an important role in the binding of the precursor to mitochondria, hence in the import. Recently, Vassarotti et al. (1987b) reported that subrogate signals in the N-terminal mature part of mitochondrial F_1-ATPase β-subunit could be generated by spontaneous mutations. The subunit, the import of which was prevented by deletion of the signal, was then imported. Effective mutations replaced different acidic residues with neutral and basic residues and generated amphiphilic helices with a less acidic character. It appeared that while an amphiphilic helix at the N terminus containing a minimum basic residues is required, all subrogate signals were acidic. They showed a reduced *in vivo* efficiency and an inefficient but detectable *in vitro* import-mediating capability.

In contrast to the case of ornithine carbamyl transferase (Horwich et al., 1986, 1987), there are reports indicating that only an N-terminal segment of the extrasequence of mitochondrial imported proteins with a sufficient length can induce uptake. Thus, Vassarotti et al. (1987a) report that deletion of residues 10–36 of the precursor of β-subunit of yeast mitochondrial F_1-ATPase does not affect import and assembly into a functional enzyme. At least 10 N-terminal residues out of the 19-residue transit sequence (see Table II) were required for the translocation to occur. Various observations on the inhibition (Furuya et al., 1987; Gillespie et al., 1985) by synthetic peptides, representing *N*-terminal segments of extra-sequences of mitochondrial migratory proteins, of the import of precursors of these proteins also favor involvement of their N-termini in the translocation process (see also Section 6.2).

Application of the gene fusion technique confirmed the crucial function of extrasequences of mitochondrial and chloroplast migratory proteins and the role of their N-terminal segment. In such experiments the amino terminal sequence of a mitochondrial or chloroplastic protein is attached to a nonexportable moiety and the localization of the hybrid is examined. A chimeric gene, the product of fusion of a part of the gene for the β-subunit of yeast F_1-ATPase coding for 350 N-terminal amino acid residues of the precursor and a large part of *E. coli* lac Z gene encoding the C terminus of the protein endowed with the enzyme activity, was found to be expressed in yeast and translocated for more than

90%. However, the respiration was blocked, presumably because of jamming of the hybrid protein at a terminal step during the transit (Douglas *et al.*, 1984). In further such experiments, Emr *et al.* (1986) constructed hybrids of yeast β-subunit and either the *E. coli* lac Z gene product (β-galactosidase) or the yeast suc 2 gene product (invertase). Such hybrids were imported *in vivo* into mitochondria in yeast. These and experiments using defective β-subunit gene deletion mutants indicated that 27 amino acids on the N terminus of the precursor were sufficient for import of a soluble protein, though the exact location of imported proteins was not elucidated. Other studies have indicated that a segment of 53 residues at the N terminus of the subunit IV of yeast cytochrome *c* oxidase (including the extrasequence of 25 residues) (Table II) was sufficient for targeting the cytosolic protein, dihydrofolate reductase, into mitochondria (Hurt *et al.*, 1984). Likewise, fusion experiments indicated that the extrasequence of the matrix enzyme ornithine carbamyl transferase could target dihydrofolate reductase into mitochondrial matrix in transfected CHO cells (Horwich *et al.*, 1985c). For certain mitochondrial proteins such as carbamylphosphate synthase I ($M_r = 165,000$), the *in vitro* translocation into isolated mitochondria is very slow. However, here again, hybrid protein experiments indicate involvement of the extrasequence: a hybrid bearing the extrasequence of 38 amino acid residues of the enzyme (Table II), 55 amino acids at the N terminus of its mature part, and the carboxyl terminal 209 amino acids of the ornithine carbamyl transferase was imported and processed in isolated mitochondria (Nguyen *et al.*, 1986). More recently, Nguyen and Shore (1987) observed translocation into mitochondria of a hybrid constructed by cDNA fusion from the signal of preornithine carbamyl trasnferase and the C terminus of the G protein of the envelope of VSV. The protein was inserted into the inner membrane, presumably in a transmembrane manner. Van Loon and Young (1986), working on three alcohol dehydrogenase isoenzymes of yeast, observed an 80–90% identity of sequences between the two cytosolic and the single mitochondrial forms. The last enzyme has a 27 residue long extrasequence (see Table II). When by the gene fusion technique the N-terminal segment of 48 residues of the mitochondrial form replaced 21 N-terminal residues of the cytosolic forms, these were translocated to the mitochondria, but when the first N-terminal 28 residues of the mitochondrial signal were deleted, the hybrid proteins could not be translocated.

As for mitochondrial imported proteins, extrasequences of those of chloroplasts are able to target soluble proteins into these organelles. Thus, the extrasequence of the small subunit of ribulose 1,5-bisphosphate carboxylase linked to various soluble proteins targeted these proteins to the stroma (Boutry *et al.*, 1987; Kuntz *et al.*, 1987; Schreier *et al.*, 1985; Van Den Broek *et al.*, 1985; Wasmann *et al.*, 1987). During these experiments it was observed that large deletions in the amino terminal or the central part of the extrasequence strongly reduced the translocation (Wasmann et al., 1987). Also, when the hybrid contained only 41 N termini out of the 57 residues of the signal, it remained mostly cytoplasmic, bound to the outer membrane, although its N terminus reached the stroma and a segment of it was cleaved (Kuntz et al., 1987). Della-Cioppa et al., (1987) report that the extrasequence of petunia chloroplast enzyme 5-enolpyruvylshikimate-3-phosphate synthase can import the E. coli soluble enzyme into coloroplasts. The experiment is of potential agronomical interest since the plant enzyme involved in production of aromatic amino acid is blocked by herbicides in which glyphosate is the active element, while the bacterial enzyme activity is not affected.

In remarkable experiments, Hurt et al. (1986a) observed that the extrasequence of a mitochondrial protein could target both the large and the small subunits of chloroplast ribulose 1,5-bisphosphate carboxylase to mitochondria. Surprisingly, however, Hurt et al. (1986b) also report that the 31 amino terminal residues of the extrasequence of the small subunit of ribulose 1,5-bisphosphate carboxylase could direct mouse dihydrofolate reductase and the mature part of the subunit IV of the mitochondrial cytochrome c oxidase into the matrix and the inner membrane of isolated yeast mitochondria, respectively. This result raises the problem of the mechanism of discrimination of proteins targeted to chloroplasts and mitochondria in plants and, in fact, other authors (Boutry et al., 1987) have recently reported that in a transgenic plant (*Nicotiana plumbaginifolia*) the transit sequence of the small subunit of ribulose 1,5-bisphosphate carboxylase directs a soluble E. coli enzyme into chloroplasts, while the same bacterial enzyme (chloramphenicol acetyltransferase) synthesized in a hybrid protein comprising N-terminally a sequence of 90 amino acids of the N-terminal end of the β-subunit of the mitochondrial ATP synthase was directed to the mitochondria of the transgenic plant cells. Furthermore a recent report by Smeekens et al. (1987) indicates that chimeric precursor proteins comprising the transit peptide of chloroplast proteins, ferredoxin, a stromal protein; or plastocyanin, an intrathylakoidal luminal protein, and the mitochondrial manganese superoxide dismutase chain, are imported to the chloroplast stroma. Such chimeric proteins were not taken up by yeast mitochondria.

Reports by Baker and Schatz (1987) and Hurt and Schatz (1987) indicate that sequences in prokaryotic or eukaryotic genome, even those directing the synthesis of segments of soluble proteins when fused to the genome of a migratory protein truncated from the part of it coding for the presequence, are able to target passenger proteins and to mediate their import into mitochondria. Such reports confirm similar observations with secretory proteins mentioned above and are in line with the previous observation of Lingappa et al. (1979) that a segment of ovalbumin, residues 234–253, having nothing to do with the signal sequence of this protein, which is located near the N terminus of the molecules (Austen, 1979; Meek *et al.*, 1982; Tabe *et al.*, 1984), could inhibit import of some secretory proteins in competitive translocation experiments.

In the case of microbodies, matrix proteins are reported to be generally devoid of an extrasequence (Geredes *et al.*, 1982; Goldman and Blobel, 1978; Goodman, 1985; Goodman *et al.*, 1984; Kruse and Kindl, 1983; Lord, 1978; Lord and Roberts, 1982; Maechina *et al.*, 1988; Osumi *et al.*, 1985; Roa and Blobel, 1983; Robbi and Lazarow, 1978; Roggenkamp *et al.*, 1984; Zimmermann and Neupert, 1980b). However, there are cases for which the presence of an extrasequence is mentioned (Arakawa *et al.*, 1987; Furuta *et al.*, 1982; Hock and Gietl, 1982; Miura *et al.*, 1984; Riezman *et al.*, 1980; Sautter *et al.*, 1987; Walk and Hock, 1978; Yamaguchi *et al.*, 1986, 1987). In one such case, that of germinating pumpkin glyoxysomal malate dehydrogenase of 33,000 Da, the extrasequence was found to be rapidly cleaved from the precursor of 38,000 Da upon its translocation, as revealed by pulse-chase experiments, so that while the precursor was cytoplasmic, only the mature form was found in the organelle (Yamaguchi *et al.*, 1987). However, in another recent report concerning *Candida tropicalis* peroxisomal acyl-CoA oxidase, Small and Lazarow (1987) indicate that only 40% of the carboxy terminal of the molecule had sufficient information to be, in part, imported in peroxisomes as revealed by protease protection. Nevertheless, further work is required to substantiate the interpretation that truncated proteins were in fact in the matrix and in any event the possibility cannot be excluded that an N terminus might still be a more efficient targeting segment in the wild-type molecule. Gould *et al.* (1987) admit the involvement of a C-terminal segment of 12 amino acids of the firefly luciferase in its targeting and translocation to peroxisomes. In addition, however, another region contained in amino acids 47 to 261 was implicated. Swinkles *et al.* (1988), comparing genes for cytosolic and glycosomal phosphoglycerate kinase of *Crithidia fasciculata*, conclude that a C-terminal extension may be involved in targeting of the microbody enzyme. A similar situation would also prevail for the *Trypanosoma brucei* enzyme.

4.2.2. Signals for Translocation to Different Organellar Compartments

The "topogenic sequence hypothesis" proposed the presence of different signal sequences and specific receptor translocators (Blobel, 1980) to explain how different organellar proteins reach different compartments. However, a discrimination for proteins of different compartments at the level of the initial interaction of percursors and membrane receptors according to their destination in mitochondrial compartments does not seem to occur. Gillepsie *et al.* (1985) have found that a peptide corresponding to the amino terminal residues 1–27 of the precursor of ornithine carbamyl transferase, a matrix protein, prevented not only the uptake of the corresponding precursor but also that of a mitochondrial inner membrane protein. Also Furuya *et al.* (1987) report that N-terminal sequences (residues 1 to 15 or 1 to 20 of the 39-residue-long extra-sequence) of cytochrome P-450$_{scc}$ prevent the import of some mitochondrial proteins of the matrix or the inner membrane. The same effect was observed with residues 1 to 14 of the N-terminal of adrenodoxin. As reported for some proteins of intermembrane compartment of mitochondria such as cytochrome b_2 (Gasser *et al.*, 1982b; Ohashi *et al.*, 1982) and cytochrome *c* peroxidase (Kaput *et al.*, 1982; Reid *et al.*, 1982), the precursors are first linked to the inner membrane, then subsequently processed and released in the intermembrane compartment. A theory was recently suggested to explain import of mitochondrial proteins into different compartments (Hurt and Van Loon, 1986). It is based on observations indicating that whole extrasequences are not always required for the uptake. Thus, proteins formed of the extrapeptide of the subunit IV of complex III of the electron transport system or of only the first 12 amino acids at its N terminus and of mouse dihydrofolate reductase were found to be targeted to the matrix of isolated yeast mitochondria *in vitro* and also during *in vivo* experiments. The N-terminal segment of less than nine residues was not sufficient (Hurt *et al.*, 1985). More recently, Van Loon *et al.* (1986, 1987) reported that the only 16 N-terminal (or less) residues of the extrasequence of cytochrome c_1 were sufficient for targeting a passenger protein to the matrix and that 13 C-terminal residues of the signal acted as a stop transfer sequence anchoring the protein (or a hybrid comprising the presequence and the dihydrofolate reductase) in the inner membrane. Also, only the nine N-terminal amino acid residues of δ-aminolevulinate, a matrix enzyme, linked to β-galactosidase, were able to import the hybrid into mitochondria. In this case the precursor is of 61,000 and the mature enzyme of 58,000 Da; hence the extrasequence should be of some 2000 Da (Keng *et al.*, 1986). Remarkably, the N terminus of an outer membrane protein of 70,000 Da was found to be involved in targeting and anchoring the protein (Riezman *et al.*, 1983b). Only 41 N-terminal residues were sufficient for targeting and anchoring hybrids constructed with *E. coli* β-galactosidase. The targeting would be mediated by a region including the 11 amino terminal residues and anchoring a stretch of uncharged segment between residues 9 and 38 (Hase *et al.*, 1984). A hybrid, comprising 21 amino terminal residues and expressed in yeast, was localized in the matrix compartment, hence the suggestion that common targeting sequences for outer membrane proteins and at least some matrix proteins exist (Hase *et al.*, 1986). Such experiments led to the generalization (Hurt and Van Loon, 1986) that amino terminal sequences of mitochondrial proteins generally possess an N-terminal "matrix targeting domain" containing a periodic array of basic amino acids involved in both attachment and transport to the matrix. Membrane-bound proteins would then have a stop transfer sequence. A long stop transfer domain with an uninterrupted stretch of uncharged amino acid residues, followed by charged residues, prevents transport through the outer or the inner membrane. The protein becomes an outer membrane protein or inner membrane protein (Figure 4). The presence, in addition, of cleavage sites would ensure that part or all of the presequence is cleaved. An intermembrane space protein is the product of cleavage of a protein first bound to the inner membrane. Further work, however, is required to understand details of translocation of proteins to different compartments. A recent proposal by Hartl et al. (1987a) for the path of translocation of an intermembrane protein, Rieseke FeS protein of the com-

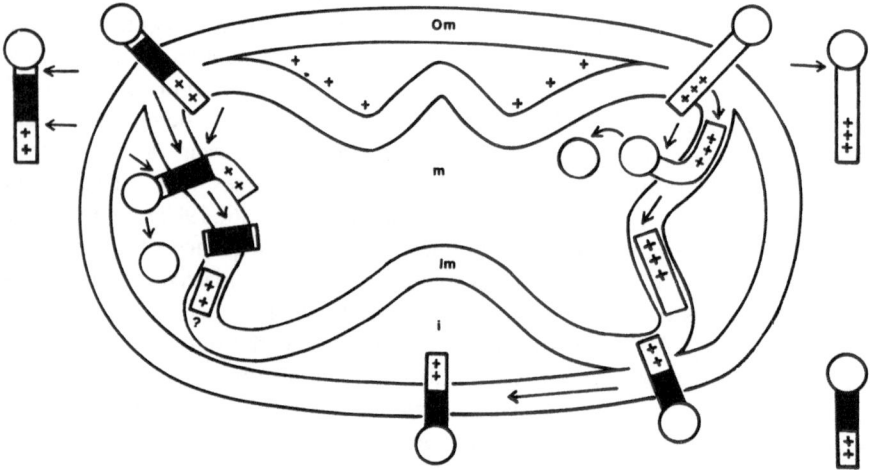

FIGURE 4. Translocation through the mitochondrial membranes of a matrix and an intermembrane compartment protein and insertion of an outer membrane protein are schematized (see Hurt and Van Loon, 1986). It is suggested that all proteins destined to mitochondria have an N-terminal segment able to target proteins to mitochondria and a "passenger" part. The mitochondria-targeting segments have in all cases a positively charged matrix-targeting segment at their N termini. In the upper right side, this matrix-targeting segment is shown followed by the passenger, appearing as a plain disc. The cleavage site of the extra sequence is indicated (arrow). The upper left drawing shows translocation and processing of the precursor of a protein destined to the intermembrane compartment (i). Such a precursor would exhibit a matrix-targeting segment, shown as a positively charged rectangle, then a segment shown as a dark rectangle, anchoring the molecule in the inner membrane. Two cleavage sites on the extra sequence exist, the matrix-targeting segment would be cleaved in the matrix and the second segment in the intermembrane compartment. For an inner membrane protein exposed to the intermembrane space, this second cleavage does not occur. In the lower right side a protein destined to the outer mitochondrial membrane is shown. This protein has, in addition to the matrix-targeting positively charged segment, an outer membrane-targeting sement shown as a dark surface. It is important to note that, for all cases, translocation through mitochondrial membranes of proteins of different compartments starts by that of their N termini. Cleavage sites are shown by arrows. Different compartments are located as in Figure 2.

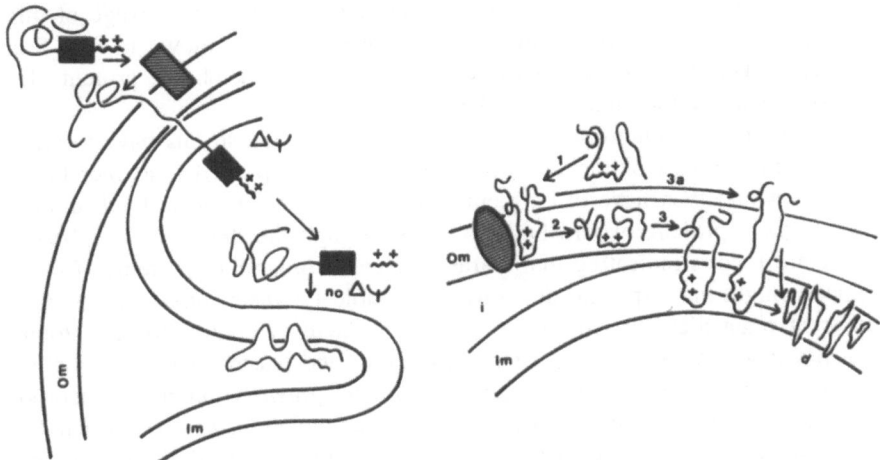

FIGURE 5. Mechanisms for incorporation of a mitochondrial intermembrane compartment protein, Rieske iron–sulfur protein (left side drawing), and of an inner membrane protein ADP-ATP carrier (right side drawing) are shown (see Hartl et al., 1987; Pfanner and Neupert, 1987). Incorporation of the precursor of iron–sulfur protein, which is cleaved of two segments from its N terminus indicated as a positively charged segment and a dark rectangle, involves interaction first with a membrane receptor (hatched rectangle). The precursor passes, at 25°C and in the presence of the membrane potential ($\Delta\psi$), through contact zones between the two mitochondrial membranes and is released in the matrix compartment where the first segment (shown with positive charges) is cleaved. The intermediate is formed in the matrix compartment and is then translocated into the intermembrane space and cleaved without requirement for $\Delta\psi$. Translocation of ADP-ATP carrier, which is devoid of a processed signal sequence, is also dependent on $\Delta\psi$; the protein interacts with a membrane receptor (striated ellipse) in order to bind specifically to the membrane (reaction 1) and at 25°C inserts into the outer membrane (reaction 2). In the presence of $\Delta\psi$ the protein passes through an intermediate stage (reaction 3) and reaches the inner membrane where it forms the dimer (d). Passage occurs through the contact zones between the two mitochondrial membranes. The initial binding to the receptor can occur at low temperature and, in the presence of $\Delta\psi$, an essential event of translocation takes place (reaction 3a). The protein on the one hand has reached the inner membrane and on the other hand is still exposed at the mitochondrial surface. Raising the temperature to 25°C leads to completion of translocation to the inner membrane and formation of the dimer (d).

plex III (Figure 5), does not comply with the above model. This protein is peripherally bound to the outer leaflet of the inner membrane. Its precursor was found to be entirely translocated into the matrix. Processing occurs in two steps with formation of an intermediate which also was located in the matrix. Van Loon and Schatz (1987) have confirmed the occurrence of the last transport pathway for iron sulfur protein. Hartl et al. (1987b) proposed that the hydrophobic stretch in the precursor does not act as stop transfer sequence but is used instead after the first cleavage in the matrix for the targeting of the intermediate to and its translocation through the inner membrane; cleavage of this sequence occurs in the intermembrane compartment. Also shown in Fig. 5 is the route taken by ADP-ATP carrier protein, an inner membrane protein, details of which do not comply with the above model.

In the case of chloroplasts, proteins imported to the stroma have to cross the two membranes of the envelope. In a further step, thylakoid membrane proteins would then have to be membrane inserted and intrathylakoid luminal compartment proteins would have to cross through the thylakoid membrane. There are reports indicating that extrasequences of migratory proteins of chloroplasts might have specific domains responsible for their correct targeting to different organellar compartments. Thus, Smeekens et al. (1986) have found that precursors of ferredoxin, a stroma protein, and plastocyanin, an intrathylakoid luminal protein of white campion (*Silene pratensis*), could be incorporated *in vitro* into pea chloroplasts. When chimeric cDNAs were constructed in which transit sequences of these two proteins were exchanged and import into isolated chloroplasts of *in vitro* synthesized hybrids was examined, ferredoxin transit peptide directed plastocyanin to the stroma. Plastocyanin transit peptide led to a processing intermediate that was arrested on its way to the lumen. A "two-domain hypothesis" was suggested, which states that the plastocyanin transit peptide comprises two domains, the first functioning in the import of proteins to chloroplasts and the second in transport across the thylakoid membrane. In the case of two luminally located thylakoid membrane proteins bound peripherally to the membrane, a two-step cleavage, one related to their passage through the envelope and another with the crossing of the thylakoid membrane, is proposed (Chia and Arntzen, 1986). Recently Kirwin et al. (1987) confirmed the two step maturation of plastocyanin and partially purified the thylakoid peptidase responsible for the second maturation step. In further studies Smeeken et al. (1987) observed that a chimeric precursor protein comprising the extrasequence of plastocyanin and the sequence of the yeast mitochondrial manganese superoxide dismutase is taken up by chloroplasts and remains in the stromal compartment in an intermediate state in which only the signal for the import into chloroplast is cleaved. This observation reveals that while the manganese superoxide dismutase section of the chimeric molecule has been imported through the two membranes of chloroplast envelope, it did not cross the thylakoid membrane. A possible explanation is that a conformational change prevents its translocation. Alternatively it might impede the correct function of the residual extra-sequence. It might also bear a segment(s) incompatible with its translocation through the thylakoid membrane.

In a recent report, Karlin-Neuman and Tobin (1986) identified blocks of homology in the transit sequences of light-harvesting chlorophyll a/b protein, a protein of the thylakoid membrane, with that of the small subunit of ribulose

1,5-bisphosphate carboxylase of the stroma compartment. Such homology blocks do not exist for signal sequences in general, including those of mitochondrial imported proteins. Located at the beginning, the middle, and the end of the extrapeptides, such sequences would constitute a common framework for a common function in protein import. Interblock segments are either dispensable or involved in import into different compartments. A recent report by Reiss *et al.* (1987) indicates that only the N- and C-termini in the transit peptide are endowed with functional importance. However, they suggest that the central part may have a role in import selectivity.

4.2.3 The Mature Part of Migratory Proteins and the Translocation Process

An important factor recently recognized to influence the import of migratory proteins into mitochondria is the protein conformation. Eilers and Schatz (1986) report that methotrexate, a folate antagonist, blocks import into mitochondria of a hybrid in which the dihydrofolate reductase sequence is fused to the extrasequence of a mitochondrial imported protein. It was suggested that as a result of the firm binding of the antagonist the enzyme moiety cannot be unfolded in a form compatible with translocation. More recently Vestweber and Schatz (1988) were able to confirm this idea by "destabilizing" the dihydrofolate reductase portion of the hybrid bearing the first 16 residues of the extrasequences of yeast cytochrome oxidase subunit IV N-terminally. Destabilization in these assays resulted from manipulating the sequence by site-directed mutagenesis and was assessed by the increased sensitivity towards proteases or by a decreased aptitude of binding methotrexate. Such experiments have shown that "destabilized" precursors were incorporated with an increased efficiency into membranes. The fact that precursors of cytochrome c peroxidase (Kaput and Blobel, 1986) and aspartate aminotransferase (Sharma and Gehring, 1986) have to be imported as apoproteins into mitochondria are also related to the requirement for an unfolded state of precursors. Chen and Douglas (1987) observed that substitution of 129 residues from the C-terminal of the 511-amino-acid-long precursor of β-subunit of yeast F_1-ATPase by yeast copper metallothionein allows the import of hybrid protein in the absence of copper. In the presence of copper, import does not occur, presumably because unfolding of the copper metallothionein part is prevented. A recent report by Ness and Weiss (1987) indicates that C-terminally truncated carbamyl phosphate synthase is imported more slowly than the full-length precursor in *N. crassa* mitochondria. It can be envisaged that, here again, an unfavorable abnormal conformation might intervene. The authors mention the alternative possibility of an involvement at the level of receptor–signal interaction. Van Steeg *et al.* (1986) have found that when a sequence of 34 amino acid residues, 26 of which constitute the extrasequence of the matrix enzyme manganese superoxide dismutase of yeast (Table II), was fused to the cytosolic yeast enzyme invertase, import into isolated mitochondria of the fusion protein did not occur and only a low binding to mitochondria was observed. Also, in yeast transformants bearing the fusion gene, some 95% of the hybrid protein remained cytoplasmic, the remainder being loosely attached to the inner membrane. Presumably, either folded domains of invertase obstruct the passage, or the membrane jamming results from passage of some hybrid molecules. Likewise, a chimeric protein comprising the extrasequence of the small subunit of ribulose 1,5-bisphosphate carboxylase, 13 N-terminal amino acids of the mature part of the enzyme, and in addition the sequence of a cytoplasmic protein (the so-called shock protein) was largely not imported. It was concluded that "a transit peptide alone may not be sufficient for efficient import" (Lubben and Keegstra, 1986). In other experiments, however, it was found that although the transit peptide of the small subunit of ribulose 1,5-bisphosphate carboxylase is sufficient for triggering translocation of the hybrid in which it was linked N terminally to neomycin phosphotransferase, a soluble cytoplasmic protein (Kuntz et al., 1987; Wasmann et al., 1987) the presence, in addition, of the 23 N-terminal residues of the mature part of the subunit favorably affected the translocation. This additional sequence would intervene through an alteration of conformation improving the recognition by the receptor (Wasmann et al., 1987). Nonetheless, others found that in the presence of these 23 residues chloroplasts would accumulate less of the hybrid, because of a competition between the rate of its uptake and that of its cytoplasmic degradation (Kuntz et al., 1987). Experiments made by Gearing and Nagley (1986) still point to a factor additional to the conformation. In these experiments a 40-amino-acid-long N-terminal segment of the precursor of subunit IV of cytochrome c oxidase, linked to subunit 8 of yeast mitochondrial ATP synthase, a highly hydrophobic protein, was unable to induce import of this protein into yeast mitochondria; while the extrasequence of the subunit 9 of $N.$ $crassa$ F_0-F_1 ATPase of 66 amino acids bound to the same subunit 8 could trigger the import and processing of the hybrid. In this interesting case, the highly hydrophobic protein with a tendency to interact with the membrane prevents a correct import unless bound to the long hydrophilic presequences of the subunit 9.

Finally a recent report by Pfanner et al. (1987d) suggested that, in addition to the targeting N-terminal extra-sequence, the hydrophobic, mature part, of the subunit 9 along with hydrophobic segments in some other mitochondrial imported proteins are functionally involved in the initial $\Delta\Psi$-independent binding, so that these proteins can be imported into mitochondria after establishment of membrane potential.

5. POST-TRANSLATIONAL PROTEIN TRANSLOCATION

5.1. Post-Translational Translocation of Secretory and Transmembrane Integral Proteins

Much controversy as to the co- or post-translational nature of translocation of exported proteins in bacteria exists. Besides the disputed case of bacteriophage f_1 (M_{13}) coat protein, other reports indicate the possibility of a post-translational translocation (Garwin and Beckwith, 1982; Koshland and Botstein, 1982). Some studies also raise only the question of a strict coupling between translation and translocation (Josephson and Randall, 1981a,b; Randall, 1983). More recently, however, the problem of post-translational import of proteins in *E. coli* inverted inner membrane vesicles was examined (Chen *et al.*, 1985; Mueller and Blobel, 1984; Rhoads *et al.*, 1984) and an energy requiring post-translational import was admitted. It is worthwhile to emphasize that the *in vitro* feasibility of such a process does not necessarily prejudge of its occurrence *in vivo*, unless it is independently shown. The case of proteins translocated through or inserted into the endoplasmic reticulum furnishes good examples for this idea. As mentioned above, such a translocation was constantly observed to occur cotranslationally, except for the case of endoproteins which insert in the membrane post-translationally. The reported occurrence of post-translational insertion of the α-subunit of $[Na^+ - K^+]$ATPase (Hiatt *et al.*, 1984) was rejected in principle (Caplan *et al.*, 1986) and could not be confirmed experimentally (Caplan *et al.*, 1986; Geering *et al.*, 1985). However, during some *in vitro* experiments the possibility of a post-translational protein translocation through the microsomal membrane, although with a reduced efficiency, was observed (Caulfield *et al.*, 1986; Mueckler and Lodish, 1986; Perara *et al.*, 1986). In these cases, generally the post-translational translocation occurred when the chains were not released from ribosomes. On the other hand, a shortened precursor largely truncated from its C-terminal segment, even if released from ribosomes, could be translocated post-translationally (Perara *et al.*, 1986). The relatively low yield generally observed in these import studies, together with the fact that nascent proteins had still to be bound to ribosomes in which translation was arrested for translocation to occur and the fact that, especially with membrane proteins, the possibility of an artifactual "reconstitution" cannot be totally excluded, were among reasons for regarding a definite conclusion as to the significance of such observations to be pending. Recent work confirming the possibility of post-translational ATP-dependent translocation of a completed secretory protein pre-pro-α-factor of yeast (Hansen *et al.*, 1986; Rothblatt and Meyer, 1986b; Waters and Blobel, 1986) support the feasibility of such a process. Very recently Wiedmann *et al.* (1988) also reported the aptitude of microsomes from the yeast *Candida maltosa* to import post-translationally the pre-pro-α-factor and, in addition, to insert an integral membrane protein, cytochrome P-450. However, such examples remain rare among eucaryotic translocated proteins. The striking instance of pre-prolactin reported to be translocated entirely post-translationally after its reduction (Maher and Singer, 1986) could not be substantiated by Ibrahimi (1987), who was able to correlate this case with those of proteins mentioned above, in which post-translational translocation mostly concerned incompleted chains still bound to ribosomes. Nevertheless, the point made by Maher and Singer (1986), the importance of the unfolding of the mature part of molecules for translocation, remains valid and was also emphasized in several studies on procaryotic exported proteins. Randall and Hardy (1986), working on translocation of the precursor of an *E. coli* exported protein, maltose-binding protein, also suggested that the protein is valid for post-translational export as far as it has not gained its final, stably folded conformation. A direct interaction either with the membrane or with a proteinaceous factor would prevent the folding of the polypeptide before the translocation process. More recently Park *et al.* (1988) reported that the presence of the signal sequence modulates (retards), *in vitro,* the folding of precursors of maltose-binding protein and ribose-binding protein as compared to the mature proteins. Other studies recognized a form of β-lactamase which is cleaved from the signal sequence but still bound to the membrane of the *E. coli* spheroplasts; this form, in contrast with the mature periplasmic enzyme, is sensitive to trypsin, hence not in its final conformation which is not cleaved by this enzyme. Thus, the form crossing the membrane reaches the final conformation only after release (Minsky *et al.,* 1986). Other authors also recognize a processed but still "immature" form for an outer membrane protein, OmpA protein. This form changes its conformation by interaction with lipopolysaccharide and is translocated to the outer membrane (Freudl *et al.,* 1986). In line with the recent developments as to the requirement of an unfolded state for protein translocation, it was suggested that a postribosomal soluble factor might act as a "denaturase" (Waters *et al.,* 1986). However, more recently, the possibility that this factor might simply correspond to a SRP-like material was raised (Fecycz and Blobel, 1987).

Further work seems to be required to unravel the exact conditions of the post-translational translocation of secretory and membrane-inserted proteins and its physiological significance, especially when the endoplasmic reticulum is involved. A point to be kept in mind is that translocation of a series of relatively short chain polypeptides (see for example Müller and Zimmermann, 1987; Rothblatt *et al.,* 1987), which concerns marginal cases among the vast majority of exported proteins, as remarked erlier (see Etémadi, 1980a), though interesting, should be considered apart. Since translation of at least some 75 residues is required for polypeptide chains to interact with the co-translational translocation machinery, those polypeptides of lengths within this range and, *a fortiori* those that have been shortened still more, cannot, in principle, be co-translationally translocated. More work seems to be required to understand the translocation of such components, regarding in particular the involvement of the general translocation machinery and also with respect to energy requirement; points on which some discrepancies exist (see Ibrahimi et al., 1986; Müller and Zimmermann, 1987; Rothblatt et al., 1987).

5.2. Post-Translational Translocation of Migratory Proteins

The demonstration of post-translational translocation of migratory proteins, in comparison with the case of secretory and integral membrane proteins, has been facile. Both the *in vivo* pulse-chase experiments (Hallermeyer and Neupert, 1976; Harmey et al., 1977) and *in vitro* experiments of uptake of some migratory proteins (Hallermeyer et al., 1977; Harmey et al., 1976) indicated such a post-translational event. Since these reports, numerous observations on post-translational translocation of mitochondrial proteins were made (e.g., see Argan et al., 1982; Chien et al., 1984; Gasser et al., 1982b; Hernandez-Yago et al., 1983; Kolansky et al., 1982; Lewin and Norman, 1983; Mihara et al., 1982; Miralles et al., 1983; Mori et al., 1982, 1985; Morita et al., 1982; Nelson et al., 1979; Niranjan et al., 1988; Ono and Tuboi, 1986; Reid and Schatz, 1982; Sakakibara et al., 1983, 1987; Suissa and Schatz, 1982; Yamamoto et al., 1981; Yoshida et al., 1985). During such studies it could generally be ascertained that post-translationally imported proteins reach the compartment where they are normally located (Campbell et al., 1982; Gasser et al., 1982a; Kalousek et al., 1984; Oda et al., 1984; Ono and Ito, 1984a,b; Ono and Tuboi, 1986). Post-translational translocation of proteins to chloroplasts was also reported in numerous cases (Chua and Schmidt, 1979; Cline et al., 1985; Della-Cioppa et al., 1986; Grossmann et al., 1980, 1982; Highfield and Ellis, 1978; Minami et al., 1986; Pfisterer et al., 1982). Precursors to chloroplast polypeptides could not generally be detected *in vivo*, presumably because of their rapid import and processing. However, post-translational translocation was shown by *in vitro* experiments. Transfer to the correct compartment was first observed during studies on the chloroplast stroma protein, the small subunit of ribulose 1,5-bisphosphate carboxylase (Chua and Schmidt, 1978), and then extended to some other nuclearly encoded chloroplast proteins (Kohorn et al., 1986; Schmidt et al., 1985).

In the case of migratory proteins of the matrix of microbodies the post-translational translocation also occurs (Becker et al., 1982; Desel et al., 1982; Fujiki and Lazarow, 1982, 1985; Fujiki et al., 1984; Gietl and Hock, 1984, 1986; Goodman et al., 1984; Kindl et al., 1980; Köller and Kindl, 1980; Kruse and Kindl, 1983; Lazarow and De Duve, 1973; Lazarow et al., 1982; Miura et *al.*, 1984; Riezman *et al.*, 1980; Robbi and Lazarow, 1978; Roggenkamp *et al.*, 1984; Teintze *et al.*, 1982; see also Crane *et al.*, 1982a,b, 1983) as shown by pulse-chase experiments or by the observation of involvement of free ribosomes in the genesis of microbody proteins. In different instances, the *in vitro* uptake of migratory proteins by microbody fractions was also shown (Becker *et al.*, 1982; Desel *et al.*, 1982; Fujiki and Lazarow, 1982, 1985; Hock and Gietl, 1982; Kruse *et al.*, 1981; Lazarow *et al.*, 1982). It can be observed that in two cases in which the demonstration of an *in vitro* uptake failed—that of the liver peroxisomal catalase (Goldman and Blobel, 1978) and that of *Ricinus communis* endosperm glyoxysomal malate synthase (Lord and Roberts, 1982)— canine pancreas microsomes were used in designing import experiments. The possibility remains that this material may not possess relevant receptors for the uptake of the mentioned microbody proteins.

5.3. Bound Ribosomes and Post-Translational Translocation of Migratory Proteins

Before the present progress in our knowledge with regard to the post-translational translation of migratory proteins, some earlier observations with mitochondrial proteins (e.g., see Bingham and Campbell, 1972; Godinot and Lardy, 1973; Gonzalez-Cadavid and Cardova, 1972; Kawajiri *et al.*, 1977) had indicated involvement of ribosomes bound to the endoplasmic reticulum in their genesis. Some of these observations might now require reevaluation. However, more recent data also reveal the involvement of a "rapidly sedimented endoplasmic reticulum" of rat liver mitochondria in the genesis of some migratory proteins (Cascarano *et al.*, 1982; Montisamo *et al.*, 1982; Parimoo *et al.*, 1982; C. Padmanaban, personal communication). There are reports on involvement of bound ribosomes and even of ribosomes tightly bound to the endoplasmic reticulum in the genesis of mitochondrial proteins (e.g., see Heinrich *et al.*, 1982; Nabi *et al.*, 1983). In addition, examples can be found in the literature in which involvement of bound ribosomes, in addition to free ribosomes, in the genesis of mitochondrial migratory proteins is recognized (see Ozasa *et al.*, 1984). In the case of chloroplasts, mention of the involvement of the rough endoplasmic reticulum in the genesis of migratory proteins is extremely limited and concerns some eukaryotic algae (Gibbs, 1979). Remarkable is also the case of *Scilla non-scripta* (Rodriguez-Garcia and Sievers, 1977) in which electron microscopic images can be interpreted as revealing provision of material from the endoplasmic reticulum to the outer membrane (A. Sievers, personal communication). In the case of microbody proteins, there are data in the literature indicating implication of bound ribosomes in some instances, in addition to free ribosomes (see references in Sugita *et al.*, 1982). In the case of peroxisomal carnitine acetyl transferase, bound ribosomes synthesize part of the enzyme despite the fact that free polysomes were five times more active than bound ribosomes (Miyazawa et al., 1983). For the synthesis of the glyoxysomal malate synthase, the involvement of ribosomes bound to the endoplasmic reticulum was rejected in the case of cucumber *(Cucumis sativus)* but could not be totally excluded in the case of castor bean *(Ricinus communis)* (Lord and Roberts, 1982). Presently, the existence of a correlation between the endoplasmic reticulum and microbodies cannot be totally excluded with regard to provision of some membrane proteins (Lord and Roberts, 1982; see also Hashimoto, 1982) and lipids (Crane et al., 1982a; Lord and Roberts, 1982). In the case of rat liver microsomal 3-hydroxy-3-methyl glutaryl-CoA reductase, the enzyme which is known to be cotranslationally inserted in the membrane of the endoplasmic reticulum, could be immunologically localized in the matrix of peroxisomes (Keller et al., 1985, 1986; see also Zaar et al., 1986). However, the correlation between the two enzymes with regard to their biogenetic story remains to be substantiated. In rat liver, part of peroxisomal acyl/alkyl dihydroxyacetone phosphate reductase is found in microsomes, and, most significantly in cases of Zellweger cerebrohepatorenal syndrome, in which peroxisomes are absent, the enzyme accumulates in the endoplasmic reticulum (Ghosh and Hajra, 1986), i.e., in a compartment which could normally furnish lipids and some proteins to peroxisomes. Much controversy has existed about the eventual glycosylated microbody proteins. The problem is of importance and not totally solved despite the fact that recent data (Beevers and Gonzalez, 1986; Harson et al., 1983) confirm the presence of asparagine-linked carbohydrate chains in glyoxysomal proteins. If so, however, the fact would support transport of material from intracellular membrane structures to microbodies since the genesis of asparagine-linked carbohydrate chains involves enzymes present in the endoplasmic reticulum and the Golgi in both animal (Kornfeld and Kornfeld, 1985) and plant (Beevers, 1979) cells (see also Etémadi, 1986b). It is clear that the problem of the correlation between the endoplasmic reticulum and microbodies deserves further research, despite the fact that the presence of a direct continuity between these compartments reported previously (see Kartenbeck and Franke, 1974) could not be confirmed by other authors (see Gorgas, 1985; Yamamoto and Fahimi, 1987; Zaar et al., 1987).

Since translocation of migratory proteins is assumed to involve, at an early step, interaction of their signal sequences with specific receptors of receptor–translocator systems, the proposal was made (Etémadi, 1985) that for some migratory proteins the endoplasmic reticulum might have assembled operative receptors able to interact with signals of migratory proteins before chain termination. It is implied that the transmembrane component(s) of receptor–translocator system(s) in such cases can be synthesized by involvement of the endoplasmic reticulum and that assembly of eventual peripheral and/or soluble factors implicated in precursor targeting can be achieved at this level. In this eventuality a "loose" binding of precursors of migratory proteins to the endoplasmic reticulum as observed in some cases (Campbell *et al.*, 1982; Kawajiri *et al.*, 1977) might be of physiological significance. The tight binding of ribosomes involved in the genesis of these proteins (Heinrich *et al.*, 1982; Nabi *et al.*, 1983) would then reveal at least the start of the translocation process and/ or binding of the apolar part of extrasequences to the endoplasmic reticulum. Figure 6 gives the description of a hypothesis envisaging different possibilities of co- or post-translational binding of the precursor of a migratory protein, here

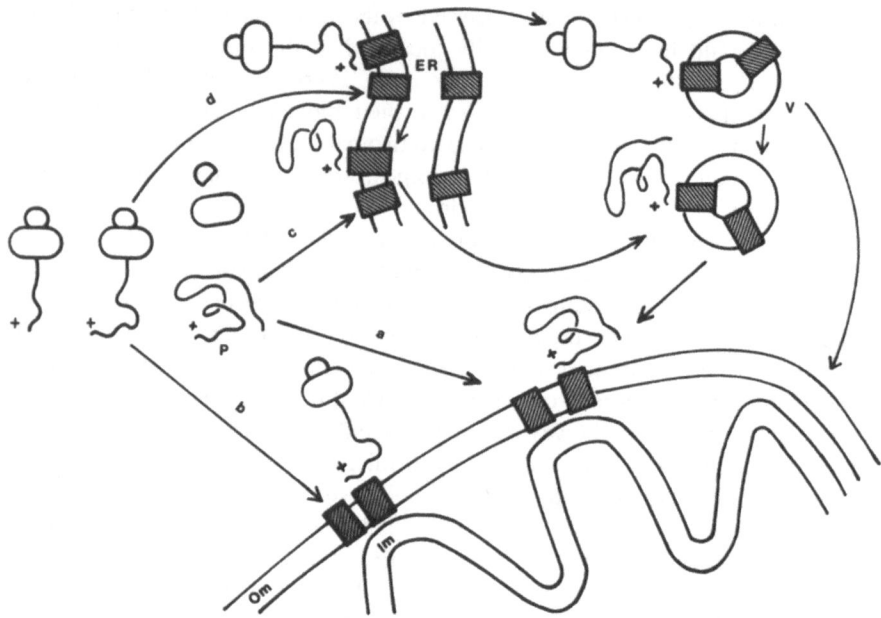

FIGURE 6. The figure is aimed at rationalizing observations on involvement of ribosomes bound not only to the outer membrane of terminal organelles, but also to the endoplasmic reticulum, in the genesis of some migratory proteins, here a mitochodrial protein. If the precursor of the protein, P, is released in the cytoplasm, it binds post-translationally to the outer membrane receptor as shown in route a. The nascent precursor chain, however, can also bind to the outer membrane receptor before chain termination, that is, cotranslationally, route b; completion of the chain and import occur subsequently. Routes through the endoplasmic reticulum (ER) are also envisaged. The completed precursor binds to the receptor on the endoplasmic reticulum and reaches the outer membrane after passage through vesicles (v) deriving from the endoplasmic reticulum, route c. Alternatively, the uncompleted chain can bind to the endoplasmic reticulum, route d, then the outer membrane through vesicles (v). As the translation continues, the uncompleted chain bound to the ER can be achieved before reaching transport vesicles. Completion of the chain can also occur on these vesicles. Binding of the precursor or uncompleted chains still linked to ribosomes to transport vesicles (not shown) and subsequent docking to the outer membrane are also possible.

a mitochondrial protein. It rationalizes, in particular, the presence of ribosomes bound to the endoplasmic reticulum in the genesis of some migratory proteins and also explains cases of the observed involvement of ribosomes bound to the terminal organelles, as, for example, that of ribosomes bound to mitochondria (Ades and Butow, 1980a,b; Kellems et al., 1974, 1975; Schatz and Butow, 1983; Suissa and Schatz, 1982). In such cases, not only a cotranslational binding but also at least for some cases a cotranslational processing, qualified as a "pseudo-cotranslational event" (Etémadi, 1985) could occur. In a recent report, the cotranslational translocation of a precursor, common to two mitochondrial enzymes (acetylglutamate kinase and acetylglutamyl phosphate reductase) was not excluded (Wandinger and Weiss, 1987).

The proposal by Fujiki et al. (1984) and Suzuki et al. (1987) of the formation of peroxisomes from preexisting peroxisomes is not in essential contradiction with this hypothesis, if it is admitted that membrane material, lipids (Crane et al., 1983; Lord and Roberts, 1982), and some proteins (Lord and Roberts, 1982) can be brought by membrane vesicles deriving from the endoplasmic reticulum; these would thus constitute some sort of "pre-microbody" structures to be "matured" by further post-translational incorporation of matrix proteins and additional membrane proteins, and/or fuse with pre-existing peroxisomes. Provision of material from the endoplasmic reticulum together with post-translational supply of proteins might well lead to fission events in "proliferating" microbodies.

5.4. The Case of Endoproteins

"Endoproteins" were thus named in order to distinguish them from "ectoproteins" (Rothman and Lenard, 1977). Since ectoproteins are integral membrane proteins having an exposure at the side remote from the cytoplasm, either with or without an exposure at the cytosolic compartment, endoproteins were defined as integral membrane proteins with only a cytoplasmic exposure. Their membrane-embedded segment, which for the best representative of these proteins, the endoplasmic reticulum membrane protein, cytochrome b_5, is the C terminus of the molecule (see Daily and Strittmatter, 1981a,b; Enoch et al., 1979), is shown as penetrating only halfway through the hydrophobic core of the membrane (see Rothman and Lenard, 1977). Recently, Thiede et al. (1986) reported the case of the rat liver endoplasmic reticulum enzyme, stearoyl-CoA desaturase, which like cytochrome b_5 is synthesized on free ribosomes. The exact mode of membrane insertion of this protein, which exhibits 62% of the hydrophobic residues, has not been elucidated. Recent work by Strittmatter et al. (1988) shows, however, that at least 33 N-terminal residues of the protein are not involved in membrane insertion. Another endoplasmic reticulum membrane protein, NADH cytochrome b_5 reductase, was thought to be C-terminally membrane anchored (Mihara *et al.*, 1978). More recently, however, binding through the N terminus was admitted for this enzyme (Kensil and Strittmatter, 1986; Kensil *et al.*, 1983). Furthermore, recent data suggest pentration of the C terminus of cytochrome b_5 deeply in the hydrophobic core, probably through the entire thickness of the membrane (Gogol and Engelman, 1984; Takagaki *et al.*, 1983a,b). More recently Rzepecki *et al.* (1986) confirmed the deep penetration of C-terminal of cytochrome b_5 and Arinc *et al.* (1987) reported the external exposure of the extreme C-terminal residues in tightly reconstituted cytochrome b_5.

Several experiments indicate that both cytochrome b_5 (see Okada *et al.*, 1982; Rachubinski *et al.*, 1980) and NADH cytochrome b_5 reductase (see Borgese and Gaetani, 1982; Okada *et al.*, 1982) are synthesized on free ribosomes and post-translationally membrane inserted. This would also be the case for stearoyl-CoA desaturase (Thiede *et al.*, 1986). These proteins thus furnish examples of integral membrane proteins bound post-translationally to the membrane, mostly to the endoplasmic reticulum but also to other membranes (for more references and details see Etémadi, 1985). Noteworthy is the fact that reconstitution by the direct insertion method leads only to a "loose" binding of cytochrome b_5 (Enoch *et al.*, 1979; Takagaki *et al.*, 1983a), small vesicles preferentially incorporating the protein (Greenhut and Roseman, 1985), while a "tight" binding, reminiscent of physiological insertion, does not require energy but the prior presence of a protein(s) or factors of membrane perturbation. The use of a detergent dialysis method (Enoch *et al.*, 1979; Takagaki *et al.*, 1983b) or incubation of cytochrome b_5 in detergent micelles with preformed vesicles (Christiansen and Carlsen, 1985), and more generally defects in membrane structure (Jain and Zakim, 1987; Scotto *et al.*, 1987), lead to a tight binding.

Recent data indicate that some proteins of mitochondrial outer membrane and of the membrane of microbodies behave as endoproteins inasmuch as they are also synthesized on free ribosomes and inserted post-translationally into the organellar membrane, without energy requirement. A mitochondrial outer membrane protein, porin (Freitag *et al.*, 1982a,b; Gasser and Schatz, 1983; Mihara and Blobel, 1982; Zwizinski *et al.*, 1984), and some other proteins of this membrane of 14,000, 45,000, and 70,000 Da (Gasser and Schatz, 1983; Riezman, 1982) were reported to be post-translationally inserted in the outer membrane, without energy requirement. It has been proposed that all mitochondrial outer membrane proteins might be inserted post-translationally (Gasser and Schatz, 1983). Recently, a major peroxisomal membrane protein of 21,700 Da was also found to be synthesized predominantly on free ribosomes (Fujiki *et al.*, 1984) and Suzuki *et al.* (1987) reported synthesis of three major peroxisomal membrane polypeptides of rat liver of 22,000, 26,000, and 70,000

Da mostly on free polysomes. Such proteins are proposed to be synthesized without an extra-sequence (Fujiki et al., 1984; Koester et al., 1986; Suzuki et al., 1987). However, further work is necessary to better understand the genesis of outer membranes of organelles through which migratory proteins are translocated. Some observations reveal the presence of a common antigenic determinant between the outer mitochondrial membrane and rough microsomes or involvement of bound ribosomes in the genesis of some outer membrane proteins (Noel et al., 1985). Also, in a recent report Hartl and Just (1987) suggest that some drug-induced rat liver peroxisomal membrane proteins of different molecular weight (42,000, 28,000, and 26,000 Da) derive by proteolysis from a common 69,000 Da membrane protein; thus, the apparent multiplicity of membrane proteins synthesized on free ribosomes might in some cases find its origin in proteolytic cleavage of a single protein.

As to whether or not the previous presence of proteinaceous material (putative receptor(s)) is required for post-translational insertion of proteins in the outer membrane of organelles, contradictory reports exist. Thus, binding of porin was not affected by protease treatment of mitochondria (Gasser and Schatz, 1983), while other reports indicate inhibition of binding of this protein after mild proteolysis (Mihara and Blobel, 1982; Ono and Tuboi, 1987; Pfaller and Neupert, 1987; Zwizinski et al., 1984). As to whether porin can (Gasser and Schatz, 1983) or cannot (Mihara and Blobel, 1982; Ono and Tuboi, 1987) bind to the isolated outer mitochondrial membrane, an agreement is not reached. There are reports (Hase et al., 1986; Mihara and Sato, 1985) indicating the similarity of the N termini of porin and the 70,000 Da outer membrane protein. It is, in addition, considered that receptors for proteins inserted in the outer membrane also interact with migratory proteins that cross the outer membrane to reach intramitochondrial compartments (Hase et al., 1986; Hurt and Van Loon, 1986; Pfaller and Neupert, 1987). If so, binding of porin to trypsin-treated mitochondria (Gasser and Schatz, 1983) would mean that either the proteinaceous material was at least not cleaved in a manner detrimental to the receptor function or, less likely, that protein receptors for many mitochondrial proteins do not exist. Clearly, the problem is strongly correlated with the very concept of the mechanism of migratory protein translocation. A recent report by Ono and Tuboi (1987) indicates that membrane potential is not required for insertion of porin which, it is suggested, occurs through the contact sites between the outer and the inner membrane before its definitive integration into the outer membrane. Studies by Kleene et al. (1987) reveal that the presence of nucleotide triphosphates is mandatory for insertion of porin and, more recently, Pfanner et al. (1988) proposed their involvement in the unfolding of porin. In fact their effect could be replaced by an acid treatment followed by a rapid alkali neutralization.

6. PROTEIN TRANSLOCATION THROUGH OR INSERTION INTO LIPID AGGREGATES

6.1. Experiments Related to Translocation of Secretory and Integral Membrane Proteins

While abundant studies to date concerned various aspects of protein translocation process by reconstitution in its general sense, not many physiologically significant experiments of reconstitution using liposomes have been carried out, despite the potential aptitude of such experiments to solve some crucial problems. Reconstitution in the restricted sense is aimed essentially and most generally at mimicking the properties of membrane proteins by their insertion in lipid bilayers. Various procedures for such insertion were developed (see Etémadi, 1985; Racker, 1979). The possibility of a *functional* reconstitution by these procedures, and particularly through the direct insertion method in which proteins or complexes are allowed to insert into preformed liposomes, led to the hope that such a procedure might be relevant to physiological protein translocation processes. A series of work (for references see Wickner, 1979) on insertion of *E. coli* phage M_{13} coat protein led to the proposal that this protein is inserted with an absolute asymmetry in liposomes formed by the detergent dilution procedure and from this it was concluded that hydrophobic forces determine the orientation of integral membrane proteins in biological membranes. However, the insertion of the coat protein with an absolute asymmetry could not be substantiated by results reported by other authors (Chamberlain *et al.*, 1978; Hu and Wiesnieski, 1979), while in the bacterial membrane the protein is asymmetrically oriented (Ohkawa and Webster, 1981). As detailed elsewhere (Etémadi, 1985), numerous examples of reconstitution clearly show that usually no unique natural orientation of membrane proteins is established by reconstitution procedures, such as detergent dilution or dialysis. Furthermore, a consensus transmembrane protein (H-$2K^K$) was reported to be inserted with a hairpin conformation and not in a transmembrane manner (Cardoza *et al.*, 1984).

The "membrane trigger hypothesis" (Wickner, 1979) considered that membrane insertion of proteins occurs post-translationally without involvement of membrane receptor–translocators, or membrane–ribosome interaction, and that no energy is required for the process; the only thermodynamics of membrane–protein interaction would allow the correct insertion of precursors into the membrane. Post-translational insertion of precoat protein in preformed phospholipid vesicles into which signal peptidase, the enzyme cleaving the signal peptide of a series of precursors of exported proteins in *E. coli* membrane, was incorporated has been reported to occur with the N terminus of the protein in the luminal side (Ohno-Iwashita and Wickner, 1983; Watts *et al.*, 1981). However, the role envisaged for the cleavage enzyme in the translocation process (Watts et al., 1981) was more recently denied (Geller and Wickner, 1985). Thus, we are left with the considerations of the membrane trigger hypothesis that the only protein–lipid interaction is at the basis of the physiological insertion of precoat protein. This, however (Rhoads et al., 1984), is contradictory to the report by Date et al. (1980), which states that membrane potential is required for transmembrane insertion of the same protein in vivo. The case of precoat protein, besides such contradictory proposals, cannot be taken as representative of most proteins translocated through or inserted into the membrane (for a discussion see Etémadi, 1985). Nevertheless, other attempts to reconstitute translocation of E. coli exported proteins into liposomes, into which the signal peptidase was inserted, were made. As a result, OmpA and maltose-binding protein were claimed to be correctly translocated and processed (Zimmermann and Wickner, 1983). Surprisingly, however, the processed proteins were found not to be membrane inserted for the case of OmpA, which is an outer membrane protein, or present in the luminal side in the case of maltose-binding protein, which is a soluble periplasmic protein, but released into the medium. The most rational interpretation would be that translocation did not occur but that some residual cleavage enzyme activity oriented toward the medium cleaved the precursors.

Some studies attempted to get insight into the interaction of signal sequences with membranes using model experiments in which synthetic signals were interacted with liposomes. In one such study (Nagaraj, 1984), the synthetic signal sequence for chicken lysozyme was labeled with the fluorescent dansyl(5-dimethyl-aminonaphtalene-1-sulfonyl) group and its interaction with phospholipid liposomes by fluorescence emission and polarization was examined. It was found that the sequence binds to the hydrophobic core of the phospholipid. In other studies, synthetic signals and their "mutated" counterparts were interacted with lipids and results were compared with the efficiency of such sequences when involved in genuine protein translocation processes. For example, Briggs and Gierasch (1984) prepared synthetic peptides with the sequence of the signal for the wild type or mutant (export defective) lambda phage receptors. A 7-residue-long segment of the wild-type signal flanked by a proline and a glycine residue (see Table I) is predicted to have an α-helical conformation, while 4 of these residues are absent in export defective mutants. Using the circular dichroism of synthetic peptides in helix-promoting environments (trifluoroethanol or SDS), the relative helicities of the wild-type and defective signals were compared. While the wild-type signal showed CD spectra characteristic of a partial helical conformation, defective polypeptides did not show such spectra. When proline or glycine in defective signals was substituted by a leucine or cysteine, respectively—substitutions that regenerate the signal function—the predicted helix promotion could be confirmed by CD data. The highest helicity, however, was associated with the wild-type signal. In further studies, Briggs et al. (1985), using wild-type, defective, and revertant signal peptides, examined their interaction with a lipid monolayer at a water–air interface, their conformational change in the presence of phospholipid vesicles, and their aptitude to promote vesicle aggregation. Such studies led to the proposal that wild-type and pseudorevertant signals interact with lipids and that this aptitude correlates with the tendency of functional signals to form the α-helical conformation. It was suggested that while the probability for interaction of signal sequences with membrane protein components exists, they may also have some contact with membrane lipid during secretion. The idea of an interaction with lipids was reinforced when it was observed (Briggs et al., 1986) by circular dichroism and Fourier transform infrared spectroscopy that the signal of the wild-type lambda phage receptor can be inserted in phospholipid monolayer and possesses an α-helical structure, while the signal sequence in the aqueous phase is unstructured and interaction, without insertion, induces a β-structure. A model for interaction of the signal sequence with the membrane through involvement of SRP and docking protein, or other factors such as the membrane potential, is presented. However, only lipids would be involved in subsequent events (Figure 7). Nevertheless, recently, such a proposal was somewhat mitigated (Chen et al., 1987) when it was observed that the peptide corresponding to the functional signal as well as two peptides corresponding to pseudorevertant types of defective mutant were able to inhibit the *in vitro* translocation of the alkaline phosphatase and the outer membrane protein A in *E. coli* inverted inner membrane vesicles. The defective mutant peptide did not show such an effect. As a result, it was not decided whether the competition occurs at the level of a receptor in the membrane or with a cytoplasmic SRP-like material. In addition, an indirect inhibitory action by interaction with lipids was not excluded.

The above experiments thus concern the early step of the binding of nascent chains to the membrane. Other reports by Fidelio et al. (1986, 1987) consider the whole process of translocation. The authors examined surface properties of some proteins (pretrypsinogen, ovalbumin) and compared them to those of signal sequences (signal sequence of pretrypsinogen, residues 21–47 of ovalbumin constituting the putative uncleaved signal of this protein and a synthetic "consensus" signal sequence). It was found that while proteins present in the subphase interact with phospholipid monolayers at the air–water interface, signal sequences do not. The reason for discrepancy with the above results of Briggs et al. (1986) is not clear, despite the remark made that in those experiments a prokaryotic signal was used. In any event, the observation was interpreted to indicate the requirement for a "signal recognition particle" in binding the signal or involvement of receptor proteins in the microsomal membrane (Fidelio et al., 1986). The proposal of a mechanism for protein translocation stands on additional observations. The surface stability of a signal

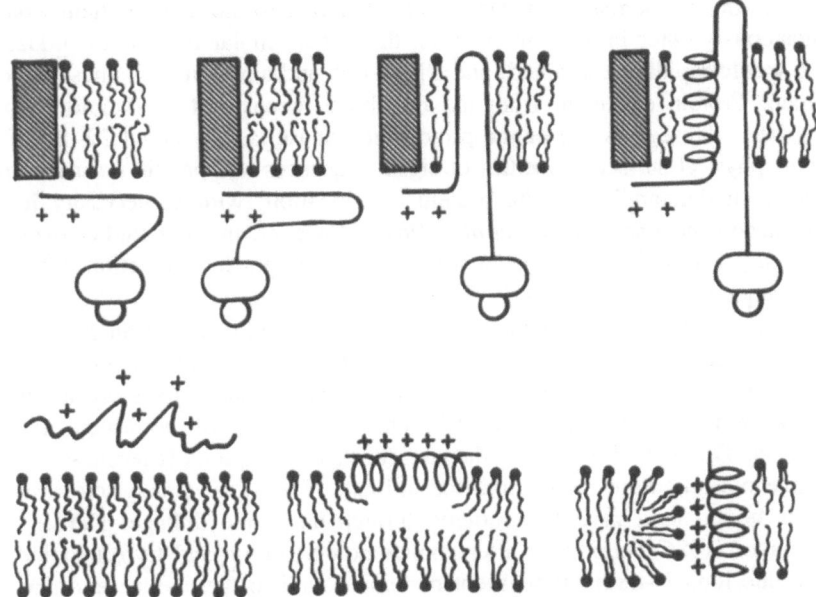

FIGURE 7. The upper drawing shows the suggested intraction of the nascent chain of a secretory or transmembrane integral protein with the membrane (Briggs et al., 1986). Proteinaceous material (hatched rectangles) or membrane potential might be involved in the initial targeting to the membrane. However, the subsequent interaction is with phospholipids only and leads from an unstructured conformation of the soluble state to the β-structure of the signal sequence. This structure is then presumably inserted into the membrane and gains the preferred α-helical conformation. No protein is involved in the actual membrane insertion. The lower drawing concerns the interaction of the N-terminal extrasequence of a mitochondrial migratory protein with the outer membrane (Roise et al., 1986; Von Heijne, 1986). It is presumed that ineraction with lipids in the cytoplasmic side (here the upper side of the bilayer) would lead to an amphipathic helical conformation of the extrasequence, which would subsequently be inserted in the membrane. Thus, proteinaceous receptors would not be involved in membrane insertion.

sequence when spread on the air–water interface alone or in mixed films with phospholipids was higher than that of proteins spread under similar conditions. Also, above a threshold of peptide or protein concentration, which was at the molar fraction of 0.2 for the signal and only of 0.0025–0.005 (depending on the lipid used) for ovalbumin, lateral immiscibility appears as revealed by observation of two collapse points. Under these conditions the signal sequence also supported higher lateral surface pressure than albumin. Based on these observations, the following translocation mechanism was suggested. After the uncleaved signal sequence of a protein such as ovalbumin is guided into the lipid core of the membrane, the part corresponding to the mature chain would spontaneously enter into the membrane, the critical molar fraction of 0.0025–0.005 would rapidly be attained, and a lipid rich and a protein-rich phase would separate. Protein extrusion from the membrane (translocation) would follow since the collapse pressure of the protein-rich phase is only about 22–24 mN/m, thus physical surface properties of signal sequences and protein chains would assist the understanding of the protein translocation, without necessity for a proteinaceous channel (Fidelio *et al.*, 1987). However, in other studies (Robinson *et al.*, 1987), involvement of a membrane protein in interaction with the signal sequence was recognized.

With regard to endoproteins, we have seen that experiment on reconstitution of cytochrome b_5 in lipid bilayers pointed to the requirement for the presence of protein or the presence or induction of membrane defects for an insertion apparently mimicking the natural membrane binding of the protein. In the case of NADH–cytochrome b_5 reductase attached through its N terminus to the endoplasmic reticulum membrane, the ultimate N-terminal amino acid residue Gly_1 is bound to a myristate moiety through an amide linkage. Kensil and Strittmatter (1986) found that after cholate solubilization and trypsinization, a 28 residue long N-terminal segment remains bound to the cholate and can be reconstituted in phospholipid vesicles. Two possible conformations—a hairpin and a transmembrane one—were envisaged for this reconstituted segment through studies on the fluorescence of Trp_{16} and energy transfer from this residue to acceptors in the head group region of the outer monolayer. The hairpin conformation was preferred to the transmembrane conformation. The actual conformation of the N terminus of the enzyme in the microsomal membrane remains, however, unclear.

6.2. Experiments on Translocation of Migratory Proteins

Before the present understanding of the requirements of the post-translational translocation of migratory proteins (i.e., the possible interaction of signal sequences of precursors with membrane receptors and energy consumption), experiments on reconstitution of protein translocation used mature proteins. Assays carried out on interaction of cytochrome *c* (holocytochrome *c*) with liposomal membranes are numerous (see Birrel and Griffith, 1976; Brown and Wütrich, 1977; Chapman and Urbina, 1971; Furuya *et al.*, 1979; Mustonen *et al.*, 1987). Generally, it is admitted that this interaction is electrostatic in nature, with phase separation and a long-range ordering effect on lipid. The occurrence in addition of a hydrophobic interaction was also admitted (Furuya *et al.*, 1979). Such experiments, while eventually having a bearing on the interaction of cytochrome *c* with the outer aspect of the inner mitochondrial membrane, do not concern the essential act of translocation through the outer membrane, which involves apocytochrome c (see below) not holocytochrome c. Other experiments concerned the mature aspartate transaminase, which was found to react electrostatically with negatively charged lipid head groups. Furthermore, the occurrence of a hydrophobic interaction was recognized. Remarkable is the observation that in these experiments the enzyme was found to bind to vesicles in such a way that it was neither cleavable by protease action nor could it interact with its substrate (Furuya et al., 1979). Such observations indicate the degree of caution with which results of reconstitution experiment should be interpreted. The insensitivity to proteolysis was often taken as revealing protein translocation, while, on the other hand, a physiologically significant translocation involves the precursor, not the mature protein used in the present experiments. Also, observations on the interaction of the mature malate dehydrogenase and lipid vesicles and interpretation given are instructive. Under conditions in which the mature enzyme is present as a monomer (see Webster et al., 1979), a conformational change as a result of interaction with liposomes occurred. This was interpreted as reflecting a hydrophobic interaction of the protein with the membrane hydrophobic core and it was thought that these observations would have a bearing on protein translocation into mitochondria since it was supposed that matrix enzymes after similar interactions with the mitochondrial membrane would somehow reach the matrix compartment where they would be trapped by interaction with substrates, coenzymes, or the pH gradient. Here again, a whole theory reminiscent of the membrane trigger hypothesis was constructed, which involves interaction of the (mature) protein with membrane lipids without requirement of proteinaceous receptor–translocator system(s) or energy consumption.

Since the above early observations, most reconstitution experiments regarding the process of translocation of migratory proteins concerned either apocytochrome c or only the initial steps of interaction of extrasequences with the membrane. Apocytochrome c was found to bind to phospholipid vesicles containing acidic phospholipids. No complete translocation occurred, however, since all bound molecules could be digested by external proteases. Molecules were bound through their C termini to vesicles. This part of apocytochrome c was protected from digestion. Remarkably, when proteases were entrapped in vesicles, most of the apocytochrome c molecules were digested to small molecules. Thus, high-affinity receptors were thought not to be strictly required for translocation (Dumont and Richards, 1984). Other studies (for references see Rietveld et al., 1986; see also Walter et al., 1986) likewise indicated interaction of apocytochrome c both electrostatically and through hydrophobic interaction with vesicles of negatively charged lipids. Interaction of apocytochrome c with phosphatidylserine vesicles within which trypsin was entrapped led to its partial degradation, indicating that part of the molecule reached the luminal compartment. In vesicles of lipid mixtures containing phosphatidylcholine and phosphatidylserine, the partial degradation of apocytochrome c molecules occurred otpimally when vesicles were enriched to an abnormally high degree (50–100%) in phosphatidylserine. A complete translocation did not occur, however. The exact part of the molecule reaching the luminal side is not clear. It was observed that as a result of cleavage from the inside no new binding sites were created for apocytochrome c, suggesting that the C terminus was not translocated (see Rietveld *et al.*, 1986). This is the end of the molecule that is reported to bind apocytochrome c to the mitochondrial membrane (Matsuura *et al.*, 1981; Stuart *et al.*, 1987). Thus, to date, a physiological translocation of apocytochrome c is not reconstituted. Likewise the object of a recent report by Pilon *et al* (1987) was the definition of conditions allowing interaction with and insertion into lipids spread on monolayer surface of apocytochrome c. These assays using phosphatidylserine entities emphasize the importance of charge interaction. Another recent report by Park *et al* (1987), indicates, however, that the import of apocytochrome c by yeast mitochondria, when mostly methylated (as ϵ-N-trimethyllysine)—as it is naturally in this case—not only was not decreased but increased 2- to 4-fold. Thus, it would be interesting to assay reconstitution using the proteinanceous receptor and the enzyme-attaching heme to apocytochrome c (Henning and Neupert, 1981; Henning *et al.*, 1983; Zwizinski *et al.*, 1984). The translocation of apocytochrome c was in fact suggested to be a concerted process involving, besides the receptor, the apocytochrome-modifying enzyme in some sort of group translocation event (Etémadi, 1985).

To get insight into the interaction of signal sequences of migratory proteins with the mitochondrial membrane, Ito *et al.* (1985) constructed synthetic peptides mimicking the structure of such sequences and assessed their effect on the import of cytochrome P-450 *scc* and adrenodoxin into adrenal cortex mitochondria. They proved to be generally inhibitory. However, binding of the precytochrome P-450 *scc* to the mitochondrial membrane was not affected. These peptides, in addition, uncoupled oxidative phosphorylation in isolated mitochondria from both rat liver and bovine adrenal cortex and at high concentrations led to the leakage of intermembrane and matrix mitochondrial enzymes, while membrane-bound enzymes were not solubilized. They were suggested to bind to the membrane, causing perturbations, possibly related to the function of extension sequences. It was proposed that "the function of the characteristic common sequences found in the extension peptide of mitochondrial enzyme precursors is not the specific recognition of mitochondrial surface, but the perturbation of the structure of mitochondrial membranes to enable the precursor peptide to translocate across the membrane." Gillepsie *et al.* (1985) synthesized a polypeptide corresponding to residues 1–27 of preornithine carbamyl transferase (see Table II) and also remarked that its interaction with mitochondria collapsed the membrane electrochemical potential. When this same peptide was added to a system (reticulocyte lysates) synthesizing the precursor of the enzyme, no collapse of electrochemical gradient occurred. Reasons for this discrepancy remain unclear. However, competition for a membrane receptor was admitted since a reversible inhibition of import occurred. Epand et al. (1986) used the same peptide and examined its interaction with various lipids. The peptide exhibited more α-helical conformation in the presence of anionic lipids such as cardiolipin or dimyristoylphosphatidylglycerol. A zwitterionic phospholipid such as dimyristoylphosphatidylcholine did not affect the conformation of the peptide. The synthetic peptide disrupted large lipid aggregates and led to discoid micelles of 30–50 nm in diameter in which it was assumed to have an amphiphilic helix conformation. These observations led to the suggestion that bilayer perturbation might play a role during protein translocation into mitochondria. Roise et al. (1986) have synthesized three peptides of 15, 25, and 33 amino acids corresponding to the N-terminal sequence of the precursor of subunit IV of yeast cytochrome c oxidase. The second of these peptides has the sequence of the extrapeptide of 25 residues (see Table II). These peptides were all soluble in aqueous buffers. They could, however, be inserted into phospholipid monolayers at the water–air interface. The two largest peptides could disrupt phospholipid liposomes. A diffusion potential negative inside enhanced and a diffusion potential positive inside decreased this effect. These two peptides also uncoupled the respiratory control of isolated yeast mitochondria. In the presence of detergent micelles the 25-residue-long peptide became partially helical as revealed by circular dichroism measurements. In addition, the sequence of peptides allowed prediction of formation of amphiphilic helices, hence the suggestion that such helices play a crucial role in protein transport into mitochondria. A scheme was presented (Figure 7, lower drawing) in which the presequence of migratory proteins would insert as an amphiphilic helix in the membrane without involvement of receptor or any other proteinaceous material. The potential of most if not all the 23 mitochondrial targeting sequences to form amphiphilic helices was also recognized by Von Heijne (1986). The fact that synthetic extra-sequences exhibit membrane perturbing activity while precursors in which these sequences are linked to "passenger" proteins do not induce such an effect, was explained (Endo and Schatz, 1988) in considering that the passenger part would mask the amphiphilic properties of extra-sequences, which would appear by the unfolding of the protein in contact with the membrane. However, several considerations should be taken into account. As remarked by Maher and Singer (1986), it cannot be excluded that in such experiments and those mentioned above for synthetic signal sequences of secretory and membrane inserted proteins, interaction between peptides and membranes would not mimic that of precursor proteins during their translocation, since synthetic peptides might exist as oligomers, while natural translocation most probably involves interaction of individual entities. Furthermore, various synthetic presequences, bound N-terminal to the mature part of a mitochondrial imported protein revealed the functionality of the synthetic prepeptides. These studies did not confirm the necessity for a potential amphiphilic α-helical conformation for functional peptides (Allison and Schatz, 1986). On the other hand Skerjanc et al. (1987) could not confirm the implication of the membrane potential in binding of the synthetic pre-sequence of ornithine carbamyl transferase to lipid vesicles.

Aoyagi et al. (1987) have recently synthesized various peptides corresponding to different segments of the extra-sequence of cytochrome P-450$_{scc}$ precursor which is of 39 residues. Two of these peptides representing residues 1–15 and 1–20 of the extra-sequence were active in inhibiting the import of some mitochondrial proteins, there were also peptides among those synthesized, that while exhibiting a random conformation in aquous buffer, showed a partially ordered conformation either α-helical or type II'β in the presence of buffer containing acidic liposomes [dipalmitoyl phosphatidylcholine plus dipalmitoyl phosphatidylglycerol (3 : 1)]. The authors suggest that precursors change their conformation when they make contact with the surface of the inner membrane. However, the primary structure of the peptides could not allow an amphiphilic helix conformation. Thus, the authors propose that a partially ordered conformation such as α-helix might be important, while amphiphilic α-helix is not required. In addition the two active peptides caused leakage of acidic liposomes, which is interpreted to confirm that membrane perturbation may be important in mitochondrial import process. Assays of these peptides performed by Furuya et al. (1987) revealed that they did not affect the intactness of mitochondria as assessed by measurement of the respiratory control ratio. They decreased the rate of import of precursors while the binding of these to the mitochrondrial surface was not affected, meaning that a blockade of internalization was occurring. Thus, these results are discordant with proposals of other authors concerning inhibition of binding of precursors to mitochondria treated with polypeptides related to extra-sequences of mitochondrial imported proteins (see Gillespie et al., 1985; Yoshida et al., 1985). Recently Roise et al. (1988) were able to confirm that the potentiality for an amphiphilic α-helix is not required for *in vitro* interaction of "artificial pre-sequences" with phospholipid monolayers or vesicles. The authors proposed, however, that such pre-sequences have to exhibit properties of amphiphilicity.

With regard to the crucial problem as to whether or not proteinaceous membrane components are implicated in binding of precursors through their extra-sequences, the significance of the above data is not clear. Again, Furuya et al. (1987) suggested a nonspecific interaction with lipids as evidenced by the fact that "active" peptides they synthesized could not only bind to mitochondria but also to microsomal membranes. However, it would remain to be clarified how such a nonspecific interaction would lead to specific organellar transport. In a recent study Myers et al. (1987) examined the interaction of the extra-sequence of ornithine carbamyl transferase with bilayer membranes of dipalmitoyl phosphatidylcholine containing small mole-percents of acidic phospholipids. A strong interaction with an enthalpy of association of -60 kcal/mol of peptide was observed. Fluorescence studies indicated, however, that the extra-sequence did not penetrate the bilayer. The authors, thus, do not exclude that further processing of precursor proteins would involve protein receptors and, finally, as discussed above, the involvement of membrane proteins, including an outer membrane protein, in the import of mitochondrial proteins was recognized (Ohba and Schatz, 1987).

Reconstitution experiments regarding chloroplast proteins also concerned the aspect of the initial binding of precursors (Pfisterer et al., 1982). Bitsch and Kloppstech (1986a,b) examined binding of such precursors to outer membrane vesicle preparations; after detergent treatment and reconstitution using sulfobetain or other detergents, the binding capacity was found in the residual particles after centrifugation and in the supernatant. Both a "partial reconstitution," that is, removal of detergent by dialysis of the whole detergent-treated vesicle preparations, and a "total reconstitution," defined as removal of detergent from the isolated supernatant, allowed an appreciable recovery of the binding activity.

7. GENERAL AND CONCLUDING REMARKS

In what precedes I have summarized the present state of the knowledge gained on physiological protein translocation processes by reconstitution experiments, reconstitution being taken in both its general and restricted senses. If the requirement for signal sequences interacting with specific receptor–translocators and the use of energy for translocation can be admitted, various processes of protein translocation with some exceptions such as that of endoproteins or apocytochrome c can be said to follow the general mechanism of "some sort of active transport" previously suggested (Etémadi, 1980a, 1985) for migratory protein translocation.

Different authors impressed by the hydrophobic nature of signal sequences of secretory and transmembrane integral proteins proposed models for translocation in which only interaction with lipids was envisaged. Such an attitude, though appearing at first sight legitimate, does not correspond to the reality, which involves proteins. However, if the implication of proteins alone in interaction with signal sequences would ultimately prove to be true, it would remain to be explained how at the molecular level the interaction with the receptor occurs. Clarification of the point would presumably await a thorough knowledge of the nature of the receptor–translocator proteins. Also, for migratory proteins an initial interaction of extrasequences with lipids was envisaged.

However, here again, as for secretory and transmembrane integral proteins, interaction with membrane proteins was ultimately admitted although much remains to be learned both on the initial steps of interaction and details of the process of translocation.

It is worthwhile at this point to mention a recent "unitary mechanism" for protein translocation processes presented by Singer et al. (1987a,b). This mechanism admits the presence of transmembrane translocator proteins. Each translocator protein is envisaged as a complex comprising n ($\sim$3–6) homologous but not identical subunits designated as α, β, γ, δ, which form a closed central aqueous channel (see Figure 8). Two classes of such translocator proteins, one for secretory and transmembrane integral proteins and the other for migratory proteins, would exist. Translocator proteins in addition to binding the signal sequence are involved in the formation of successive subdomains of

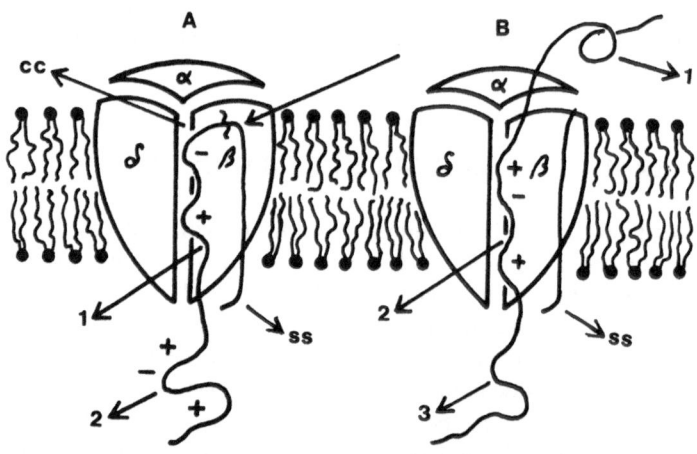

FIGURE 8. The unitary hypothesis for the mechanism of translocation of proteins (Singer et al., 1987a). The translocator protein is formed of subunits α, β, γ, δ, ... One of the interfaces, where interaction between subunits is weaker, say the interface between β and γ subunits, is destined to allow translocation. Subunit γ is not shown, so that the central aqueous channel (cc) appears. Translocation is suggested to be initiated by interaction of the signal sequence (ss) with β or γ subunit in such a way that its N-terminal remains on the cis (cytoplasmic) side. The rest of the signal sequence interacts with the β–γ interface, so that its site of cleavage (arrow in Figure 8A) reaches the trans (extracytoplasmic) face. Thus, necessarily, part of the mature chain enters the interface. As a result of interaction of the signal and the first segment of the mature chain with the interface, the central channel (cc) opens; subsequently, charged groups are located in contact with the aqueous environment of the channel, while hydrophobic residues face the hydrophobic core of the membrane. Translocation involves successive transfer of subdomains (indicated as 1 and 2 in Figure 8A and 1, 2, and 3 in Figure 8B).

20–40 residues and in their energy-dependent translocation. Subdomains have secondary structures which could contain stretches of amphiphilic helix, or β-structure, or both. The first subdomain would be translocated as a result of an extensive quanternary rearrangement in translocator protein resulting from energy-dependent conformational changes in β and/or γ subunits. Translocation of the first subdomain pulls the following sequence into the interface where seeding of the second subdomain, followed by translocation, occurs. The process can thus continue until the whole protein is translocated. Translocation of secretory proteins without a cleaved signal is explained by the same mechanism except that the signal sequence, if it does not have too strong an affinity for the translocator protein, would form part of the last subdomain and be translocated. For a transmembrane protein with its C terminus located cytoplasmically, after the hydrophobic membrane anchoring segment is pulled in the interface, it slips laterally and forms a transmembrane helix; the β–γ interface would close up, thus the hydrophobic segment acts as a stop transfer sequence. For proteins with their N termini located cytoplasmically and their C termini extracytoplasmically, it is admitted that the membrane-anchoring sequence is the uncleaved signal. For a protein with the N terminus located cytoplasmically but crossing the membrane thickness several times, the process of translocation begins as for proteins with uncleaved signals, then successive stop transfer and internal signals would be involved. At no stage is part of the protein sequence inserted directly into the lipid bilayer. A recent interesting report by Fikes and Bassford (1987) can be interpreted as supporting the idea that, depending on its affinity for the membrane, a signal sequence can anchor the protein in the membrane or be translocated with the mature part of the molecule. However, the problem of the chronology of the translocation of the uncleaved N-terminal signal as compared to that of the bulk of the protein was not addressed in this work.

While the essence of the "unitary mechanism" for protein translocation presently seems acceptable and in accordance with the generalization of the involvement of "active transport" mechanisms (Etémadi, 1980a, 1985, 1987, 1988), the strict limitations placed as to the subunit structure of translocator proteins and the length of subdomains to be translocated would require substantiation. There are also aspects of the theory that would require caution. The orientation of the signal sequence restricted to the cytoplasmic exposure of its N terminus does not account for the presence of transmembrane proteins with extracytoplasmic location of their N termini (see Etémadi, 1986a) that do not have a cleaved extrasequence. Examples of such proteins are encountered both in eukaryotic and prokaryotic cells. Thus, bovine retinal opsin (Ovchinnikov *et al.*, 1983; Schechter *et al.*, 1979), rat cytochrome P-450 (Sakaguchi *et al.*, 1987) and *E. coli* signal peptidase I (Moore and Miura, 1987) have their uncleaved membrane-anchored N-terminal sequences oriented to the extracytoplasmic side. In addition, the orientation of prebacteriorhodopsin in protoplasts of *Halobacterium halobium* is remarkable and does not comply with the proposal of Singer *et al.* (1987a,b). The protein normally has a cleaved extrasequence in the cells, which in protoplasts remains unprocessed and exposed on the *trans* side of the membrane (Seehara and Khorana, 1984). Remarkably, it was recently suggested (Von Heijne and Segrest, 1987) that the 13-residue-long extrasequence of the protein has the information to form an amphiphilic helix with a large nonpolar face and a narrow negatively charged polar face and that it would be inserted in the membrane contributing to one of the four helical hairpins. However, biological membranes being asymmetrical structures (see references in Etémadi, 1980a,b,c, 1985; Bergelson and Barsukov, 1978; Op Den Kamp, 1979; Van Deenen, 1981), only one orientation for the N terminus of membrane proteins and their precursors can be accepted. The proposal of Von Heijne and Segrest (1987) would require that the N terminus of the first hairpin be cytoplasmic, which is incompatible with the data of Seehara and Khorana (1984) on the localization of the N-terminal of the precursor of bacteriorhodopsin. However, if a reverted insertion of the first hairpin into the membrane is admitted—which, in passing, is not possible unless a concomitant insertion of a third hydrophobic segment is accepted—then both the cleavage site of the precursor and the N-terminal of the membrane protein would be cytoplasmically located. Both these corollaries are contradicted, the first by the general localization of the active site of the cleavage enzyme that is extracytoplasmic and the second by the orientation of the N terminus of the mature protein in this case (Ovchinnikov, 1982). Thus, it would seem that neither a loop model nor a hair-pin model for insertion of the precursor of bacteriorhodopsin can account for the reality. With regard to migratory proteins, reservations as to the generality of the mode of insertion as proposed by the unitary hypothesis can also be expressed. Different authors (e.g., see Daum *et al.*, 1982; Gasser *et al.*, 1982b; Hurt and Van Loon, 1986; Kaput *et al.*, 1982) suggested that the N termini of proteins of mitochondrial intermembrane space, inner membrane, and the matrix compartment are oriented extracytoplasmically. Schleyer and Neupert (1985) and Pfanner *et al.* (1987) have shown that the N termini of precursors of some migratory proteins reach the matrix and the extrasequence can be cleaved even under conditions in which part of the molecule is still cytoplasmic. A similar situation was also recently encountered during experiments on translocation of a hybrid protein to chloroplasts (see Kuntz *et al.*, 1987). However, the above proposals and the reported observations can still be accommodated with the mechanism suggested by Singer *et al.* (1987a,b), since the localization of the cleaved sequence is not determined. In fact, in the above cases, it was not demonstrated that cleaved sequences were released in the mitochondrial matrix or the chloroplast stroma, so that a loop-like model as suggested by Singer *et al.* (1987a) still has a chance to be applicable. Nonetheless, the recent demonstration by Hartl et al. (1987a,b) and Pfanner et al. (1987a) that, in the case of a mitochondrial intermembrane compartment protein, the entire precursor is first released in the matrix before being processed (see Figure 5a) cannot be explained on the basis of the suggested mechanism. Also, the path of translocation of ADP-ATP carrier protein (see Figure 5b) does not comply with the unitary mechanism. With regard to chloroplast proteins, the uncleaved precursor of light harvesting chlorophyll a/b binding protein can also be imported, as reported by Chitnis et al. (1986), and even inserted in the thylakoid membrane, partly in an unprocessed precursor form. Chitnis et al. (1987) find in addition that this insertion requires a developmentally regulated stromal factor. Despite these reservations the mechanism proposed by Singer et al. (1987a,b) confirms in essence the view that some sort of active transport is involved, in general, in protein translocation processes and strongly supports the statement made previously (Etémadi, 1985) that not a single case of physiologically significant reconstitution in its restricted sense was realized to date since in such experiments only interaction between the membrane lipids and translocated proteins was sought. In the light of our present knowledge, the minimal requirements formulated previously for a significant reconstitution of migratory proteins (Etémadi, 1985) can be extended to cases of secretory and transmembrane integral proteins. These requirements were emphasized to be the use of precursors of proteins, the translocation of which is examined, the prior reconstitution of receptor–translocators in the liposomal membrane in such a way that they exhibit at least in part a right-side orientation, provision of energy in an appropriate form, and soluble cofactors. In addition, it turns out that the proposal to use an *in vitro* translation system for reconstitution of translocation of secretory and transmembrane integral proteins remains valid, since such systems provide ATP used not only for translation but also for translocation (see in particular Chen and Tai, 1987b). In addition, recent reports regarding the feasibility of the *in vitro* post-translational translocation of such proteins in yeast (Hansen et al., 1986; Rothblatt and Meyer, 1986b; Waters and Blobel, 1986; Wiedmann et al., 1988) encourage designing *in vitro* translocation assays using completed precursors in order to determine the limits of such a process, which, as mentioned above, was questioned (Ibrahimi, 1987) for a case and which presumably should remain exceptional, at least under physiological *in vivo* conditions.

The generalized unitary mechanism of Singer et al. (1987a,b) obviously does not account for translocation of endoproteins, of proteins like apocytochrome c, and of the proteins of the thylakoid membrane and of the luminal compartment of chloroplast thylakoids. It also dismisses the genesis and membrane insertion of transmembrane proteins crossing the membrane thickness once or several times and lacking cleaved N-terminal sequences, but, bound to the membrane with their N-termini located in the extracytoplasmic compartment.

ACKNOWLEDGMENTS. I wish to thank my wife, Raymonde, for her indispensable assistance during the preparation of this manuscript. Financial support during this work was given by the Centre National de la Recherche Scientifique.

8. REFERENCES

Abrahamsen, L., Moks, T., Nilsson, B., Hellman, U., and Uhlen, M., 1985, Analysis of signals for secretion in the staphylococcal protein A gene. *EMBO J.* **4**:3901–3906.

Adams, G. A., and Rose, J. K., 1985a, Structural requirements of a membrane spanning domain for protein anchoring and cell surface transport. *Cell* **41**:1007–1015.

Adams, G. A., and Rose, J. K., 1985b, Incorporation of a charged amino acid into the membrane spanning region blocks cell surface transport but not membrane anchoring of a viral glycoprotein. *Mol. Cell. Biol.* **5**:1442–1448.

Adelman, M. R., Sabatini, D. D., and Blobel, G., 1973, Ribosome–membrane insertion. Nondestructive disassembly of rat liver rough microsomes into ribosomal and microsomal components. *J. Cell Biol.* **56**:206–229.

Ades, I. Z., and Butow, R. A., 1980a, The product of mitochondria-bound cytoplasmic polysomes in yeast. *J. Biol. Chem.* **255**:9918–9924.

Ades, I. Z., and Butow, R. A., 1980b, The transport of proteins into yeast mitochondria, kinetics and pools. *J. Biol. Chem.* **255**:9925–9935.

Adler, L. A., and Arvidson, S., 1987, Correlation between the rate of exoprotein synthesis and the amount of multiprotein complex or membrane-bound ribosome (MBRP-complex) in *Staphylococcus aureus*. *J. Gen. Microbiol.* **133**:803–813.

Adrian, G. S., McCammon, M. T., Montgomory, D. L., and Douglas, M. G., 1986, Sequences required for delivery and localization of ADP/ATP translocator to the mitochondrial inner membrane, *Mol. Cell. Biol.* **6**:626–634.

Ainger, K. J., and Meyer, D. I., 1986, Translocation of nascent secretory proteins across membranes can occur late in translation. *EMBO J.* **5**:951–955.

Akiyama, Y., and Ito, K., 1985, The Sec Y membrane component of the bacterial protein export machinery: Analysis by new electrophoretic methods for integral membrane proteins. *EMBO J.* **4**:3351–3356.

Akiyama, Y., and Ito, K., 1987, Topology of the *Sec* Y protein an integral membrane protein involved in protein export in *Escherichia coli*, *EMBO J.* **6**:3465–3470.

Allison, D. S., and Schatz, G., 1986, Artificial mitochondrial presequences. *Proc. Natl. Acad. Sci. U.S.A.* **83**:9011–9015.

Amar-Costesec, A., Todd, J. A., and Kreibich, G., 1984, Segregation of the polypeptide translocation apparatus to regions of the endoplasmic reticulum containing ribophorin and ribosomes. I. Functional tests on rat liver microsomal subfranctions. *J. Cell Biol.* **99**:2247–2253.

Anderson, D. J., Walter, P., and Blobel, G., 1982, Signal recognition protein is required for the integration of acetylcholine receptor δ subunit, a transmembrane glycoprotein, into the endoplasmic reticulum. *J. Cell Biol.* **93**:501–506.

Anderson, L., 1981, Identification of mitochondrial proteins and some of their precursors in two-dimensional electrophoretic maps of human cells. *Proc. Natl. Acad. Sci. U.S.A.* **78**:2407–2411.

Aoyagi, H., Lee, S., Kanmera, T., Mihara, H., and Kato, T., 1987, Interaction of synthetic fragments of the extension peptide of cytochrome P-450 *(scc)* precursor with phospholipid bilayer, *J. Biochem.* **102**:813–820.

Arakawa, H., Takiguchi, M., Amaya, Y., Nagata, S., Hayashi, H., and Mori, M., 1987, cDNA-derived amino acid sequence of rat mitochondrial 3-oxoacyl-CoA thiolase with no transient presequence: Structural relationship with peroxisomal isoenzyme. *EMBO J.* **6**:1361–1366.

Argan, C., and Shore, G. C., 1985, The precursor of ornithine carbamyl transferase is transported to mitochondria as a 5 S complex containing an import factor. *Biochem. Biophys. Res. Commun.* **131**:289–298.

Argan, C., Lusty, C., and Shore, G., 1982, Nature of surface components on mitochondria which are required for import of ornithine carbamyl transferase. *J. Cell Biol.* **95**(2, Pt. 2): 391a (Abstract).

Argan, C., Lusty, C. J., and Shore, G. C., 1983, Membrane and cytosolic components affecting transport of the precursor of ornithine carbamyl transferase into mitochondria. *J. Biol. Chem.* **258**:6667–6670.

Arinc, E., Rzepecki, L. M., and Strittmatter, P., 1987, Topography of the C terminus of cytochrome b_5 tightly bound to dimyristoyl phosphatidylcholine vesicles, *J. Biol. Chem.* **262**:15563–15567.

Armstrong, J., Nieman, H., Smeekens, S., Rottier, P., and Warren, G., 1984, Sequence and topology of a model intracellular membrane protein E_1 glycoprotin from coronavirus. *Nature (London)* **308**:751–752.

Arpin, M., Matsuura, S., Margoliash, E., Sabatini, D. D., and Morimoto, T., 1980, A segment of cytochrome c containing information for the uptake of newly synthesized cytochrome polypeptides by mitochondria. *Eur. J. Cell Biol.* **22**:152 (Abstract M 451).

Audigier, Y., Friedlander, M., and Blobel, G., 1987, Multiple topogenic sequences in bovine opsin, *Proc. Natl. Acad. Sci. USA* **84**:5783–5787.

Austen, B. M., 1979, Predicted secondary structures of amino-terminal extension sequences on secreted proteins. *FEBS. Lett.* **103**:308–313.

Austen, B. M., and Ridd, D. H., 1983, Studies on the binding of a synthetic signal peptide to pancreatic rough microsomal vesicles. *Biochem. Soc. Trans.* **11**:160–161.

Austen, B. M., Hermon-Taylor, J., Kaderbhai, M. A., and Ridd, D. H., 1984, Design and synthesis of a consensus signal sequence that inhibits protein translocation into rough microsomal vesicles. *Biochem. J.* **224**:317–325.

Bacallao, R., Crooke, E., Shiba, K., Wickner, W., and Ito, K., 1986, The Sec Y protein can act post-translationally to promote bacterial protein export. *J. Biol. Chem.* **261**:12907–12910.

Baker, A., and Schatz, G., 1987, Sequences from a prokaryotic genome on the mouse dihydrofolate reductase gene can restore the import of a truncated precursor protein into yeast mitochondria. *Proc. Natl. Acad. Sci. U.S.A.* **84**:3117–3121.

Bakker, E. P., and Randall, L. L., 1984, The requirement for energy during export of β-lactamase in *Escherichia coli* is fulfilled by the total protonmotive force. *EMBO J.* **3**:895–900.

Bancroft, C. F., Sussman-Berger, P., and Dobner, P. R., 1980, Biosynthesis of rat growth hormone and its messenger RNA. *Ann. N.Y. Acad. Sci.* **343**:56–68.

Bankaitis, V. A., and Bassford, P. J. Jr., 1985a, Proper interaction between at least two components is required for efficient export of proteins to *Escherichia coli* cell envelope. *J. Bacteriol.* **161**:169–178.

Bankaitis, V. A., and Bassford, P. J. Jr., 1985b, Sequences within the mature maltose-binding protein of *Escherichia coli* may be actively involved in initiating the export process. *Ann. Inst. Pasteur Microbiol.* **136B**:3–7.

Bankaitis, V. A., Rasmussen, B. A., and Bassford, P. J. Jr., 1984, Intragenic suppressor mutations that restore export of maltose-binding protein with a truncated signal peptide. *Cell* **37**:243–252.

Bar-Nun, S., Kreibich, G., Adesnik, M., Alterman, L., Negishi, M., and Sabatini, D. D., 1980, Synthesis and insertion of cytochrome P-450 into endoplasmic reticulum. *Proc. Natl. Acad. Sci. U.S.A.* **77**:965–969.

Bassford, P. J. Jr., Silhavy, T. J., and Beckwith, J., 1979, Use of gene fusion to study secretion of maltose-binding protein in *E. coli* periplasm. *J. Bacteriol.* **139**:19–31.

Bassuener, R., Wobus, U., and Rapoport, T. A., 1984, Signal recognition particle triggers the translocation of storage globulin polypeptides from field beans *(Vicia faba L.)* across mammalian endoplasmic reticulum membrane. *FEBS Lett.* **166**:314–320.

Becker, W. M., Riezman, H., Weir, E. M., Titus, D. E., and Leaver, C. J., 1982, In vitro synthesis and compartmentation of glyoxysomal enzymes from cucumber. *Ann. N.Y. Acad. Sci.* **386**:329–349.

Bedbrook, J. R., Smith, S. M., and Ellis, R. J., 1980, Molecular cloning and sequencing of cDNA encoding the precursor to the small subunit of chloroplast ribulose-1,5-bisphosphate carboxylase. *Nature (London)* **287**:692–697.

Bedouelle, H., Bassford, P. J. Jr., Fowler, A. V., Zabin, I., Beckwith, J., and Hofnung, M., 1980, Mutations which alter the function of the signal sequence of the maltose binding protein of *Escherichia coli*. *Nature (London)* **285**:78–81.

Beevers, H., 1979, Microbodies in higher plants. *Annu. Rev. Plant Physiol.* **30**:159–193.

Beevers, H., and Gonzalez, E., 1987, Proteins and phospholipids of glyoxysomal membranes from castor bean, *Methods Enzymol.* **148**:528–532.

Behra, R., and Christen, P., 1986, In vitro import into mitochondria of the precursor of mitochondrial aspartate aminotransferase. *J. Biol. Chem.* **261**:257–263.

Bell, G. I., Swain, W. F., Pictet, R., Cordell, B., Goodman, H. M., and Rutter, W. J., 1979, Nucleotide sequence of a cDNA clone encoding human preproinsulin. *Nature (London)* **282**:525–527.

Bergelson, L. D., and Barsukov, L. I., 1977, Topological asymmetry of phospholipids in membranes. The distribution of phospholipids in biological membranes is related to that in bilayer membranes of small vesicles. *Science* **179**:224–230.

Bernstein, M., Hoffmann, W., Ammerer, G., and Schekman, R., 1985, Characterization of a gene product (Sec 53 p) required for protein assembly in yeast endoplasmic reticulum. *J. Cell Biol.* **101**:2374–2382.

Bhat, N. K., and Avadhani, N. G., 1985, Transport of proteins into hepatic and nonhepatic mitochondria: Specificity of uptake and processing of precursor form of carbamyl phosphate synthetase. *Biochemistry* **24**:8107–8113.

Bickman, E. R., Oliver, D. B., Garwin, J. L., Kumamoto, C., and Beckwith, J., 1984, The use of extragenic suppressors to define genes involved in protein export in *Escherichia coli*. *Mol. Gen. Genet.* **196**:24–27.

Bielinska, R. M., Rogers, G., Rucinsky, T., and Boime, I., 1979, Processing in vitro of placental peptide hormones by smooth microsomes. *Proc. Natl. Acad. Sci. U.S.A.* **76**:6152–6156.

Bingham, R. W., and Campbell, P. B., 1972, Studies on the biosynthesis of mitochondrial malate dehydrogenase and the location of its synthesis in liver cell of the rat. *Biochem. J.* **126**:211–215.

Birrel, G. B., and Griffith, O. H., 1976, Cytochrome *c* induced lateral phase separation in a diphosphatylglycerol-steroid spin-label model membrane. *Biochemistry* **95**:2925–2929.

Bitsch, A., and Kloppstech, K., 1986a, Reconstitution of the solubilized envelope receptors for nuclear-coded precursor proteins. *Plant Biol.* **2**:241–246.

Bitsch, A., and Kloppstech, K., 1986b, Transport of proteins into chloroplasts. Reconstitution of the binding capacity for nuclear-coded precursor proteins after solubilization of envelope with detergents. *Eur. J. Cell Biol.* **40**:160–166.

Blachly-Dyson, E., and Stevens, T. H., 1987, Yeast carboxypeptidase Y can be translocated and glycosylated without its aminoterminal signal sequence. *J. Cell Biol.* **104**:1183–1191.

Black, S. D., French, J. S., William, C. H. Jr., and Coon, M. J., 1979, Role of hydrophobic polypeptide in the N-terminal region of NADPH–cytochrome P-450 reductase in complex formation with P-450 LM. *Biochem. Biophys. Res. Commun.* **91**:1528–1535.

Blobel, G., 1980, Intracellular protein topogenesis. *Proc. Natl. Acad. Sci. U.S.A.* **77**:1496–1500.

Blobel, G., and Dobberstein, B., 1975, Transfer of proteins across membranes. II. Reconstitution of functional rough microsomes from heterologous components. *J. Cell Biol.* **67**:852–862.

Blobel, G., and Sabatini, D. D., 1971, Ribosome–membrane interaction in eukaryotic cells, in *Biomembranes* (L. A. Manson, ed.), Vol. 2, pp. 193–195, Plenum Press, N.Y.

Boeke, J. D., and Model, P., 1982, A prokaryotic membrane anchor sequence: Carboxyl terminus of bacteriophage f_1 gene III protein retains it in the membrane. *Proc. Natl. Acad. Sci. U.S.A.* **79**:5200–5204.

Boime, I., Szczesna, E., and Smith, D., 1977, Membrane dependent cleavage of the human placental lactogen precursor to its native form in ascites cell-free extract. *Eur. J. Biochem.* **73**:515–520.

Bonatti, S., and Blobel, G., 1979, Absence of a cleavable signal sequence in Sindbis virus glycoprotein PE_2. *J. Biol. Chem.* **254**:12261–12264.

Bonatti, S., Migliaccio, G., Blobel, G., and Walter, P., 1984, Role of signal recognition particle in the membrane assembly of Sindbis viral glycoproteins. *Eur. J. Biochem.* **140**:499–502.

Borgese, N., and Gaetani, S., 1982, Biosynthesis and post-translational insertion into membranes of rat liver NADH–cytochrome b_5 reductase. *J. Cell Biol.* **95**(2, Pt. 2):399a (abstract).

Bos, T. J., Davis, A. R., and Nayak, D. P., 1984, NH_2 terminal hydrophobic region of influenza virus neuraminidase provides the signal function in translocation. *Proc. Natl. Acad. Sci. U.S.A.* **81**:2327–2331.

Boutry, M., Nagy, F., Poulsen, C., Aoyagi, K., and Chua, N. H., 1987, Targeting of bacterial chloramphenicol acetyltransferase to mitochondria in transgenic plants. *Nature (London)* **328**:340–342.

Bremer, E., Cole, S., Hindennach, I., Henning, U., Beck, E., Kurz, C., and Schaller, H., 1982, Export of a protein into the outer membrane of *Escherichia coli*. Stable incorporation of OmpA protein requires less than 193 amino terminal acid residues. *J. Biochem.* **122**:223–231.

Bretscher, M. S., 1973, Membrane structure: Some general principles. *Science* **181**:622–629.

Briggs, M. S., and Gierasch, L. M., 1984, Exploring the conformational roles of sequences: Synthesis and conformational analysis of λ receptor protein wild-type and mutant signal peptides. *Biochemistry* **23**:3111–3114.

Briggs, M. S., Gierasch, L. M., Zlotnick, A., Lear, J. D., and De Grado, W. F., 1985, In vivo function and membrane binding properties are correlated for *Escherichia coli* Lam B signal peptides. *Science* **228**:1096–1099.

Briggs, M. S., Cornell, D. G., Dluhy, R. A., and Gierasch, L. M., 1986, Conformation of signal peptides induced by lipids suggest initial steps in protein export. *Science* **233**:206–208.

Brown, D. A., and Simoni, R. S., 1984, Biogenesis of 3-hydroxy-3-methyl glutaryl coenzyme A reductase, an integral glycoprotein of the endoplasmic reticulum. *Proc. Natl. Acad. Sci. U.S.A.* **81**:1674–1678.

Brown, L. R., and Wütrich, K., 1977, NMR and ESR studies of the interactions of cytochrome *c* with mixed cardiolipin–phosphatidylcholine vesicles. *Biochim. Biophys. Acta* **468**:389–410.

Brown, P. A., Halvorson, H. O., Raney, P., and Perlman, D., 1984, Conformational alterations in the proximal portion of the yeast invertase signal peptide do not block secretion. *Mol. Gen. Genet.* **197**:351–357.

Brusilow, W. S. A., Gunsalus, R. P., Hardeman, E. C., Decker, K. P., and Simoni, R. D., 1981, In vitro synthesis of F_0 and F_1 components of the proton translocating ATPase of *Escherichia coli*. *J. Biol. Chem.* **256**:3141–3144.

Büchel, D. E., Gronenbord, B., and Müller-Hill, B., 1980, Sequences of the lactose permease gene. *Nature (London)* **283**:541–545.
Burns, D., and Lewin, A., 1986, Inhibition of the import of mitochondrial proteins by RNase. *J. Biol. Chem.* **261**:6155–6163.
Burr, F. A., and Burr, B., 1981, In vitro uptake and processing of prezein and other maize proteins by maize membranes. *J. Cell Biol.* **90**:427–434.
Burstein, V., and Schechter, I., 1978, Primary structure of N-terminal peptide segment linked to the variable and constant regions of immunoglobulin light chain precursors: Implication of the organization and controlled expression of immunoglobulin genes. *Biochemistry* **17**:2392–2400.
Campbell, M. T., Sutton, R., and Pollak, J. K., 1982, Import of carbamoylphosphate synthetase into mitochondria from fetal rat liver. *Eur. J. Biochem.* **125**:401–406.
Cancedda, R., and Schlesinger, M. J., 1974, Localization of polyribosomes containing alkaline phosphatase nascent polypeptides on membranes of *Escherichia coli*. *J. Bacteriol.* **117**:290–301.
Caplan, M. J., Palade, G. E., and Jamieson, J. D., 1986, Newly synthesized sodium–potassium ATPase α-subunit has no cytosolic intermediate in MDCK cells. *J. Biol. Chem.* **261**:2860–2865.
Cardoza, J. D., Kleinfeld, A. M., Stallcup, K. C., and Mescher, M. F., 1984, Hairpin configuration of H-2K^k in liposomes formed by detergent dialysis. *Biochemistry* **23**:4401–4409.
Carlson, M., and Botstein, D., 1982, Two differently regulated mRNAs with different 5′ ends encode secreted and intracellular forms of yeast invertase. *Cell* **28**:145–154.
Cascarano, J., Montisano, D. F., Pickett, C. B., and Jones, T. W., 1982, Rough endoplasmic reticulum–mitochondrial complexes from rat liver. *Exp. Cell Res.* **139**:39–50.
Cashmore, A. R., 1983, Nuclear gene encoding the small subunit of ribulose-1,5-bisphosphate carboxylase, *Basic Life Sci.* **26**:29–38.
Caulfield, M. P., Tai, P. C., and Davis, B. D., 1983, Association of penicillin-binding proteins and other enzymes with the ribosome-free membrane fraction of *Bacillus subtilis*. *J. Bacteriol.* **156**:1–5.
Caulfield, M. P., Horiuchi, S., Tai, P. C., and Davis, B. D., 1984, The 64-kilodalton membrane protein of *Bacillus subtilis* is also present as a multicomplex on membrane free ribosomes. *Proc. Natl. Acad. Sci. U.S.A.* **81**:7772–7776.
Caulfield, M. P., Furlong, D., Tai, P. C., and Davis, B. D., 1985, Secretory S complex of *Bacillus subtilis* forms a large organized structure when released from ribosomes. *Proc. Natl. Acad. Sci. U.S.A.* **82**:4031–4035.
Caulfield, M. P., Duong, L. T., and Rosenblatt, M., 1986, Demonstration of post-translational secretion of human placental lactogen by a mammalian *in vitro* translation system. *J. Biol. Chem.* **261**:10953–10956.
Chamberlain, B. K., Nozaki, Y., Tanford, C., and Webster, R. E., 1978, Association of the major coat protein of fd bacteriophage with phospholipid vesicles. *Biochim. Biophys. Acta* **510**:18–37.
Chapman, D., and Urbina, J., 1971, Phase transition and bilayer structure of *Mycoplasma laidlawii* B. *FEBS Lett.* **12**:169–172.
Chen, L., and Tai, P. C., 1985, ATP is essential for protein translocation into *Escherichia coli* membrane vesicles. *Proc. Natl. Acad. Sci. U.S.A.* **82**:4384–4388.
Chen, L., and Tai, P. C., 1987a, Effects of antibodies and other inhibitors on ATP-dependent protein translocation into membrane vesicles. *J. Bacteriol.* **169**:2373–2379.
Chen, L., and Tai, P. C., 1987b, Evidence for the involvement of ATP in co-translational protein translocation. *Nature (London)* **328**:164–166.
Chen, L., Rhoads, D., and Tai, P. C., 1985, Alkaline phosphatase and OmpA protein can be

translocated post-translationally into membrane vesicles of *Escherichia coli. J. Bacteriol.* **161:**973–980.
Chen, L., Tai, P. C., Briggs, M. S., and Gierasch, L. M., 1987, Protein translocation into *Escherichia coli* membrane vesicles is inhibited by functional synthetic signal peptides. *J. Biol. Chem.* **262:**1427–1429.
Chen, W. J., and Douglas, M. G., 1987a, Phosphodiester bond cleavage outside mitochondria is required for the completion of protein import into mitochondrial matrix. *Cell* **49:**651–658.
Chen, W. J., and Douglas, M. G., 1987b, The role of protein structure in the mitochondrial import pathway. Unfolding of mitochondrially bound precursors is required for membrane translocation, *J. Biol. Chem.* **262:**15605–15609.
Cheng, M. Y., Pollock, R. A., Hendrick, J. A., and Horwich, A. L., 1987, Import and processing of human ornithine transcarbamylase precursor by mitochondria from *Saccharomyces cerevisiae. Proc. Natl. Acad. Sci. U.S.A.* **84:**4063–4067.
Chia, C. P., and Arntzen, C. J., 1986, Evidence for two step processing of nuclear encoded chloroplast protein during membrane assembly. *J. Cell Biol.* **103:**725–731.
Chien, S. M., and Freeman, K. B., 1983, Synthesis and uptake of precursor forms of mammalian mitochondrial proteins. *Fed. Proc.* **42:**2125 (Abstract).
Chien, S. M., Patel, H. V., and Freeman, K. B., 1984, Import of rat liver mitochondrial malate dehydrogenase. Binding of the precursor to mitochondria, an intermediate stage in import. *J. Biol. Chem.* **259:**13633–13636.
Chiocchia, K. B., and Dirckamer, K., 1984, Direct evidence for the transmembrane orientation of the hepatic glycoprotein receptors. *J. Biol. Chem.* **259:**15440–15446.
Chitnis, P. R., Hartl, E., Kohorn, B. L., Tobin, E. M., and Thornber, J. P., 1986, Assembly of the precursor and processed light-harvesting chlorophyll a/b protein of *Lemna* into the light-harvesting complex II of barley etioplasts, *J. Cell. Biol.* **102:**982–988.
Chitnis, P., Nechushtai, R., and Thornber, J. P., 1987, Insertion of the precursor of the light-harvesting chlorophyll a/b protein into the thylakoid requires the presence of a developmentally regulated stromal factor, *Plant Mol. Biol.* **10:**3–11.
Christiansen, K., and Carlsen, J., 1985, Reconstitution of cytochrome b_5 into lipid vesicles in a form which is nonsusceptible to attack by carboxypeptidase Y. *Biochim. Biophys. Acta* **815:**215–222.
Chu, T. W., Grand, P. M., and Strauss, A. W., 1987a, Mutation of a neutral amino acid in the transit peptide of rat mitochondrial malate dehydrogenase abolishes binding and import, *J. Biol. Chem.* **262:**15759–15764.
Chu, T. W., Grant, P. M., and Strauss, A. W., 1987b, The role of arginine residues in the rat mitochondrial malate dehydrogenase transit peptide, *J. Biol. Chem.* **262:**12806–12811.
Chua, N. H., and Schmidt, G. W., 1978, Post-translational transport into intact chloroplasts of a precursor to the small subunit of ribulose 1,5-bisphosphate carboxylase. *Proc. Natl. Acad. Sci. U. S.A.* **75:**6110–6114.
Chua, N. H., and Schmidt, G. W., 1979, Transport of proteins into mitochondria and chloroplasts. *J. Cell Biol.* **81:**461–483.
Chyn, T. L., Martonosi, A. N., Morimoto, T., and Sabatini, D. D., 1979, In vitro synthesis of the Ca^{2+} transport ATPase by ribosomes bound to sarcoplasmic reticulum membranes. *Proc. Natl. Acad. Sci. U.S.A.* **76:**1241–1245.
Claesson, L., Larhammer, D., Rask, L., and Peterson, P. A., 1983, cDNA clone for human invariant γ chain of class II histocompatibility and its implications for protein structure. *Proc. Natl. Acad. Sci. U.S.A.* **80:**7395–7399.
Cline, K., 1986, Import of proteins into chloroplasts. Membrane integration of a thylakoid precursor protein reconstituted in chloroplast lysates. *J. Biol. Chem.* **261:**14804–14810.

Cline, K., Werner-Washburne, M., Lubben, T. H., and Keegstra, K., 1985, Precursors to two nuclear-encoded chloroplast proteins bind to the outer envelope membrane before being imported into chloroplasts. *J. Biol. Chem.* **260**:3691-3696.

Coleman, J., Inukai, M., and Inouye, M., 1985, Dual functions of signal peptide in protein transfer across the membrane. *Cell* **43**:351-360.

Collier, D. N., Bankaitis, V. A., Weiss, J. B., and Bassford, P. J., Jr., 1988, The antifolding activity of Sec B promotes the export of *E. coli* maltose-binding protein, *Cell* **53**:273-283.

Cornwell, K. L., and Keegstra, K., 1987, Evidence that a chloroplast surface protein is associated with a specific binding site for the precursor to the small subunit of ribulose 1,5-bisphosphate carboxylase, *Plant. Physiol.* **85**:760-785.

Coruzzi, G., Brogli, R., Cashmore, A., and Chua, N. H., 1983, Nucleotide sequences of two pea cDNA clones encoding the small subunit of ribulose 1,5-bisphosphate carboxylase and the major chlorophyll a/b-binding thylakoid membrane polypeptide. *J. Biol. Chem.* **258**:1399-1402.

Côté, C., and Boulet, D., 1985, Differential import and processing of the precursors to F_1-ATPase β-subunit and ornithine carbamyl transferase by liver, spleen, heart and kidney mitochondria. *Biochem. Biophys. Res. Commun.* **129**:240-247.

Cover, W. H., Ryan, J. P., Bassford, P. J. Jr., Walsh, K. A., Bollinger, J., and Randall, L. L., 1987, Suppression of a signal sequence mutation by an amino acid substitution in the mature portion of the maltose-binding protein. *J. Bacteriol.* **169**:1794-1800.

Crane, D., Holmes, R., and Masters, C. J., 1982a, Synthesis and incorporation of phospholipid by peroxisomes of mouse liver. *Biochim. Biophys. Acta* **712**:57-64.

Crane, D. I., Holmes, R. S., and Masters, C. J., 1982b, Proteolytic modification of mouse liver catalase. *Biochem. Biophys. Res. Commun.* **104**:1567-1572.

Crane, D. I., Holmes, R. S., and Masters, C. J., 1983, On the synthesis and incorporation of catalase and urate oxidase into the peroxisomes of mouse liver. *Int. J. Biochem.* **15**:1429-1437.

Crimaudo, C., Hortsch, M., Gausephol, H., and Meyer, D. I., 1987, Human ribophorins I and II: the primary structure and membrane topology of two highly conserved rough endoplasmic reticulum-specific glycoproteins, *EMBO J.* **6**:75-82.

Culter, D. F., and Garoff, H., 1986a, Mutants of the membrane binding region of Semliki forest virus E_2 protein. I. Cell surface transport and fusogenic activity. *J. Cell Biol.* **102**:889-901.

Culter, D. F., Melancon, P., and Garoff, H., 1986b, Mutants of membrane binding region of Semliki forest virus E_2 protein. II. Topology and membrane binding. *J. Cell Biol.* **102**:902-910.

Dailey, H. A., and Strittmatter, P., 1981a, The role of COOH-terminal anionic residues in binding cytochrome b_5 to phospholipid vesicles and biological membranes. *J. Biol. Chem.* **256**:1677-1680.

Dailey, H. A., and Strittmatter, P. 1981b, Orientation of the carboxyl and NH_2 termini of the membrane-binding segment of cytochrome b_5 on the same side of phospholipid bilayers. *J. Biol. Chem.* **256**:3951-3955.

Dalbey, R. E., and Wickner, W., 1986, The role of the polar, carboxyl terminal domain of *Escherichia coli* leader peptidase in its translocation across the plasma membrane. *J. Biol. Chem.* **261**:13844-13849.

Dalbey, R. E., and Wickner, W., 1988, Characterization of the internal signal anchor domain of *Escherichia coli* leader peptidase, *J. Biol. Chem.* **263**:404-408.

Daniels, C. J., Bole, D. G., Quay, S. C., and Oxender, D. L., 1981, Role of membrane potential in the secretion of proteins into the periplasm of *Escherichia coli*. *Proc. Natl. Acad. Sci. U.S.A.* **78**:5396-5400.

Date, T., Goodman, J. M., and Wickner, W. T., 1980, Procoat, the precursor of M_{13} coat protein, requires an electrochemical potential for membrane insertion. *Proc. Natl. Acad. Sci. U.S.A.* **77**:4669–4673.

Daum, G., Gasser, S. M., and Schatz, G., 1982, Import of proteins into mitochondria. Energy dependent, two step processing of the intermembrane space enzyme cytochrome b_2 by isolated yeast mitochondria. *J. Biol. Chem.* **257**:13075–13080.

Davis, B. D., and Tai, P. C., 1980, The mechanisms of protein secretion across membranes. *Nature (London)* **283**:433–438.

Davis, N. G., and Model, P., 1985, An artificial anchor domain hydrophobicity suffices to stop transfer. *Cell* **41**:607–614.

Davis, N. G., Boeke, J. D., and Model, P., 1985, Fine structure of membrane anchor domain. *J. Mol. Biol.* **181**:111–121.

De Geus, P., Vehreij, H. M., Riegman, N. H., Hoekstra, W. P. M., and de Haas, G. P., 1984, The pro- and mature forms of the *Escherichia coli* K-12 outer membrane phospholipase A are identical. *EMBO J.* **3**:1799–1802.

De Haan, M., Van Loon, A. P. G. M., Kreike, J., Vaessen, R. T. M. J., and Grivell, L. A., 1983, The biosynthesis of the ubiquinol–cytochrome c reductase complex in yeast. DNA sequence analysis of the nuclear gene coding for the 14 kDa subunit. *Eur. J. Biochem.* **138**:169–177.

De Lemos-Chiarandini, C., Frey, A. B., Sabatini, D. D., and Kreibich, G., 1987, Determination of membrane topology of the phenobarbital-induced cytochrome P-450 isoenzyme PB-4 using site specific antibodies, *J. Cell. Biol.* **104**:209–219.

Della-Cioppa, G., Bauer, S. C., Klein, B. K., Shah, D. M., Fraley, R. T., and Kishore, G. M., 1986, Translocation of the precursor of 5-enoylpyruvoylshikimate-3-phosphate synthase into chloroplasts of higher plants *in vitro*. *Proc. Natl. Acad. Sci. U.S.A.* **83**:6873–6877.

Della-Cioppa, G., Bauer, S. C., Taylor, M. L., Rochester, D. E., Klein, B. K., Shah, D. M., Fraley, R. T. and Kishore, G. M., 1987, Targeting a herbicide-resistant enzyme from *Escherichia coli* to chloroplasts of higher plants. *Bio/Technology* **5**:579–584.

Desel, H., Zimmermann, R., Janes, M., Miller, F., and Neupert, W., 1982, Biosynthesis of glyoxysomal enzymes in *Neurospora crassa*. *Ann. N.Y. Acad. Sci.* **386**:377–388.

Deshaies, R. J., and Schekman, R., 1987, A yeast mutant defective at an early stage in import of secretory protein precursors into the endoplasmic reticulum, *J. Cell. Biol.* **105**:633–645.

Devillers-Thiery, A., Kindt, T., Scheele, G., and Blobel, G., 1975, Homology in amino-terminal sequence of precursors to pancreatic secretory proteins. *Proc. Natl. Acad. Sci. U.S.A.* **72**:5016–5020.

DiRienzo, J. M., and Inouye, M., 1979, Lipid fluidity dependent biosynthesis and assembly of outer membrane proteins of *E. coli*. *Cell* **17**:155–161.

DiRienzo, J. M., Nakamura, K., and Inouye, M., 1978, The membrane proteins of gram negative bacteria: Biosynthesis assembly and function. *Annu. Rev. Biochem.* **47**:481–552.

Distel, B., Veenhuis, M., and Tabak, H. F., 1987, Import of alcohol oxidase into peroxisomes of *Saccharomyces cerevisiae*, *EMBO J.* **6**:3111–3116.

Dorbani, L., Janczic, V., Linden, M., Leterrier, J. F., Nelson, B. D., and Rendon, A., 1987, Subfractionation of the outer membrane of rat brain mitochondria: Evidence for the existence of a domain containing the porin hexokinase complex. *Arch. Biochem. Biophys.* **252**:188–196.

Douglas, M. G., Geller, B. L., and Emr, S. D., 1984, Intracellular targeting and import of an F_1-ATPase-subunit-β-galactosidase hybrid protein into yeast mitochondria, *Proc. Natl. Acad. Sci. U.S.A.* **81**:3983–3987.

Douma, A. C., Veenhuis, M., Sulter, G. J., and Harder, W., 1987, A proton translocating adenosine triphosphatase is associated with the peroxisomal membrane of yeasts. *Arch. Microbiol.* **147**:42–47.

Dumont, M. E., and Richards, F. M., 1984, Insertion of apocytochrome c into lipid vesicles. *J. Biol. Chem.* **259**:4147–4156.

Duong, L. T., Caulfield, M. P., and Rosenblatt, M., 1987, Synthetic signal peptide and analogs display different activities in mammalian and plant *in vitro* secretion synthesis. *J. Biol. Chem.* **262**:6328–6333.

Ehring, R., Beyreuther, K., Wright, J. K., and Overath, P., 1980, *In vitro* and *in vivo* products of *E. coli* lactose permease gene are identical. *Nature (London)* **283**:537–540.

Eilers, M., Hwang, S., and Schatz, G., 1988, Unfolding and refolding of a purified precursor protein during import into isolated mitochondria, *EMBO J.* **7**:1139–1145.

Eilers, M., and Schatz, G., 1986, Binding of a specific ligand inhibits import of a purified precursor protein into mitochondria. *Nature (London)* **382**:228–232.

Eilers, M., Opplinger, W., and Schatz, G., 1987, Both ATP and energized inner membrane are required to import a purified precursor protein into mitochondria. *EMBO J.* **6**:1073–1077.

Emr, S. D., and Bassford, P. J. Jr., 1982, Localization and processing of outer membrane and periplasmic proteins in *Escherichia coli* strains harboring export specific suppressor mutations. *J. Biol. Chem.* **257**:5852–5860.

Emr, S. D., and Silhavy, T. J., 1982, Molecular components of the signal sequence that function in the initiation of protein export. *J. Cell Biol.* **95**:689–696.

Emr, S. D., and Silhavy, T. J., 1983, Importance of secondary structure in the signal sequence for protein secretion. *Proc. Natl. Acad. Sci. U.S.A.* **80**:4599–4603.

Emr, S. D., Hall, M. N., and Silhavy, T. J., 1980, Mechanism of protein localization: The signal hypothesis and bacteria. *J. Cell Biol.* **86**:701–711.

Emr, S. D., Hanley-Way, S., and Silhavy, T. J., 1981, Suppressor mutations which restore export of a protein with a defective signal sequence. *Cell* **23**:79–88.

Emr, S., Vassarotti, A., Garret, J., Geller, B., Takeda, M., and Douglas, M., 1986, The amino terminus of the yeast F_1-ATPase β-subunit precursor functions as a mitochondrial import signal. *J. Cell Biol.* **102**:523–533.

Endo, T., and Schatz, G., 1988, Latent membrane perturbation activity of a mitochondrial precursor protein is exposed by unfolding, *EMBO J.* **7**:1153–1158.

Enequist, H. G., Hirst, T. R., Harayama, S., Hardy, S. J. S., and Randall, L. L., 1981, Energy is required for maturation of exported proteins in *Escherichia coli*. *Eur. J. Biochem.* **116**:227–233.

Engelman, D. M., and Steitz, T. A., 1981, The spontaneous insertion of proteins into and across membranes; the helical hairpin hypothesis. *Cell* **23**:411–422.

Enoch, H. G., Fleming, P. J., and Strittmatter, P., 1979, The binding of cytochrome b_5 to phospholipid vesicles and biological membranes. Effect of orientation on intermembrane transfer and digestion by carboxypeptidase Y. *J. Biol. Chem.* **254**:6483–6488.

Enosawa, S., and Ohashi, A., 1986, Localization of enzyme for heme attachment to apocytochrome c in yeast mitochondria. *Biochem. Biophys. Res. Commun.* **141**:1145–1150.

Epand, R. M., Hui, S. M., Argan, C., Gillespie, L. L., and Shore, G. C., 1986, Structural analysis and amphiphilic properties of a chemically synthesized mitochondrial signal peptide. *J. Biol. Chem.* **261**:10017–10020.

Erickson, A. H., Walter, P., and Blobel, G., 1983, Translocation of a lysosomal enzyme across the microsomal membrane requires signal recognition particles. *Biochem. Biophys. Res. Commun.* **115**:275–280.

Eskridge, E. M., and Shields, D., 1982, Cell-free processing of small polypeptide hormones is a co-translational event. *J. Cell Biol.* **95**:394a (Abstract).

Etémadi, A.-H., 1980a, Membrane asymmetry. A survey and critical appraisal of the methodology. I. Methods for assessing the asymmetric orientation and distribution of proteins. *Biochim. Biophys. Acta* **604**:347–422.

Etémadi, A.-H., 1980b, Membrane asymmetry. A survey and critical appraisal of the methodology. II. Methods for assessing the unequal distribution of lipids. *Biochim. Biophys. Acta* **604:**423–475.

Etémadi, A.-H., 1980c, Tendances organisationnelles des constituants des membranes biologiques et les problèmes de l'asymétrie de leur distribution. *Biochimie* **62:**111–134.

Etémadi, A.-H., 1985, Functional and orientational features of protein molecules in reconstituted lipid membranes. *Adv. Lipid Res.* **21:**281–428.

Etémadi, A.-H., 1986a, Mechanism of biogenesis and translocation of cell surface membrane proteins, in *Handbook of Plasma Membranes*, CRC Press, Boca Raton, FL.

Etémadi, A.-H., 1986b, Intracellular path and maturation of cell surface proteins, in *Handbook of Plasma Membranes*, CRC Press, Boca Raton, FL.

Etémadi, A.-H., 1987, Reconstitution and cellular protein translocation processes, in: *International Symposium Membranes Lipids, Metabolism Organization*, Varna, Abstract lecture 1-5.

Etémadi, A.-H., 1988, Overview on protein translocation and membrane insertion processes and conditions for their physiologically significant reconstitution, in: *International Conference on Biomembranes in Health and Disease*, Lucknow, Abstract Lecture 43.

Evans, E. A., Gilmore, R., and Blobel, G., 1986, Purification of microsomal signal peptidase as a complex. *Proc. Natl. Acad. Sci. U.S.A.* **83:**581–585.

Fecycz, I. T., and Blobel, G., 1987, Soluble factors stimulating secretory protein translocation in bacteria and yeast can substitute for each other. *Proc. Natl. Acad. Sci. U.S.A.* **84:**3723–3727.

Fenton, W. A., Hack, A. M., Helfgott, D., and Rosenberg, L. E., 1984, Biogenesis of mitochondrial enzyme methylmalonyl CoA mutase. Synthesis and processing of a precursor in a cell-free system and in cultured cells. *J. Biol. Chem.* **259:**6616–6621.

Ferro-Novick, S., Novick, P., Field, C., and Schekman, R., 1984a, Yeast secretory mutants that block formation of active cell surface enzymes. *J. Cell Biol.* **98:**35–43.

Ferro-Novick, S., Hausen, W., Sehaver, I., and Schekman, R., 1984b, Genes required for completion of import of proteins into the endoplasmic reticulum in yeast. *J. Cell Biol.* **98:**44–53.

Fidelio, G. D., Austen, B. M., Chapman, D., and Lucy, J. A., 1986, Properties of signal sequence peptides at an air–water interface. *Biochem. J.* **238:**301–304.

Fidelio, G. D., Austen, B. M., and Chapman, D., and Lucy, J. A., 1987, Interaction of ovalbumin and its putative signal sequence with phospholipid monolayers. Possible importance of differing lateral stabilities in protein translocation. *Biochem. J.* **244:**295–301.

Fields, S., Winter, G., and Brownlee, G. G., 1981, Structure of neuraminidase gene in human influenza virus A/PR/8/34. *Nature (London)* **230:**213–217.

Fikes, J. D., Bankaitis, V. A., Ryan, J. P., and Bassford, P. J., Jr., 1987, Mutational alterations affecting the export competence of a truncated but fully functional maltose-binding protein signal peptide, *J. Bacteriol.* **169:**2345–2351.

Fikes, J. D., and Bassford, P. J. Jr., 1987, Export of unprocessed precursor maltose-binding protein to the periplasm of *Escherichia coli* cells. *J. Bacteriol.* **169:**2352–2359.

Finidori, J., Rizzolo, L., Gonzalez, A., Kreibich, G., Adesnik, M., and Sabatini, D. D., 1987, The influenza hemagglutinin insertion signal is not cleaved and does not halt translocation when presented to the endoplasmic reticulum membrane as part of a translocating polypeptide. *J. Cell Biol.* **104:**1705–1714.

Firguira, F. A., Hendrick, J. P., Kalousek, F., Kraus, J. P., and Rosenberg, L. E., 1984, RNA required for import of precursor proteins into mitochondria. *Science* **226:**1319–1322.

Fitts, R., Reuveny, Z., Van Amsterdam, J., Mulholland, J., and Botstein, D., 1987, Substitution of tyrosine for either cysteine in β-lactamase prevents release from the membrane during secretion, *Proc. Natl. Acad. Sci. USA* **84:**8540–8543.

Fluegg, V. I., and Hinz, G., 1986, Energy dependence of protein translocation into chloroplasts. *Eur. J. Biochem.* **160:**563–570.

Freitag, H., Janes, M., and Neupert, W., 1982a, Biosynthesis of mitochondrial porin and insertion into the outer mitochondrial membrane of *Neurospora crassa*. *Eur. J. Biochem.* **126:**197–202.

Freitag, H., Neupert, W., and Benz, R., 1982b, Purification and characterization of a pore protein of the outer mitochondrial membrane from *Neurospora crassa*. *Eur. J. Biochem.* **123:**629–639.

Freudl, R., Braun, G., Hindennach, I., and Henning, U., 1985, Lethal mutations in the structural gene of an outer membrane protein (OmpA) of *Escherichia coli* K-12. *Mol. Gen. Genet.* **210:**76–81.

Freudl, R., MacIntyre, S., Degen, M., and Henning, U., 1988, Alterations to the signal peptide of an outer membrane protein (OmpA) of *Escherichia coli* K_{12} can promote either the co-translational or the post-translational mode of processing, *J. Biol. Chem.* **262:**344–349.

Freudl, R., Schwartz, H., Stierhof, Y. D., Gamon, K., Hindennach, I., and Henning, U., 1986, An outer membrane protein (OmpA) of *Escherichia coli* K-12 undergoes a conformational change during export. *J. Biol. Chem.* **261:**11355–11361.

Friedlander, M., and Blobel, G., 1985, Bovine opsin has more than one signal sequence, *Nature* **318:**338–343.

Fujiki, Y., and Lazarow, P. B., 1982, Post-translational uptake of an *in vitro* synthesized peroxisomal polypeptide of rat liver. *J. Cell Biol.* **95**(2, Pt. 2): 398a (Abstract).

Fujiki, Y., and Lazarow, P. B., 1985, Post-translational import of fatty acyl CoA oxidase and catalase into peroxisomes. *J. Biol. Chem.* **260:**5603–5609.

Fujiki, Y., Rachubinski, R. A., and Lazarow, P. B., 1984, Synthesis of a major integral membrane polypeptide of rat liver peroxisomes on free polysomes. *Proc. Natl. Acad. Sci. U.S.A.* **81:**7127–7131.

Furuta, S., Hashimoto, T., Miura, S., Mori, M., and Tatibana, M., 1982, Cell free synthesis of enzymes of peroxisomal β-oxidation. *Biochem. Biophys. Res. Commun.* **105:**639–646.

Furuya, S., Okada, M., Ito, A., Aoyagi, H., Kanmera, T., Kato, T., Sagara, Y., Horiuchi, T., and Omura, T., 1987, Synthetic partial extension peptides of P-450 *(scc)* and adrenodoxin precursors: effects on the import of mitochondrial enzyme precursors, *J. Biochem. (Tokyo)* **102:**821–832.

Furuya, E., Yoshida, Y., and Tagawa, K., 1979, Interaction of aspartate aminotransferase with negatively charged lecithin liposomes. *J. Biochem. (Tokyo)* **85:**1157–1163.

Garcia, P. D., and Ghrayeb, J., Inouye, M., and Walter, P., 1987, Wild type and mutant signal peptide of *Escherichia coli* outer membrane lipoprotein interact with equal efficiency with mammalian signal recognition particle. *J. Biol. Chem.* **262:**9463–9468.

Gardel, C., Benson, S., Hunt, J., Michaelis, S., and Beckwith, J., 1987, sec D, a new gene involved in protein export in *Escherichia coli, J. Bacteriol.* **169:**1285–1290.

Garnier, J., Gaye, P., Mercier, J. C., and Robson, B., 1980, Structural properties of signal peptides and their membrane insertion. *Biochimie* **62:**231–239.

Garwin, J. L., and Beckwith, J., 1982, Secretion and processing of ribulose-binding protein in *Escherichia coli*. *J. Bacteriol.* **149:**789–792.

Gasser, S. M., and Schatz, G., 1983, Import of proteins into mitochondria. *In vitro*, studies on the biogenesis of the outer membrane. *J. Biol. Chem.* **258:**3427–3430.

Gasser, S. M., Daum, G., and Schatz, G., 1982a, Import of proteins into mitochondria. Energy dependent uptake of precursors by isolated mitochondria. *J. Biol. Chem.* **257:**13034–13041.

Gasser, S. M., Ohashi, A., Daum, G., Böhni, P. C., Gibson, J., Reid, G. A., Yonetani, T., and Schatz, G., 1982b, Imported mitochondrial proteins cytochrome b_2 and cytochrome c_1 are processed in two steps. *Proc. Natl. Acad. Sci. U.S.A.* **79:**267–271.

Gearing, D. P., and Nagley, P., 1986, Yeast mitochondrial ATPase subunit 8, normally a mitochondrial gene product, expressed *in vitro* and imported back into the organelle. *EMBO J.* **5**:3651–3655.

Geering, K., Meyer, D., Paccolat, M. P., Kraehenbuhl, J. P., and Rossier, B. C., 1985, Membrane insertion of α- and β-subunits of Na^+, K^+-ATPase. *J. Biol. Chem.* **260**:5154–5160.

Geller, B. L., and Wickner, W., 1985, M_{13} procoat inserts into liposomes in absence of other membrane proteins. *J. Biol. Chem.* **260**:13281–13285.

Geller, B. L., Movva, N. R., and Wickner, W., 1986, Both ATP and the electrochemical potential are required for optimal assembly of proOmpA into *Escherichia coli* inner membrane vesicles. *Proc. Natl. Acad. Sci. U.S.A.* **83**:4219–4222.

Geredes, H. H., Behrends, W., and Kindl, H., 1982, Biosynthesis of microbody matrix enzyme in greening cotyledons. Glycolate oxidase synthesized *in vivo* and *in vitro*. *Planta* **156**:572–578.

Gething, M. J., and Sambrook, J., 1982, Construction of influenza hemagglutinin genes that code for intracellular and secreted forms of the protein. *Nature (London)* **300**:593–603.

Ghersa, P., Huber, P., Semenza, G., and Wacker, H., 1986, Cell free synthesis, membrane integration and glycosylation of pro-sucrase-isomaltase. *J. Biol. Chem.* **261**:7969–7974.

Ghosh, M. K., and Hajra, A. K., 1986, Subcellular distribution and properties of acyl/alkyl dihydroxy acetone phosphate reductase in rodent livers. *Arch. Biochem. Biophys.* **245**:523–530.

Ghrayeb, J., and Inouye, M., 1984, Nine amino acid residues at the amino terminal of lipoprotein are sufficient for its modification, processing and localization in the outer membrane of *Escherichia coli*. *J. Biol. Chem.* **259**:463–467.

Ghrayeb, J., Lunn, C. A., Inouye, S., and Inouye, M., 1985, An alternate pathway for the processing of the prolipoprotein signal peptide in *Escherichia coli*. *J. Biol. Chem.* **260**:10961–10965.

Giam, C. Z., Chai, T., Hayashi, S., and Wu, H. C., 1984, Prolipoprotein modification and processing in *Escherichia coli*. A unique secondary structure in prolipoproteins signal sequence for the recognition by glyceryltransferase. *Eur. J. Biochem.* **141**:331–337.

Gibbs, S. P., 1979, The route of entry of cytoplasmically, synthesized proteins into chloroplasts of algae possessing chloroplast ER. *J. Cell Sci.* **34**:253–266.

Gietl, C., and Hock, B., 1984, Import of *in vitro* synthesized glyoxysomal malate dehydrogenase into isolated watermelon glyoxysomes. *Planta* **162**:261–267.

Gietl, C., and Hock, B., 1986, Import of glyoxysomal malate dehydrogenase precursor into glyoxysomes: A heterologous *in-vitro* system. *Planta* **167**:87–93.

Gillepsie, L. L., 1987, Identification of an outer mitochondrial membrane protein that interacts with a synthetic signal peptide. *J. Biol. Chem.* **262**:7939–7942.

Gillepsie, L. L., Argan, C., Taneja, A. T., Hodges, R. S., Freeman, K. B., and Shore, G. C., 1985, A synthetic signal peptide blocks import of precursor proteins destined for the mitochondrial inner membrane matrix. *J. Biol. Chem.* **260**:16045–16048.

Gilmore, R., and Blobel, G., 1983, Transient involvement of signal recognition particle and its receptor in the microsomal membrane prior to protein translocation. *Cell* **35**:667–685.

Gilmore, R., and Blobel, G., 1985, Translocation of secretory proteins across the microsomal membrane occurs through an environment accessible to aqueous perturbants. Translocation through proteinaceous channel. *Cell* **42**:497–505.

Gilmore, R., Walter, P., and Blobel, G., 1982, Protein translocation across the endoplasmic reticulum. II. Isolation and characterization of the signal recognition particle receptor. *J. Cell Biol.* **95**(2, Pt. 1): 470–477.

Godinot, C., and Lardy, H. A., 1973, Biosynthesis of glutamate dehydrogenase in rat liver. Demonstration of its microsomal localization and hypothetical mechanism of transfer to mitochondria. *Biochemstry* **12**:2051–2061.

Gogol, E. P., and Engelman, D. M., 1984, Neutron scattering shows that cytochrome b_5 penetrates deeply the lipid bilayer. *Biophys. J.* **46**:491–495.
Goldman, B. M., and Blobel, G., 1978, Biosynthesis of peroxisomes: Intracellular site of synthesis of catalase and uricase. *Proc. Natl. Acad. Sci. U.S.A.* **75**:5066–5070.
Gonzalez, E., 1986, Glycoproteins in the matrix of glyoxysomes in endosperm of castor bean seedlings. *Plant Physiol.* **80**:950–955.
Gonzalez, F. J., and Kasper, C. B., 1980, In vitro translation of epoxide hydratase messenger RNA. *Biochem. Biophys. Res. Commun.* **93**:1254–1258.
Gonzalez-Cadavid, N. F., and Cardova, C. S., 1974, Role of membrane bound and free polysomes in the synthesis of cytochrome c in rat liver. *Biochem. J.* **140**:157–167.
Goodman, J. M., 1985, Dihydroxyacetone synthase is an abundant constituent of the methanol-induced peroxisomes of *Candida boidinii*. *J. Biol. Chem.* **260**:7108–7113.
Goodman, J. M., Scott, C. W., Donahue, P. N., and Atherthon, J. P., 1984, Alcohol dehydrogenase assembles post-translationally into the peroxisomes of *Candida boidinni*. *J. Biol. Chem.* **259**:8485–8493.
Gordon, J. I., Alpers, D. H., Schonfeld, G., Andy, R., Smith, D. P., and Strauss, A., 1981, The primary translation product of rat intestinal apo A_1 mRNA is a preprotein. *Fed. Proc.* **40**:1635.
Gorgas, K., 1985, Serial section analysis of mouse hepatic peroxisomes. *Anat. Embryol.* **172**:21–32.
Gorin, M. B., Yancey, S. B., Cline, J., Revel, J. P., and Horwitz, J., 1984, The major intrinsic protein (MIP) of the bovine lens fiber membrane: Characterization and structure based on cDNA cloning. *Cell* **39**:49–59.
Gould, S. J., Keller, G. A., and Subramani, S., 1987, Identification of a peroxisomal targeting signal at the carboxy terminus of firefly luciferase, *J. Biol. Chem.* **105**:2923–2931.
Greenhut, S. F., and Roseman, M. A., 1985, Distribution of cytochrome b_5 between sonicated phospholipid vesicles of different sizes. *J. Biol. Chem.* **260**:5883–5886.
Grossman, A., Bartlett, S. G., and Chua, N. H., 1980, Energy dependent uptake of cytoplasmically synthesized polypeptides by chloroplasts. *Nature (London)* **285**:625–628.
Grossman, A. R., Bartlett, S. G., Schmidt, G. W., Mullet, J. E., and Chua, N. H., 1982, Optimal conditions for post-translational uptake of proteins by isolated chloroplasts. In vitro synthesis and transport of plastocyanin, ferredoxin-NADP$^+$ oxidoreductase and fructose 1,6-diphosphatase, *J. Biol. Chem.* **257**:1558–1563.
Guan, J. L., and Rose, J. K., 1984, Conversion of a secretory protein into a transmembrane protein results in its transport to the Golgi complex but not to the cell surface. *Cell* **37**:779–787.
Gundelfinger, E. D., Krause, E., Melli, M., and Dobberstein, B., 1983, The organization of 7 SL RNA in the signal recognition particle. *Nucleic Acids Res.* **11**:7363–7374.
Halegoua, S., and Inouye, M., 1979, Translocation and assembly of outer membrane proteins of *Escherichia coli*. Selective accumulation of precursor and novel assembly intermediates caused by phenethyl alcohol. *J. Mol. Biol.* **130**:30–61.
Halegoua, S., Hirashima, A., Sekizawa, J., and Inouye, M., 1976, Protein synthesis in toluene-treated *Escherichia coli*. Exclusive synthesis of membrane protein. *Eur. J. Biochem.* **69**:163–167.
Halegoua, S., Sekizawa, J., and Inouye, M., 1977, A new form of structural lipoprotein of outer membrane of *Escherichia coli*. *J. Biol. Chem.* **252**:2324–2330.
Hallermeyer, G., and Neupert, W., 1976, Studies on the synthesis of mitochondrial proteins in the cytoplasm and on their transport in the mitochondria, in *Genetics and Biogenesis of Chloroplasts and Mitochondria* (T. Bücher, W. Neupert, W. Sebald, and S. Werner, eds.), pp. 807–812, North-Holland/Elsevier, Amsterdam.

Hallermeyer, G., Zimmerman, R., and Neupert, W., 1977, Kinetic studies on the transport of cytoplasmically synthesized proteins into mitochondria in intact cells of *Neurospora crassa*. *Eur. J. Biochem.* **81**:523–532.

Hampsey, D. M., Lewin, A. S., and Kohlhaw, G. B., 1983, Submitochondrial localization, cell free synthesis and mitochondrial import of 2-isopropylmalate synthase of yeast. *Proc. Natl. Acad. Sci. U.S.A.* **80**:1270–1274.

Hansen, W., Garcia, P. D., and Walter, P., 1986, *In vitro* protein translocation across the yeast endoplasmic reticulum. ATP dependent post-translational translocation of the prepro-α-factor. *Cell* **45**:397–406.

Harmey, M. A., Hallermeyer, G., and Neupert, W., 1976, *In vitro* synthesis and transport into mitochondria of cytoplasmically translated proteins, in *Genetics and Biogenesis of Chloroplasts and Mitochondria* (T. Bücher, W. Neupert, W. Sebald, and S. Werner, eds.), pp. 813–818, North-Holland/Elsevier, Amsterdam.

Harmey, M. A., Hallermeyer, G., Korb, H., and Neupert, W., 1977, Transport of cytoplasmically synthesized proteins into mitochondria in a cell free system from *Neurospora crassa*. *Eur. J. Biochem.* **81**:533–544.

Harnik-Ort, V., Prakash, K., Marcantonio, E., Colman, D. R., Rosenfeld, M. G., Adesnik, M., and Sabatini, D. D., 1987, Isolation and characterization of cDNA clones for rat ribophorin. I: Complete coding sequence and *in vitro* synthesis and insertion of the encoded product into endoplasmic reticulum. *J. Cell Biol.* **104**:855–863.

Harold, F. M., and Van Brunt, J., 1977, Circulation of H^+ and K^+ across the plasma membrane is not obligatory for bacterial growth. *Science* **197**:372–373.

Harson, M. M., Conder, M. J., and Lord, J. M., 1983, Endoplasmic reticulum and glyoxysomal membranes from castor bean endosperm: Interaction between membrane glycoproteins and organelle matrix proteins. *Planta* **157**:143–149.

Hartl, F. U., and Just, W. W., 1987, Integral membrane polypeptides of rat liver peroxisomes: Topology and response to different metabolic states. *Arch. Biochem. Biophys.* **255**:109–119.

Hartl, F. U., Pfanner, N., and Neupert, W., 1987a, Translocation intermediates on the import pathway of proteins into mitochondria. *Biochem. Soc. Trans.* **15**:96–97.

Hartl, F. U., Ostermann, J., Guiard, B., and Neupert, W., 1987b, Successive translocation into and out of the mitochondrial matrix: targeting of proteins in the intermembrane space by bipartite signal peptide, *Cell* **51**:1027–1037.

Hase, T., Mueller, U., Riezman, H., and Schatz, G., 1984, A 70-kd protein of the yeast mitochondrial outer membrane is targeted and anchored via its extreme amino terminus. *EMBO J.* **3**:3157–3164.

Hase, T., Nakai, M., and Matsubara, H., 1986, The N-terminal 21 amino acids of the 70 kd protein of the yeast mitochondrial outer membrane direct *E. coli* β-galactosidase into mitochondrial matrix space in yeast cells. *FEBS Lett.* **197**:199–203.

Hashimoto, T., 1982, Individual proxisomal β-oxidation enzymes. *Ann. N.Y. Acad. Sci.* **385**:5–12.

Heinemann, F. S., and Ozols, J., 1984, The covalent structure of hepatic microsomal epoxide hydrolase, II. The complete amino acid sequence. *J. Biol. Chem.* **259**:797–804.

Heinrich, P. C., Schmelzer, E., Northmann, C. K., and Witt, I., 1982, Rat liver cytochrome c oxidase subunits IV and V; cell free synthesis as larger molecular weight precursors, mRNA sizes and sites of synthesis. *Prog. Clin. Biol. Res.* **102B**:149–159.

Henning, B., and Neupert, W., 1981, Assembly of cytochrome c. Apocytochrome c is bound to specific sites on mitochondria before its conversion to holocytochrome c. *Eur. J. Biochem.* **121**:203–212.

Henning, B., Koehler, H., and Neupert, W., 1983, Receptor sites involved in post-translational transport of apocytochrome c into mitochondria. Specificity, affinity and number of sites. *Proc. Natl. Acad. Sci. U.S.A.* **80**:4963–4967.

Henning, U., Cole, S. T., Bremer, E., Hindennach, I., and Schaller, H., 1983, Gene fusion using OmpA gene coding for a major outer membrane protein of *Escherichia coli*. *Eur. J. Biochem.* **136**:233–240.

Hernandez-Yago, J., and Grisolia, S., 1987, Apocytochrome c competes with pre-ornithine carbamyl transferase for transport into mitochondria, *Biochem. Biophys. Res. Commun.* **146**:1318–1323.

Hernandez-Yago, J., Knecht, E., Felipo, V., Miralles, V., and Grisolia, S., 1983, Exit of proteins and fragments thereof from mitochondria is accelerated by the import of cytosolic synthesized protein. *Biochem. Biophys. Res. Commun.* **113**:199–204.

Hiatt, A., McDonough, A., and Edelman, I. S., 1984, Assembly of the $(Na^+ + K^+)$-adenosine triphosphatase, post-translational membrane integration of the α-subunit. *J. Biol. Chem.* **259**:2629–2635.

Higgins, C. F., Haag, P. D., Nikaido, K., Ardelshir, F., Garcia, G., and Ames, G. F. L., 1982, Complete nucleotide sequence and identification of membrane components of histidine transport operon of *S. typhimurium*. *Nature (London)* **298**:723–727.

Highfield, P. E., and Ellis, R. J., 1978, Synthesis and transport of the small subunit of chloroplast ribulose bisphosphate carboxylase. *Nature (London)* **271**:420–424.

Hock, B., and Gietl, C., 1982, Cell-free synthesis of watermelon glyoxysomal malate dehydrogenase: A comparison with the mitochondrial isoenzyme. *Ann. N.Y. Acad. Sci.* **386**:350–361.

Holland, E. C., Leung, J. O., and Dirckamer, K., 1984, Rat liver asialoglycoprotein receptor lacks a cleavable NH_2-terminal signal sequence. *Proc. Natl. Acad. Sci. U.S.A.* **81**:7338–7342.

Horiuchi, S., Marty-Mazars, D., Tai, P., and Davis, B. D., 1983a, Localization and quantitation of proteins characteristic of the complex membrane of *Bacillus subtilis*. *J. Bacteriol.* **153**:1215–1221.

Horiuchi, S., Tai, P., and Davis, B. D., 1983b, A 64-kilodalton membrane protein of *Bacillus subtilis* covered by secreting ribosomes. *Proc Natl. Acad. Sci. U.S.A.* **80**:3287–3291.

Hortin, G., and Boime, I., 1980a, Inhibition of preprotein processing in ascites tumor lysates by incorporation of leucine analog. *Proc. Natl. Acad. Sci. U.S.A.* **77**:1356–1360.

Hortin, G., and Boime, I., 1980b, Pre-prolactin accumulates in rat pituitary cells incubated with threonine analog. *J. Biol. Chem.* **255**:7051–7053.

Hortsch, M., and Meyer, D. I., 1988, The human docking protein does not associate with the membrane of the rough endoplasmic reticulum via a signal or insertion sequence-mediated mechanism, *Biochem. Biophys. Res. Commun.* **150**:111–117.

Hortsch, M., Avossa, A., and Meyer, D. I., 1985, A structural and functional analysis of docking protein: Characterization of active domains by proteolysis and specific antibodies. *J. Biol. Chem.* **260**:9137–9145.

Hortsch, M., Avossa, D., and Meyer, D. I., 1986, Characterization of secretory protein translocation: Ribosome–membrane interaction in endoplasmic reticulum. *J. Cell Biol.* **103**:241–253.

Horwich, A. L., Fenton, W. A., Williams, K. R., Kalousek, F., Kraus, J. P. Doolittle, R. F., Konigsberg, W., and Rosenberg, L. E., 1984, Structure and expression of a complementary DNA for the nuclear coded precursor of human mitochondrial ornithine transcarbamylase. *Science* **224**:1068–1074.

Horwich, A. L., Fenton, W. A., Firgaira, F. A., Fox, J. E., Kolansky, D., Mellman, I. S., and Rosenberg, L. E., 1985a, Expression of amplified DNA sequences for ornithine transcarbamylase in HeLa cells: Arginine residues may be required for mitochondrial import of enzyme precursor. *J. Cell Biol.* **100:**1515–1521.

Horwich, A. L., Kalousek, F., and Rosenberg, L. E., 1985b, Arginine in the leader peptide is required for both import and proteolytic cleavage of a mitochondrial precursor. *Proc. Natl. Acad. Sci. U.S.A.* **82:**4930–4933.

Horwich, A. L., Kalousek, F., Mellman, I., and Rosenberg, L. E., 1985c, A leader peptide is sufficient to direct mitochondrial import of a chimeric protein. *EMBO J.* **4:**1129–1135.

Horwich, A. L., Kalousek, F., Fenton, W. A., Pollock, R. A., and Rosenberg, L. E., 1986, Targeting of pre-ornithine transcarbamylase to mitochondria: Definition of critical regions and residues in the leader peptide. *Cell* **44:**451–459.

Horwich, A. L., Kalousek, F., Fenton, W. A., Furtak, K., Pollock, R. A., and Rosenberg, L. E., 1987, The ornithine transcarbamylase leader peptide directs mitochondrial import through both midportion structure and net positive charge. *J. Cell Biol.* **105:**669–677.

Hu, V. W., and Wiesnieski, B. J., 1979, Photoactive labeling of M_{13} coat protein in model membranes by use of glycolipid probe. *Proc. Natl. Acad. Sci. U.S.A.* **76:**5460–5464.

Hurt, E. C., and Schatz, G., 1987, A cytosolic protein contains a cryptic mitochondrial targeting signal. *Nature (London)* **325:**499–503.

Hurt, E. C., and Van Loon, A. P. G. M., 1986, How proteins find mitochondria and intramitochondrial components. *Trends Biochem. Sci.* **11:**204–207.

Hurt, E. C., Pesold-Hurt, B., and Schatz, G., 1984, The amino terminal region of an imported mitochondrial precursor polypeptide can direct cytoplasmic dehydrofolate reductase into the mitochondrial matrix. *EMBO J.* **3:**3149–3156.

Hurt, E. C., Mueller, U., and Schatz, G., 1985, The first twelve amino acids of a yeast mitochondrial outer membrane protein can direct a nuclear encoded cytochrome oxidase subunit to the mitochondrial inner membrane. *EMBO J.* **4:**3509–3518.

Hurt, E. C., Goldschmidt-Clermont, M., Pesold-Hurt, B., Rochaix, J. D., and Schatz, G., 1986a, A mitochondrial presequence can transport a chloroplast encoded protein into yeast mitochondria. *J. Biol. Chem.* **261:**11440–11443.

Hurt, E. C., Soltanifar, N., Goldschmidt-Clermont, M., Rochaix, J. D., and Schatz, G., 1986b, The cleavable pre-sequence of an imported chloroplast protein directs attached polypeptides into yeast mitochondria. *EMBO J.* **5:**1343–1350.

Ibrahimi, I., 1987, Dithiothreitol and the translocation of preprolactin across mammalian endoplasmic reticulum, *J. Cell. Biol.* **105:**1555–1560.

Ibrahimi, I., and Gentz, R., 1987, A functional interaction between the signal peptide and the translation apparatus is detected by the use of a single point mutation which blocks translocation across mammalian endoplasmic reticulum. *J. Biol. Chem.* **262:**10189–10194.

Ibrahimi, I., Culter, D., Stueber, D., and Bujard, H., 1986, Determinants for protein translocation across mammalian endoplasmic reticulum. Membrane insertion of truncated and full-length prelysozyme molecule, *Eur. J. Biochem.* **155:**571–576.

Ichihara, S., Hussain, M., and Mizushima, S., 1982, Mechanism of export of outer membrane lipoproteins through the cytoplasmic membrane in *Escherichia coli*. Binding of lipoprotein precursors to the peptidoglycan layer. *J. Biol. Chem.* **257:**495–500.

Ikeda, Y., Keese, S. M., Fenton, W. A., and Tanaka, K., 1987, Biosynthesis of four rat liver mitochondrial acyl-CoA dehydrogenases: *In vitro* synthesis, import into mitochondria and processing of their precursors in a cell free system and in cultured cells. *Arch. Biochem. Biophys.* **252:**662–674.

Innis, M. A., Tokunaga, M., Williams, M. E., Loranger, J. M., Chang, S. Y., Chang, S., and Wu, H. C., 1984, Nucleotide sequence of *Escherichia coli* prolipoprotein signal peptidase (1 sp) gene. *Proc. Natl. Acad. Sci. U.S.A.* **81:**3708–3712.

Inouye, H., and Beckwith, J., 1977, Synthesis and processing of an *Escherichia coli* alkaline phosphatase precursor *in vitro*. *Proc. Natl. Acad. Sci. U.S.A.* **74**:1440–1444.

Inouye, H., Barnes, W., and Beckwith, J., 1982, Signal sequence of alkaline phosphatase of *E. coli*. *J. Bacteriol.* **149**:434–439.

Inouye, M., and Halegoua, S., 1980, Secretion and membrane localizaiton of proteins of *Escherichia coli*. *CRC Crit. Rev. Biochem.* **7**:339–371.

Inouye, M., DiRienzo, J., Maeda, T., Movva, R., Nakamura, K., Lee, A. G., Pirtle, R., and Pirtle, I., 1980, Secretion of outer membrane proteins of *Escherichia coli* across the cytoplasmic membrane. *Ann. N.Y. Acad. Sci.* **343**:362–367.

Inouye, S., Soberon, X., Franceschini, T., Nakamura, K., Itakura, K., and Inouye, M., 1982, Role of positive charge on the amino terminal region of the signal peptide in protein secretion across the membrane. *Proc. Natl. Acad. Sci. U.S.A.* **79**:3438–3441.

Inouye, S., Hsu, C. P. S., Itakura, K., and Inouye, M., 1983a, Requirement for signal peptide cleavage of *Escherichia coli* lipoprotein. *Science* **221**:59–61.

Inouye, S., Franceschini, T., Sato, M., Itakura, K., and Inouye, M., 1983b, Prolipoprotein signal peptidase of *Escherichia coli* requires a cysteine residue at the cleavage site. *EMBO J.* **2**:87–91.

Inouye, S., Duffaud, G., and Inouye, M., 1986, Structural requirement at the cleavage site for efficient processing of the lipoprotein secretory precursor of *Escherichia coli*. *J. Biol. Chem.* **261**:10970–10975.

Inukai, M., and Inouye, M., 1983, Association of the prolipoprotein accumulated in the presence of globomycin with the outer membrane of *Escherichia coli*. *Eur. J. Biochem.* **130**:27–32.

Ito, A., Ogishima, T., Ou, W., Omura, T., Aoyagi, H., Lee, S., Mihara, H., and Izumiya, N., 1985, Effects of synthetic model peptides resembling the extension peptides of mitochondrial enzyme precursors on import of the precursors into mitochondria. *J. Biochem. (Tokyo)* **98**:1571–1582.

Ito K., 1984, Identification of the sec Y (prl A) gene product involved in protein export in *Escherichia coli*. *Mol. Gen. Genet.* **197**:204–208.

Ito, K., Bassford, P. J. Jr., and Beckwith, J., 1981, Protein localization in *E. coli*. Is there a common step in insertion of periplasmic and outer membrane proteins? *Cell* **24**:707-717.

Ito, K., Cerretti, D. P., Nashimoto, H., and Nomura, M., 1984, Characterization of an amber mutation in the structural gene for ribosomal protein L_{15}; which impairs the expression of the protein export gene, sec Y, in *Escherichia coli*. *EMBO J.* **3**:2319–2324.

Jain, M., and Zakim, D., 1987, The spontaneous incorporation of proteins into preformed bilayers, *Biochim. Biophys. Acta* **906**:33–68.

Jaussi, R., Cotton, B., Juretic, N., Christen, P., and Schuemperli, D., 1985, The primary structure of the precursor of chicken mitochondrial aspartate aminotransferase. Cloning and sequence analysis of cDNA. *J. Biol. Chem.* **260**:16060–16063.

Johnson, L. V., Walsh, M. L., Bockus, B. J., and Chen, L. B., 1981, Monitoring of relative mitochondrial membrane potential in living cells by fluorescence microscopy. *J. Cell Biol.* **88**:526–535.

Josefsson, L. G., and Randall, L. L., 1981a, Processing *in vivo* of precursor maltose-binding protein in *Escherichia coli* occurs post-translationally as well as co-translationally. *J. Biol. Chem.* **256**:2504–2507.

Josefsson, L. G., and Randall, L. L., 1981b, Different exported proteins in *Excherichia coli* show differences in the temporal mode of processing *in vivo*. *Cell* **25**:151–157.

Joste, V., Berrez, J. C., Latruff, N., and Nelson, B. D., 1987, An import factor in reticulocyte lysates which stimulates processing of several precursors destined for the rat liver mitochondrial inner membrane, *Acta Chem. Scand. Ser. B.* **B 41**:770–772.

Kadonaga, J. T., Gantier, A. E., Strauss, D. R., Charles, A. D., Edge, M. D., and Knowles, J.

R., 1984, The role of the β-lactamase signal sequence in the secretion of proteins by *Escherichia coli*. *J. Biol. Chem.* **259**:2149–2154.

Kaiser, C. A., Preuss, D., Grisafi, P., and Botstein, D., 1987, Many random sequences funtionally replace the secretion signal sequences of yeast invertase. *Science* **235**:312–317.

Kalousek, F., Orsulak, M. D., and Rosenberg, L. E., 1984, Newly processed transcarbamylase subunits are assembled to trimers in rat liver mitochondria. *J. Biol. Chem.* **259**:5392–5395.

Kamisaki, Y., Sakakibara, R., Horio, Y., and Wada, H., 1982, Low isoelectric point of precursor of mitochondrial glutamic oxaloacetate transaminase isoenzyme synthesized *in vitro*. *Biochem. Int.* **4**:289–296.

Kaput, J., and Blobel, G., 1986, Binding of the heme ligand inhibits translocation of *in vitro* synthesized cytochrome *c* peroxidase into yeast mitochondria. *Yeast* **2**:S182.

Kaput, J., Goltz, S., and Blobel, G., 1982, Nucleotide sequence of the yeast nuclear gene for cytochrome *c* peroxidase precursor. Functional implications of the presequence for protein transport into mitochondria. *J. Biol. Chem.* **257**:15054–15058.

Karlin-Neumann, G. A., and Tobin, E. M., 1986, Transit peptides of nuclear-encoded chloroplast proteins share a common amino acid framework. *EMBO J.* **5**:9–13.

Kartenbeck, J., and Franke, W. W., 1974, Membrane relationship between endoplasmic reticulum and peroxisomes in rat hepatocytes and Morris hepatoma cells. *Cytobiology* **10**:152–156.

Kasper, C. B., and Porter, T. D., 1985, Coding nucleotide sequence of rat NADPH-cytochrome P-450 oxidoreductase cDNA and identification of flavin binding domain, *Proc. Natl. Acad. Sci. U.S.A.* **82**:973–977.

Katz, F. N., Rothman, J. E., Lingappa, V. R., Blobel, G., and Lodish, H. F., 1977, Membrane assembly *in vitro:* Synthesis, glycosylation and asymmetric insertion of transmembrane proteins. *Proc. Natl. Acad. Sci. U.S.A.* **74**:3278–3282.

Kawajiri, K., Harano, T., and Omura, T., 1977, Biosynthesis of the mitochondrial matrix enzyme, glutamate dehydrogenase, in rat liver cells. II. Significance of binding of glutamate dehydrogenase to microsomal membrane. *J. Biochem. (Tokyo)* **82**:1417–1423.

Kellems, R. E., Allison, V. F., and Butow, R. A., 1974, Cytoplasmic type 80 S ribosomes associated with yeast mitochondria. II. Evidence for the association of cytoplasmic ribosomes with the outer mitochondrial membrane. *in situ. J. Biol. Chem.* **249**:3297–3303.

Kellems, R. E., Allison, V. F., and Butow, R. A., 1975, Cytoplasmic type 80 S ribosomes associated with yeast mitochondria. IV. Attachment of ribosomes to the outer membrane of isolated mitochondria. *J. Cell Biol.* **65**:1–14.

Keller, G. A., Barton, M. C., Shapiro, D. J., and Singer, S. J., 1985, 3-Hydroxy-3-methyl glutaryl CoA reductase is present in peroxisomes in rat liver cells. *Proc. Natl. Acad. Sci. U.S.A.* **82**:770–774.

Keller, G. A., Pazirandeh, M., and Krisans, S., 1986, 3-Hydroxyl-3-methyl glutaryl coenzyme A reductase localization in rat liver peroxisomes and microsomes of control and cholestyramine-treated animals: Quantitative biochemical and immunoelectron microscopical analyses. *J. Cell Biol.* **103**:875–886.

Keller, G. A., Gould, S., Deluca, M., and Subramani, S., 1987, Firefly luciferase is targeted to peroxisomes in mammalian cells. *Proc. Natl. Acad. Sci. U.S.A.* **84**:3264–3265.

Kendall, D. A., Bock, S. C., and Kaiser, E. T., 1986, Idealization of the hydrophobic segment of the alkaline phosphatase signal peptide. *Nature (London)* **321**:706–708.

Keng, T., Alani, E., and Guarente, L., 1986, The nine aminoterminal residues of δ-aminolevulinate synthase direct-β-galactosidase into the mitochondrial matrix. *Mol. Cell. Biol.* **6**:355–364.

Kensil, C. R., and Strittmatter, P., 1986, Binding and fluorescence properties of the membrane domain of NADH–cytochrome b_5 reductase. Determination of the depth of Trp-16 in the bilayer. *J. Biol. Chem.* **261**:7316–7321.

Kensil, C. R., Hediger, M. A., Ozols, J., and Strittmatter, P., 1983, Isolation and partial characterization of the amino-terminal membrane binding domain of NADH–cytochrome b_5 reductase. *J. Biol. Chem.* **258**:14656–14663.

Kiino, D. R., and Silhavy, T. J., 1984, Mutation Prl F_1 relieves the lethality associated with export of β-galactosidase hybrid proteins in *Escherichia coli*. *J. Bacteriol.* **158**:878–883.

Kindl, H., Köller, W., and Frevert, J., 1980, Cytoplasmic precursor pool during glyoxysome biosynthesis. *Hoppe-Seyler's Z. Physiol. Chem.* **361**:465–467.

Kinoshita, N., Unemoto, T., and Kobayashi, H., 1984, Proton motive force is not obligatory for growth of *Escherichia coli*. *J. Bacteriol.* **160**:1074–1077.

Kirwin, P. M., Elderfield, P. D., and Robinson, C., 1987, Transport of proteins into chloroplasts, Partial purification of thylakoidal processing peptidase involved in plastocyanin biogenesis, *J. Biol. Chem.* **262**:16386–16390.

Kleene, R., Pfanner, N., Pfaller, R., Link, T. A., Sebald, W., Neupert, W., and Tropschug, M., 1987, Mitochondrial porin of *Neurospora crassa:* cDNA cloning, *in vitro* expression and import into mitochondria, *EMBO J.* **6**:2627–2633.

Kloppstech, K., and Bitsch, A., 1986, Cross-linking of envelope proteins presumably involved in binding of nuclear coded chloroplast precursor proteins. *Plant Biol.* **2**:235–240.

Koester, A., Heisig, M., Heinrich, P. C., and Just, W. W., 1986, *In vitro* synthesis of peroxisomal membrane polypeptides. *Biochem. Biophys. Res. Commun.* **137**:628–632.

Kohorn, B. D., Harel, E., Chitnis, P. R., Thornber, J. P., and Tobin, E. M., 1986, Functional and mutational analysis of the light-harvesting chlorophyll a/b protein of thylakoid membranes. *J. Cell Biol.* **102**:972–981.

Kolansky, D. M., Conboy, J. G., Fenton, W. A., and Rosenberg, P. E., 1982, Energy dependent translocation of precursor of ornithine transcarbamylase by isolated rat liver mitochodria. *J. Biol. Chem.* **257**:8467–8471.

Köller, W., and Kindl, H., 1980, Cytosolic malate synthethase. A small pool characterized by rapid turnover. *Hoppe-Seyler's Z. Physiol. Chem.* **361**:1437–1444.

Kopito, R. R., and Lodish, H. F., 1985, Primary structure and transmembrane orientation of the murine anion exchange protein. *Nature (London)* **316**:234–238.

Kornfeld, R., and Kornfeld, S., 1985, Assembly of asparagine-linked oligosaccharides. *Annu. Rev. Biochem.* **54**:631–664.

Koshland, D., and Botstein, D., 1980, Secretion of β-lactamase requires the carboxy end of the protein. *Cell* **20**:749–760.

Koshland, D., and Botstein, D., 1982, Evidence for post-translational translocation of β-lactamase across the bacterial inner membrane. *Cell* **30**:893–902.

Koshland, D., Sauer, R. T., and Botstein, D., 1982, Diverse effects of mutations in the signal sequence on the secretion of β-lactamase in *Salmonella typhimurium*. *Cell*. **30**:903–914.

Kreibich, G., Ullrich, B. L., and Sabatini, D. D., 1978a, Proteins of rough microsomal membranes related to ribosome binding. I. Identification of ribophorins I and II, membrane proteins characteristic of rough microsomes. *J. Cell Biol.* **77**:464–487.

Kreibich, G., Freienstein, C. M., Pereyra, P. N., Ullrich, B. L., and Sabatini, D. D., 1978b, Proteins of rough microsomal membranes related to ribosome binding. II. Cross-linking of bound ribosomes to specific membrane proteins exposed at the binding site. *J. Cell Biol.* **77**:488–506.

Kreibich, G., Csako-Graham, M., Grebeneau, R., and Sabatini, D. D., 1980, Functional and structural characteristics of endoplasmic reticulum membrane proteins associated with ribosome binding sites. *Ann N.Y. Acad. Sci.* **343**:17–33.

Kreibich, G., Ojakian, G., Rodriguez-Boulan, E. J., and Sabatini, D. D., 1982, Recovery of ribophorins in "inverted rough" vesicles derived from liver rough microsomes. *J. Cell Biol.* **93**:111–121.

Krieg, U. C., Walter, P., and Johnson, A. E., 1986, Photo cross-linking of the signal sequence of nascent preprolactin to the 54-kilodalton polypeptide of the signal recognition particle. *Proc. Natl. Acad. Sci. U.S.A.* **83**:8604–8608.
Kruse, C., and Kindl, H., 1983, Oligomerization of malate synthase during glyoxysome biosynthesis. *Arch. Biochem. Biophys.* **223**:629–638.
Kruse, C., Frevert, J., and Kindl, H., 1981, Selective uptake by glyoxysome of *in vitro* translated malate synthase. *FEBS. Lett.* **129**:36–38.
Kuhn, A., Wickner, W., and Kreil, G., 1986, The cytoplasmic terminus of M_{13} procoat is required for the membrane insertion of its central domain. *Nature (London)* **322**:335–339.
Kumamoto, C. A., and Beckwith, J., 1983, Mutations in a new gene, sec B, cause defective protein localization in *Escherichia coli*. *J. Bacteriol.* **154**:253–260.
Kumamoto, C. A., Oliver, D. B., and Beckwith, J., 1984, Signal sequence mutations disrupt feedback between secretion of an exported protein and its synthesis in *E. coli*. *Nature (London)* **308**:863–864.
Kumamoto, T., Morohashi, K., Ito, A., and Omura, T., 1987, Site-directed mutagenesis of basic amino acid residues in the extension peptide of P-450 (scc) precursor: effects on import of precursor into mitochondria, *J. Biochem. (Tokyo)* **102**:833–834.
Kuntz, M., Simons, A., Schell, J., and Schreier, P. H., 1987, Targeting of protein to chloroplasts in transgenic tobacco by fusion to mutated transit peptide. *Mol. Gen. Genet* **205**:454–460.
Kurzchalia, T. V., Wiedmann, M., Girshovich, A. S., Bochkareva, E. S., Bielka, H., and Rapoport, T. A., 1986, The signal sequence of nascent preprolactin interacts with the 54 K polypeptide of signal recognition particle. *Nature (London)* **320**:634–636.
Lai, J. S., Sarvas, M., Brammar, W. J., Neugebauer, K., and Wu, H. C., 1981, *Bacillus licheniformis* penicillinase synthesized in *Escherichia coli* contains covalently linked fatty acid and glyceride. *Proc. Natl. Acad. Sci. U.S.A.* **78**:3506–3510.
Lamppa, G. K., Morelli, G., and Chua, N. H., 1985, Structure and developmental regulation of a wheat gene encoding the major chlorophyll a/b-binding polypeptide. *Mol. Cell. Biol.* **5**:1370–1378.
Lauffer, L., Garcia, P. D., Harkins, R. N., Coussens, L., Axel, U., and Walter, P., 1985, Topology of signal recognition particle receptor in endoplasmic reticulum membrane. *Nature (London)* **318**:334–338.
Lazarow, P. B., and De Duve, C., 1973, The synthesis and turnover of rat liver peroxisomes. V. Intracellular pathway of catalase synthesis. *J. Cell Biol.* **59**:507–524.
Lazarow, P. B., Robbin, M., Fujiki, Y., and Wong, L., 1982, Biogenesis of peroxisomal proteins *in vivo* and *in vitro*. *Ann. N.Y. Acad. Sci.* **386**:285–300.
Lee, C. A., and Beckwith, J., 1986, Suppression of growth and protein secretion defects in *Escherichia coli* sec A mutants by decreasing protein synthesis. *J. Bacteriol.* **166**:878–883.
Lee, C. A., Fournier, M. J., and Beckwith, J., 1985, *Escherichia coli* 6S RNA is not essential for growth of protein secretion. *J. Bacteriol.* **161**:1156–1161.
Leskes, A., Siekevitz, P., and Palade, G. E., 1971a, Differentiation of endoplasmic reticulum in hepatocytes. I. Glucose-6-phosphatase distribution in situ. *J. Cell Biol.* **49**:264–287.
Leskes, A., Siekevitz, P., and Palade, G. E., 1971b, Differentiation of endoplasmic reticulum in hepatocytes. II. Glucose-6-phosphatase in rough microsomes. *J. Cell Biol.* **49**:288–302.
Lewin, A. S., and Norman, D. K., 1983, Assembly of F_1-ATPase in isolated mitochondria. *J. Biol. Chem.* **258**:6750–6755.
Liao, M. J., and Khorana, H. G., 1984, Removal of the carboxyl terminal peptide does not affect refolding or function of bacteriorhodopsin as a light-dependent proton pump. *J. Biol. Chem.* **259**:4194–4199.
Lin, J. J. C., Kanazawa, H., Ozols, J., and Wu, H. C., 1978, An *Escherichia coli* mutant with

an amino acid alteration within the signal sequence of outer membrane prolipoprotein. *Proc. Natl. Acad. Sci. U.S.A.* **75**:4891–4895.

Lin, J. J. C., Kanazawa, H., and Wu, H. C., 1980, Assembly of outer membrane proteins in *Escherichia coli* mutant, with a single amino acid replacement within the signal sequence of prolipoprotein. *J. Bacteriol.* **141**:550–557.

Lingappa, V. R., Katz, F. N., Lodish, H. F., and Blobel, G., 1978, A signal sequence for the insertion of a membrane glycoprotein. Similarity to the signals of secretory proteins in a primary structure and function. *J. Biol. Chem.* **253**:8667–8670.

Lingappa, V. R., Lingappa, J. R., and Blobel, G., 1979, Chicken ovalbumin contains an internal signal sequence. *Nature (London)* **281**:117–121.

Lingelbach, K. R., Graf, L. J., Dunn, A. R., and Hoogenraad, N. J., 1986, Effect of deletion within the leader peptide of pre-ornithine transcarbamylase on mitochondrial import. *Eur. J. Biochem.* **161**:19–23.

Lipp, J., and Dobberstein, B., 1986a, Signal recognition particle-dependent membrane insertion of mouse invariant chain: A membrane-spanning protein with cytoplasmically exposed amino terminus. *J. Cell Biol.* **102**:2169–2175.

Lipp, J., and Dobberstein, B., 1986b, The membrane-spanning segment of invariant chain (Iγ) contains a potentially cleavable signal sequence. *Cell* **46**:1103–1112.

Lord, J. M, 1978, Evidence that proliferation of endoplasmic reticulum precedes formation of glyoxysomes and mitochondria in germinating castor bean endosperm. *J. Exp. Bot.* **29**:13–23.

Lord, J. M., and Roberts, L. M., 1982, Glyoxysome biosynthesis via endoplasmic reticulum in castor bean endosperm. *Ann. N.Y. Acad. Sci.* **386**:362–376.

Lory, S., Tai, P. C., and Davis, B. D., 1983, Mechanism of protein secretion by gram-negative bacteria: *Pseudomonas aeruginosa* exotoxin A. *J. Bacteriol.* **156**:695–702.

Lubben, T. H., Bansberg, J., and Keegstra, K., 1987, Stop-transfer regions do not halt translocation of proteins into chloroplasts, *Science* **238**:1112–1114.

Lubben, T. H., and Keegstra, K., 1986, Efficient *in vitro* import of a cytosolic heat shock protein into pea chloroplasts. *Proc. Natl. Acad. Sci. U.S.A.* **83**:5502–5506.

Lubin, I. M., Wu, L. N. Y., Wuthier, R. E., and Fisher, R. R., 1987, Rhodamine 123 inhibits import of rat liver mitochondrial transhydrogenase. *Biochem. Biophys. Res. Commun.* **144**:477–483.

Lunn, C. A., and Inouye, M., 1987, Effects of prolipoprotein signal peptide mutations on secretion of hybrid prolipo-β-lactamase in *Escherichia coli*. *J. Biol. Chem.* **262**:8318–8324.

Lycett, G. W., Delauney, A. J., Catehouse, J. A., Gilroy, J., Croy, R. R. D., and Boutler, D., 1983, The vicilin gene family of pea *(Pisum sativum* L.) a complete cDNA sequence for preprovicilin. *Nucleic Acids Res.* **11**:2367–2380.

Maarse, A. C., Van Loon, A. P. G. M., Riezman, H., Gregor, I., Schatz, G., and Grivell, L. A., 1984, Subunit IV of yeast cytochrome oxidase: Cloning and nucleotide sequencing of the gene and partial amino acid sequencing of the mature protein. *EMBO J.* **3**:2831–2837.

Maccecchini, M. L., Rudin, Y., Blobel, G., and Schatz, G., 1979a, Import of proteins into mitochondria: Precursor forms of the extramitochondrially made F_1-ATPase subunits in yeast. *Proc. Natl. Acad. Sci. U.S.A.* **76**:343–347.

Maccecchini, M. L., Rudin, Y., and Schatz, G., 1979b, Transport of proteins across the mitochondrial outer membrane. A precursor form of the cytoplasmically made intermembrane enzyme, cytochrome *c* peroxidase. *J. Biol. Chem.* **254**:7468–7471.

Maeshima, M., Yokoi, H., and Asahi, T., 1988, Evidence for no proteolytic processing during transport of isocitrate lyase into glyoxysomes in castor bean endosperm, *Plant Cell Physiol.* **29**:381–384.

Maher, P. S., and Singer, S. J., 1986, Disulfide bonds and the translocation of proteins across membranes. *Proc. Natl. Acad. Sci. U.S.A.* **83:**9001–9005.

Majzoub, J. A., Rosenblatt, M., Fennick, B., Mannus, R., Kronenberg, H. M., Potts, J. T. Jr., and Habener, J. F., 1980, Synthetic pre-pro-parathyroid hormone leader sequence inhibits cell-free processing of placental, parathyroid and pituitary prehormones. *J. Biol. Chem.* **255:**11478–11484.

Marcantonio, E., Amar-Costesec, A., and Kreibich, G., 1984, Segregation of polypeptide translocation apparatus to regions of the endoplasmic reticulum containing ribophorin and ribosomes. II. Rat liver microsomal subfractions contain equimolar amounts of ribophorins and ribosomes. *J. Cell Biol.* **99:**2254–2259.

Martial, J. A., Hallwell, R. A., Baxter, J. D., and Goodman, H. M, 1979, Human growth hormone: Complementary DNA cloning and expression in bacteria. *Science* **205:**602–607.

Marty-Mazars, D. M., Horiuchi, S., Tai, P. C., and Davis, B. D., 1983, Proteins of ribosome-bearing and free-membrane domains in *Bacillus subtilis*. *J. Bacteriol.* **154:**1381–1388.

Matocha, M. F., and Waterman, M. R., 1984, Discriminatory processing of the precursor forms of cytochrome P-450 *scc* and adrenodoxin by adrenocortical and heart mitochondria. *J. Biol. Chem.* **259:**8672–8678.

Matocha, M. F., and Waterman, M. R., 1986, Import and processing of P-450 *scc* and P-450 11β precursors by corpus luteal mitochondria: A processing pathway recognizing homologous and heterologous precursors. *Arch. Biochem. Biophys.* **250:**456–460.

Matsuura, S., Arpin, M., Hannun, C., Margoliash, E., Sabatini, D. D., and Morimoto, T., 1981, *In vitro* synthesis and post-translational uptake of cytochrome *c* into isolated mitochondria: Role of a specific addressing signal in the apocytochrome *c*. *Proc. Natl. Acad. Sci. U.S.A.* **78:**4368–4372.

McIntyre, S., Freudl, R., Degen, M., Hindennach, I., and Henning, U., 1987, The signal sequence of an *E. coli* outer membrane protein can mediate translocation of a not normally secreted protein across the plasma membrane. *J. Biol. Chem.* **262:**8416–8422.

McKean, D. J., and Maurer, R. A., 1978, Complete amino acid sequence of the precursor region of rat prolactin. *Biochemistry* **17:**5215–5219.

MacLennan, D. H., Brandl, C. J., Korczak, B., and Green, N. M., 1985, Amino-acid sequence of $Ca^{2+} + Mg^{2+}$-dependent ATPase from rabbit muscle sarcoplasmic reticulum, deduced from its complementary DNA sequence. *Nature* **316:**696–700.

Meek, R. L., Walsh, K. A., and Palmiter, R. D., 1982, The signal sequence of ovalbumin is located near the aminoterminus. *J Biol. Chem.* **257:**12245–12251.

Meyer, D. I., 1982, Translocation of secretory and membrane proteins across the endoplasmic reticulum: The requirement for two specific receptors. *J. Cell Biol.* **95**(2, Pt. 2) 388a (Abstract).

Meyer, D. I., 1985, Signal recognition particle does not mediate a translational arrest of nascent secretory proteins in mammalian cell free systems. *EMBO J.* **4:**2031–2033.

Meyer, D. I., and Dobberstein, B., 1980a, A membrane component essential for vectorial translocation of nascent proteins across the endoplasmic reticulum: Requirement for its extraction and reassociation with the membrane. *J. Cell Biol.* **87:**498–502.

Meyer, D. I., and Dobberstein, B., 1980b, Identification and characterization of a membrane component essential for the translocation of nascent proteins across the membrane of the endoplasmic reticulum. *J. Cell Biol.* **87:**503–508.

Meyer, D. I., Krause, E., and Dobberstein, B., 1982a, Secretory protein translocation across membranes. The role of the "docking protein." *Nature (London)* **297:**647–650.

Meyer, D. I., Louvard, D., and Dobberstein, B., 1982b, Characterization of molecules involved in protein translocation using a specific antibody. *J. Cell Biol.* **92:**579–582.

Michaelis, S., Guarente, L., and Beckwith, J., 1983a, In vitro construction and characterization of Pho A-lac Z gene fusions in *Escherichia coli. J. Bacteriol.* **154**:356-365.

Michaelis, S., Inouye, H., Oliver, D., and Beckwith, J., 1983b, Mutations that alter the signal sequences of alkaline phosphatase in *Escherichia coli. J. Bacteriol.* **154**:366-374.

Mihara, K., and Blobel, G., 1982, In vitro synthesis and integration into mitochondria of porin, a major protein of the outer mitochondrial membrane of *Saccharomyces cerevisiae, Proc. Natl. Acad. Sci. U.S.A.* **79**:7012-7016.

Mihara, K., and Sato, R., 1985, Molecular cloning and sequencing of cDNA from yeast porin, an outer mitochodrial membrane protein: A search for targeting signal in primary structure. *EMBO J.* **4**:769-774.

Mihara, K., Sato, R., Sakakibara, R., and Wada, H., 1978, Reduced nicotinamide adenine dinucleotide-cytochrome b_5 reductase: Location of the hydrophobic membrane binding region at the carboxyl terminal and the masked amino terminus. *Biochemistry* **17**:2829-2834.

Mihara, K., Omura, T., Harano, T., Brenner, S. C., Fleischer, S., Rayagoplan, K. V., and Blobel, G., 1982, Rat liver L-glutamate dehydrogenase, malate dehydrongenase, D-β-hydroxybutyrate dehydrogenase and sulfite oxidase are each synthesized as larger precursors by cytoplasmic free polysomes. *J. Biol. Chem.* **257**:3355-3358.

Milstein, C., Brownlee, G. G., Harrison, T. M., and Mathews, M. B., 1972, A possible precursor of immunoglobulin light chains. *Nature New Biol.* **239**:117-120.

Minami, E., Shinohara, K., Kuwabara, T., and Watanabe, A., 1986, In vitro synthesis and assembly of photosystem II proteins of spinach chloroplasts. *Arch. Biochem. Biophys.* **244**:517-527.

Minsky, A., Summers, R. G., and Knowles, J. R., 1986, Secretion of β-lactamase into the periplasm of *Escherichia coli:* Evidence for a distinct release step associated with a conformation change. *Proc. Natl. Acad. Sci. U.S.A.* **83**:4180-4184.

Mirales, V., Felipo, V., Hernandez-Yago, J., and Grisolia, S., 1983, Transport of the precursor for rat liver glutamate dehydrogenase into mitochondria in vitro. *Biochem. Biophys. Res. Commun.* **110**:449-503.

Mishkind, M. L., Wessler, S. R., and Schmidt, G. W., 1985, Functional determinants in transit sequence: Import and partial maturation by vascular plant chloroplasts of the ribulose 1,5-bisphosphate carboxylase small subunit of *Chlamydomonas. J. Cell Biol.* **100**:226-234.

Miura, S., Mori, M., Morita, T., and Tatibana, M., 1982, A high isolectric point of mitochondrial ornithine carbamyltransferase precursor and inhibition of its processing in vitro by basic proteins. *Biochem. Int.* **4**:201-208.

Miura, S., Mori, M., and Tatibana, M., 1983, Transport of ornithine carbamyltransferase precursor into mitochondria. Stimulation by potassium ion, magnesium ion and reticulocyte cytosolic protein(s). *J. Biol. Chem.* **258**:6671-6674.

Miura, S., Mori, M., Takiguchi, M., Tatibana, M., Furuta, S., Miyazawa, S., and Hashimoto, T., 1984, Biosynthesis and intracellular transport of enzymes of peroxisomal β-oxidation. *J. Biol. Chem.* **259**:6397-6402.

Miura, S., Amaya, Y., and Mori, M., 1986, A metalloprotease involved in the processing of mitochondrial precursor proteins. *Biochem. Biophys. Res. Commun.* **134**:1151-1159.

Miyazawa, S., Ozasa, H., Furuta, S., Osumi, T., Hashimoto, T., Miura, S., Mori, M., and Tatibana, M., 1983, Biosynthesis and turnover of carnitine acetyltransferase in rat liver. *J. Biochem. (Tokyo)* **93**:453-459.

Montisano, D. F., Cascarano, J., Pickett, C. B., and James, T. W., 1982, Association between mitochondria and rough endoplasmic reticulum in rat liver, *Anat. Rec.* **203**:441-450.

Moore, K. E., and Miura, S., 1987, A small hydrophobic domain anchors leader peptidase to the cytoplasmic membrane of *Escherichia coli. J. Biol. Chem.* **262**:8806-8813.

Mori, M., Morita, T., Ikeda, F., Amaya, Y., Tatibana, M., and Cohen, P. P., 1981, Synthesis, intracellular transport and processing of the precursor for mitochondrial ornithine transcarbamylase and carbamyl phosphate synthetase I in isolated hepatocytes. *Proc. Natl. Acad. Sci. U.S.A.* **78**:6056–6060.

Mori, M., Miura, S., Morita, T., and Tatibana, M., 1982, Synthesis and intracellular transport of mitochondrial carbamyl phosphate synthetase I and ornithine transcarbamylase. *Adv. Exp. Med. Biol.* **153**:267–276.

Mori, M., Matsue, H., Miura, S., Tatibana, M., and Hashimoto, T., 1985, Transport of matrix proteins into mitochondrial matrix. Evidence suggesting a common pathway for 3-ketoacyl-CoA thiolase and enzymes having presequences. *Eur. J. Biochem* **149**:181–186.

Morita, T., Mori, M., Ikeda, F., and Tatibana, M., 1982, Transport of carbamyl phosphate synthetase I and ornithine transcarbamylase into mitochondria. Inhibition by rhodamine 123 and accumulation of enzyme precursors in isolated hepatocytes. *J. Biol. Chem.* **257**:13075–13080.

Mueckler, M., and Lodish, H. F., 1986, Post-translational insertion of a fragment of the glucose transporter into microsomes requires phosphoanhydride bond cleavage. *Nature (London)* **322**:549–552.

Mueckler, M., Caruso, C., Baldwin, S. A., Panico, M., Blench, I., Morris, H. R., Allard, W. J., Lienhard, G. E., and Lodish, H. F., 1985, Sequence and structure of a human glucose transporter. *Science* **229**:941–945.

Mueller, M., and Blobel, G., 1984, Protein export in *Escherichia coli* requires a soluble activity. *Proc. Natl. Acad. Sci. U.S.A.* **81**:7737–7741.

Mueller, M., Ibrahimi, I., Chang, C. N., Walter, P., and Blobel, G., 1982, A bacterial secretory protein requires signal recognition particle for translocation across mammalian endoplasmic reticulum. *J. Biol. Chem.* **257**:11860–11863.

Müller, G., and Zimmermann, R., 1987, Import of honeybee prepromelittin into the endoplasmic reticulum: structural basis of independence of SRP and docking protein, *EMBO J.* **6**:2099–2107.

Murén, E. M., and Randall, L. L., 1985, Export of α-amylase by *Bacillus amyloliquefaciens* requires proton motive force. *J. Bacteriol.* **164**:712–716.

Mustonen, P., Virtanen, J. A., Somerharju, P. J., and Kinnunen, P. K. J., 1987, Binding of cytochrome c to liposomes as revealed by the quenching of fluorescence from pyrene-labeled phospholipids. *Biochemistry* **26**:2991–2997.

Myers, M., Mayorga, D. L., Emtage, J., and Friere, E., 1987, Thermodynamic characterization of interactions between ornithine transcarbamylase leader peptide and phospholipid bilayer membranes, *Biochemistry* **26**:4309–4315.

Nabi, N., Ishikawa, T., Ohashi, M., and Omura, T., 1983, Contribution of cytoplasmic free and membrane-bound ribosomes to the synthesis of NADPH–adrenodoxin of bovine adrenal cortex mitochondria. *J. Biochem. (Tokyo)* **94**:1505–1515.

Nagaraj, R., 1984, Interaction of synthetic signal sequence fragments with model membranes. *FEBS Lett* **165**:79–82.

Nagata, S., Tsunetsuga-Yokota, Y., Naito, A., and Kaziro, Y., 1983, Molecular cloning and sequence determination of the nuclear gene coding for mitochondrial elongation factor Tn of *Saccharomyces cerevisiae. Proc. Natl. Acad. Sci. U.S.A.* **80**:6182–6196.

Nakagawa, T., Maeshima, M., Muto, H., Kajiura, H., Hattori, H., and Asahi, T., 1987, Separation, amino-terminal sequence and cell free synthesis of the smallest subunit of sweet potato cytochrome c oxidase. *Eur. J. Biochem.* **165**:303–307.

Natsoulis, G., Hilger, F., and Fink, G. R., 1986, The HTS gene encodes both the cytoplasmc and mitochondrial histidine tRNA synthetases of *S. cerevisiae. Cell* **46**:235–243.

Nelson, N., Maccecchini, M. L., Rudin, Y., and Schatz, G., 1979, Import of proteins into mitochondria, in *Biological Function of Proteinases* (H. Holzer and H. Tschesche, eds.), pp. 109–119, Springer, Berlin.

Nesmeyanova, M. A., 1982, On the possible participation of acid phospholipids in the translocation of secreted proteins through the bacterial cytoplasmic membrane, *FEBS Lett.* **142**:189–193.
Ness, S. A., and Weiss, R. L., 1987, Carboxyl-terminal sequences influence the import of mitochondrial protein precursors *in vivo*, *Proc. Natl. Acad. Sci. USA* **84**:6692–6696.
Neupert, W., and Schatz, G., 1981, How proteins are transported into mitochondria. *Trends Biochem. Sci.* **6**:1–4.
Nguyen, M., and Shore, G. C., 1987, Import of hybrid vesicular stomatitis G protein to the mitochondrial inner membrane. *J. Biol. Chem.* **262**:3929–3931.
Nguyen, M., Argan, C., Lusty, C. J., and Shore, G. C., 1986, Import and processing of hybrid proteins by mammalian mitochondria *in vitro*. *J. Biol. Chem.* **261**:800–805.
Nguyen, M., Argan, C., Sheffield, W. P., Bell, A. W., Shields, D., and Shore, G. C., 1987, A signal sequence domain essential for processing, but not import, of mitochondrial pre-ornithine carbamyltransferase. *J. Cell Biol.* **104**:1193–1198.
Nicholson, D. W., Koehler, H., and Neupert, W., 1987, Import of cytochrome *c* into mitochondria. Cytochrome *c* heme lyase. *Eur. J. Biochem.* **144**:147–157.
Niranjan, B. G., Raza, H., Shayiq, R. M., Jefcoate, C. R., and Avadhani, N. G., 1988, Hepatic mitochondrial cytochrome P-450 system. Identificatin and characterization of a precursor form of mitochondrial cytochrome P-450 induced by 3-methylcholanthrene, *J. Biol.Chem.* **262**:575–580.
Noda, M., Shimizu, S., Tanabe, T., Takai, T., Kayano, T., Ireka, T., Takahashi, H., Nakayama, H., Hanaoka, Y., Inayama, S., Hayashiba, H., Miyata, T., and Numa, S., 1984, Primary structure of *Electrophorus electricus* sodium channel deduced from cDNA sequence. *Nature (London)* **312**:121–127.
Noel, C., Nicolau, Y., Argan, C., Rachubinski, R. A., and Shore, G. C., 1985, *In vitro* synthesis and assembly of a 68 k Da outer membrane protein from rat liver. *Biochim. Biophys. Acta* **814**:35–42.
Oda, T., Ichiyama, A., Miura, S., and Mori, M., 1984, Uptake and processing of serine: Pyruvate aminotransferase precursor by rat liver mitochondria *in vitro* and *in vivo*. *J. Biochem. (Tokyo)* **95**:815–824.
Ogishima, T., Okada, Y., and Omura, T., 1985, Import and processing of the precursor of cytochrome P-450 *(scc)* by bovine adrenal cortex mitochondria. *J. Biochem.* **98**:781–791.
Ohashi, A., Gibson, J., Gregor, I., and Schatz, G., 1982, Import of proteins into mitochondria. The precursor of cytochrome c_1 is processed in two steps one of them heme dependent. *J. Biol. Chem.* **257**:13042–13047.
Ohba, M., and Schatz, G., 1987, Disruption of the outer membrane restores protein import to trypsin treated yeast mitochondria. *EMBO J.* **6**:2117–2122.
Ohkawa, I., and Webster, R. E., 1981, The orientation of the major coat protein of bacteriophage f_1 in the cytoplasmic membrane of *Escherichia coli*. *J. Biol. Chem.* **256**:9951–9958.
Ohno-Iwashita, Y., and Wickner, W., 1983, Reconstitution of rapid and asymmetrric assembly of M_{13} procoat protein into liposomes which have bacterial leader peptidase. *J. Biol. Chem.* **258**:1895–1900.
Okada, Y., Frey, A. B., Guenthner, T. M., Oesh, F., and Sabatini, D. D., 1982, Sites of synthesis and mode of insertion of cytochrome b_5, cytochrome b_5 reductase, cytochrome P-450 reductase and epoxide hydrolase. *Eur. J. Biochem.* **122**:393–402.
Oliver, D. B., 1985, Identification of five new essential genes involved in the synthesis of a secreted protein in *Escherichia coli*. *J. Bacteriol.* **161**:285–291.
Oliver, D. B., and Beckwith, J., 1981, *E. coli* mutant pleiotropically defective in the export of secreted proteins. *Cell* **25**:765–772.
Oliver, D. B., and Beckwith, J., 1982a, Identification of a new gene (sec A) and gene product

involved in the secretion of the envelope proteins in *Escherichia coli. J. Bacteriol.* **150:**686–691.
Oliver, D. B., and Beckwith, J., 1982b, Regulation of a membrane component is required for protein secretion in *Escherichia coli. Cell* **30:**311–319.
Oliver, D. B., and Liss, L. R., 1985, PrlA-mediated suppression of signal sequence mutations is modulated by the sec A gene product of *Escherichia coli* K-12. *J. Bacteriol.* **161:**817–819.
Ono, H., and Ito, A., 1984a, Transport of the precursor of sulfite oxidase into intermembrane space of liver mitochondria: Characterization of import and processing activities. *J. Biochem. (Tokyo)* **95:**345–352.
Ono, H., and Ito, A., 1984b, Transport of the precursor for sulfite oxidase into intermembrane space of liver mitochondria: Binding of the precursor to outer mitochondrial membrane. *J. Biochem. (Tokyo)* **95:**353–358.
Ono, H., and Tuboi, S., 1985, Partial purification of the receptor protein for import of preornithine aminotransferase into mitochondria. *Biochem. Int.* **10:**351–357.
Ono, H., and Tuboi, S., 1986, Translocation of proteins into rat liver mitochondria. The precursor polypeptides of a large subunit of succinate dehydrogenase and ornithine aminotransferase and their import into their own locations of mitochondria. *Eur. J. Biochem.* **155:**543–549.
Ono, H., and Tuboi, S., 1988, The cytosolic factor required for import of precursors of mitochondrial proteins into mitochondria, *J. Biol. Chem.* **263:**3188–3193.
Ono, H., Yoshimura, N., Sato, M., and Tuboi, S., 1985, Translocation of proteins into rat liver mitochondria. Existence of two different precursor polypeptides of liver fumarase and import of the precursor into mitochondrria. *J. Biol. Chem.* **260:**3402–3407.
Op den Kamp, J. A. F., 1979, Lipid asymmetry in membranes, *Ann. Rev. Biochem.* **48:**47–71.
Osumi, T., Ishii, N., Hijikata, M., Kamijo, K., Ozasa, H., Furuta, S., Miyazawa, S., Kondo, K., and Inoue, K., 1985, Molecular cloning and nucleotide sequence of cDNA for rat peroxisomal enoyl CoA: Hydrarase-3-hydroxyacyl-CoA dehydrogenase bifunctional enzyme. *J. Biol. Chem.* **260:**8905–8910.
Otha, S., and Schatz, G., 1984, A purified precursor polypeptide requires a cytosolic protein fraction for import into mitochondria. *EMBO J.* **3:**651–657.
Ou, W., Ito, A., Morohashi, K., Fujii-Kuriyama, Y., and Omura, T., 1986, Processing-independent *in vitro* translocation of cytochrome P-450 *(scc)* precursor across mitochondrial membranes. *J. Biochem. (Tokyo)* **100:**1287–1296.
Ovchinnikov, Yu. A., 1982, Rhodopsin and bacteriorhodopsin: Structure function relationship, *FEBS Lett.* **148:**179–191.
Ovchinnikov, Yu. A., Abdulaev, N. G., Feigina, M. Y., Artamonov, I. D., Bogachuk, A. S., Zolotarev, A. S., Egamian, E. R., and Kostetsky, P. V., 1983, Visual rhodopsin III. Total amino acid sequence and arrangement in the membrane, *Bioorg. Khim* **9:**1331–1340.
Ozasa, H., Furuta, S., Miyazawa, S., Osumi, T., Hashimoto, T., Mori, M., Miura, S., and Tatibana, M., 1984, Biosynthesis of enzymes of rat liver mitochondrial β-oxidation. *Eur. J. Biochem.* **144:**453–458.
Pain, D., and Blobel, G., 1987, Protein import into chloroplasts requires a chloroplast ATPase. *Proc. Natl. Acad. Sci. U.S.A.* **84:**3288–3292.
Pain, D., Kanwar, Y. S., and Blobel, G., 1988, Identification of a receptor for protein import into chloroplasts and its localization to envelope contact zones, *Nature* **311:**232–237.
Palade, G. E., 1975, Intracellular aspects of the process of protein synthesis. *Science* **189:**347–358.
Palade, G. E., and Siekevitz, P., 1956a, Liver microsomes. An integrated morphological and biochemical study. *J. Biophys. Biochem. Cytol.* **2:**171–200.
Palade, G. E., and Siekevitz, P., 1956b, Pancreatic microsomes. An integrated morphological and biochemical study. *J. Biophys. Biochem. Cytol.* **2:**671–690.

Palmiter, R. D., Gagnon, J., and Walsh, K. A., 1978, Ovalbumin, a secreted protein without a transient hydrophobic leader sequence. *Proc. Natl. Acad. Sci. U.S.A.* **75**:94–98.

Palva, E. T., Hirst, T. R., Hardy, S. J. S., Holmgren, J., and Randall, L. L., 1981, Synthesis of a precursor of the B subunit of heat-labile enterotoxin in *Escherichia coli*. *J. Bacteriol.* **146**:325–330.

Parimoo, S., Rao, N., and Padmanaban, C., 1982, Cytochrome *c* oxidase is preferably synthesized in the rough endoplasmic reticulum–mitochondria complex in rat liver. *Biochem. J.* **208**:505–507.

Park, K. S., Frost, B., Tuck, M., Ho, L. L., Kim, S., and Paik, W. K., 1987, Enzymatic metabolism of *in vitro* synthesized apocytochrome c enhances its transport into mitochondria, *J. Biol. Chem.* **262**:14702–14708.

Park, S., Liu, G., Topping, T. B., Cover, W. H., and Randall, L. L., 1988, Modulation of folding pathways of exported proteins by the leader sequence, *Science* **239**:1033–1035.

Paul, D. L., and Goodenough, D. A., 1983, *In vitro* synthesis and membrane insertion of bovine MP 26, an integral protein from lens fiber plasma membrane. *J. Cell Biol.* **96**:633–638.

Perara, E. and Lingappa, R., 1986, A former aminoterminal signal sequence engineered to an internal location directs translocation of both flanking protein domains. *J. Cell Biol.* **101**:2292–2301.

Perara, E., Rothman, R. E., and Lingappa, R., 1986, Uncoupling translocation from translation implication for transport of proteins across membrane. *Science* **232**:348–352.

Perlman, D., and Halvorson, H. O., 1983, A putative signal peptidase recognition site and sequence in eukaryotic and prokaryotic signal peptides. *J. Mol. Biol.* **167**:633–638.

Pfaller, R., and Neupert, W., 1987, High affinity binding sites involved in the import of porin into mitochondria. *EMBO J.* **6**:2635–2642.

Pfanner, N., and Neupert, W., 1985, Transport of proteins into mitochondria: a potassium diffusion potential is able to drive the import of ADP/ATP carrier. *EMBO J.* **4**:2819–2825.

Pfanner, N., and Neupert, W., 1986, Transport of F_1-ATPase subunit β into mitochodria depends on both a membrane potential and nucleotide triphosphates. *FEBS Lett.* **209**:152–156.

Pfanner, N., and Neupert, W., 1987, Distinct steps in the import of ADP/ATP carrier into mitochondria. *J. Biol. Chem.* **262**:7528–7536.

Pfanner, N., Tropschug, M., and Neupert, W., 1987a, Mitochondrial protein import: Nucleotide triphosphates are involved in conferring import competence to precursors. *Cell* **49**:815–823.

Pfanner, N., Hartl., F. U., Bernard, G., and Neupert, W., 1987b, Mitochondrial precursor proteins are imported through a hydrophilic membrane environment, *Eur. J. Biochem.* **169**:289–293.

Pfanner, N., Hoeben, P., Tropschug, M., and Neupert, W., 1987c, The carboxyl-terminal two-third of the ADP/ATP carrier polypeptide contains sufficient information to direct translocation into mitochondria, *J. Biol. Chem.* **262**:14851–14854.

Pfanner, N., Müller, H. K., Harmey, M. A., and Neupert, W., 1987d, Mitochondrial protein import: involvement of the mature part of a cleavable precursor protein in the binding to receptor sites, *EMBO J.* **6**:3449–3454.

Pfanner, N., Pfaller, R., Kleene, R., Ito, M., Tropschug, M., and Neupert, W., 1988, Role of ATP in mitochondrial protein import. Conformational alteration of a precursor protein can substitute for ATP requirement, *J. Biol. Chem.* **263**:4049–4051.

Pfisterer, J., Lachman, P., and Kloppstech, K., 1982, Transport of proteins into chloroplasts. Binding of nuclear-coded chloroplast proteins to the chloroplast envelope. *Eur. J. Biochem.* **126**:143–148.

Pilon, M., Jordi, W., De Kruijff, B., and Domel, R. A., 1987, Interaction of mitochondrial precursor protein apocytochrome c with phosphatidylserine in model membranes, *Biochim.*

Biophys. Acta **902**:207–216.
Pollit, S., and Zalkin, H., 1983, Role of primary structure and disulfide bound formation in β-lactamase secretion. *J. Bacteriol.* **153**:27–32.
Poulis, M. I., Sham, D. C., Campbell, H. D., and Young, I. G., 1981, In vitro synthesis of respiratory NADH dehydrogenase of *Escherichia coli*. Role of UUG as initiation codon. *Biochemistry* **20**:4178–4185.
Prehn, S., Tsamaloukas, A., and Rapoport, T. A., 1980, Demonstration of specific receptors of the rough endoplasmic membrane for the signal sequence of carp preproinsulin. *Eur. J. Biochem.* **107**:185–195.
Prehn, S., Nürnberg, P., and Rapoport, T. A., 1981, A receptor for signal segments of secretory proteins in rough endoplasmic reticulum membrane. *FEBS Lett.* **123**:79–84.
Prehn, S., Wiedmann, M., Rapoport, T. A., and Zwieb, C., 1987, Protein translocation across wheat germ microsomal membranes requires an SRP-like component. *EMBO J.* **6**:2093–2097.
Rachubinski, R. A., Verma, D. P. S., and Bergeron, J. J. M., 1980, Synthesis of rat liver microsomal cytochrome b_5 by free ribosomes. *J. Cell Biol.* **85**:705–716.
Racker, E., 1979, Reconstitution of membrane processes, in *Methods in Enzymology* (S. P. Colowick and D. Kaplan, eds.), Vol. 55, pp. 699–711, Academic Press, New York.
Randall, L. L., 1983, Translocation of domains of nascent periplasmic proteins across the cytoplasmic membrane is independent of elongation. *Cell* **33**:231–240.
Randall, L. L., and Hardy, S. J. S., 1977, Synthesis of exported proteins by membrane bound polysomes from *Escherichia coli*. *Eur. J. Biochem.* **75**:43–53.
Randall, L. L., and Hardy, S. J. S., 1986, Correlation of competence for export with lack of tertiary structure of the mature species: A study *in vivo* of maltose-binding protein in *E. coli*. *Cell* **46**:921–928.
Randall, L. L., Hardy, S. J. S., and Josefsson, L. G., 1978, Precursors of three exported proteins in *Escherichia coli*. *Proc. Natl. Acad. Sci. U.S.A.* **75**:1209–1212.
Rasmussen, B. A., and Bassford, P. J. Jr., 1985, Both linked and unlinked mutations can alter the intracellular site of synthesis of exported proteins of *Escherichia coli*. *J. Bacteriol.* **161**:258–264.
Rasmussen, B. A., and Silhavy, T. J., 1987, The first 28 amino acids of mature lam B are required for rapid and efficient export from the cytoplasm, *Genes Dev.* **1**: 185–196.
Redman, C. M., and Sabatini, D. D., 1966, Vectorial discharge of peptides released by puromycin from attached ribosomes. *Proc. Natl. Acad. Sci. U.S.A.* **56**:608–615.
Rees-Jones, R., and Alawqati, O., 1984, Proton-translocating adenosine triphosphatase in rough and smooth microsomes from rat liver. *Biochemistry* **23**:2236–2240.
Reid, G. A., and Schatz, G., 1982, Import of proteins into mitochondria. Extramitochondrial pool and post-translational import of mitochondrial proteins precursors *in vivo*. *J. Biol. Chem.* **257**:13062–13067.
Reid, G. A., Yonetani, T., and Schatz, G., 1982, Import of proteins into mitochondria. Import and maturation of mitochondrial intermembrane space enzymes cytochrome b_2 and cytochrome c peroxidase in intact yeast cells. *J. Biol. Chem.* **257**:13068–13074.
Reiss, B., Wasmann, C. C., and Bohnert, H. J., 1987, Regions in the transit peptide of SSU essential for transport into chloroplasts, *Mol. Gen. Genet.* **209**:116–121.
Rhoads, D. B., Tai, P. C., and Davis, B. D., 1984, Energy requiring translocation of OmpA protein and alkaline phosphatase of *Escherichia coli* into inner membrane vesicle. *J. Bacteriol.* **159**:63–70.
Richter, J. D., Lorenz, L. J., and Andet, R. G., 1985, Membrane-bound mRNAs are recruited from preinitiated ribonucleoprotein particles in injected *Xenopus* oocytes. *J. Biol. Chem.* **260**:4448–4454.

Rietveld, A., Jordi, W., and De Kruijff, B., 1986, Studies on the lipid dependency and mechanism of the translocation of the mitochondrial precursor protein apocytochrome *c* across model membranes. *J. Biol. Chem.* **261**:3846–3856.

Riezman, H., 1982, Binding of precursors of cytoplasmically-synthesized mitochondrial proteins to isolated outer membranes of yeast mitochondria. *Prog. Clin. Res.* **102b**:161–170.

Riezman, H., Weir, E. M., Leaver, C. J., Titus, D. E., and Becker, W. M., 1980, Regulation of glyoxysomal enzymes during germination of cucumber. *Plant Physiol.* **65**:40–46.

Riezman, H., Hay, R., Witte, C., Nelson, N., and Schatz, G., 1983a, Yeast mitochondrial outer membrane specifically binds cytoplasmically synthesized precursors of mitochondrial proteins. *EMBO J.* **2**:1113–1118.

Riezman, H., Hase, T., Van Loon, A. P. G. M., Grivell, L. A., Suda, K., and Schatz, G., 1983b, Import of proteins into mitochondria: A 70-kilodalton outer membrane protein with a large carboxy-terminal deletion is still transported to the outer membrane. *EMBO J.* **2**:2161–2168.

Riggs, P. D., Derman, A. I., and Beckwith, J., 1988, A mutation affecting the regulation of a sec A-lac Z fusion defines a new sec gene, *Genetics* **118**:1571–579.

Rizzolo, L. J., Finidori, J., Gonzalez, A., Arpin, M., Ivanov, I. E., Adesnik, M., and Sabatini, D. D., 1985, Biosynthesis and intracellular sorting of growth hormone-viral envelope glycoprotein hybrids. *J. Cell Biol.* **101**:1351–1356.

Roa, M., and Blobel, G., 1983, Biosynthesis of peroxisomal enzymes in the methylotropic yeast *Hansenula polymorpha*. *Proc. Natl. Acad. Sci. U.S.A.* **80**:6872–6876.

Robbi, M., and Lazarow, P. B., 1978, Synthesis of catalase in two cell free protein synthesizing systems in rat liver. *Proc. Natl. Acad. Sci. U.S.A.* **75**:4344–4348.

Robinson, A., Kaderbhai, M. A., and Austen, B. M., 1987, Identification of signal binding proteins integrated into the rough endoplasmic reticulum membrane. *Biochem. J.* **242**:767–777.

Rodriguez-Garcia, M. F., and Sievers, A., 1977, Membrane contacts of the endoplasmic reticulum with plastids and with the plasmalemma in the endothecium of *Scilla non-scripta*. *Cytobiologie* **15**:85–95.

Roggenkamp, R., Janiwicz, Z., Stanikowski, B., and Hollenberg, C. P., 1984, Biosynthesis and regulation of the peroxisomal methanol oxidase from methylotropic yeast *Hansenula polymorpha*. *Mol. Gen. Genet.* **194**:489–493.

Roise, D., Horvath, S. J., Tomich, J. M., Richards, J. H., and Schatz, G., 1986, A chemically synthesized pre-sequence of an imported mitochondrial protein can form an amphiphilic helix and perturb natural and artificial phospholipid bilayers. *EMBO J.* **5**:1327–1334.

Roise, D., Theiler, F., Horvath, S., Tomich, J. M., Richards, J. H., Allison, D. S., and Schatz, C., 1988, Amphiphilicity is essential for mitochondrial presequence function, *EMBO J.* **7**:649–653.

Rose, J. K., and Bergman, J. E., 1982, Expression from cloned cDNA of cell surface and secreted forms of the glycoprotein of vesicular stomatitis virus in eukaryotic cells. *Cell* **30**:753–762.

Rosenblatt, M., Habener, J. F., Tyler, F. A., Shepard, G. L., and Potts, J. T. Jr., 1979, Chemical synthesis of precursor-specific region of pre-proparathyroid hormone. *J. Biol. Chem.* **254**:1414–1421.

Rosenblatt, M., Beaudett, N. V., and Fasman, G. D., 1980, Conformational studies on the synthetic precursor specific region of pre-proparathyroid hormone. *Proc. Natl. Acad. Sci. U.S.A.* **77**:3983–3987.

Rosenkrantz, A., Alam, T., Kim, K. S., Clark, B., Srere, P. A., Garante, L. P., 1986, Mitochondrial and nonmitochondrial citrate synthases in *Saccharomyces cerevisiae* are encoded by distinct homologous genes, *Mol. Cell. Biol.* **6**:4509–4515.

Rothblatt, J. A., and Meyer, D. I., 1986a, Secretion in yeast: Reconstitution of translocation and glycosylation of α-factor and invertase in homolgous cell free systems. *Cell* **44**:619–628.

Rothblatt, J. A., and Meyer, D. I., 1986b, Secretion in yeast: Translocation and glycosylation of pre-pro α-factor *in vitro* can occur via an ATP dependent post-translational mechanism. *EMBO J.* **5**:1031–1036.

Rothblatt, J. A., Webb, J. R., Ammerer, G., and Meyer, D. I., 1987, Secretion in yeast: structural features influencing the post-translational translocation of prepro α-factor *in vitro*, *EMBO J.* **6**:3455–3468.

Rothman, J. E., and Lenard, J., 1977, Membrane asymmetry, the nature of membrane asymmetry provides clues to the puzzle of how membranes are assembled. *Science* **195**:743–753.

Rothstein, S. J., Lazarus, C. M., Smith, W. E., Baulcombe, D. C., and Gatenby, A. A., 1984, Secretion of a wheat α-amylase expressed in yeast. *Nature (London)* **308**:662–665.

Rothstein, S. J., Gatenby, A. A., Wiley, D. L., and Gray, J. C., 1985, Binding of pea cytochrome f to the inner membrane of *Escherichia coli* requires the bacterial sec A gene product. *Proc. Natl. Acad. Sci. U.S.A.* **82**:7955–7959.

Rottier, P., Armstrong, J., and Meyer, D. I., 1985, Signal recognition particle-dependent insertion of coronavirus E_1, an intracellular membrane glycoprotein. *J. Biol. Chem.* **260**:4648–4652.

Russel, M., and Model, P., 1981, A mutant downstream from the signal peptidase cleaving site affects cleavage but not membrane insertion of phage coat protein. *Proc. Natl. Acad. Sci. U.S.A.* **78**:1717–1721.

Ryan, J. P., Duncan, M. C., Bankaitis, Y. A., and Bassford, P. J. Jr., 1986, Intragenic reversion mutations that improve export of maltose-binding protein in *Escherichia coli* mal E signal sequence mutants. *J. Biol. Chem.* **261**:3389–3395.

Rzepecki, L. M., Strittmatter, P., and Herbette, L. G., 1986, X-ray diffraction analysis of cytochrome b_5 reconstituted in egg phosphatidylcholine vesicles, *Biophys. J.* **49**:829–838.

Sakaguchi, M., Mihara, K., and Sato, R., 1984, Signal recognition particle is required for cotranslational insertion of cytochrome P-450 into microsomal membrane. *Proc. Natl. Acad. Sci. U.S.A.* **81**:3361–3364.

Sakaguchi, M., Mihara, K., and Sato, R., 1987, A short amino terminal segment of microsomal cytochrome P-450 functions both as an insertion signal and as a stop-transfer sequence. *EMBO J.* **6**:2425–2431.

Sakakibara, R., Horio, Y., Ishiguro, M., Kanagawa, K., Matsuo, H., and Wada, H., 1987, An antibody and anti-idiotypic antibody against the extra signal peptide of pre-aspartate aminotransferase, *Biochem. Biophys. Res. Commun.* **148**:979–988.

Sakakibara, R., Kamisaki, Y., Horio, Y., and Wada, H., 1983, Precursor of mitochondrial glutamic oxaloacetic transaminase isoenzyme exits as dimer. *Biochem. Int.* **6**:231–238.

Santos, E., Kung, H., Young, I. G., and Kaback, H. R., 1982, *In vitro* synthesis of membrane bound D-lactate dehydrogenase of *Escherichia coli*. *Biochemistry* **21**:2085–2091.

Sautter, C., Keller, G., and Hock, B., 1988, Glyoxysomal citrate synthase from watermelon cotyledons: immunocytochemical localization and heterologous translation in *Xenopus* oocytes, *Planta* **173**:289–297.

Schatz, G., 1979, How mitochondria import proteins from the cytoplasm. *FEBS Lett.* **103**:203–211.

Schatz, G., and Butow, R. A., 1983, How are proteins imported into mitochondria. *Cell* **32**:316–318.

Schechter, I., Burstein, Y., Zemell, R., Ziv, E., Kantor, F., and Papermaster, D. S., 1979, Messenger RNA for opsin from bovine retina: Isolation and partial sequence of the *in vitro* translation product. *Proc. Natl. Acad. Sci. U.S.A.* **76**:2654–2658.

Scheele, G. A., Jacoby, R., and Carne, T., 1980, Mechanism of compartmentation of secretory proteins: Transport of exocrine pancreatic proteins across the microsomal membrane. *J. Cell Biol.* **87**:611–628.

Schleyer, M., and Neupert, W., 1985, Transport of proteins into mitochondria: Translational intermediates spanning contact sites between outer and inner membranes. *Cell* **43:**339–350.

Schleyer, M., Schmidt, B., and Neupert, W., 1982, Requirements of a membrane potential for the post-translational transfer of proteins into mitochondria. *Eur. J. Biochem.* **125:**109–116.

Schmidt, B., Henning, B., Zimmermann, R., and Neupert, W., 1983, Biosynthetic pathway of mitochondrial ATPase subunit 9 in *Neuropsora crassa. J. Cell Biol.* **96:**248–255.

Schmidt, G. W., Devillers-Thiery, A., Desruisseaux, H., Blobel, G., and Chua, N. H., 1979, Amino terminal amino acid sequences of precursor and mature forms of the ribulose 1,5-bisphosphate carboxylase small subunit from *Chlamydomonas reinhardtii. J. Cell Biol.* **83:**615–623.

Schmidt, R. J., Gillham, N. W., and Boynton, J. E., 1985, Processing of the precursor to a chloroplast ribosomal protein made in the cytosol occurs in two steps, one of which depends on a protein made in the chloroplast. *Mol. Cell. Biol.* **5:**1093–1099.

Schreier, P., Seftor, E. A., Schell, J., and Bohnert, H. J., 1985, The use of nuclear-encoded sequences to direct the light-regulated synthesis and transport of a foreign protein into plant chloroplasts. *EMBO J.* **4:**25–32.

Schwaiger, M., Herzog, V., and Neupert, W., 1987, Characterization of translocation contact sites involved in the import of mitochondrial proteins. *J. Cell Biol.* **105:**235–246.

Scotto, A. W., Goodwyn, D., and Zakim, D., 1987, Reconstitution of membrane proteins: Sequential incorporation of integral membrane proteins into preformed lipid bilayers. *Biochemistry* **26:**833–839.

Scoulica, E., Krause, E., Meese, K., and Dobberstein, B., 1987, Disassembly and domain structure of the proteins in the signal recognition particle. *Eur. J. Biochem.* **163:**519–528.

Seeburg, P. H., Shine, J., Martial, J. A., Ivarie, R. D., Morris, J. A., Ullrich, A., Baxter, J. D., and Goodman, H. M., 1978, Synthesis of growth hormone by bacteria. *Nature (London)* **276:**795–798.

Seehara, J. S., and Khorana, H. G., 1984, Bacteriorhodopsin precursor characterization and its integration into the purple membrane. *J. Biol. Chem.* **259:**4187–4193.

Sekizawa, J., Inouye, S., Halegoua, S., and Inouye, M., 1977, Precursors of major outer membrane proteins of *Escherichia coli. Biochem. Biophys. Res. Commun.* **77:**1126–1133.

Severina, I. I., and Skulachev, V. P., 1984, Ethylrhodamine as a fluorescent penetrating cation and a membrane potential sensitive probe in cyanobacterial cells. *FEBS Lett.* **165:**61–71.

Sharma, C. P., and Gehring, H., 1986, The precursor of mitochondrial aspartate aminotransferase is translocated into mitochondria as apoprotein. *J. Biol. Chem.* **261:**11146–11149.

Shiba, K., Ito, K., Yura, T., and Cerretti, D. P., 1984, A defined mutation in the protein export gene within the spc ribosomal protein operon of *Escherichia coli:* Isolation and characterization of a new temperature-sensitive sec Y mutant. *EMBO J.* **3:**631–635.

Shiba, K., Ito, K., and Yura, T., 1986, Suppressors of the sec Y 24 mutation: Identification and characterization of additional ssy genes in *Escherichia coli. J. Bacteriol.* **166:**849–856.

Sidhu, A., and Beattie, D., 1983, Kinetics of assembly of complex III into the yeast mitochondrial membrane. Evidence for a precursor to the iron-sulfur protein. *J. Biol. Chem.* **256:**10649–10656.

Siegel, V., and Walter, P., 1985, Elongation arrest is not a prerequisite for secretory protein translocation across the microsomal membrane. *J. Cell Biol.* **100:**1913–1921.

Siegel, V., and Walter, P., 1986, Removal of the Alu structural domain from signal recognition particle leaves its protein translocation activity intact. *Nature (London)* **320:**81–84.

Siegel, V., and Walter, P., 1988a, Each of the activities of signal recognition particle (SRP) is contained within a distinct domain: Analysis of biochemical mutants of SRP, *Cell* **52:**39–49.

Siegel, V., and Walter, P., 1988b, The affinity of signal recognition particle for presecretory

proteins is dependent on nascent chain length, *EMBO J.* **7:**1769–1775.
Simon, K., Perara, E., and Lingappa, V. R., 1987, Translocation of globin fusion proteins across the endoplasmic reticulum membrane in *Xenopus laevis* oocytes. *J. Cell Biol.* **104:**1165–1172.
Singer, S. J., 1977, The fluid mosaic model of membrane structure, in *Structure of Biological Membranes* (S. Abrahamsson and I. Pascher, eds.), pp. 443–461, Plenum Press, New York.
Singer, S. J., Maher, P. A., and Yaffe, M., 1987a, On the translocation of proteins across membranes. *Proc. Natl. Acad. Sci. U.S.A.* **84:**1015–1019.
Singer, S. J., Maher, P. A., and Yaffe, M. P., 1987b, On the transfer of integral proteins into membranes. *Proc. Natl. Acad. Sci. U.S.A.* **84:**1960–1964.
Skerjanc, I. S., Shore, G. C., and Silvius, J. R., 1987, The interaction of a synthetic mitochondrial signal peptide with lipid membranes is independent of transbilayer potential, *EMBO J.* **6:**3117–3123.
Small, G. M., and Lazarow, P. B., 1987, Import of the carboxy-terminal portion of acyl-CoA oxidase into peroxisomes of *Candida tropicalis*. *J. Cell Biol.* **105:**247–250.
Smeekens, S., Bauerle, C., Hageman, J., Keegstra, K., and Wasbeek, P., 1986, The role of the transit peptide in the routing of precursors towards different chloroplast comparttments. *Cell* **46:**365–375.
Smeekens, S., van Steeg, H., Bauerle, C., Bettenbroek, H., Keegstra, K., and Weisbeek, P., 1987, Import into chloroplasts of a yeast mitochondrial protein directed by ferredoxin and plastocyanin transit peptides, *Plant Mol. Biol.* **9:**377–388.
Smith, H., Bron, S., Van Ee, J., and Venema, G., 1987, Construction and use of signal sequence selection vectors in *Escherichia coli* and *Bacillus subtilis*. *J. Bacteriol.* **169:**3321–3328.
Smith, W. P., 1980, Co-translational secretion of diphtheria toxin and alkaline phosphatase *in vitro:* Involvement of membrane protein(s). *J. Bacteriol.* **141:**1142–1147.
Smith, W. P., Tai, P. C., and Davis, B. D., 1978a, Nascent peptides as sole attachment of polysomes to membranes in bacteria. *Proc. Natl. Acad. Sci. U.S.A.* **75:**814–817.
Smith, W. P., Tai, P. C., and Davis, B. D., 1978b, Interaction of secreted nascent chains with surrounding membrane in *Bacillus subtilis*. *Proc. Natl. Acad. Sci. U.S.A.* **75:**5921–5925.
Smith, W. P., Tai, P. C., and Davis, B. D., 1979, Extracellular labeling of growing secreted polypeptide chains with surrounding membrane with diazoiodosulfanilic acid. *Biochemistry* **18:**198–202.
Smith, W. P., Tai, P. C., and Davis, B. D., 1981, *Bacillus licheniformis* penicillinase: Cleavage and attachment of lipid during co-translational secretion. *Proc. Natl. Acad. Sci. U.S.A.* **78:**3501–3505.
Soellner, T., Pfanner, N., and Neupert, W., 1988, Mitochondrial protein import: differential recognition of various transport intermediates by antibodies, *FEBS. Lett.* **229:**25–29.
Spiess, M., and Lodish, H. F., 1985, The sequene of a second human asialoglycoprotein receptor: Conservation of two receptors during evolution. *Proc. Natl. Acad. Sci. U.S.A.* **82:**6465–6469.
Spiess, M., and Lodish, H. F., 1986, An internal signal sequence; the asialoglycoprotein receptor membrane anchor is necessary. *Cell* **44:**177–185.
Spiess, M., and Handschin, C., 1987, Deletion analysis of the internal signal-anchor domain of the human asialoglycoprotein receptor H. *EMBO J.* **6:**2683–2691.
Spiess, M., Schwartz, A. L., and Lodish, H. F., 1985, Sequence of human asialoglycoprotein receptor cDNA, an internal signal sequence for membrane insertion. *J. Biol. Chem.* **260:**1979–1982.
Strauch, K. L., Kumamoto, C. A., and Beckwith, J., 1986, Does sec A mediate coupling between secretion and translation in *Escherichia coli?* *J. Bacteriol.* **166:**505–512.
Strittmatter, P., Thiede, M. A., Hackett, C. S., and Ozols, J., 1988, Bacterial synthesis of active

rat stearyl-CoA desaturase lacking the 26-residue amino-terminal amino acid sequence, *J. Biol. Chem.* **262**:2532–2535.

Strom, M. S., and Lory, S., 1987, Mapping of export signals of *Pseudomonas aeruginosa* pili with alkaline phosphatase fusions. *J. Bacteriol.* **169**:3181–3188.

Strubin, M., Mach, B., and Long, E. O., 1984, The complete sequence of the mRNA for HLA-DR associated invariant chain reveals a polypeptide chain with an unusual transmembrane polarity. *EMBO J.* **3**:869–872.

Stuart, R. A., Neupert, W., and Tropschug, M., 1987, Deficiency in mRNA splicing in a cytochrome *c* mutant of *Neurospora crassa:* Importance of carboxy terminus for import of apocytochrome *c* into mitochondria. *EMBO J.* **6**:2131–2137.

Sugimoto, K., Sugisaki, H., Okamoto, T., and Takanami, M., 1977, Studies on the bacteriophage fd DNA. IV. The sequence of messenger RNA for the major coat protein gene. *J. Mol. Biol.* **111**:487–507.

Sugita, H., Tobe, T., Sakamoto, T., and Higashi, T., 1982, Immature precursor catalase in subcellular fractions of rat liver. *J. Biochem. (Tokyo)* **92**:509–515.

Suissa, M., and Schatz, G., 1982, Import of proteins into mitochondria. Translatable mRNA for imported mitochondrial proteins are present in free as well as mitochondrial-bound cytoplasmic polysomes. *J. Biol. Chem.* **257**:13048–13055.

Sutcliffe, J. G., 1978, Nucleotide sequence of the ampicillin resistance gene of *Escherichia coli* plasmide PBR 322. *Proc. Natl. Acad. Sci. U.S.A.* **75**:3737–3741.

Suzuki, Y., Orii, T., Takiguchi, M., Mori, M., Hijikata, M., and Hashimoto, T., 1987, Biosynthesis of membrane polypeptides of rat liver peroxisomes. *J. Biochem.* **101**:491–496.

Sveda, M. M., Markoff, L. J., and Lai, C. J., 1982, Cell surface expression of the influenza virus hemagglutinin requires the hydrophobic carboxy terminal sequences. *Cell* **30**:649–656.

Swinkels, B. W., Evers, R., and Borst, P., 1988, The topogenic signal of glycosomal (microbody) phosphoglycerate kinase of *Crithidia fasciaculata* resides in a carboxy-terminal extension, *EMBO J.* **7**:1159–1165.

Sztul, E. S., Hendrick, J. P., Kraus, J. P., Wall, D., Kalousek, F., and Rosenberg, L. E., 1987, Import of rat ornithine transcarbamylase precursor into mitochondria: two-step processing of the leader peptide, *J. Cell. Biol.* **105**:2631–2639.

Tabe, L., Krieg, P., Strachan, R., Jackson, D., Wallis, E., and Colman, A., 1984, Segregation of mutant ovalbumins and ovalbumin-globin proteins in *Xenopus* oocytes. *J. Mol. Biol.* **180**:645–666.

Tadashi, J., Hisayuki, N., Shuichiro, M., Kazunori, S., and Yoshimasa, M., 1985, Cloning and sequence analysis of a cDNA encoding porcine mitochondrial aspartate aminotransferase precursor, *Proc. Natl. Acad. Sci. USA* **82**:6065–6069.

Tajima, S., Lauffer, L., Rath, V. L., and Walter, P., 1986, The signal recognition particle receptor is a complex that contains two distinct polypeptide chains. *J. Cell Biol.* **103**:1167–1178.

Takagaki, Y., Radhakrishnan, R., Gupta, C. M., and Khorana, H. G., 1983a, The membrane embedded segment of cytochrome b_5 as studied by cross-linking with photoactivable phospholipids. I. Transferable form. *J. Biol. Chem.* **258**:9128–9135.

Takagaki, Y., Radhakrishnan, R., Wirtz, K. W. A., and Khorana, H. G., 1983b, The membrane embedded segment of cytochrome b_5 as studied by cross-linking with photoactivable phospholipids. II. Nontransferable form. *J. Biol. Chem.* **258**:9136–9143.

Takiguchi, M., Miura, S., Mori, M., and Tatibana, M., 1983, Transport of proteins into mitochondria: A high conservation of precursor uptake and processing systems. *Comp. Biochem. Physiol. B* **75B**:227–231.

Taldmage, K., Stahl, S., and Gilbert, W., 1980a, Eukaryotic signal sequence transports insulin antigen in *Escherichia coli*. *Proc. Natl. Acad. Sci. U.S.A.* **77**:3369–3373.

Talmadge, K., Kaufman, S., and Gilbert, W., 1980b, Bacteria mature preproinsulin to proinsulin. *Proc. Natl. Acad. Sci. U.S.A.* **77**:3988-3992.

Taldmadge, K., Brosius, J., and Gilbert, W., 1981, An "internal" signal sequence directs secretion and processing of proinsulin in bacteria. *Nature (London)* **294**:176-178.

Teintze, M., Slaughter, M., Weiss, H., and Neupert, W., 1982, Biogenesis of mitochondrial ubiquinol: Cytochrome creductase (cytochrome bc_1 complex). Precursor proteins and their transfer into mitochondria. *J. Biol. Chem.* **257**:10364-10371.

Thiede, M. A., Ozols, J., and Strittmatter, P., 1986, Construction and sequence of cDNA for rat liver stearyl coenzyme A desaturase. *J. Biol. Chem.* **261**:13230-13235.

Thibodeau, S. N., Lee, D. C., and Palmiter, R. D., 1978, Identical precursors for serum transferrin and egg white conalbumin. *J. Biol. Chem.* **253**:3771-3774.

Turn, N. J., and Silhavy, T. J., 1987, Characterization and *in vivo* cloning of *prl* C, a suppressor of signal sequence mutation in *Escherichia coli, Genetics* **116**:513-521.

Tweten, R. K., and Iandolo, J. J., 1983, Transport and processing of staphylococcal enterotoxin B. *J. Bacteriol.* **153**:297-303.

Valusek, G. P., Inouye, S., and Inouye, M., 1984, Effects of replacing serine and threonine residues within the signal peptide on the secretion of the major outer membrane lipoprotein of *Escherichia coli. J. Biol. Chem.* **259**:6195-6200.

Van Deenen, L. L. M., 1981, Topology and dynamics of phosopholipids in membranes, *FEBS Lett.* **123**:3-15.

Van Den Broek, G., Timko, M. P., Kausch, A. P., Cashmore, A. R., Montagu, M. V., and Herrera-Estrella, L., 1985, Targeting of a foreign protein to chloroplasts by fusion to the transit peptide from the small subunit of ribulose 1,5-bisphosphate carboxylase. *Nature (London)* **313**:358-363.

Van Loon, A. P. G. M., and Schatz, G., 1987, Transport of proteins to mitochondrial intermembrane space: The "sorting" domain of cytochrome c_1 presequence is a stop-transfer sequence specific for mitochondrial inner membrane. *EMBO J.* **6**:2441-2448.

Van Loon, A. P. G. M., and Young, E. T., 1986, Intracellular sorting of alcohol dehydrogenase isoenzymes in yeast: A cytosolic location reflects absence of an aminoterminal targeting sequence for the mitochondria. *EMBO J.* **5**:161-165.

Van Loon, A. P. G. M., De Groot, R. J., De Haan, M., Dekker, A., and Grivell, L. A., 1984, The DNA sequence of the nuclear gene coding for the 17 kd subunit VI of the yeast ubiquinol-cytochrome *c* reductase: A protein with an extremely high content of acidic amino acid. *EMBO J.* **3**:1039-1043.

Van Loon, A. P. G. M., Brändli, A. W., and Schatz, G., 1986, The presequence of two imported mitochondrial proteins contains information for intracellular and intramitochondrial sorting. *Cell* **44**:801-812.

Van Loon, A. P. G. M., Brädli, A. W., Pesold-Hurt, B., Blank, D., and Schatz, G., 1987, Transport of proteins into the mitochondrial intermembrane space: "Matrix targeting" and the sorting domains in cytochrome c_1 presequence. *EMBO J.* **6**:2433-2439.

Van Steeg, H., Oudshoorn, P., Van Hell, B., Polman, J. E. M., and Grivell, L. A., 1986, Targeting efficiency of a mitochondrial presequence is dependent on the passenger protein. *EMBO J.* **5**:3643-3650.

Vassarotti, A., Chen, W. J., Smagula, C., and Douglas, M. G., 1987a, Sequences distal to mitochondrial targeting sequence are necessary for the maturation of F_1-ATPase β-subunit precursor in mitochondria. *J. Biol. Chem.* **262**:411-418.

Vassarotti, A., Stroud, R., and Douglas, M., 1987b, Independent mutations at the amino terminus of a protein act as surrogate signals for mitochondrial import. *EMBO J.* **6**:705-711.

Verner, K., and Schatz, G., 1987, Import of an incompletely folded precursor protein into isolated

mitochondria requires an energized inner membrane, but no added ATP. *EMBO J.* **6:**2449–2456.
Vestweber, D., and Schatz, G., 1988, Point mutations destabilizing a precursor protein enhance its post-translational import into mitochondria, *EMBO J.* **7:**1147–1151.
Viebrock, A., Perz, A., and Sebald, W., 1982, The imported preprotein of the proteolipid subunit of the mitochondrial ATP synthase from *Neurospora crassa*. Molecular cloning and sequencing of the mRNA. *EMBO J.* **1:**565–571.
Von Heijne, G., 1980, Transmembrane translocation of proteins. A detailed physicochemical analysis. *Eur. J. Biochem.* **103:**431–438.
Von Heijne, G., 1984, How signal sequences maintain cleavage specificity. *J. Mol. Biol.* **173:**243–251.
Von Heijne, G., 1985, Signal sequences. The limits of variation. *J. Mol. Biol.* **134:**99–105.
Von Heijne, G., 1986, Mitochondrial targeting sequences may form amphiphilic helices. *EMBO J.* **5:**1335–1342.
Von Heijne, G., and Blomberg, C., 1979, Transmembrane translocation of proteins. The direct transfer model. *Eur. J. Biochem.* **97:**175–181.
Von Heijne, G., and Segrest, J. P., 1987, The leader peptides from bacteriorhodopsin and halorhodopsin are potential membranes spanning amphiphilic helixes. *FEBS Lett.* **213:**238–240.
Walk, R. A., and Hock, B., 1978, Cell-free synthesis of glyoxysomal malate dehydrogenase. *Biochem. Biophys. Res. Commun.* **81:**636–645.
Walter, A., Margolis, D., Mohan, R., and Blumenthal, R., 1986, Apocytochrome *c* induces pH-dependent vesicle fusion. *Membr. Biochem.* **6:**217–237.
Walter, P., and Blobel, G., 1980, Purification of a membrane-associated protein complex required for translocation across the endoplasmic reticulum. *Proc. Natl. Acad. Sci. U.S.A.* **77:**7112–7116.
Walter, P., and Blobel, G., 1981a, Translocation of proteins across the endoplasmic reticulum. II. Signal recognition protein (SRP) mediates the selective binding to microsomal membranes of *in vitro* assembled polysomes synthesizing secretory proteins. *J. Cell Biol.* **91:**551–556.
Walter, P., and Blobel, G., 1981b, Translocation of proteins across the endoplasmic reticulum. III. Signal recognition protein (SRP) causes signal sequence dependent and site specific arrest of chain elongation that is released by microsomal membranes. *J. Cell Biol.* **91:**557–561.
Walter, P., and Blobel, G., 1983, Disassembly and reconstitution of signal recognition particle. *Cell* **34:**525–533.
Walter, P., Jackson, R. C., Marcus, M. M., Lingappa, V. R., and Blobel, G., 1979, Tryptic dissection of receptor(s) for the translocation of presecretory proteins across the microsomal membrane and reconstitution of translocation activity. *Proc. Natl. Acad. Sci. U.S.A.* **76:**1795–1799.
Walter, P., Ibrahimi, I., and Blobel, G., 1981, Translocation of proteins across the endoplasmic reticulum. I. Signal recognition protein (SRP) binds to *in vitro* assembled polysomes synthesizing secretory protein. *J. Cell Biol.* **91:**545–551.
Wandinger-Ness, A. U., and Weiss, R. L., 1987, A single precursor protein for two separate mitochondrial enzymes in *Neurospora crassa*, *J. Biol. Chem.* **262:**5823–5830.
Wanner, B. L., Sarthy, A., and Beckwith, J. R., 1979, *Escherichia coli* periplasmic mutant that reduces amounts of several periplasmic and outer membrane proteins. *J. Bacteriol.* **140:**229–239.
Warren, T. G., and Dobberstein, B., 1978, Protein transfer across microsomal membrane reassembled from separated membrane components, *Nature* **273:**569–571.
Wasmann, C., C., Reis, B., Bartlett, S. G., and Bohnert, H., 1987, The importance of the transit peptide and the transported protein for import into chloroplasts, MGG. *Mol. Gen. Genet.* **205:**445–453.

Watanabe, K., and Kubo, S., 1982, Mitochondrial adenylate kinase from chicken liver. Purification, characterization and its cell-free synthesis. *Eur. J. Biochem.* **123**:587–592.

Watanabe, M., Hunt, J. F., and Blobel, G., 1986, In vitro synthesized outer membrane protein is integrated into bacterial inner membrane but translocated across microsomal membranes. *Nature* **323**:71–73.

Waters, M. G., and Blobel, G., 1986, Secretory protein translocation in a yeast cell free system can occur post-translationally and requires ATP hydrolysis. *J. Cell Biol.* **102**:1543–1550.

Waters, M. G., Chirico, W. J., and Blobel, G., 1986, Protein translocation across yeast microsomal membranes is stimulated by a soluble factor. *J. Cell Biol.* **103**:2629–2636.

Watson, E. E., 1984, Compilation of published signal sequences. *Nucleic Acids Res.* **12**:5145–5164.

Watts, C., Silver, P., and Wickner, W., 1981, Membrane assembly from purified components. II. Assembly of M_{13} procoat into liposomes reconstituted with purified leader peptides. *Cell* **25**:347–353.

Webster, K. A., Patel, H. V., Freeman, K. B., and Papahadjopoulos, D., 1979, Interaction of mitochondrial malate dehydrogenase monomer with phospholipid vesicles. *Biochem. J.* **178**:147–158.

White, J. A., and Scandalios, J. G., 1987, In vitro synthesis and processing of Mn-superoxide dismutase (SOD-3) into maize mitochondria, *Biochim. Biophys. Acta* **226**:16–25.

Wickner, W., 1979, The assembly of proteins into biological membranes. The membrane trigger hypothesis. *Annu. Rev. Biochem.* **49**:23–45.

Wiedmann, M., Huth, A., and Rapoport, T. A., 1986a, A signal sequence is required for the functions of the signal recognition particle. *Biochem. Biophys. Res. Commun.* **134**:790–796.

Wiedmann, M., Huth, A., and Rapoport, T. A., 1986b, Internally transposed signal sequence of carp preproinsulin retains its function with the signal recognition particle. *FEBS Lett.* **194**:139–145.

Wiedmann, M., Kurzchalia, T. V., Bielka, H., and Rapoport, T. A., 1987, Direct probing of the interaction between the signal sequence of nascent preprolactin and the signal recognition particle by specific cross-linking. *J. Cell Biol.* **104**:201–208.

Wiedmann, M., Kurzchalia, T. V., Hartmann, E., and Rapoport, T. A., 1987, A signal sequence receptor in the endoplasmic reticulum membrane, *Nature* **328**:830–833.

Wiedmann, M., Wiedmann, B., Voigt, S., Wachter, E., Müller, H. G., and Rapport, T. A., 1988, Post-translational transport of proteins into microsomal membranes of *Candida maltosa*, *EMBO J.* **7**:1763–1768.

Wolfe, P. B., and Wickner, W., 1984, Bacterial leader peptidase, a membrane protein without a leader peptide uses the same export pathway as presecretory proteins, *Cell* **36**:1067–1072.

Wolfe, P. B., Wickner, W., and Goodman, J. M., 1983, Sequence of the leader peptidase gene of *Escherichia coli* and the orientation of leader peptidase in the bacterial envelope. *J. Biol. Chem.* **258**:12073–12080.

Wu, H. C., Hou, C., Lin, J. J. C., and Yem, D. W., 1977, Biochemical characterization of a mutant lipoprotein of *Escherichia coli*. *Proc. Natl. Acad. Sci. U.S.A.* **74**:1388–1392.

Yaffe, M. P., and Schatz, G., 1984, Two nuclear mutations that block mitochondrial protein import in yeast. *Proc. Natl. Acad. Sci. U.S.A.* **81**:4819–4823.

Yamaguchi, J., Mori, H., and Nishimura, M., 1987, Biosynthesis and intracellular transport of glyoxysomal malate dehydrogenase in germinating pumpkin cotyledons. *FEBS Lett.* **213**:329–332.

Yamaguchi, J., Nishimura, M., and Akazawa, T., 1986, Purification and characterization of heme-containing low-affinity form of catalase from seeding cotyledons, *Eur. J. Biochem.* **159**:315–322.

Yamamoto, K., and Fahimi, H. D., 1987, Three-dimensional reconstruction of a peroxisomal reticulum in regenerating rat liver. *J. Cell Biol.* **105**:713–722.

Yamamoto, M., Hayashi, N., an Kikuchi, G., 1981, Regulation of synthesis and intracellular translocation of β-aminolevulinate synthase by heme and its relation to the heme saturation of tryptophan pyrolase in rat liver, *Arch. Biochem Biophys.* **209**:451–459.

Yamane, K., Ichihara, S., and Mizushima, S., 1987, Transloction of protein across *Escherichia coli* membrane vesicles requires both the proton motive force and ATP. *J. Biol. Chem.* **262**:2358–2362.

Yazgu, M., Shiota-Niiya, S., Shimamoto, T., Kanazawa, H., Futai, M., and Tsuchiya, T., 1984, Nucleotide sequence of mel B gene and characteristics of deduced amino acid sequence of melibiose carrier in *Escherichia coli. J. Biol. Chem.* **259**:4320–4326.

Yoshida, Y., Hashimoto, T., Kimura, H., Sakakibara, S., and Tagawa, K., 1985, Interaction with mitochondrial membranes of a synthetic peptide with a sequence common to extra peptides of mitochondrial precursor protein. *Biochem. Biophys. Res. Commun.* **128**:775–780.

Yost, S. C., Hedgpeth, J., and Lingappa, V. R., 1983, A stop transfer sequence confers predictable transmembrane orientation to a previously secreted protein in cell free systems. *Cell* **34**:759–766.

Young, E. T., and Pilgrim, D., 1985, Isolation and DNA sequence of ADH_3, a nuclear gene encoding the mitochondrial isoenzyme of alcohol dehydrogenase in *Saccharomyces cerevisiae. Mol. Cell. Biol.* **5**:3024–3034.

Yu, L. M., Merchant, S., Theg, S. M., and Selman, B. R., 1988, Ioslation of cDNA clone for the γ subunit of the chloroplast ATP synthase of *Chlamydomonas reinhardtii:* import and cleavage of the precursor, *Proc. Natl. Acad. Sci. USA* **85**:1369–1373.

Zaar, K., Völkl, A., and Fahimi, H. D., 1986, Isolation and characterization of peroxisomes from the renal cortex of beef, sheep, and cat. *Eur. J. Cell Biol.* **40**:16–24.

Zaar, K., Völkl, A., and Fahimi, H. D., 1987, Association of isolated bovine kidney cortex peroxisomes with endoplasmic reticulum. *Biochim. Biophys. Acta* **897**:135–142.

Zerial, M., Melancon, P., Schneider, C., and Garoff, H., 1986, The transmembrane segment of the human transferrin receptor functions as a signal peptide. *EMBO J.* **5**:1543–1550.

Zerial, M., Huylebroeck, D., and Garoff, H., 1987, Foreign transmembrane peptides replacing the internal signal sequences of transferrin receptor allow its translocation and membrane binding. *Cell* **48**:147–155.

Zimmermann, R., and Neupert, W., 1980a, Transport of proteins into mitochondria. Post-translational transfer of ADP/ATP carrier into mitochondria *in vitro. Eur. J. Biochem.* **109**:217–229.

Zimmermann, R., and Neupert, W., 1980b, Biogenesis of glyoxysomes. Synthesis and intracellular transfer of isocitrate lyase. *Eur. J. Biochem.* **112**:225–233.

Zimmermann, R., and Neupert, W., 1981, Different transport pathways of individual precursor proteins in mitochondria. *Eur. J. Biochem.* **116**:455–460.

Zimmermann, R., and Wickner, W., 1983, Energetics and intermediates of the assembly of protein OmpA into the outer membrane of *Escherichia coli. J. Biol. Chem.* **258**:3920–3925.

Zimmermann, R., Watts, C., and Wickner, W., 1982, The biosynthesis of membrane bound M_{13} coat protein. Energetics and assembly intermediates. *J. Biol. Chem.* **257**:6529–6536.

Zwizinski, C., and Neupert, W., 1983, Precursor proteins are transported into mitochondria in the absence of proteolytic cleavage of the additional sequences. *J. Biol. Chem.* **258**:13340–13346.

Zwizinski, C., Schleyer, M., and Neupert, W., 1983, Transport of proteins into mitochondria. Precursor to the ADP/ATP carrier binds to receptor sites on isolated mitochondria. *J. Biol. Chem.* **258**:4071–4074.

Zwizinski, C., Schleyer, M., and Neupert, W., 1984, Proteinaceous receptor for the import of mitochondrial precursor proteins. *J. Biol. Chem.* **259**:7850–7856.

Index

Absorption spectrometry
  lipid deuteration, 34
  polymorphic behavior of lipids, 35
  Raman spectroscopy, 35
Acetylcholine, 105
Acetylcholine receptor, 129–131, 339–359
  cation-selective channel, 339
  comparison of reconstituted and native, 354
  from electric tissue, 350–353
  of electrocytes, 340
  electrophysiology, 341, 342
  gating models, 350, 356–358
  glycoprotein subunits of, 130, 340
  incorporation into lipid bilayers, 130, 346–350
  liposomal, 131
  molecular structure of, 130
  of muscle, 340
  nicotinic pharmacology of, 340
  number of active channels, 354
  oligomeric structure, 358
  parameters of, 353
  pharmacological properties, 130, 352–355
  properties of reconstituted channel, 350–356
  purification by affinity chromatography, 343
  reconstitution into lipid bilayers, 339–359
    technical aspects, 342–350
  from SDS-PAGE, 355
  sensitivity to organic solvents, 346
  single channel current
    recording of, 347–350
  in soybean lipid vesicles, 342, 343
  tissue distribution
  of *Torpedo* electroplax, 130, 131
  two types of channel, 358
  of vertebrate brain, 340
  in *Xenopus* oocytes, 358

Actinomyces naeslandii
  glycolipid binding, 264
Affinity chromatography, 235, 261
  of acetylcholine receptor, 343
  respiratory complex, 289
Alamethicin, 207, 208
Aldosterone, 133, 134
Amiloride
  diuretic, 134
  inhibition of Na^+ channel, 134
Amphiphilic molecules, 1
Anabaena variabilis, 302, 303
Anacystis nidulans
  chilling damage, 79
  DSC of lipids, 72, 73
  freeze fracture, 70, 71
Antimycin A, 285
Apocytochrome *c*, 441, 442
Ascorbate, 286, 293
Asialoglycoprotein receptor, 410
  of hepatocytes, 262
  oligomeric complex of, 263
Asolectin, 289
ATPase
  proton-translocating, 279, 280
  radiation inactivation, 288
ATP synthase, 310
  reconstituted, 287

B. Caldontenax, 306
Bacteria
  aerobic respiration, 280
  anaerobic respiration, 282
  anaerobic electron transfer systems, 308–310
  cytochrome oxidase, 291–293

Bacteria (cont.)
  electron transfer chains
    reconstitution of, 280
  exported proteins, 383
  H^+/O ratio, 282–286
  membrane vesicles, 286
  NADH dehydrogenases, 305, 306
  oxidative phosphorylation, 287
  proton pump, 293–296
  respiratory burst, 283
  respiratory chains, 280–282
  terminal oxidants, 282
  transhydrogenase, 310–311
Bacterial
  cytochrome bc_1 complexes, 299–305
    properties of, 299–303
  cytochrome oxidase
    aa_3-type, 296, 297
    d-type, 297–299
    o-type, 297–299
  reconstitution of proton pump, 293–296
Bacterial respiratory complexes
  H^+/O ratio: See Mitochondria, 312
  hydrophobic nature, 289
  purification of, 289
  reconstitution in liposomes, 287–291
  reconstitution methods, 289–291
    detergent dialysis, 290
    detergent dilution, 290
    freeze-thaw-sonication, 290
    incorporation, 290
  requirement for polar lipids, 289
Bacteriorhodopsin, 408
Bacillus caldolyticus, 294, 295
Bacillus megaterium, 283
Bacillus stearothermophilus, 306
  acyl chain phase changes, 148
Bacillus subtilis, 283
  NADH dehydrogenase, 305, 306
  protein translocation, 391
  X-ray diffraction of lipids, 29
Band 3 protein
  reconstitution into BLM, 133
Basophil
  Fcε receptor, 321, 322
  leukemia cells, 325
Batrachotoxin
  action on reconstituted Na^+ channels, 109, 111, 112

Bilayer
  lipid composition, 3
Bilayer fusion
  Ca^{2+}-induced, 66
Bilayer lipid membranes (BLM), 97–137
  Ca^{2+} channels, 112–119
    effect of cholesterol on Ca^{2+} conductance, 120
    effect of phosphatidylserine on Ca^{2+} conductance, 120
  incorporation of
    acetylcholine receptor, 346–350
    erythrocyte band 3 protein, 133
    kidney proteolipid, 134, 135
    muscle Na^+ channel, 111
    native channels, 108
    polypeptides, 107
    sarcolemma Ca^{2+} channels, 117–119
    sarcoplasmic reticulum K^+ channel, 123
  interaction with microtubules, 115, 116
  properties of nerve Ca^{2+} channel, 113–117
  reconstitution of
    chloride channel, 126, 127
    ciliary membrane Ca^{2+} channel, 135
    K^+ channel from synaptosomes, 119
    mitochondrial anion-selective channel, 136
    Na^+ channel, 108, 109
  transport mechanisms, 107–112
Bilayer phases
  acyl chain length, 44, 45
  lyotrophic phases, 46
Biological membranes, 1
  damage at subzero temperatures, 78–80
  extrinsic proteins of, 3
  intrinsic proteins of, 3, 4
  mosaic of lipid and protein, 3
  lipid constituents, 25
  permeability barrier, 2
  physical properties of, 3
  structural unit of, 2
Bleomycin, 373
Blood platelet
  surface glycoproteins of, 239
Boltzmann distribution, 172
Bone marrow macrophage
  liposomal uptake, 374
Brij 58, 16–18
α-Bungarotoxin, 343

Index

Butylated hydroxytoluene (BHT)
  potentiation of PEG-induced fusion, 197

Ca^{2+}
  binding to acidic phospholipids, 246, 249
  effect on phospholipid phase separation, 158
  fusion of
    liposomes containing acidic P-lipids, 158
    phosphatidylserine liposomes, 196
  fusion mechanisms, 158
  in virally-induced membrane fusion, 211, 212
  intracellular concentration and fusion, 193
  lectin-induced agglutination, 244–249
  role in membrane fusion, 199, 200, 250–257
Ca^{2+}-ATPase, 77
Ca^{2+}-channel, 102, 103
  action of verapamil, 118, 119
  activation by ATP, 119
  in BLMs, 112–119
  block by Cd^{2+}, 114
  of ciliary membranes, 135
  comparison of
    ER and SR channels, 115
    native and reconstituted channels, 114
    cardiac and skeletal muscle channel, 118
  effect of tubulin, 115, 116
  kinetic characteristics of, 103, 105
  reconsitituted from
    brain ER, 114, 115
    muscle membranes, 117–119
    nerve membranes, 112–117
  regulation by microtubule proteins, 117
  ruthenium red-induced inhibition, 104, 115, 119
  stimulation of, 106
  structural diversity of, 118
  sarcoplasmic reticulum, 13
Candida tropicalis
  peroxisomal acyl-CoA oxidase, 420
Calpastatin
  in human erythrocytes, 194, 198
Carbamylcholine, 351, 354, 355
Carbocyanine
  probe for membrane fusion, 201
Carboxyfluorecein
  liposomal, 365–373

Cardiolipin, 66, 443
  cytochrome *c* interaction, 67
Cell surface
  glycoproteins, 229, 230
  -mediated responses, 229, 230
  receptors, 230
Cerebroside, 250
  fatty acyl chain length
    antibody interaction, 254
Ceremide, 237
  carbohydrate head group hydration, 254
Chemiosmotic theory, 279, 287
Chicken erythrocyte,
  fusion of, 250
Chlamydomonas reinhardtii
  ribulose 1, 5-bisphosphate carboxylase, 395, 401
Chloroplast
  cytochrome b_5, 299–301
  destacking of thylakoid membranes, 75
  ferredoxin, 424
  heat-stressed, 74, 76
  lipids of, 64
  membrane-associated particles, 76, 77
  migratory proteins, 384, 393, 400–403
  photophosphorylation, 279
  photosynthetic membranes, 67
  plastocyanin, 424
  post-translational translocation of proteins, 429
  protease treatment, 395, 396
  ribulose 1, 5-bisphosphate carboxylase
Cholate, 287, 323
Cholesterol
  effects on lipid fluidity, 150
  interaction with
    filipin, 148
    polymyxin B, 148
  lipid domain formation, 158–161
  liposomal, 364, 366, 367
  quantitation of lipid heterogeneity, 159
  stabilizing role in liposomes, 365, 366
Cholesteryl oleate, 370
Ciliary membranes
  reconstituted Ca^{2+} channel, 135
Cl^- channel, 126–127
  dimeric complex, 127
  erythrocyte exchange system, 127
Colchicine, 115

Colloid osmotic swelling
  role in membrane fusion, 191
Concanavalin A (Con A), 231, 232, 237, 244, 245, 249, 260, 261, 264, 266
  agglutination
    electrostatic forces in, 244
    mannose binding, 237
Coronene
  fluorescence probe, 168–171
Crithidia fasciculata
  glycosomal phosphoglycerate kinase, 420
Critical micelle concentration (CMC), 5–7, 14, 16–18, 20, 25, 323
  of polar lipids, 39
Cromolyn
  Ca^{2+} chelation, 328
  -binding protein, 328
Croton tiglium lectin, 236, 245
Cryopreservatives
  lipid polymorphic behavior, 30
Cytochromes, 282
Cytochrome bc_1 complexes, 299–305
  reconstitution of, 303–305
Cytochrome *c*, 66
Cytochrome *c* oxidase
  subunit organization, 288, 289
  radiation inactivation, 288
Cytochrome oxidase, 291–299
  subunit composition, 291
Cytochrome P450, 67
Cytoskeleton, 148

Dehydroascorbate, 285
Dehydrogenases, 280
Deoxycholate, 19, 287
Desialation
  of glycoproteins, 262
Detergent
  adsorption of hydrophobic beads, 14–19
  aggregation number, 7
  amphiphilic properties, 5
  bile salts, 6
  heterogeneity, 7
  interaction energy, 9
  interaction with
    diacyl lipid bilayers, 8, 9
    proteins, 7, 8
  inverted cone shape, 38
  ionic, 6

Detergent (*cont.*)
  -lipid-protein
    mixed micelles, 2
  in membrane reconstitution, 1–23
  micelle, 2, 5, 9
    sizes, 7
  nonionic, 6
  polyoxyethylene derivatives, 6, 7
  properties of, 5–7
  protective nature of, 22
  pure and mixed micelles, 9
  removal of, 10, 11
  self-association, 5
  solubilization of intrinsic proteins, 8
Deuterium-NMR
  fatty acyl chain motion, 167
Dialysis
  detergent removal, 14, 19
Dicetylphosphate, 364
Dielectrophoresis
  and membrane fusion, 204
Dictyostelium discoideum
  glycoproteins and cell adhesion, 264
  lectin of, 265
Differential scanning calorimetry (DSC), 61–64
  of *Anacystis nidulans* lipids, 72, 73
  cholesterol-induced heterogeneity, 159
  of chloroplast membranes, 74
  of 1, 2-dioleoylphosphatidylcholine, 61, 62
  of 1, 2-dipalmitoylphosphatidylcholine, 59, 60
  of 1, 2-distearoylphosphatidylcholine, 61, 62
  endothermic transition, 36
  of glycolipids, 241, 242
  lipid phase transitions, 35–37, 149
  kinetics of pretransition, 52
  of lamellar crystalline phases, 48–50
  phase mixing, 36
  of phase-separated domains, 71
  of phospholipid domains, 147–149
  transition enthalpy, 37
Dilatometry
  and lipid phase transitions, 37, 38
Dilauroylphosphatidylcholine, 45
  liposomal, 367
Dimyristoylphosphatidylcholine, 44
  domains of, 64
  kinetics of pretransition, 52–54

Dimyristoylphosphatidylcholine (cont.)
　liposomal, 367
Dimyristoylglycerophosphoric acid, 63
　cochleate phase, 64
Dimyristoylphosphatidylglycerol, 443
Dioleoylphosphatidylcholine, 42, 66
　liposomal, 367
Dioleoylphosphatidylethanolamine, 42
Dipalmitoylphosphatidylcholine, 44, 49, 50
　dilatometry of, 37, 38
　DSC scans of, 36, 37
　electron density profile, 54
　enantomers of, 58, 59
　kinetics of $P_{\beta'}$ and $L_{\beta'}$ phases, 51, 53, 54
　interaction with orinithine carbamyltransferase, 445
Dipalmitoylphosphatidylglycerol, 53
Dipalmitoylphosphatidylethanolamine, 59, 60
Distearoylphosphatidylcholine, 46, 371, 373
Distearoylphosphatidylethanolamine
　DSC scans of, 36, 37
Dodecane, 42
Dilichos biflorus lectin, 232, 234

EDTA
　reversal of lectin-induced agglutination, 246
Electric eel
　electroplax, 350
Electron microscopy
　direct visualization of lipidic particles, 30
　freeze-fracture of lipid cubic phases, 56, 57
　freeze-fracture replicas, 30
　fusion of phospholipid vesicles, 209, 210
　and lipid polymorphism, 30, 31
　of *Neurospora* mitochondria, 135
　of PEG-induced fusion, 202
　of phospholipid domains, 147, 148
　of reconstituted Fcε, 332
　of thylakoid membranes, 75
EPR spectroscopy, 146
　of phospholipid domains, 147
ESR spin-label
　fatty acyl chain motion, 167
Electron transfer chain
　of bacteria, 280–282
Electrophysiology
　of acetylcholine receptor, 341, 342
　analysis of single-channel parameters, 107
　analysis of reconstituted membranes, 101–107

Electrophysiology (cont.)
　of bilayer lipid membranes, 98
　concentration gradients, 101
　current amplitude histograms, 103, 104
　ion channel kinetics, 107
　ion transport processes, 102, 103
　of K^+-channels, 119–126
　of muscle, 111
　of reconstituted Ca^{2+}-channels, 102, 112–119
　single channel current, 103
　of sodium channel, 105
Endocytic pathway, 262
Endocytosis
　receptor-mediated, 263
Endoplasmic reticulum
　cytochrome b_5, 433
　NADH-Cytochrome b_5 reductase, 433, 440
　reconstituted Ca^{2+}-channels, 115
　stearoyl-CoA desaturase, 433
Endoproteins, 433–435
Endosome, 190
Energy-transducing complexes
　of bacterial respiratory chains, 279–313
　molecular weight of, 288
Epithelial membrane
　Na^+ transport, 133, 134
Erythrocyte
　acyl chain phase changes, 148
　electrically-induced fusion, 205, 206
　fusion by
　　benzyl alcohol, 193, 194
　　chlorpromazine, 193
　　heat, 191, 192
　　oleoylglycerol, 192
　　Sendai virus, 210, 211
　lipid agglutination, 245
　lipid asymmetry, 146, 148
　lysis inhibition by La^{3+}, 196
　lysis by Sendai virus, 211
　membrane, 67
　　sialoglycoproteins, 133
　　skeleton disruption, 192
　　skeleton and electrical fusion, 208
　　structure of band 3 protein, 133
Escherichia coli, 285, 286, 297, 298, 308, 310, 311
　alkaline phosphatase, 407
　asparagine synthetase, 401
　ATP-dependent protein translocation, 400

Escherichia coli (cont.)
  binding to glycolipid, 253, 254
  exported proteins of, 398
  β-galactosidase, 392, 418, 421
  lac Z gene, 417
  β-lactamase, 428
  lipoprotein, 406
  maltose-binding protein, 428
  membrane and lipid conformational transitions, 29
  membrane lipid asymmetry, 146
  membrane vesicles, 391
  mutant defective in protein export, 391, 392
  outer membrane protein, 412
  phage M_{13} coat protein, 436
  phage T_4, 412
  protein carboxy termini, 409
  protein translocation, 399
  respiratory enzymes, 280, 281
Excitation-contraction coupling, 111
  voltage-gating by K^+ conductance, 123
Exocytosis
  in biological systems, 214–218
  and Ca^{2+}-activated proteases, 218
  and chromaffin granule release, 214–216
  and cytoskeletal modifications, 218
  fusion pore, 216
  fusion of secretory vesicles, 213
  inhibition of, 215
  model for, 209
  model lipid-membranes, 212-214
  role of Ca^{2+}, 215
  role of osmotic forces, 216
  shared bilayer, 215

Fatty acyl chain, 3
  flexibility of, 145
Fibrinogen, 239
Fibroblast
  fusion by oleylamine, 191
Fibronectin, 239
Firefly luciferase, 395
Fluorescence
  membrane dynamics and heterogeneity, 145–177
  depolarization
    fatty acyl chain motion, 168
  photobleaching
    phospholipid lateral diffusion, 54, 55

Fluorescent probes, 62, 147, 149
  anisotropy decay-associated spectra, 152–157
  decay, 150
  detection of acyl chain phase changes, 148
  membrane disordering, 168
  membrane fusion, 201, 202
  partition between phases, 150
  pyrene excimer fluorescence, 160, 161
Fluorescence polarization
  anisotropic rotation, 151
  steady-state, 150
  time-dependent, 150
Fluorescence spectroscopy
  of phospholipid domains, 147, 149
Formate dehydrogenase, 308–310
Forssman antigen, 246
Fourier transform, 43
  IR spectroscopy, 438
Freeze-fracture, 62, 64, 65, 242: *see also* Electron microscopy
  of *Anacystis* plasma membranes, 70, 71
  of chloroplasts, 74, 75, 81
  of cholesterol-induced lipid heterogeneity, 159
  of membranes treated with filipin, 148
Fumarate reductase, 308–310
Fusion
  influence of temperature, 66
  of vesicles, 66
Fusogenic lipids, 190–197
  activation of cellular proteases, 192–195
  osmotic effects, 190–192
  osmotic cell swelling, 195–197

Galactolipids, 48
Ganglioside, 64, 241, 249, 367
  and bilayer phase separation, 255
  and endogenous protein phosphorylation, 238
  and motion of carbohydrate head group, 252
  sialic acid content of, 250
  and transmembrane signaling, 238
Gel filtration chromatography
  of Fcε proteoliposomes, 332, 333
Globoside, 249, 250
Glycolipid
  acyl chain of, 241

Index

Glycolipid (cont.)
  carbohydrate head group
    protrusion of, 251, 252, 260
    and membrane fusion, 250–252
  and cell activation by lectin, 261
  epitopes of, 255
  and intermembrane interactions, 249, 250
  lateral distribution in bilayer, 242
  as lectin receptor, 237–239
  localization in plasma membrane, 238
  and membrane fluidity, 240
  phase separation, 241
  and properties of membranes, 240–242
  synthetic, 243
Glycophorin, 256
  -wheat germ agglutinin interaction, 256, 257
Glycoprotein
  as lectin receptor, 237–239
  mannosylated, 265, 266
  oligosaccharide chains of, 238
  protein shielding by carbohydrate, 239
Glycosphingolipid, 237
  micelles of, 240, 241
Glyoxosome
  malate dehydrogenase, 420
Golgi, 190, 263
Gramicidin, 107
  channel in BLM, 134
  effect on Ca^{2+}-channel, 117

Halobacterium halobium
  cubic phases of membrane lipids, 57
  prebacteriorhodopsin, 447, 448
Hen erythrocyte
  fusogenic studies on, 190, 191, 197
  PED-induced fusion, 197
  polysialoganglioside-induced fusion
Hexagonal-II phase, 30
  cytochrome *c*-induced, 67
  of dioleoylphosphatidylethanolamine, 68
  of glycolipids, 64, 65
  induced by metal ions, 80
Hexamethonium, 353, 354
High affinity IgE receptor
  detergent sensitivity, 322–324
  dissocation of, 323
  ester-linked fatty acids of, 325
  mechanism of action, 327
  monomeric IgE binding, 322

High affinity IgE receptor (cont.)
  in planar bilayers, 328–330
  protective effect of lipids, 324, 325
  purification of, 322–327
  purification protocol, 325–327
  receptor aggregation, 322, 327
  reconstitution of, 321–335
  tetrameric structure of, 324
  in vesicles, 330–334
Hybridoma
  electrically induced, 205
  PEG induced, 197
Hydrophobic beads
  adsorption properties of, 14–19

Import factors, 397
Influenza virus
  hemagglutinin, 409
  sialic acid binding by, 264, 265
Insulin
  secretory vesicles, 216
Iodoacetamide
  inhibition of fusion by, 193
Ionophore A23187, 329
  and cellular Ca^{2+} concentration, 193, 195
Isoenzymes, 412

K^+ channel
  blockers of, 119, 120
  Ca^{2+}-activation of, 121, 122
  delayed rectifier channel, 120
  effect of
    Cs^+, 123, 125
    decamethonium, 123, 124
    hexamethonium, 123, 124
    membrane surface charge, 125
    phospholipid charge, 124, 125
    succinyl choline, 126
  of intracellular membranes, 123–126
  ion selectivity of, 119, 121
  of muscle cells, 121
  of sarcoplasmic reticulum
    electrogenic Ca^{2+} flux, 124
  two open states of, 126
  variety of, 119
  varying states
    in cardiac membranes, 124
  voltage dependence of, 121

K⁺ channel (cont.)
  voltage-dependent inhibition
    by mersalyl, 123
K⁺ transport
  across cytoplasmic membranes, 119–122
Kidney plasma membranes
  reconstitution of Na⁺ channels, 134, 135

La³⁺
  fusion of phosphatidylserine liposomes, 196
  inhibition of erythrocyte lysis, 196
Lactosylceramide, 251
Lauryl maltoside, 289
Lecithin, 8, 10, 12, 16–19, 21: see also Phosphatidylcholine
Lecithin-cholesterol acyltransferase, 366, 367
Lectin
  use in affinity chromatography, 235, 261
  affinity constants, 235
  agglutination of transformed cells, 260
  binding to terminal saccharide, 234, 235
  biosynthesis of, 234
  carbohydrate binding
    rate, 235
    specificity, 231, 234–237
  carbohydrate content, 231
  carbohydrate mapping, 259
  in cytochemistry, 260
  endogenous, 262–266
    biological effects, 261, 262
  exogenous, 259–261
  -glycolipid interaction, 237–239, 243, 244
  glycoprotein binding, 237–239
  and intermembrane interactions, 249, 250
  limited diagnostic value of, 260
  metal ion binding by, 231
  mitogenic stimulation by, 261
  molecular properties of, 232, 233
  and phase separation, 255
  -receptor interactions, 230
  sialic acid binding by, 265
  structural features of, 231–234
  subunits of, 231
  and vesicle fusion, 249
Lectin-carbohydrate interactions, 229–266
  in biological systems, 259–266
  Ca²⁺ and agglutination, 244–246
  and cell differentiation, 265
  competing sugar, 244
  in model systems, 240–259

Lectin-induced agglutination
  dehydration of phospholipid head group, 246–249
  effect of Ca²⁺, 244–246
  effect of fatty acid, 253, 254
  lateral clustering, 254–256
Lentil lectin, 232, 234
Light-harvesting apparatus
  photosystem-II, 74–78
  chlorophyll a/b complexes, 76
Limulus polyphemus
  amoebocyte degranulation, 216
Lipid-protein interactions, 25–82
  exclusion of proteins from bilayer gel phases, 69–73
  high temperature stress, 74–78
  role of acidic lipids, 80–82
  at subzero temperatures, 78–80
Lipid vesicles
  and reconstitution, 11, 12
Liposomes
  acidic, 444
  carboxyfluorescein marker of, 365–373
  carriers of drugs, 363–375
  cholesterol content, 364–368, 370, 371
  cholesterol enrichment of, 366
  circulating in blood, 366–373
  circumlatory half-life, 370, 371
  clearance, 364
    and size, 369
    and surface charge, 369
  cytochrome c interaction, 440
  destabilization by HDLs, 364
  distribution in tissues, 373–375
  entrapment of
    albumin, 364
    carboxyfluorescein, 365
    penicillin, 364
    β-fructofuranosidase, 364
  intravascular survival, 368
  in vivo studies, 366
  ligand-bearing, 365, 374
  loss of small polar solutes, 367
  multilamellar vesicles, 364
  negatively charged, 364
  and plasma HDLs, 364–366
  pore formation, 368
  and protein translocation processes, 436–445
  and reconstitution, 386

Index

Liposomes (cont.)
  removal by RES, 365, 368, 372, 374
  retention of drugs, 364
  small unilamellar vesicles, 365, 366
  targeting to cells, 374
Lotus tetragonolobus lectin, 236
Lymphoma cells
  electrically induced fusion of, 205
Lysophospholipids
  hexagonal-I phase, 38
  hexagonal-II phase, 38
  inverted cone shape, 38
Lysosomal enzymes, 236
Lysozyme
  effect on phosphatidylserine bilayers, 157

Mast cell
  activation of, 327
  Ca^{2+} entry, 327
  exocytosis, 217
  Fcϵ receptor, 321, 322
  histamine release, 328
Membrane: see also Biological membranes
  environment of, 2–5
  protein solubilization, 1, 2
  lipid components of, 2
  proteins, 3–5
    activation by phospholipids, 146
Membrane fusion
  Ca^{2+}-induced lipid phase separation, 209
  chemically induced, 190
  and cytoskeleton, 190, 207, 208
  disruption of membrane skeleton, 192
  electrically induced, 190, 204–208
    dielectrophoresis and high-voltage pulse, 204
    potentiation by pronase, 206
  endocytosis, 190
  exocytosis, 190
  fertilization, 189
  fluorescein labels, 205
  fusogenic agents and osmotic forces, 189–218
  heat-induced, 191, 192
  hybridoma production, 189
  inhibition by protease inhibitors, 193
  molecular models of, 208–210
  of myoblasts, 189, 190
  PEG-induced, 197–204
  PEG and phospholipid hydration, 199–201

Membrane fusion (cont.)
  shared bilayer model, 209, 210
  role of Ca^{2+}, 199, 200
  syncytial trophoblast, 189
  virally induced, 190, 195, 210–212
    cell swelling, 211
    involvement of viral protease, 212
    viral swelling, 210, 211
Membrane polar lipids
  absorption spectroscopy, 135
  bilayer asymmetry of, 146
  bilayer phases, 2, 6, 42–48
  biophysical characterization of polymorphism, 27–38
  Brownian motion of molecules, 34
  cubic phases, 26, 28
  and EM methods, 30, 31
  factors determining lipid phase, 38–42
  hexagonal-I phase, 28
  hexagonal-II phase, 26, 27, 30, 55, 56, 64, 65, 67, 68, 88
  hydrophobic to hydrophilic balance, 39, 40
  immobility of headgroup, 49
  lateral diffusion in bilayer, 54, 58
  lipid crystallization and protein aggregation, 71
  molecular arrangements of, 28
  nonbilayer phases, 26
  phase behavior, 25–82
    of mixed lipid systems, 57–66
    of pure lipid systems, 42–55
  phase and molecular shape, 38
  polymorphism, 26–57
  pure lipid
    hexagonal phases, 55, 56
    cubic phases, 56, 57
    lamellar crystalline phases, 48–51
    lamellar gel phases, 51–53
    lamellar liquid-crystalline phases, 53–55
    in nonaqueous solvents, 54
  X-ray diffraction, 27–30: see also X-ray diffracton
Membrane reconstitution, 9–22
  detergent-induced, 14–19
    alterations of vesicle size, 21
  detergent–lipid–protein interactions, 10, 11
  detergent removal, 14–19
  experimental requirements, 11–13
  incorporation of intrinsic proteins, 22
  kinetic control of vesicle size, 11, 19–22

Membrane reconstitution (cont.)
  and membrane potentials, 13
  solubilization, 14
  vesicle size, 20
Metalloendoprotease
  and mast cell exocytosis, 195
  and myoblast fusion, 194, 195
Methotrexate, 425
Methylolphilus methylotrophus, 298
  methanol dehydrogenase of, 299
Micelle, 1
  detergent, 38, 443
  discoidal, 443
  of glycolipid, 240
  inverted, 56, 68, 208, 209
  of lysophospholipids, 38
  and membrane fusion, 208, 209
  mixed, 2, 4, 9, 22
    of Brij 58 and lecithin, 16, 17
  number of lipid molecules, 9
Microbody, 403, 420
  post-translational translocation of proteins, 429, 430
Micropipette
  and Ca^{2+}-channel recording, 113
  patch-clamp bilayers, 99, 102, 346
Microsomes, 67
  Ca^{2+} channels, 103, 104
  cytochrome P-450, 409
  membrane proteins of, 386–390
Microtubule
  -associated proteins, 117
  depolymerization, 117
  influence of Ca^{2+} on, 117
Migratory proteins
  primary structure, 414, 415
Mitochondria, 384
  acyl chain phase changes, 148
  apocytochrome *c*, 442
  aspartate aminotransferase, 397
  ATP synthase, 419, 426
  carbamylphosphate synthetase, 394, 395
  cytochrome *b*, 300, 301
  cytochrome b_2, 421
  cytochrome *c* oxidase, 419
  cytochrome *c* peroxidase, 421
  digitonin treatment, 395
  dihydrofolate reductase, 418, 421, 425
  F_1-ATPase, 402
  malate dehydrogenase, 416

Mitochondria (cont.)
  manganese superoxide dismutase, 424
  matrix enzymes, 401
  migratory proteins, 400–403, 412–426
  mild proteolysis of, 393, 435
  ornithine carbamyl transferase, 416, 417, 420
  outer membrane
    porin, 434, 435
    reconstitution of, 136
  oxidative phosphorylation, 279
  post-translational translocation of proteins, 429
  respiratory enzymes, 280, 281
  transhydrogenase, 311
Monazomycin, 107
Monensin, 215
Monoclonal antibody
  to lactylceramide, 254
Monogalactosyldiacylglycerol, 76, 77
  kinetics of phase transition, 29
Multilamellar vesicles, 10
  repulsive forces between bilayers, 41, 42
  surface-induced polarization density, 42
Muscle
  action potential, 124
  Ca^{2+}-channel from, 117–119
  excitation of transverse tubule membrane, 124
  Na^+-channels of, 110–112
  Sarcoplasmic reticulum, 111
    Ca^{2+} permeability of, 124, 125
  transverse tubule system, 111
Mycobacterium phlei, 286
Myelin
  glycolipid content, 238
Myeloma cells
  electrically induced fusion of, 205
Myoblast, 189
  Ca^{2+}-activated proteinase and fusion, 194, 195
  fusion of, 265
    Ca^{2+} requirement, 266
  myotube formation, 229

Na^+ channels
  of muscle membranes, 110–112
  of nerve membranes, 108–110
  polypeptide subunits of, 110

Index

Na⁺ channels (cont.)
  reconstituted from epithelial membranes, 134
  structure of, 110
Na⁺,K⁺-ATPase
  association with lipids, 128
  inhibitors of, 128, 129
  regulation by phosphorylation, 128
  subunits of, 128
Na⁺,K⁺ pump, 13
  and action potential, 128
  and cellular resting potential, 128
  of plasma membrane, 128
N-acetyl-neuraminic acid (NANA), 237
NADH and NADPH, 280
NADH dehydrogenase
  proteolytic digestion of, 306
NADH: quinone oxidoreductase
  Na⁺ pumping, 307
  proton pumping, 306, 307
Nitrobacter agilis, 293–295
Nematocyst
  discharge of, 216
Neuraminidase
  treatment of erythrocytes, 208
Neurospora crassa, 394
  F_0-F_1-ATPase, 426
  mitochondrial carbamyl phosphate synthase, 425
  mitochondrial proteins, 385
  outer mitochondrial membrane arrays, 136
Neurotoxins, 109
  radiolabeled, 110
  voltage-dependent, 111, 112
Nitroxide
  spin probe, 241
Nuclear magnetic resonance (NMR)
  acyl chain length
    and bilayer thickness, 46
  determination of molecular displacements, 32
  deuterium spectra, 33
  deuterium substitution, 32, 33
  hexagonal-II phase, 32–35
  of lamellar lipid phase, 32–34
  of lipid phase changes, 32, 33
  and lipid polymorphism, 32–35
  ³¹P-NMR spectra, 33, 34
  pulsed field gradient, 32
  rotational diffusion coefficient, 32

Nuclear magnetic resonance (NMR) (cont.)
  static and dynamic data from, 32
  translational diffusion coefficient, 32

Octyl glucoside, 8, 15, 18–20, 323
Olfactory receptors
  in BLM, 132, 133
  lack of effect of
    odorous compounds on conductance, 132
Ornithine carbamyl transferase, 395
Osmotic forces
  and cell fusion, 204
Osmotic swelling of cells
  electrically induced, 205
  membrane skeleton disruption, 195, 196
Osteoclast, 189
Ouabain
  inhibition of Na⁺K⁺-ATPase, 128

Palmitoylphosphatidylcholine, 332
Paracoccus denitrificans, 296, 298, 299, 306, 307, 311
  apocytochrome c, 395
  cytochromes, 300
  electron transfer chain, 280, 281
  H⁺/O ratio, 283–286
Paramecium aureola
  Ca²⁺ channels of, 135
  reconstitution of
    mitochrondrial anion-selective channel, 135
  trichocyst membranes, 217
Parathyroid hormone
  release of, 216
Patch-clamp bilayers, 97–137
Pea lectin, 234
Peanut agglutinin, 233, 236
Peroxisome
  acyl/alkyl dihydroxyacetone phosphate reductase, 431
  carnitine acetyl transferase, 430
Penicillin
  liposomial, 364
Phosphatidylcholine, 9, 10, 43, 47–49, 63, 100, 241, 244, 246, 249, 324, 364, 441
  acyl chain length, 254
  bilayers, 4
  calorimetric study of, 48
  ²H-NMR of saturated lipid, 52

Phosphatidylcholine (*cont.*)
  hexagonal-II phase
    when mixed with fatty acids, 41
Phosphatidylethanolamine, 48, 66, 100, 113, 119, 121; 213, 244, 246
  ^{2}H-NMR of saturated lipid, 52
  interaction with intrinsic proteins, 67
  lamellar to hexagonal-II phase, 29
  phase transition kinetics of, 29
  WGA-induced phase transition, 256–258
Phosphatidylglycerol, 48
Phosphatidylglycerol phosphate
  diphytanyl derivatives, 57
Phosphatidylinositol, 2
Phosphatidylserine, 2, 63, 66, 113, 119, 121, 256, 441
  Ca^{2+} binding and hydration, 199, 200
  Ca^{2+}-induced phase changes, 47
  vesicle aggregation by Mg^{2+}, 212
Phosphodiesterase
  Ca^{2+} stimulation of, 195
Phospholipase A_2
  digestion of chloroplasts, 81
Phospholipid
  hexagonal-I phase, 28, 38, 55
  hexagonal-II phase, 26, 27, 30, 31, 33, 34, 38, 39, 41, 42, 55–57
    curvature of lipid tubes, 42
    hydration factors, 56
  lamellar crystalline phases, 48–51
  lateral separations, 146
  racemic mixtures, 59
  vesicle fusion, 21
Phospholipid bilayers
  acyl chain
    structural order of, 151
  anisotropy value, 151, 152
  crystallization and protein exclusion, 69
  cubic phase, 26, 32, 33
  diffusion studies, 145
  domain formation, 145–167
  electrostatic repulsion, 199
  fluidity of, 145
  hydration repulsion of, 199
  phases of, 42–48
  thickness, 44
Phospholipid domains, 145–167
  chemically induced
    decay-associated spectra, 162–167

Phospholipid (*cont.*)
  extrinsic effectors, 157, 158
  fluorescent lifetime, 149, 150
  fluorescence polarization, 150–152
  influence of cholesterol, 158–161
  methods of detection, 147, 148
  relaxation times, 167–174
  thermal effects on, 149
Phospholipid phase behavior, 25–82
  bilayer-nonbilayer phase separations, 64–65
  Ca^{2+}-induced phase separation, 63
  electrostatic effects, 65, 66
  influence of high salt concentrations, 68
  influence of protein, 66, 67
  lateral miscibility, 61–64
  of lipid enantomers, 58–61
Phospholipid phase transition,
  lamellar to hexagonal-II, 39
  temperature dependence of, 39
Photosynthetic membranes
  nonbilayer lipids of, 73
Photosystem-I, 75
Photosystem-II, 74, 75, 78
Phytohemagglutinin (PHA), 233, 261
Phytophthora palmivora
  secretion of zoospores, 209
Planar lipid membranes
  formation of, 344–346
Plant cell organelles
  effects of high temperatures, 74–78
Plant protoplasts
  fusion by PEG, 197, 204, 209
Platelets
  serotonin secretion by, 216
Polycations, 66
Polyethyleneglycol (PEG)
  Ca^{2+} protease and membrane fusion, 198
  contaminant metal ions and membrane fusion, 198
  dehydration and cell fusion, 199–201
  differences between cell and lipid vesicle fusion, 200
  fusion enhancement by rehydration, 201–204
  intramembrane particle distribution, 198
  and membrane fusion, 197–204
  purity and fusogenic ability, 197, 198
Polylysine, 66
  effect on negatively charged bilayers, 158

Polyoxyethylene, 21
Postsynaptic membrane
  conductance change, 339
Protease
  activation in PEG-induced fusion, 198
  activation by Ca^{2+}, 192–195
  degradation of erythrocyte membrane proteins, 192, 193
  inhibitors and fusion inhibition, 193
  lateral movement of IMPs, 193, 194
Protein
  denaturation, 10
Protein-lipid interactions, 66–82
  phase separation of nonbilayer lipids in membranes, 73, 74
Protein translocation processes, 379–449
  and active transport, 385
  bacterial membrane protein, 391–393
  of chloroplasts, 396, 397
  endoproteins, 433–435
  energy dependence, 400–403
    for migratory proteins, 398–400
    for secretory proteins, 398–400
    for transmembrane proteins, 398–400
  in lipid aggregates, 436–445
  to organellar compartments, 420–425
  membrane-anchoring signals, 403–426
  membrane-protein interactions, 380
  membrane trigger hypothesis, 381
  microsomal docking protein, 388
  microsomal membrane proteins, 386–390
  in mitochondria, 394–396
  organelle membrane proteins, 393–397
  post-translational, 427–435
  reconstitution of, 386, 436–445
  signal recognition particle, 386–389
  signal sequences, 403–408
  unitary mechanism of, 446, 447
Proteoglycan, 265
Proteolipid, 110
Proteoliposomes, 291
  containing cytochrome bc_1, 303
  multicomponent, 310
  phosphatidylserine-containing, 212
Proteolysis, 441
Pseudomonas aeruginosa pilin of, 408
Pseudomonas AM1, 283, 285, 294, 295
Puromycin, 383
Pyrene

Pyrene (*cont.*)
  emission spectra of, 164
  excimer fluorescence, 160, 161
  lateral diffusion of, 160

Radiation-inactivation
  of enzymes, 288
Raman spectroscopy, 151
Receptor
  incorporation into phospholipid, 230
  interactions, 230
  specificity of, 230
Reconstitution
  of active transport, 128, 129
  formation of bilayers, 98–101
  of chloride channel, 126, 127
  of dopamine receptor, 132
  of epithelial Na^+ transport, 133, 134
  of erythrocyte band 3, 133
  of K^+ transport, 119–126
  of kidney plasma membrane, 134, 135
  of membrane molecular mechanisms, 97–137
  of mitochondrial outer membrane, 136
  of Na^+ transport, 108–112
  of sensory mechanisms, 132, 133
  of synaptic events, 129–132
Red blood cell membrane: *see* Erythrocyte
Reticuloendothelial system, 365, 368, 372
  phagocytic cells of, 374
Reticulocyte lysate, 386, 387
  import factors of, 398
Retinal rod
  outer segment disc membranes, 67, 132
  dehydration of, 74
Rhodamines, 416
Rhodobacter sphaeroides, 293, 296, 298, 299, 301–304
  cytochrome c oxidase purification, 289
Ribophorins, 382
  and ribosome binding, 389, 390
Ribosomes
  endoplasmic reticulum-bound, 383
  and post-translational translocation, 430
  free, 383
Ricinus communis
  glyoxysomal malate synthase, 430
Ricinus communis lectin, 230, 233, 234, 236, 244, 245, 248, 255, 257

Rough endoplasmic reticulum, 190
  bound ribosomes of, 383
  proteolysis of, 388

Saccharomyces cerevisiae: see also Yeast
  alcohol oxidase, 395
  histidine tRNA synthetase, 412
  invertase, 413
  mutant defective in protein translocation, 389
Sanguinarine, 133
Sarcoplasmic reticulum
  acyl chain phase changes, 148
  Ca^{2+} channels of, 103
  conductance of, 103
  K^+ channel of, 123
Saxitoxin, 109
Semliki forest virus
  P62 protein, 411
Sendai virus, 264
  binding protein HN, 264, 265
  cell lysis in absence of Ca^{2+}, 211
  M protein, loss of, 210
  and membrane fusion, 210, 211
Sialic acid, 256, 266
Signal hypothesis, 379, 381
Signal peptidases, 406
  uncouplers of, 399
Signal recognition particle, 386
Signal sequence, 403–408
  deletion of, 408
  hydrophobic segment of, 407
  and membrane anchoring domain, 409–411
  for migratory proteins, 412–426
  N-terminal segment, 382
  primary structures, 405, 414, 415
Sodium cholate, 10
Sodium dodecylsulfate (SDS), 7
  as denaturant, 8
  polyacrylamide gel electrophoresis, 1, 355
Solubilization
  of membrane proteins, 1
Sonication, 19, 20, 64
Soybean lectin, 233, 234, 245, 248, 249, 255
Spermine, 248
Sphingomyelin, 324
  liposomal, 367
Staphylococcus aureus
  membrane-bound ribosomes, 391

Salmonella typhimurium
  secretion-defective mutants, 412
Streptomyces fecalis, 399
Subcellular membranes
  nonbilayer lipids of, 73
Suberyldicholine, 354
Succinylcholine, 126
Sulpholobus sulphataricus
  membrane lipids of, 57
Suzukacillin, 107
Synaptosome
  Ca^{2+}-channels of, 113, 114
Syncytial trophoblast, 189
Synexin, 215

Tetraethylammonium, 120
Tetrahymena pyriformis
  acyl chain phase changes, 148
  mucocyst secretion, 216
Tetrodotoxin, 132
Thermophilic bacterium PS3, 287, 288, 294–297, 299
Thermus thermophilus, 291, 293, 296
  proton pump, 306
Torpedo
  electric tissue, 350–352
  α-Toxin binding protein, 343
Transferrin receptor, 409, 411
Transmembrane proteins
  anchoring domain, 408
Trimethyloxonium
  action on single ion channels, 110
Triton X-100, 7, 289, 323
  extraction of cytochrome o-type oxidase, 297
  solubilization of bacterial membranes, 287
d-Tubocurarine, 353, 354
Tubulin, 115
  association with lipids, 115
Turbidity
  and vesicle agglutination, 243

Ubiquinol
  oxidation of, 297
Ubiquinone, 280, 281
Ultracentrifugation
  of reconstructed Fcϵ, 334

Index

Valinomycin, 282, 287, 291
Valonia utricularis
   electrical breakdown potential, 205
Van der Waals forces
   between bilayers, 47
   between lipid hydrocarbon chains, 39
Vasopressin, 134
   aggregation of Na^+ channel subunits, 135
Vectorial transport, 379
Very low density lipoproteins, 367
Vesicular stomatitis virus
   G protein, 381
Vibrio alginolyticus, 307
Virus
   attachment, 229, 264
   cell entry, 229, 264
   cell fusion, 195
   hydrophobic domains, 195
Visual receptors
   in BLMs, 132
Von Willebrand factor, 239

Wheat germ agglutinin (WGA), 233, 235, 249, 255–257, 261
   -glycophorin interaction, 256, 257
Winged bean lectin, 235
Wolionella succinogenes, 308, 310

Xenopus laevis
   endoplasmic reticulum, 407

Xenopus laevis (cont.)
   oocytes, 389
X-ray diffraction, 27–30, 46, 60, 61, 64
   agreement with EM, 30
   Bragg reflections, 27, 43
   of cholesterol-phospholipid interactions, 159
   cubic phases, 56, 57
   diffuse scattering of lamellar liquid-crystalline phase, 53
   electron density distributions, 43
   of lamellar crystalline phases, 48, 49
   low angle, 27, 71
   of multilamellar systems, 27
   of $P_{\beta'}$ phase, 52, 53
   of phosphatidylcholine, 44
   and phospholipid domains, 147
   of pretransition, 52
   scattering intensities, 47
   of stacked bilayers, 42, 43
   synchron radiation, 28, 29
   time-resolved methods, 29

Yeast: *see also Saccharomyces cerevisiae*
   alcohol dehydrogenase isoenzymes, 418
   cytochrome oxidase, 425
   cytosolic invertase, 407, 418, 425, 426
   F_1-ATPase, 425
   mitochondrial complex III, 413
   secretory invertase, 399